S0-CEE-512

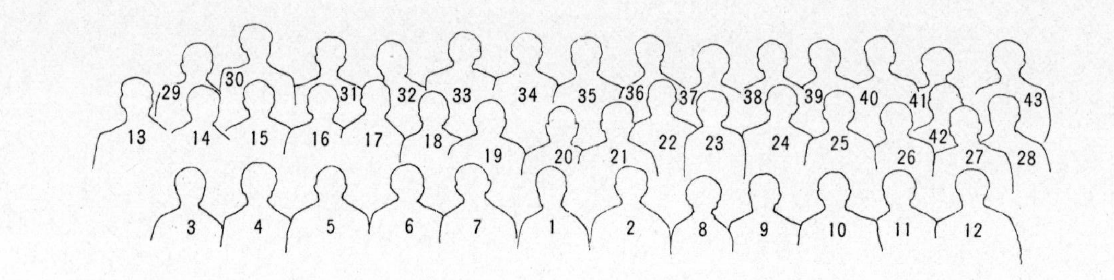

1. H.I.H. Princess Takamatsu
2. H.I.H. Prince Takamatsu
3. Mrs. Ohno
4. Dr. Silagi
5. Dr. Weber
6. Dr. Barski
7. Dr. G. Schapira
8. Mrs. Weinhouse
9. Dr. Weinhouse
10. Dr. Paul
11. Mrs. Johnson
12. Dr. Johnson
13. Dr. Kaya
14. Dr. Ono
15. Dr. Ohno
16. Dr. Rutter
17. Dr. Potter
18. Mrs. Tomkins
19. Dr. Higuchi
20. Dr. Kurokawa
21. Dr. Kamahora
22. Dr. Ostertag
23. Mrs. Ostertag
24. Dr. Shall
25. Dr. Pierce
26. Dr. Prasad
27. Dr. Nakahara
28. Dr. Ikawa
29. Dr. Sugimura
30. Mr. Silagi
31. Dr. Katunuma
32. Dr. Dube
33. Dr. Tomkins
34. Dr. Thompson
35. Dr. Isaka
36. Dr. Koyama
37. Dr. Tanaka
38. Dr. Tsuiki
39. Dr. Hozumi
40. Dr. Okada
41. Dr. Ichikawa
42. Dr. Holtzer
43. Dr. Sugano

DIFFERENTIATION AND CONTROL OF MALIGNANCY OF TUMOR CELLS

Proceedings of the 4th International Symposium of
The Princess Takamatsu Cancer Research Fund, Tokyo, 1973

DIFFERENTIATION AND CONTROL OF MALIGNANCY OF TUMOR CELLS

Edited by
WARO NAKAHARA, TETSUO ONO,
TAKASHI SUGIMURA, and HARUO SUGANO

UNIVERSITY PARK PRESS
Baltimore • London • Tokyo

UNIVERSITY PARK PRESS
Baltimore • London • Tokyo

Library of Congress Cataloging in Publication Data
Main entry under title:

Differentiation and control of malignancy of tumor cells.

1. Cancer cells—Congresses. I. Nakahara, Warō, 1894– ed. II. Takamatsu no Miya Hi Gan Kenkyū Kikin. [DNLM: 1. Cell differentiation—Congresses. 2. Cell transformation, Neoplastic—Congresses. 3. Neoplasms—Congresses. QZ200 D569 1973]
RC269.D53 616.9'94'07 74-28499
ISBN 0-8391-0804-4

UTP 3047-68090-5149
Printed in Japan.

Originally published by
UNIVERSITY OF TOKYO PRESS

Princess Takamatsu Cancer Research Fund

Organizing Committee of the 4th International Symposium

Waro NAKAHARA
National Cancer Center Research Institute, Tsukiji, Tokyo, Japan
Tetsuo ONO
Cancer Institute, Japanese Foundation for Cancer Research, Toshima-ku, Tokyo, Japan
Takashi SUGIMURA
National Cancer Center Research Institute, Tsukiji, Tokyo, Japan
Haruo SUGANO
Cancer Institute, Japanese Foundation for Cancer Research, Toshima-ku, Tokyo, Japan

Participants

BARSKI, G.
Labaratoire de Virologie et de Culture de Tissus, Institut Gustave-Roussy (CNRS. E.R. N° 38), 94-Villejuif, France

DUBE, S. K.
Abteilung Molekulare Biologie, Max-Planck-Institut für experimentelle Medizin, Hermann-Rein-Str. 3 Ruf. 3031, 3400-Göttingen, West Germany

HOLTZER, H.
Department of Anatomy, School of Medicine, University of Pennsylvania, Philadelphia, Pa. 19104, U.S.A.

HOZUMI, M.
National Cancer Center Research Institute, Tsukiji 5-1-1, Chuo-ku, Tokyo 104, Japan

IKAWA, Y.
Cancer Institute, Japanese Foundation for Cancer Research, Kami-Ikebukuro 1-37-1, Toshima-ku, Tokyo 170, Japan

ISAKA, H.
Sasaki Institute, Sasaki Foundation, 2-2 Surugadai, Kanda, Chiyoda-ku, Tokyo 101, Japan

JOHNSON, G. S.
Laboratory of Molecular Biology, National Cancer Institute, National Institutes of Health, Bethesda, Md. 20014, U.S.A.

KATUNUMA, N.
Institute for Enzyme Research, School of Medicine, Tokushima University, Kuramoto-cho 3-18-15, Tokushima, Japan

Koyama, H.
Cancer Institute, Japanese Foundation for Cancer Research, Kami-Ikebukuro 1-37-1, Toshima-ku, Tokyo 170, Japan

Ohno, S.
Department of Biology, City of Hope Medical Center, 1500 East Duarte Road, Duarte, Calif. 91010, U.S.A.

Okada, Y.
Research Institute for Microbial Diseases, Osaka University, Yamada-kami, Suita, Osaka 565, Japan

Ono, T.
Cancer Institute, Japanese Foundation for Cancer Research, Kami-Ikebukuro 1-37-1, Toshima-ku, Tokyo 170, Japan

Ostertag, W.
Abteilung Molekulare Biologie, Max-Planck-Institut für experimentalle Medizin, Hermann-Rein-Str. 3 Ruf. 3031, 3400-Göttingen, West Germany

Paul, J.
Beatson Institute for Cancer Research, 132 Hill Street, Glasgow G3 6UD, U. K.

Pierce, G. B.
Department of Pathology, University of Colorado Medical Center, 4200 East Ninth Avenue, Denver, Colo. 80220, U.S.A.

Potter, V. R.
McArdle Laboratory for Cancer Research, University of Wisconsin, Madison, Wis. 53706, U.S.A.

Prasad, K. N.
Department of Radiology, University of Colorado Medical Center, 4200 East Ninth Avenue, Denver, Colo. 80220, U.S.A.

Rutter, W. J.
Department of Biochemistry and Biophysics, University of California, San Francisco, Calif. 94122, U.S.A.

Schapira, F.
Institut de Pathologie Moléculaire, Université de Paris, 24, Rue du Faubourg-Saint-Jacques, Paris 14e, France

Schapira, G.
Institut de Pathologie Moléculaire, Université de Paris, 24, Rue du Faubourg-Saint-Jacques, Paris 14e, France

Shall, S.
Biochemistry Laboratory, University of Sussex, Falmer, Brighton BN1 9QG, Sussex, U. K.

Silagi, S.
Department of Obstetrics and Gynecology, Cornell University Medical College, 525 East 68th Street, New York, N.Y. 10021, U.S.A.

SUGANO, H.
Cancer Institute, Japanese Foundation for Cancer Research, Kami-Ikebukuro 1-37-1, Toshima-ku, Tokyo 170, Japan

SUGIMURA, T.
National Cancer Center Research Institute, Tsukiji 5-1-1, Chuo-ku, Tokyo 104, Japan

TANAKA, T.
Department of Nutrition and Physiological Chemistry, Osaka University Medical School, 33 Joan-cho, Kita-ku, Osaka 530, Japan

THOMPSON, E. B.
Biosynthesis Section, Laboratory of Biochemistry, National Cancer Institute, National Institutes of Health, Bethesda, Md. 20014, U.S.A.

TOMKINS, G. M.
Department of Biochemistry and Biophysics, School of Medicine, University of California, San Francisco, Calif. 94122, U.S.A.

TSUIKI, S.
Research Institute for Tuberculosis, Leprosy and Cancer, Tohoku University, 4-12 Hirose-machi, Sendai 980, Japan

WEBER, G.
Department of Pharmacology, Indiana University School of Medicine, 1100 West Michigan Street, Indianapolis, Indiana 46202, U.S.A.

WEINHOUSE, S.
Fels Research Institute, School of Medicine, Temple University, Philadelphia, Pa. 19140, U.S.A.

Observers

TAKEHIKO AMANO, Mitsubishi-Kasei Institute of Life Sciences, Tokyo
GORO CHIHARA, National Cancer Center Research Institute, Tokyo
SETSURO FUJII, Institute for Enzyme Research, Tokushima University, Tokushima
FUMIKO FUKUOKA, National Cancer Center Research Institute, Tokyo
MITSURU FURUSAWA, Faculty of Science, Osaka City University, Osaka
OSAMU HAYAISHI, Faculty of Medicine, Kyoto University, Kyoto
HIDEMATSU HIRAI, Hokkaido University School of Medicine, Sapporo
TAKESHI HIRAYAMA, National Cancer Center Research Institute, Tokyo
AKIRA ICHIHARA, Institute for Enzyme Research, Tokushima University, Tokushima
YASUO ICHIKAWA, Institute for Virus Research, Kyoto University, Kyoto
KOICHI IWAI, Institute of Endocrinology, Gunma University, Maebashi
TAKEO KAKUNAGA, Research Institute for Microbial Diseases, Osaka University, Osaka
JUNTARO KAMAHORA, Research Institute for Microbial Diseases, Osaka University, Osaka
ATUNORI KASHIWAGI, Osaka University School of Medicine, Osaka
TSUNEHIKO KATSUNUMA, Institute for Enzyme Research, Tokushima University, Tokushima
TAKASHI KAWACHI, National Cancer Center Research Institute, Tokyo
SHIGEKO KIJIMOTO, Hokkaido University School of Medicine, Sapporo
SEIICHIRO KINOSHITA, Faculty of Science, University of Tokyo, Tokyo
KIKUKO KOGURE, National Cancer Center Research Institute, Tokyo
KIYOHIDE KOJIMA, Aichi Cancer Center Research Institute, Nagoya
TOSHIO KUROKAWA, Cancer Institute Hospital, Tokyo
SAJIRO MAKINO, Chromosome Research Unit, Faculty of Science, Hokkaido University, Sapporo
TAIJIRO MATSUSHIMA, Institute of Medical Science, University of Tokyo, Tokyo
TAIJU MATSUZAWA, Research Institute for Tuberculosis, Leprosy and Cancer, Tohoku University, Sendai
OSAMU MIDORIKAWA, Faculty of Medicine,

Kyoto University, Kyoto
YOSHIAKI MIURA, Chiba University School of Medicine
TAEKO MIYAGI, Research Institute for Tuberculosis, Leprosy and Cancer, Tohoku University, Sendai
MICHIKO MURA, National Cancer Center Research Institute, Tokyo
MASAMI MURAMATSU, Tokushima University School of Medicine, Tokushima
SUSUMU NISHIMURA, National Cancer Center Research Institute, Tokyo
TAKUZO ODA, Cancer Institute, Okayama University, Okayama
SHIGEYOSHI ODASHIMA, National Institute of Hygienic Sciences, Tokyo
MOCHIHIKO OHASHI, Tokyo Metropolitan Institute of Gerontology, Tokyo
ATSUSHI OIKAWA, National Cancer Center Research Institute, Tokyo
YOSHIO SAKURAI, Cancer Chemotherapy Center, Japanese Foundation for Cancer Research, Tokyo
YUKIHIRO SANADA, Institute for Enzyme Research, Tokushima University, Tokushima
KOZO SASAJIMA, National Cancer Center Research Institute, Tokyo
HARUO SATO, Research Institute for Tuberculosis, Leprosy and Cancer, Tohoku University, Sendai
KIYOMI SATO, Hirosaki University School of Medicine, Hirosaki
MAKOTO SEIJI, Tohoku University School of Medicine, Sendai
TADASHI SUGAWA, Osaka City University School of Medicine, Osaka
TADASHI YAMAMOTO, Institute of Medical Science, University of Tokyo, Tokyo
ISAO YAMANE, Research Institute for Tuberculosis, Leprosy and Cancer, Tohoku University, Sendai
YUKIO SUGINO, Takeda Biological Research Laboratories, Osaka
TAKETOSHI SUGIYAMA, Kobe University School of Medicine, Kobe
TAKEHIKO TACHIBANA, National Cancer Center Research Institute, Tokyo
SHOZO TAKAYAMA, Cancer Institute, Japanese Foundation for Cancer Research, Tokyo
HIROSHI TERAYAMA, Faculty of Science, University of Tokyo, Tokyo
AKIRA TOKUNAGA, National Cancer Center Research Institute, Tokyo
REIKO TOKUZEN, National Cancer Center Research Institute, Tokyo
MIKIO TOMIDA, Cancer Institute, Japanese Foundation for Cancer Research, Tokyo
IKUKO TOMINO, Jikei University School of Medicine, Japanese Foundation for Cancer Research, Tokyo
SHIGERU TSUKAGOSHI, Cancer Chemotherapy Center, Japanese Foundation for Cancer Research, Tokyo
KEMPO TSUKAMOTO, National Cancer Center, Tokyo
TADASHI UTAKOJI, Cancer Institute, Japanese Foundation for Cancer Research, Tokyo
MINRO WATANABE, Research Institute for Tuberculosis, Leprosy and Cancer, Tohoku University, Sendai

Opening Address

H.I.H. Princess Kikuko Takamatsu

The first of the international symposia of the Cancer Research Fund which bears my name was held in 1970 on the subject of Virology and Immunology of Human Tumors. The second and the third symposia were held in 1971 and 1972 on Chemical Carcinogenesis and Cancer Epidemiology, respectively. The successful issue of these past symposia has been sufficient encouragement for undertaking the present symposium which is the fourth in the series.

As the subject for this fourth symposium my scientific advisors suggested the problem of Cell Differentiation, with special reference to malignancy. Particular topics included are the regulation of gene expression, cell differentiation in tissue culture, involving methods for suppression and induction of differentiation. Experiments on cell fusion will be considered and also the possibility of reversion of tumor cells to normal ones. These complex topics are related to the latest advance in molecular biology and are of fundamental importance not only to basic cancer research but also to the study of life science in general.

I am glad to act upon this suggestion and believe that discussion of these subjects by the international group of competent scientists engaged in studies in the forefront of these researches is bound to be invaluable.

On the occasion of the opening meeting of this symposium, it is my real pleasure to welcome all the participants, especially those from abroad, and extend to them my best wishes for a successful conclusion of this symposium.

My cordial thanks are due to the organizing committee for the thought, time, and energy they have expended in preparation.

Dr. Waro Nakahara

The old Chinese scholar Confucius opened his "Analects" with two famous sentences: "Is it not pleasurable to learn with constant perseverance and application?" and "Is it not pleasurable to have friends coming from distant countries?"

The second of these sentences, written half a century B.C., if brought up-to-date and to suit ourselves, would be: "Friends in science coming from distant parts of the world for scientific discussion"; that is to hold an international symposium.

Since the Confucian sentence was put in interrogative form, I would decidedly answer: "It is most pleasurable indeed to hold the 4th International Symposium of the Princess Takamatsu Cancer Research Fund on Differentiation and Control of Malignancy of Tumor Cells!" I am sure you will all agree with me and with Confucius.

Prof. John Paul

Your Imperial Highness, it is my great honour to be privileged to reply to your welcoming address. In doing so I should like to take the opportunity to state what I believe to be the greater significance of these symposia.

When we of the twentieth century look back across the ages to the ancient civilizations of China, Egypt, or Greece, what do we see to admire in them? Not the battles they fought nor their conquests of other peoples but surely their creative achievements—in art, in literature, in architecture, in engineering, and in science. If our civilization can survive its recurrent crises, what will the people of a future millenium remember of us? Not, I submit, the wars in Vietnam or Korea or the Middle East but, more probably, those great scientific achievements which have annihilated distance, taken man into space and unravelled some of nature's most intricate riddles. We scientists who tackle problems of Cell Differentiation are humble members of a much larger company which is engaged in a series of epic endeavours of a kind the world never saw until this century. These are endeavours in which some of the most brilliant minds of the entire world unite to push back the frontiers of knowledge and this symposium, in which scientists from every corner of the earth are brought together, is a manifestation of them.

Our own task is perhaps a special one. Not only does it involve a daunting intellectual challenge: it has also a great humanitarian aim. For it is clear to all of us here that the only long-term hope of abolishing the scourge of cancer from mankind lies in research, particularly in research which is directed to understanding the nature of that discipline which maintains constant harmony among all the cells of an individual, a discipline which sometimes fails and results in cancer.

Often, in the day-to-day routine of our experiments and in the frustrations and disappointments which commonly accompany them, we forget that, minor figures though we all are as individuals, we are, neverthelss, principal actors in the high drama of human history. Your Imperial Highness has clearly recognised, in bestowing your patronage, that in bringing scientists together from every corner of the world to exchange ideas, you are helping to further progress towards our common aims.

The subject of this year's symposium is well chosen. We now know many of the factors that can cause cancer—viruses, carcinogenic chemicals, physical agents such as X-rays—but we do not know how they cause it. In particular we have not been albe to identify the crucial point at which all these agents converge. However, the wave of molecular biology, which started with the elucidation of the genetic code, and the solving of the problems of protein synthesis in bacteria, has now been impinging on our understanding of the working of mammalian cells for some years. Our ideas are rapidly clarifying, as this symposium will show, and there is a feeling of confidence in the air that soon we will understand rather well the normal workings of animal cells and the ways in which carcinogenic factors disrupt them.

The Princess Takamatsu Cancer Symposia have established a reputation for high standards and we confidently expect that the present symposium will maintain these. Your organising committee has gone to great pains to bring together a distinguished company and has made all arrangements with impeccable efficiency. On behalf of the foreign guests, I wish, therefore, to express our appreciation and thanks to you and to them. To your Imperial Highness we must all express our very special gratitude for extending your gracious hospitality to us and creating this opportunity for us to meet together in your beautiful country.

Prof. Sidney Weinhouse

Your Imperial Highness, it is a great honor to respond to your gracious welcome to the participants of this international symposium. I have the special privilege of thanking you on behalf of the foreign guests, and to express to you our appreciation for your support and encouragement. The reading of research reports in scientific publications is a necessary, but sometimes tedious enterprise; and it is never a satisfactory substitute for communicating directly and personally; with the spoken as well as the written word. This symposium is a welcome and timely opportunity to share the excitement of new discoveries and ideas; to make and to renew friendships; and to come to know our Japanese colleagues, not just as names in books or journals, but as fellow-workers toward a common goal, the control of a disease whose burden of misery and grief is incalculable. Your Imperial Highness has sponsored previous international symposia on virology and immunology, chemical carcinogenesis, and cancer epidemiology. Each represents a spearhead in a worldwide assault on the cancer problem. These symposia have each broken new ground, and the proceedings thus far published represent milestones of progress in each of these fields. Cell differentiation, the subject of this fourth symposium, presents a subtler but no less promising approach to the control of malignancy. Researchers have long recognized that cancer is associated with a disorder of differentiation. With the rapidly burgeoning knowledge of the molecular biology and enzymology of the cancer cell, this hitherto elusive feature of neoplasia is now being dissected at the molecular level. One can envision the possibility that some day this disorder of differentiation, when better understood, can be prevented or corrected.

As in other fields of cancer research, Japanese scientists are at the forefront of these studies. It should be a source of great satisfaction to you that the universally acclaimed leadership of Japan in cancer research is due not only to the brilliance of your scientists but in no small measure, I feel sure, to the generous support provided by the Princess Takamatsu Cancer Research Fund. We thank you for this privilege of being here, and of sharing in your equally unsurpassed warmth, friendship, and hospitality.

Contents

Phenotypes of Tumor Cells

Control of Differentiation: Induction of Differentiation

Control of Differentiation: Suppression of Differentiation

Tumor Reversal

REGULATION OF GENE EXPRESSION

Hormonal Regulation of Gene Expression

Gordon M. Tomkins
Department of Biochemistry and Biophysics, University of California, San Francisco, California, U.S.A.

Abstract: Adrenal glucocorticoids induce specific enzymes in cultured hepatoma cells and kill cultured immunocytes. In the former system, we have established that the events leading to enzyme induction involve (1) the association of the hormone with specific allosteric cytoplasmic receptors, (2) following conformational changes, the association of the receptor-steroid complex with specific high affinity nuclear sites containing DNA, and (3) the accumulation of mRNA active in coding for the synthesis of specific induced protein. Additional gene products regulate inactivation of the messenger which appears to be under specific physiological control as well.

Cultured lymphoma cells also contain receptors for glucocorticoids which when complexed with steroids migrate to the nucleus and set in motion a series of events resulting in cell death. This phenomenon permits the selection of steroid-resistant variants which provides genetic as well as biochemical methods for the analysis of cell-steroid interactions.

Cultured lymphoma cells are also killed by cyclic AMP, and variants resistant to this agent can also be selected. These cells therefore facilitate the study, with chemical and biological methods of the actions of individual hormones as well as the interaction between different classes of hormones.

Not surprisingly, in view of their important biological role, hormones interact with membranes and chromatin, the cellular structures primarily responsible for coordinating regulation and differentiation. In this context, it is of interest that important aspects of malignant transformation may be characterized by aberrations of hormonal control mechanisms. This conclusion is based on the observations that transformed cells require less serum for growth than do untransformed cells, and that the serum factors are hormonal in nature (*4*, *8*).

Effectors like polypeptide growth factors and antigens which interact with specific membrane receptors can have quite general effects on cellular function, for example, antigen-membrane receptor interactions stimulate cellular proliferation and differentiation and, ultimately, antibody production. The latter, however, is not the immediate result of effector-receptor combination. Other classes of effectors, however, appear to affect gene expression more directly. These agents, for example, the steroids and, very likely, thyroxin, combine with intracellular receptors; and the receptor-hormone complexes thus formed interact with specific nuclear acceptor sites associated with chromatin (*10*).

Mechanism of the Action of Glucocorticoid

I shall limit the present communication to an outline of some of our experimental work on regulation by the adrenal glucocorticoids. We have worked primarily with 2 lines of steroid-sensitive cultured cells: rat hepatoma (HTC) (*17*) and mouse lymphoma (S49) (*9*). In the former cells, glucocorticoids induce the accumulation of the mRNA, and therefore augment the synthesis of, tyrosine aminotransferase (TAT) (*17*). In the latter cell lines, steroids induce macromolecules which inhibit cell growth and ultimately lead to cell death.

In both cell types, biologically active steroids penetrate the cell membrane (*11*) and combine with specific cytoplasmic receptor proteins (*1*, *12*). The strength of this association, which is non-covalent but of high affinity, depends on the structure of the steroid molecule (*12*). Based on experiments comparing the effects of a large number of steroids on enzyme induction, we have inferred that glucocorticoid receptor (GR) molecules equilibrate between "active" and "inactive" conformations. We assume that "optimal inducers" associate only with the active conformation, "anti-inducers" only with the inactive form, and "sub-optimal inducers" with both forms of the receptors (*14*). There are 10,000–20,000 cytoplasmic receptor sites per cell which sediment at 6S in low salt and at 4S at higher ionic strengths (*1*).

Following the reaction of receptors with biologically active inducing steroids, a temperature- or salt-dependent "activation" of GR complex takes place allowing subsequent localization of the complexes in the nucleus (*2*, *5*).

Cell-free experiments have suggested that HTC, S49, and rat uterine cell nuclei contain specific binding sites for GR complexes. These acceptor sites are destroyed when nuclei are treated with pancreatic DNase prior to reaction with the complexes (*2*, *6*).

Uterine nuclei also contain distinct sites for the binding of estradiol-receptor (E_2R) complexes, whereas HTC cell nuclei do not contain E_2R acceptors (*6*). The E_2R acceptor sites in uterine nuclei, unlike the GR acceptors do not appear to be DNase-sensitive (*6*). We therefore infer the existence of distinct classes of nuclear acceptor sites specific for different receptor-steroid complexes (*6*).

Nature of Nuclear Acceptor

The DNase sensitivity of the nuclear acceptor sites indicates that DNA plays a role in the structure of GR acceptor activity. In support of this idea, we find that soluble GR complexes bind to purified DNA (*2*). This reaction, however, appears to be only partially specific for the polynucleotide since double-stranded DNA from HTC cells, *Escherichia coli* and bacteriophages λ or T7 all react with "activated" GR complexes. On the other hand, single-stranded DNA forms a weaker association, and the complexes do not associate with RNA, suggesting at least some element of specific protein-polynucleotide recognition, rather than simple ion exchange (*13*).

Indeed, the central enigma in steroid hormone action seems to reside in the proper identification of the nuclear acceptor sites. At the moment, however, significant obstacles prevent the resolution of this question. For instance, experiments with intact uterine cells have shown that the nuclear E_2R acceptor sites do not become saturated as the concentration of E_2R complex increases (*18*). Similar results have been obtained with intact HTC cells (*3*). Second, when the nuclear acceptor sites for E_2R in uterus or GR in HTC cells are saturated by exposure of intact cells to estradiol or dexamethasone, there is no dimunition of the ability of isolated nuclei to bind the corresponding receptor-steroid complex (*7*). Despite these apparent paradoxes, we imagine that the acceptor sites detected in cell-free experiments somehow represent those involved in the biological action of the steroids. In intact cells, however, for unknown reasons, more sites are available and they have a lower affinity for GR than corresponding sites in isolated nuclei.

Steroid-resistant Variants

The resolution of this question will obviously require further investigation. Moreover, the resolving power of biochemical analysis may not be adequate for this task. For this reason we have begun a study of steroid action using another cell line. Although cultured S49 mouse lymphoma cells are killed on exposure to adrenal glucocorticoids, steroid-resistant variants appear at random at a rate of about 3×10^{-6}/cells/generation (*15, 16*). Steroid resistance is stable and heritable, and its frequency is increased by a variety of mutagens. We imagine, therefore, that the transition from steroid sensitivity to resistance is the result of a mutation (*15*). Whatever their origin, however, the variants can be used to analyze cell-steroid interactions (*16*). From the general model of hormone action, we predict that 3 resistant phenotypes should exist: those with defective steroid binding activity (r-), those which retain receptor activity but where nuclear transfer of GR complexes does not occur (nt-); and those where the first 2 reactions take place, but in which cell death does not follow ("deathless" or d-).

Experiments in intact cells and in cell-free extracts were carried out to screen steroid-resistant variant, and all 3 predicted phenotypes have been detected (*16*). Approximately 80% of the spontaneously arising variants are of the r- type whereas the remaining 20% are divided equally between the nt- and d- cell types.

Preliminary experiments have shown that receptor-steroid complexes prepared from several steroid-resistant nt- clones do not associate in cell-free transfer experiments with "wild-type" or "mutant" nuclei or with purified DNA. These results lead to a number of conclusions. First, since steroid binding activity is retained while nuclear and DNA binding are diminished the variant receptor molecules must be chemically altered. This suggests that in the case of this clone at least, the transition from steroid sensitive to steroid-resistant is the result of a mutation. Second, the experiments show that formation of the receptor-steroid complex, although necessary is not sufficient for hormone action. Third, the results show that the receptor has 2 separate domains: a hormone-binding site and a nuclear binding site. Other clones, of the d- phenotype bind greater than wild type amounts of receptor steroid complex to nuclei in intact cells. Cell-free DNA binding experiments show that receptor-glucocorticoid complexes isolated from these cells associate more strongly than wild-type complexes with DNA.

CONCLUDING REMARKS

These results, considered together with those obtained with nt- complexes greatly strengthen the idea that DNA plays a major role in the nuclear binding of receptor-glucocorticoid complexes. Whether specific base sequences are involved in this association and how this binding produces biological effects remains to be determined.

ACKNOWLEDGMENTS

This work was supported by Grants Nos. GM 17239 and GM 19527 from the National Institute of General Medical Sciences, NIH, and Contract No. CP 33332 within the Virus Cancer Program of the National Cancer Institute, NIH.

REFERENCES

1. Baxter, J. D. and Tomkins, G. M. Specific cytoplasmic glucocorticoid hormone receptors in hepatoma tissue culture cells. Proc. Natl. Acad. Sci. U.S., *68*: 932–937, 1971.
2. Baxter, J. D., Rousseau, G. G., Benson, M. C., Garcea, R. L., Ito, J., and Tomkins, G. M. Role of DNA and specific cytoplasmic receptors in glucocorticoid action. Proc. Natl. Acad. Sci. U.S., *69*: 1892–1896, 1972.
3. Baxter, J. D. Unpublished.
4. Dulbecco, R. Topoinhibition and serum requirement of transformed and untransformed cells. Nature, *227*: 802–806, 1970.
5. Higgins, S. J., Rousseau, G. G., Baxter, J. D., and Tomkins, G. M. Early events in glucocorticoid action: activation of the steroid receptor and its subsequent specific nuclear binding studied in a cell-free system. J. Biol. Chem., *248*: 5866–5872, 1973.
6. Higgins, S. J., Rousseau, G. G., Baxter, J. D., and Tomkins, G. M. Nature of nuclear acceptor sites for glucocorticoid- and estrogen-receptor complexes. J. Biol. Chem., *248*: 5873–5879, 1973.
7. Higgins, S. J., Rousseau, G. G., Baxter, J. D., and Tomkins, G. M. Nuclear binding of steroid receptors: comparison in intact cells and cell-free systems. Proc. Natl. Acad. Sci. U.S., *70*: 3415–3418, 1973.
8. Holley, R. W. and Kiernan, J. A. "Contact inhibition" of cell division in 3T3 cells. Proc. Natl. Acad. Sci. U.S., *60*: 300-304, 1968.
9. Horibata, K. and Harris, A. W. Mouse myelomas and lymphomas in culture. Exp. Cell Res., *60*: 61–77, 1970.

10. King, R. J. B. and Mainwaring, W. I. P. (eds.), Steroid-Cell Interactions, University Park Press, Baltimore, 1974.
11. Levinson, B. B., Baxter, J. D., Rousseau, G. G., and Tomkins, G. M. Cellular site of glucocorticoid-receptor complex formation. Science, *175*: 189–190, 1972.
12. Rousseau, G. G., Baxter, J. D., and Tomkins, G. M. Glucocorticoid receptors; relations between steroid binding and biological effects. J. Mol. Biol., *67*: 99–115, 1972.
13. Rousseau, G. G. Unpublished.
14. Samuels, H. H. and Tomkins, G. M. Relation of steroid structure to enzyme induction in hepatoma tissue culture cells. J. Mol. Biol., *52*: 57–74, 1970.
15. Sibley, C. H. and Tomkins, G. M. Unpublished.
16. Sibley, C. H., Gehring, U., Bourne, H., and Tomkins, G. M. Hormonal control of cellular growth. *In;* B. Clarkson and R. Baserga (eds.), Control of Proliferation in Animal Cells, pp. 115–124, Cold Spring Harbor Laboratory, New York, 1974.
17. Thompson, E. B., Tomkins, G. M., and Curran, J. F. Induction of tyrosine α-ketoglutarate transaminase by steroid hormones in a newly established tissue culture cell line. Proc. Natl. Acad. Sci. U.S., *56*: 296–303, 1966.
18. Williams, D. and Gorski, J. Kinetic and equilibrium analysis of estradiol in uterus: a model of binding-site distribution in uterine cells. Proc. Natl. Acad. Sci. U.S., *69*: 3464–3468, 1972.

Discussion of Paper by Dr. Tomkins

DR. PAUL: I understood you to say that there is no specificity in the binding of steroid receptors to DNA. Does this imply that they bind equally well to prokaryotic and eukaryotic DNA?

DR. TOMKINS: Yes, unfortunately, but not to RNA. Receptor-steroid complexes bind to λ or T7 DNA as well as hepatoma DNA. However, I don't believe this is simply ion exchange since the complex does not bind to RNA and also because an essential -SH group is required for receptor-steroid complex binding to DNA.

DR. G. SCHAPIRA: DNA has something to do with nuclear receptors. If you add DNA to cytoplasmic receptor, what will occur?

DR. TOMKINS: Receptor-steroid complexes bind rather non-specifically to isolated DNA (but not to RNA). We found that the cytoplasmic receptor-steroid complex migrates to the nucleus and associates with specific nuclear acceptor sites. These sites are composed of DNA and other nuclear factors, we imagine. Our main job is to define the nature of the nuclear sites.

DR. BARSKI: Are there any data on the behavior of steroid hormone receptor in the somatic cell hybrids obtained by crossing of steroid hormones receptor containing and non-containing cells?

DR. TOMKINS: Yes. Receptor-less (r^-) cells do not extinguish the receptor activity in hybrids between r^- and r^+ cells.

DR. DUBE: Could one then say that receptors are cytoplasmic proteins and not, for example, chromat inlinked? Of course, hormone-receptor complex interacts with DNA in the chromatin as you showed.

DR. TOMKINS: It seems as though free receptors are cytoplasmic while receptor-steroid complexes are nuclear. We imagine that receptors are "non-histone proteins" which reversibly attach to chromatin in the presence of steroid effectors.

Simplicity of Mammalian Regulatory Systems as Exemplified by the X-linked *Tfm* Locus of the Mouse

Susumu Ohno

Department of Biology, City of Hope National Medical Center, Duarte, California, U.S.A.

Abstract: Taking the nature of natural selection into account, I have argued that, while the construction of the complex mammalian body form from a single fertilized egg no doubt requires a large number of regulatory systems, each mammalian regulatory system, when singled out and analyzed, should turn out to be surprisingly simple.

A steroid hormone appears to function as an inducer in the sense of Jacob and Monod. Each target cell cytoplasma contains a small number (2,000 subunits or 500 tetrameric molecules) of a specific receptor protein having a high binding affinity to a specific steroid (K_m in the region of 0.5 to 1.0×10^{-9} M). The act of inducer binding apparently creates an acceptor-binding site *via* allosteric effect. Thus, the inducer-bound receptor protein moves into the nucleus and associates with a finite number of acceptor-binding sites in the chromatin.

Utilizing *noninducible* mutation of the androgen-receptor protein locus, we have shown that the mouse genome, and thus most likely all other mammalian genomes, contains only one androgen-receptor protein (*Tfm*) locus on the X chromosome. Inasmuch as the mammalian embryo has an inherent tendency to develop as a female, male development must be induced by an androgen. Therefore, it can be said that the X-linked *Tfm* locus is the one-and-only regulatory locus which determines the sexual phenotypes of mammals.

Today, everyone seems to be interested in the nature of the genetic regulatory systems that operate in mammals. While *Escherichia coli* and other prokaryotes proved themselves to be magnificent tools for elucidation of the principles of genetic regulatory mechanisms, our ultimate interest obviously lies in knowing what makes us and our immediate relatives tick. Such renewed and widespread interest is always accompanied by the hope of finding new biological principles. It has often been stated that genetic regulatory mechanisms of so complicated an organism as a mammal must be fundamentally different from those that operate in simple prokaryotes. Is the apparent complexity of mammalian regulatory systems one of quantity or quality? The organization of the complex mammalian body form, from a single fertilized egg, undoubtedly requires interplay between a very large number of genetic regulatory systems. Indeed, any given somatic cell type is under the control of not one, but several independent regulatory systems. Liver cells, for example, are endowed with machinery to respond to a number of different steroid as well as peptide hormones.

I have pointed out that the nature of natural selection is such that during evolution from unicellular prokaryotes to multicellular eukaryotes of increasing complexity, a necessary increase in the number of regulatory systems had to be compensated by simplifying each regulatory system. Thus, when any particular mammalian regulatory system is singled out and analyzed, it will turn out to be surprisingly simple (*15*).

In this paper, I will argue that the nuclear-cytosol steroid hormone-receptor protein essentially functions like well-elucidated regulatory proteins of *E. coli* and that the manifestation of mammalian sexual phenotypes is under the control of a single X-linked gene locus which specifies the androgen-receptor protein.

Nuclear-cytosol Steroid-receptor Proteins as the True Regulatory Proteins

Thanks to the work of many outstanding groups, we now know that steroid hormones essentially function like inducers of bacterial regulatory systems (*19*).

Each target-cell cytoplasma (not membrane, but supernatant fraction) of any steroid hormone contains a limited number of receptor protein molecules having specific binding affinity to that particular steroid hormone. The receptor proteins of different steroid hormones (estrogen, progesterone, androgenic steroids, *etc.*) share many common characteristics. The receptor subunit has a molecular weight of about 70,000, which indicates that there are about 700 amino acid residues in the peptide chain and the subunit is 5 times as long as it is wide. In androgen target cells of mice which we study, receptors appear to exist in a tetrameric form, and each target cell cytoplasma contains about 500 such molecules. The receptor protein

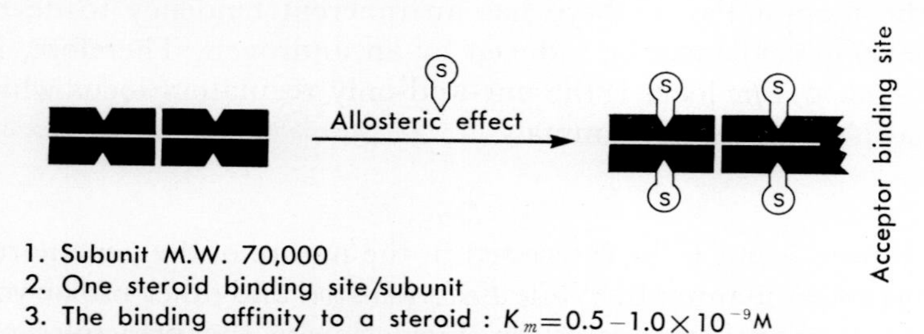
S
Allosteric effect
S
S
S
S
Acceptor binding site
1. Subunit M.W. 70,000
2. One steroid binding site/subunit
3. The binding affinity to a steroid : K_m=0.5–1.0×10^{-9}M
4. 500 molecules (2,000 subunits)/cells

FIG. 1. A schematic illustration of steroid-receptor proteins in the supernatant (cytosol) fraction of the target-cell cytoplasma. Each subunit apparently has one binding site to a specific steroid. The androgen-receptor protein in the mouse kidney appears to exist in the tetrameric form *in vitro*. The act of binding to a specific steroid hormone creates the acceptor binding site on the receptor molecule *via* allosteric effect.

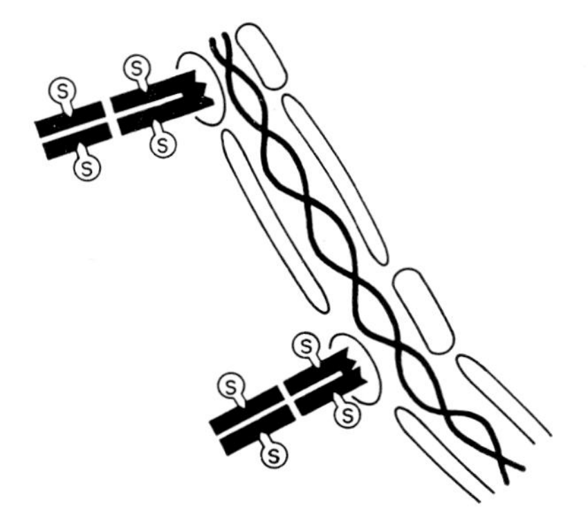

1. Assuming receptor protein remains tetrameric,
250 acceptor sites/haploid chromatin
2. Acceptor sites
DNA ?
organ specific acidic proteins ?
heterogeneous nuclear RNA ?
(messenger precursor)

FIG. 2. The steroid-bound receptor proteins which move into the nucleus and associate with specific sites on the chromatin *via* the acceptor-binding site. Whether they associate directly with a number of *operator* base sequences of DNA, or *via* organ-specific nonhistone acidic proteins is not clear. It is also possible that they associate with freshly transcribed heterogeneous nuclear RNA's of specific structural genes, thereby exercising post-transcriptional control.

shows a very respectable binding affinity to specific steroid, almost always having K_m in the vicinity of 0.5 to 1.0×10^{-9} M. The most important characteristic of steroid-receptor proteins is that the act of binding to a specific steroid creates an acceptor-binding site within that molecule via allosteric effect (Fig. 1).

Steroid-bound receptor proteins then move into the nucleus where they associate with the chromatin (Fig. 2). The only thing that we do not know is whether the steroid-bound receptor binds directly to DNA (*13*), or indirectly through organ-specific, nonhistone acidic protein (*14*), or to heterogeneous nuclear RNA's which are the precursors of mRNA's (*9*). The problem has been that binding to all 3 components of the chromatin mentioned above can be demonstrated under *in vitro* conditions.

Nevertheless, there is a general agreement that the binding between nuclear chromatin and steroid-bound receptors is specific in that only the target-cell nuclear chromatin contains a fixed number of acceptor-binding sites. In our system, that is to say dihydrotestosterone-receptor protein in the mouse-kidney proximal tubules, assuming that the receptor protein remains tetrameric, we find that there are about 250 acceptor sites per haploid amount of chromatin. A figure of 250 acceptor binding sites per haploid nucleus is a very satisfying number in that it indicates that in any given target-cell nucleus, steroid-bound receptors associate with 30 or so specific structural genes or their immediate transcription products.

Here I will draw an analogy between steroid target cells and what is known in the *lac* operon system of *E. coli*. *E. coli* is a haploid organism containing one *lac* operon, and therefore, one *operator* base sequence of 24 base pairs long and about 10 molecules of *lac*-repressor regulatory protein. What Riggs and his colleagues in our department have shown is that of the 10 regulatory molecules, one is of course specifically bound to an *operator*; among the 9 remaining molecules, however, only one is free, and 8 are bound nonspecifically to other parts of the *E. coli* genome; *nonoperator* DNA binding (*10*). It then follows that of 250 or so acceptor-binding sites per genome of steroid target cells, only one-eighth, 30 or so, might actually represent meaningful regulatory bindings. Thirty is quite an agreeable number in that an androgenic steroid indeed appears to cause the target cell-specific induction of about that many proteins and enzymes.

If we talk about the androgen target cells of mice, one of the target-cell types which we study is the kidney proximal tubule. In these cells, the administration of testosterone causes 100-fold induction of alcohol dehydrogenase and β-glucuronidase (*2*). Since several other enzymes are also known to be inducible, it would not be surprising if the total number of inducible enzymes and proteins in this target-cell type amounts to about 30.

Similarly, testosterone induces nerve growth factor (NGF), epithelial growth factor (EGF), and several peptidases in another target organ, namely, submaxillary salivary glands (*8*, *12*).

It appears that the steroid-receptor protein has all the necessary qualities to be a true regulatory protein in the sense of Jacob and Monod. It then follows that the gene locus, which specifies nuclear-cytosol receptor protein for a specific steroid hormone, should be regarded as the regulatory locus of that system. Needless to say, the gene locus in the genome can only be localized through a mutation.

Mouse X-linked Androgen-receptor Protein (Tfm) Mutation as a Noninducible Mutation of the Androgen-receptor Locus

What kinds of mutations should we expect of the steroid-receptor locus? What Jacob and Monod termed i^s (*noninducible*) mutation renders mutant target cells incapable of responding to either internal or externally administered steroid hormone. Such a mutation can come about if there is a mutational reduction, say by one order of magnitude, of the receptor's binding affinity for a steroid hormone.

Conversely, a mutational amino acid substitution near the steroid-binding site might hinder the manifestation of allosteric effect. Such mutant receptor proteins bind to steroid as strongly as the wild type, but the act of steroid binding does not create the acceptor-binding site, thus precluding an interaction between the androgen-bound receptor and the genome (Fig. 3).

Assuming that either of the above 2 types of mutations occurs, what would be the phenotype expected of an individual bearing the mutation? As far as ubiquitous hormones such as corticosteroids are concerned, i^s-type mutations are expected to behave as embryonic lethals, thus precluding any usefulness.

How about sex hormones? The affected individual in this case leads a healthy,

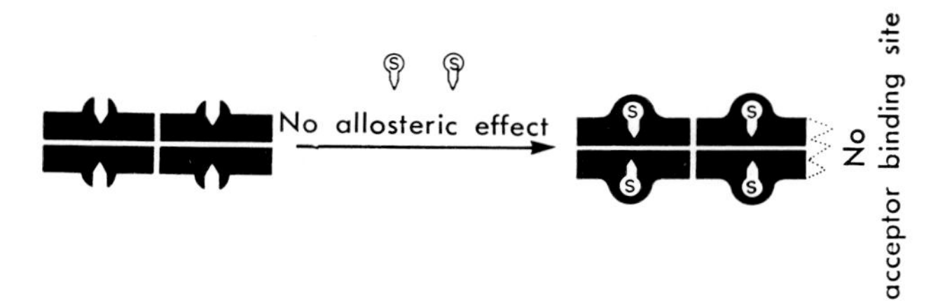

1. A mutational reduction in the binding affinity to an androgen:
   for example, a change in K_m from $0.5-1.0 \times 10^{-9}$M to 1.0×10^{-8} M
2. A mutation impairment of allosteric effect: the acceptor site not created.

FIG. 3. Two alternative forms of mutational changes in androgen-receptor protein which give the *noninducible* (*Tfm*) phenotype to the affected individual. Top: a mutational reduction in the binding affinity to effective androgens; 5α-dihydrotestosterone, testosterone, and 5α-androstan-3α, 17β-diol. Bottom: mutational hindrance in allosteric effect. Thus, the binding to a steroid does not create an effective acceptor-binding site.

although somewhat joyless, postnatal life. In fact, because of Jost's embryological work (*7*), one can fairly well predict the phenotype expected of the mutant which sustains an i^s-type mutation of the androgen-receptor locus.

Alfred Jost has shown that the mammalian embryo has an inherent tendency to develop as a female. This is simply because the differentiation of the embryonic Mullerian duct to Fallopian tubules and the uterus, of the urogenital sinus to the vagina, and the establishment of a connection between the nipple and underlying mammary glands are automatic processes that do not require any inducer. In sharp contrast, the differentiation of the Wolffian duct to epididymis, vas deferens, and seminal vesicles, of the urogenital sinus to prostates and penis, and of the hypothalamus-pituitary axis to a male behavioral direction, can occur only in the presence of androgenic steroids. This is the reason for the statement: femaleness and maleness represent the *noninduced* state and the *induced* state of one and the same regulatory system, and an inducer of this system must be testosterone or an effective metabolite of it.

Based on Jost's finding, one does not expect to see much effect of i^s mutation of the androgen-receptor locus on the female (*18*). However, the XY zygote will be profoundly affected. Since in such an XY, another regulatory system which determines the fate of indifferent gonads is not affected, he will develop testes and the testis will synthesize androgenic steroids, yet the target cells of androgens are not able to respond. Thus, he or she will manifest an externally female phenotype, and internally no Wolffian duct derivatives will be seen beyond the vagina, which ends as a blind sac.

This is a perfect description of the well-known *testicular feminization* syndrome in man, and an apparently homologous mutation has been described in cattle, rats, and dogs. In all these species, however, the X linkage of the *testicular* mutation locus has been suspected but not proven.

TABLE 1. Androgen Target-cell Types Affected by the X-linked *Tfm* Mutation

Embryonic Wolffian duct	Disappears despite externally administered androgen (Ref. *5*).
Embryonic urogenital sinus	No masculinization despite externally administered androgen (Ref. *5*).
Hypothalamus-pituitary axis	No neonatal imprinting.
	No negative-feedback inhibition by androgens of LH release (Ref. *6*).
Breast development	The connection between the nipple and mammary glands is not severed by androgens (Ref. *11*).
Kidney proximal tubules	No induction of specific enzymes.
	No hypertrophy by androgens (Ref. *2*).
Submaxillary salivary glands	No induction of specific proteins.
	No hypertrophy by androgens (Ref. *12*).

Only in the mouse was Lyon able to establish X linkage of the *testicular feminization* locus by its close linkage to 2 well-known X-linked coat-marker genes, Tabby and Blochy (*11*). In view of the evolutionary conservation *in toto* of the mammalian X chromosome (*16*, *17*), X linkage of the *testicular feminization* locus in the mouse means that the corresponding gene will be X linked in all other mammalian species including man.

Now that a desirable mutation that can be subjected to experimental analysis has been found, the first question is whether or not the *Tfm* locus is the only androgen-receptor protein locus in the genome. If it is, this apparent point mutation of the single X-linked locus should render all the divergent target-cell types in the body nonresponsive to administered androgens. On the other hand, if there is more than one receptor locus, some, but not all of the target-cell types should be affected by this mutation. As the list shows (Table 1), the *Tfm* mutation apparently affects every conceivable target-cell type in the body of affected *Tfm*/Y. Thus, it can be concluded that the mouse genome contains no other androgen-receptor protein locus.

Inasmuch as the entire body of the affected *Tfm*/Y totally ignores androgens, internally produced as well as externally administered, their testes are functioning as a supplier of estrogen. Testosterone is invariably converted to estrogen by peripheral tissues. This explains why the maintenance of the feminine phenotype by affected *Tfm*/Y is dependent upon the presence of testes. The name *testicular feminization* is a very appropriate name indeed. James H. Kan of our group is on the verge of inducing lactation in *Tfm*/Y by the administration of testosterone instead of estrogen together with progesterone, cortisol, and prolactin.

The next question is, how did the *Tfm* mutation affect the androgen-receptor protein? Initial observations (*1*, *4*) are disappointing in that the only difference that we can detect between the wild-type target cells of normal males as well as females and the target cells of affected *Tfm*/Y is in the amount of androgen-receptor protein. *Tfm*/Y target cells contained only 20 to 30% of the normal amount of androgen-receptor protein. The binding affinity to 5α-dihydrotestosterone, that is to say K_m, is apparently not appreciably affected by the *Tfm* mutation, remaining at about 0.5 to 1.0×10^{-9}M. This result is disappointing because a mere difference in amount does not prove that the androgen-receptor protein has mutated in *Tfm*/Y. It might equally well be that the *Tfm* locus specifies a gene product which regulates the production of androgen-receptor protein rather than the androgen receptor itself.

Because of this disappointment, we went through about a year of uncertainty (*3*). Fortunately, further studies by Barbara Attardi of this laboratory, have finally shown that the kinetic properties of the androgen-recepetor protein specified by affected *Tfm*/Y have indeed changed. Within target cells, testosterone is metabolized to 5α-dihydrotestosterone and then to 5α-androstan-3α(β), 17β-diol. Of the above 3 kinds of androgenic steroids, present within target cells, the wild-type receptor shows highest binding affinity for dihydrotestosterone and least affinity for androstandiol. The *Tfm* receptor protein, on the other hand, shows increased binding affinity to androstandiol. This then is the first indication that the *Tfm* mutation caused an amino acid substitution in the vicinity of the androgen-binding site of receptor molecules.

It was further found that mutated *Tfm* receptor protein does not release dihydrotestosterone as readily as does the wild-type receptor protein. When dihydrotestosterone-bound wild-type receptor protein is subjected to thorough dialysis, it retains only 10% of bound dihydrotestosterone, whereas mutated *Tfm* receptor protein retains as much as 40 to 50% of bound dihydrotestosterone. This peculiar property of mutated *Tfm* receptor protein has misled us in the past (*3*).

Thus, it appears that the *noninducible* mutation sustained by the X-linked *Tfm* locus of the mouse is not the simple first type, which represents a mutational reduction in the binding affinity to effective androgens. Rather it belongs to the more complicated second type. Mutated *Tfm* receptor protein binds as strongly to dihydrotestosterone as the wild type but there is a mutational hindrance of allosteric effect, so that the act of binding to dihydrotestosterone does not readily create an acceptor-binding site (Fig. 3). Indeed, our study of 1972 (*3*) indicated that dihydrotestosterone-bound *Tfm* receptor protein binds to the chromatin fraction less tightly than the wild-type receptor protein.

In conclusion, I can state unequivocally that the mouse genome, and most likely all other mammalian genomes, contains only one androgen-receptor protein locus on the X chromosome and this is the one-and-only regulatory locus which determines sexual phenotypes (*15*).

ACKNOWLEDGMENT

This work was supported in part by a contract #NO1-CB-33907 from the National Cancer Institute, U. S. Public Health Service.

REFERENCES

1. Bullock, L. P., Bardin, C. W., and Ohno, S. The androgen-insensitive mouse: absence of intranuclear androgen retention in the kidney. Biochem. Biophys. Res. Commun., *44*: 1537–1543, 1971.
2. Dofuku, R., Tettenborn, U., and Ohno, S. Testosterone-"regulon" in the mouse kidney. Nature New Biol., *232*: 5–7, 1971.
3. Drews, U., Itakura, H., Dofuku, R., Tettenborn, U., and Ohno, S. Nuclear DHT receptor in *Tfm*/Y kidney cells. Nature New Biol., *238*: 216–217, 1972.

4. Gehring, U., Tomkins, G. M., and Ohno, S. Effect of the androgen-insensitivity mutation on a cytoplasmic receptor for dihydrotestosterone. Nature New Biol., *232*: 106–107, 1971.
5. Goldstein, J. L. and Wilson, J. D. Studies on the pathogenesis of the pseudohermaphroditism in the mouse with *testicular feminization*. J. Clin. Invest., *51*: 1647–1655, 1972.
6. Itakura, H. and Ohno, S. The effect of the mouse X-linked *testicular feminization* mutation on the hypothalamus-pituitary axis. I. Paradoxical effect of testosterone upon pituitary gonadotrophs. Clin. Genet., *4*: 91–97, 1973.
7. Jost, A. The role of fetal hormones in prenatal development. The Harvey Lectures. Series 55, Academic Press, New York, 1961.
8. Levi-Montalcini, R. and Angeletti, P. U. Growth control of the sympathetic system by a specific protein factor. Quart. Rev. Biol., *36*: 99–108, 1961.
9. Liao, S., Liang, T., and Tymoczko, J. L. Ribonucleoprotein binding of steroid-"receptor" complexes. Nature New Biol., *241*: 211–212, 1973.
10. Lin, S. Y. and Riggs, A. D. *Lac*-repressor binding to DNA not containing the *lac* operator and to synthetic poly dAT. Nature, *228*: 1184–1186, 1970.
11. Lyon, M. F. and Hawkes, S. G. An X-linked gene for *testicular feminization* of the mouse. Nature, *227*: 1217–1219, 1970.
12. Lyon, M. F., Hendry, I., and Short, R. V. The submaxillary salivary glands as test organs for response to androgen in mice with *testicular feminization*. J. Endocrinol., *58*: 357–362, 1973.
13. Mainwaring, W. I. P. and Mangan, F. R. The specific binding of steroid-receptor complexes to DNA: evidence from androgen receptors in rat prostates. Adv. Biosci., *7*: 165–172, 1971.
14. O'Malley, B. W., Sherman, M. R., Toft, D. O., Spelsberg, T. C., Schrader, W. T., and Steggles, A. W. A specific oviduct target-tissue receptor for progesterone: identification, characterization, partial purification, intercompartmental transfer kinetics, and specific interaction with the genome. Adv. Biosci., *7*: 213–231, 1971.
15. Ohno, S. Simplicity of mammalian regulatory systems inferred by single-gene determination of sex phenotypes. Nature, *234*: 134–137, 1971.
16. Ohno, S. Ancient linkage groups conserved in human chromosomes and the concept of frozen accidents. Nature, *244*: 259, 1973.
17. Ohno, S., Becak, W., and Becak, M. L. X autosome ratio and behavior pattern of individual X chromosomes in placental mammals. Chromosoma, *15*: 14–19, 1964.
18. Ohno, S., Christian, L., and Attardi, B. Role of testosterone in normal female function. Nature New Biol., *243*: 119–120, 1973.
19. See all the papers in the Shering Workshop on Steroid Hormone Receptors. Adv. Biosci., *7*: 1971.

Discussion of Paper by Dr. Ohno

DR. SUGIMURA: You assumed the presence of 250 receptor sites per haploid chromatin sets. What is the basis of the assumption?

DR. OHNO: The chromatin fraction isolated from the mouse kidney was mixed with ^{3}H-dihydrotestosterone labeled near 8S receptor protein and passed through a Sephadex G-200 column. The count at the void volume peak is regarded as the receptor-chromatin complex and that near the 8S peak is regarded as free receptors. Making proper allowance for the number of target-cell nuclei per kidney, *etc.*, the number of acceptor binding sites per haploid chromatin was calculated.

DR. WEBER: Is your view related to the simplification analysis already embodied in the pleiotropy concept elaborated by Charles Darwin nearly a century ago?

DR. OHNO: Yes and no. Charles Darwin had no idea of the gene. See his idea of pangenes in the body fluid.

DR. OSTERTAG: You argue on the basis of mutational load and mutation rate per gene locus that there are only some 10^4 genes in mammals. I agree with that conclusion. But you go on that on the basis of entire gene size you must have 98% "junk DNA." If the gene is very large, much larger than you estimate, you could possibly reduce the amount of "junk DNA" to minimal proportions—that a lot of DNA in the chromomeres is functional is shown by the presence of lethal mutations all over the chromomeres. There is good evidence that many of these lethals are not due to frameshift.

DR. OHNO: Your argument is somewhat along the same line as that for the necessity of spacer sequences. The spacer sequence between ribosomal RNA genes can be long as in *Xenopus laevis* or short as in *Xenopus mullerii.* This shows that most of the long spacer sequence is dispensable. If the DNA base sequence in front of the structural gene corresponding to the 5′-end nontranslatable part of heterogeneous nuclear RNA contributes to the mutation load, one would expect that most of the deficient mutations of the structural genes are not associated with amino acid substitutions. The fact is that nearly all the known enzyme deficiencies of man and other mammals are associated with amino acid substitutions. It is for this reason I cannot buy your idea.

DR. TOMKINS: The mutation rate per rad increases with increasing genome size (Wolff and Abrahamson). Doesn't this mean that there is no "junk" DNA?

DR. OHNO: What they have shown is merely that the larger genome offers a larger target, and thus, sustains more hits. If functional parts (genes) and nonfunctional parts (junk) are interspaced, this is exactly the result that one expects.

DR. PAUL: In relation to Dr. Tomkins' comment, a simple explanation of Wolff's finding might be that the size of transcriptional units is in proportion to the size of the genome. Mutations might then occur as a result of polarity of transcriptions. Chovmk's work suggests this strongly.

DR. TOMKINS: Do you think all mutations in drosophola are polar? If not, there are very few genes in this organism.

DR. PAUL: In the rosy locus most mutations are null mutations until one end of the locus is reached. Then structural mutations of xanthine dehydrogenase emerge. This argues very strongly that many mutations are of a polar nature.

DR. OHNO: I do not buy the idea of polarity, for the reason stated in my answer to the question by Dr. Ostertag.

DR. MURAMATSU: It may be a matter of definition of the word "junk." However, when you say that the nonmessenger portion of these heterogeneous nuclear RNA's does not have any function, do you believe that a cell is continuously making some 10 to 20 times longer polynucleotide sequences per mRNA without any advantage? It does not seem reasonable that a cell is wasting this much energy on macromolecular synthesis just for nothing.

DR. OHNO: *E. coli* divides every so many minutes. Cells of multicellular organisms, on the other hand, are designed to divide twice a day or so. Some slowdown would surely be desirable.

Chromosomal Non-histone Proteins and the Transcriptional Unit

John PAUL, R.S. GILMOUR, I. HICKEY, P.R. HARRISON, N. AFFARA, and J. WINDASS

Beatson Institute for Cancer Research, Glasgow, U. K.

Abstract: Evidence relating to the structure of the transcriptional unit in eukaryotes is reviewed. In confirmation of earlier experiments it has now proved possible to demonstrate organ-specific behaviour in chromatin. Using DNA copied from globin mRNA by reverse transcriptase as a probe, it has been shown that RNA, transcribed by *Escherichia coli* RNA polymerase from erythroid cell chromatin, contains globin mRNA whereas RNA, transcribed similarly from brain chromatin, does not. It has also been shown that non-histone proteins from erythroid tissue are instrumental in promoting transcription from the globin gene. Based on this information and evidence from other laboratories, a structure for the transcriptional unit is proposed. This envisages a fundamental structure in which the DNA coding for a functional RNA is preceded by DNA containing regulatory sequences. Some of these are postulated to bind non-histone proteins which cause a local unwinding of supercoiled nucleohistone and thereby permit attachment of RNA polymerase. By reduplication and mutation of this fundamental structure, other, more complex, structures could arise which might give rise, on the one hand, to multiple copies of the same gene or, on the other, to a large transcriptional unit with a single protein-specifying sequence. It is shown that induction of haemoglobin synthesis in the Friend cell involves co-ordinate transcriptional and translational control of several genes. Somatic cell genetic studies implicate both positive and negative controlling elements. Model control circuits suggested by these findings are discussed.

The transcriptional unit in prokaryotes is the operon. What is the corresponding structure in eukaryotes? Although we have little certain knowledge, we have amassed sufficient experimental evidence to make some interesting speculations. I believe that there is now very good evidence that each chromosome is basically a single DNA double helix coated with histones, forming a nucleohistone thread which is highly supercoiled in such a way as to prevent the binding of large molecules such as RNA polymerase. We have much evidence to show that non-histone proteins are bound, possibly intermittently, along this thread and that they have the effect of causing local unwinding of the supercoil so that RNA polymerase molecules can become attached. I have postulated that these regions, which I call address sites, have a special configuration and that address sites therefore correspond quite closely to bacterial promoters except that the predominant control is positive rather than negative (*23*).

It seems likely that, despite the very large size of most eukaryotic genomes, the number of cistrons is relatively small (*18*). Most DNA in these large genomes is probably derived by tandem reduplication of simple transcriptional units. In consequence, many genes may persist as multiple copies but others may drift due to the accumulation of mutations until, in extreme instances, only a single functional copy remains. This communication will be devoted to discussing the evidence for the above statements.

Template Specificity of Chromatin

First, I wish to consider some aspects of the control of transcription in eukaryotes. Whereas the genome in prokaryotes is essentially naked DNA, in eukaryotes the DNA is associated with proteins to form chromatin. In some of our early work, we (*25, 26*) obtained evidence that chromatin was organ-specific in that RNA transcribed from a sample of chromatin by bacterial RNA polymerase, was characteristic of the organ from which the chromatin was derived, as judged by RNA/DNA hybridisation. These findings were verified by other investigators (*31*). The only techniques available at that time were the so called low C_0t (concentration $\times$ time) hybridisation methods; the evidence was therefore not complete because it was possible to measure only RNA transcribed from repetitive DNA. Recently, however, we (*15*) have obtained results which seem to prove this point conclusively (similar findings have been obtained independently by Axel *et al.* (*1*)). In these experiments, transcription of the globin gene was studied by exploiting the discovery (*19, 30, 34*) that viral RNA dependent DNA polymerase (reverse transcriptase) could be used to make a copy of globin mRNA. It is possible to obtain deoxyribonucleotide triphosphates of very high specific activity and consequently highly labelled complementary DNA (cDNA) can be prepared; we commonly use it with a specific activity of from $10–50 \times 10^6$ dpm/μg. By hybridising this cDNA to a specimen of RNA, presumed to contain globin messenger sequences, the cDNA can be used as a specific probe for these; because of its very high specific activity, it is very sensitive and can easily be used to measure $10^{-5}–10^{-4}$ μg of globin mRNA.

In these recent experiments, RNA was transcribed from chromatin with *Escherichia coli* polymerase and globin cDNA was used as a probe to determine whether this transcript contained globin mRNA sequences. Chromatin was prepared from erythropoietic tissue (foetal liver) and also from non-erythropoietic tissue (brain). As shown in Fig. 1, the RNA isolated from the incubation mixture with brain chromatin had no detectable globin mRNA whereas the RNA isolated from the incubation mixture containing foetal liver chromatin had. One possible defect in these experiments is that chromatin from erythropoietic tissue might al-

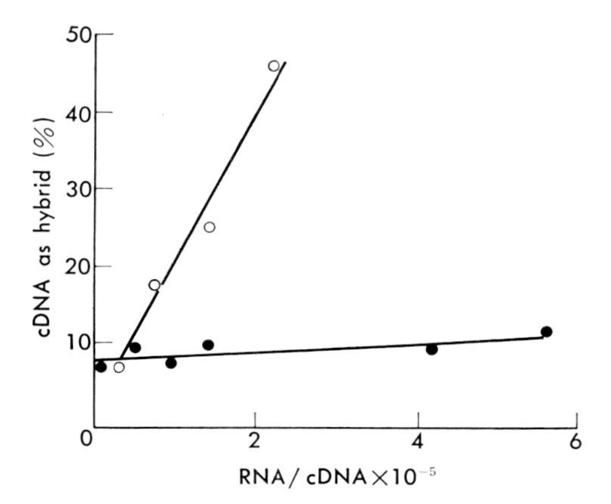

FIG. 1. Measurement of the globin mRNA content of RNA transcribed from mouse foetal liver chromatin or brain chromatin. Reverse transcriptase was used to make single-stranded ^{3}H-cDNA from a template of purified globin mRNA. RNA was transcribed from each chromatin template with *E. coli* RNA polymerase and purified. A constant amount of cDNA was hybridised to increasing amounts of each RNA in conditions which would lead to all reactions being complete. The amount of cDNA hybridised was measured by its resistance to S1 nuclease, which specifically digests single-stranded DNA but not double-stranded DNA or RNA-DNA hybrid. ○ hybrid formed with RNA transcribed from foetal liver template; ● hybrid formed with RNA from brain chromatin template.

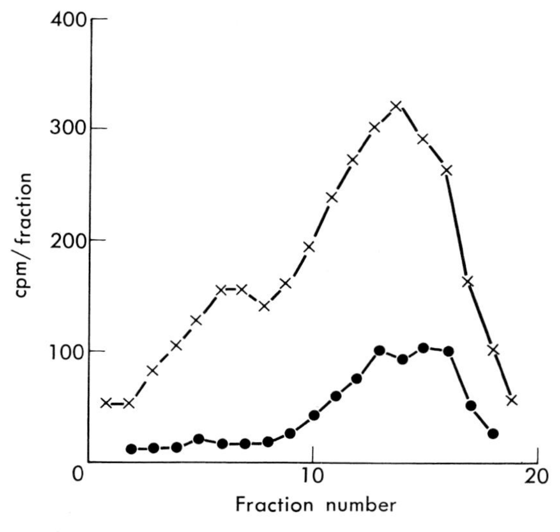

FIG. 2. Demonstration of a hybrid between cDNA and newly synthesised RNA. RNA was synthesised on a template of foetal liver chromatin, as described for Fig. 1, but ^{32}P-labelled nucleotide triphosphates were included in the reaction mixture. After incubation with ^{3}H-cDNA, the presumptive hybrid was banded in CsCl. Both ^{3}H-cDNA (X) and ^{32}P-RNA (●) are found at the buoyant density position of an RNA-DNA hybrid molecule.

ready contain some mRNA. To investigate this, we performed a number of control experiments. For example, we prepared incubation mixtures from which nucleotides were missing and were unable then to demonstrate globin mRNA sequences in the RNA isolated from the reaction mixture. However, we felt that even this criterion was inadequate since it was possible that during active transcription, partly transcribed sequences were displaced. We needed clear proof that a hybrid was formed between cDNA and newly made RNA and therefore performed an experiment in which RNA was made in the presence of ^{32}P-labelled nucleotides and then hybridised to ^{3}H-labelled cDNA. The hybrid material was isolated and banded on cesium chloride. As shown in Fig. 2, most of the material banded with a buoyant density of 1.78 which is characteristic of a hybrid. Only a small amount of the tritium labelled DNA could be seen at a buoyant density of 1.71 which is the density of cDNA itself. Moreover, a peak of ^{32}P counts appeared in exactly the same position as the hybrid. Hence, the hybrid contained both ^{3}H-labelled DNA and ^{32}P-RNA, an observation which leaves little doubt that the globin gene was actually transcribed in the course of the reaction.

Regulatory Function of Non-histone Proteins

What is the reason for the tissue specificity of chromatin? It is not due to differences in DNA and we must therefore assume that it is due to associated molecules, probably proteins. In our earlier studies (*14*, *26*) we obtained direct evidence for this and presumptive evidence for the separate roles of histones and non-histone proteins. In some of these experiments, we simply measured the rate of RNA synthesis on a given template in standard conditions. When all samples contained the same amount of DNA, we observed that chromatin had only about one fifth of the template activity of DNA. If DNA were combined with histones (by dissolving these components in 5M urea: 2M NaCl and then progressively reducing the salt and urea content during 18 hr), the resulting nucleohistone had a much lower template activity than either. However, if a similar experiment were conducted with the addition of non-histone proteins during the reconstitution then this material gave the same value as chromatin. These observations suggested that some components of the non-histone proteins neutralised the inhibitory effect of histones and did so to the same quantitative degree as occurred in native chromatin. A further observation made in these experiments was that the addition of extra histones to chromatin did not significantly reduce its template activity (*27*). These results suggested that non-histone proteins somehow antagonised the inhibitory action of histones.

The results were only quantitative and the question remained; do non-histone proteins antagonise the inhibition of histones in a specific way? We had already demonstrated this (*14*, *26*) in experiments similar to those described above except that RNA from the reaction was isolated and examined by molecular hybridisation to DNA. The product transcribed from DNA was found to hybridise with a large fraction of DNA while the product from nucleohistone hybridised with a very small amount of DNA. On the other hand, the product transcribed from chromatin hybridised to an intermediate extent. Most important, the product transcribed from reconstituted chromatin, made from DNA, histones and non-histone proteins, hybridised to exactly the same amount. This argued that the activating effect of non-histone proteins was specific and the conclusion was borne out by competition experiments in which we studied RNA transcribed from DNA, native chromatin, chromatin reconstituted from DNA, histones, and non-histone proteins and a 'pseudo-chromatin' made from DNA, histones, and bovine serum albumin. All the templates were prepared from calf thymus; the RNA transcribed from them was hybridised in competition with natural RNA isolated from calf thymus nuclei. We found that, whereas very poor competition was obtained with the RNA transcribed from either DNA or the 'pseudo-chromatin,' quite effective competition was obtained against the RNA transcribed on the one hand from native chromatin and on the other from reconstituted chromatin. These experiments argued very strongly that precise reconstitution was obtained and that this depended on the presence of non-histone proteins. They also suffered from the defect that they were performed by low C_0t hybridisation.

We have therefore carried out some experiments in which we have used globin cDNA. We first showed that RNA transcribed from reconstituted brain chromatin, like native brain chromatin, did not contain globin mRNA sequences, whereas RNA transcribed from reconstituted foetal liver chromatin, like native liver chromatin, did contain globin mRNA sequences. We then prepared reconstituted chromatin using brain chromatin to which we added non-histone proteins from mouse foetal liver. When RNA was transcribed from the resulting material, it was possible to demonstrate that it contained globin mRNA sequences. These results then seem to show that histones non-specifically inhibit transcription and that non-histone proteins are capable of reversing this inhibition in a specific way.

A simple model based on these observations would propose that non-histone proteins associate with the entire stretch of nucleohistone to be transcribed but I consider this model unsatisfactory because the specificity of the non-histone effect suggests that it is dependent on DNA configuration and it seems unlikely that sequences coding for protein structure could also be restricted in the way this would demand. Accordingly, I prefer an alternative model (*23*) which proposes that the informational sequences are preceded by one or more address loci to which non-histone proteins can attach.

Structure of Transcriptional Units

In the simplest eukaryotic chromosomes which presumably occur in organisms with very small genomes, such as *Dictyostelium discoides*, the transcriptional units may be of precisely the structure proposed and recent results (*21*) strongly suggest this. However, most organisms have much larger genomes. Consider the genome of *Drosophila*. This contains sufficient DNA for approximately 10^5 genes. However, most of our current evidence suggests that the actual number of genes is much smaller (*2*, *18*).

In these studies, a very close correlation has been shown between the number of complementation groups and the number of bands in different parts of the genome and since there are about 5,000 bands in the giant chromosomes of most Diptera, this strongly suggests that the number of complementation groups, which is likely to approximate to the number of genes, if of the order of 5,000. O'Brien (*22*) has pointed out that this conclusion should be accepted only with great caution since the studies depend on the use of lethal mutations and many mutations studied in *Drosophila* are not lethal. There is a paradox here which has not been resolved. Assuming that the data are correct, however, there are only two possible explanations. One is that the ratio of lethal to non-lethal mutations is very high. The alternative is that essential and non-essential genes are distributed in such a way that each band contains one essential gene and a number of non-essential ones. Most people favour the former interpretation and this would lead to the conclusion that about 95% of the DNA in the *Drosophila* genome does not code for proteins.

Many geneticists find it difficult to accept the notion that so much DNA is non-functional in the sense of providing information for proteins. Moreover, from a number of studies, we know that most of the DNA in the genomes of many organisms is transcribed some time in the life cycle. Whether the transcript is translated or not is not clear but there are strong indications that a great deal of it is not. We are now in a position to consider possible ways in which the extra DNA might arise. There is no doubt that some genes are reduplicated many times. The classical example is ribosomal genes and it may be remembered that these consist of spacer regions alternating with regions from which ribosomal precursor molecules are transcribed. This structure could readily be explained by the alternation of address sites and coding regions. Other examples of structures of this type of arrangement are probably the genes for tRNA, 5S RNA, and histones.

The number of genes constructed in this way is probably quite small. Studies of the re-annealing kinetics of DNA from many species show that a large part of the genome of eukaryotes is unique implying that there are single or only a few copies of most genes (*5*). It could be argued that the 'unique' DNA has little biological significance but this argument can no longer be upheld since it has been demonstrated quite convincingly that some well-known genes are unique. For example, Suzuki *et al.* (*33*) have shown that the silk fibroin gene is unique in the silkworm. Moreover, we and others (*3, 4, 16*) have shown the same for the globin gene in the mouse and the duck. The technique we used in our experiments was to employ a minute amount of globin cDNA to trace the re-annealing of the globin gene during a C_0t type experiment. We were able to show that re-annealing of the globin gene occurred during re-annealing of the unique sequences of DNA. It is of interest, by the way, that we found that DNA from erythropoietic tissue behaved in the same way as DNA from sperm, indicating that there is no gene amplification in erythropoietic tissue.

Hence, some and possibly most, genes are unique. Moreover, there is a great deal of strongly suggestive evidence now that the transcriptional units containing these are very large. Chovnick and his colleagues (*8*) have presented genetic evidence to this effect for the gene representing xanthine dehydrogenase in diptera. We, ourselves (*36*) and also Scherrer's group (*17*) have shown that the precursor of globin mRNA is very large; presumably, therefore, the transcriptional unit is large. In our experiments, we injected high molecular weight RNA from mouse foetal liver (an erythropoietic tissue) into *Xenopus* oocytes and demonstrated the synthesis of mouse haemoglobin in them. Direct observation of certain chromosomes, for example, the lampbrush chromosomes of amphibia, also seems to indicate that some transcripts from these can be very large indeed. Some of the lampbrush loops, which are apparently transcribed in their entire length, can be more than 100 μm long. The implication from these and other observations would therefore seem to be either that these transcripts are polycistronic messengers or that much of the DNA and RNA is not structural.

Much of the RNA from these large transcripts is certainly degraded. Just how much is not used to specify polypeptides is not known, but if we assume that the figure of 5,000 functional genes is right for *Drosophila* and that only a few of these are present as multiple copies, it could be well over 90% in this organism.

There is another strong argument which indicates that we should seriously consider the likelihood that much RNA, though transcribed, is not translated. This argument derives from what Callan calls the C-paradox (*7*). The genomes of quite closely related organisms can vary enormously and a particularly well-studied example is the amphibia in which some organisms can have 20 times as much DNA per cell as morphologically similar organisms. As a particular example, *Ambystoma* has a genome 8 times larger than that of *Xenopus* and Strauss (*32*), has shown that it is correspondingly more complex. It is unthinkable that *Ambystoma* should require 8 times as much DNA as *Xenopus* and we must therefore infer that a high proportion of its DNA is non-functional. Indeed it may be remarked that the genome of *Ambystoma* is about 250 times larger than that of *Drosophila* and I believe that most people would consider is extremely unlikely that 250 times as much information is required to make a newt than to make a fruitfly.

Recently some evidence has been obtained about the arrangement of DNA in the eukaryotic genome which is probably relevant to the fine structure of the transcriptional unit. It has been recognised for some time that there are three kinetic classes of DNA in the eukaryotic genome. One class, which anneals very rapidly, consists of a very large number of copies of very short sequences. This kind of DNA is clustered, can frequently be isolated as a satellite in buoyant density gradients and apparently occurs mainly in the chromomeres. Walker refers to it as simple sequence DNA (*35*). The second category is moderately repetitive DNA which consists of a smaller number of repeats of much longer sequences, on the average about 300 nucleotide pairs. Finally, there is unique DNA. Recent evidence suggests that the unique and moderately repetitive DNAs are intimately intermixed and indeed probably alternate in such a way that after about 1,000 nucleotide pairs of unique DNA, there appears a block of repetitive sequences (*10*).

We can now list some of the general points that must be taken into consideration in proposing any model of the transcriptional unit.

1) Eukaryotic chromosomes contain, besides DNA, histones and non-histone proteins. Histones non-specifically inhibit transcription whereas non-histone proteins specifically facilitate transcription.
2) Most of the DNA in the eukaryotic genome is not used directly in protein synthesis. The greater part of it is transcribed but most of the transcription product is degraded.

3) Some genes are present as multiple copies in which an untranscribed spacer alternates with a transcribed region several hundred times. Although some of the transcription product is degraded, most of it appears as functional RNA in the cytoplasm.
4) Some other genes are present as single copies. These are transcribed as much larger transcriptional units and most of the transcriptional product is degraded.

I believe that one can propose a compound transcriptional unit derived by tandem reduplication from the basic unit. I have already proposed which explains all these features (Fig. 3) (*23*). Tandem reduplication of simple transcriptional units would by itself provide an alternating model which is immediately reminiscent of the ribosomal gene. If strong selective pressures existed to maintain a large number of copies of these genes (as seems to be the case for both ribosomal and histone genes) then, of course, these structures would be expected to persist. On the other hand, if selective pressure demanded no more than a single copy of the gene

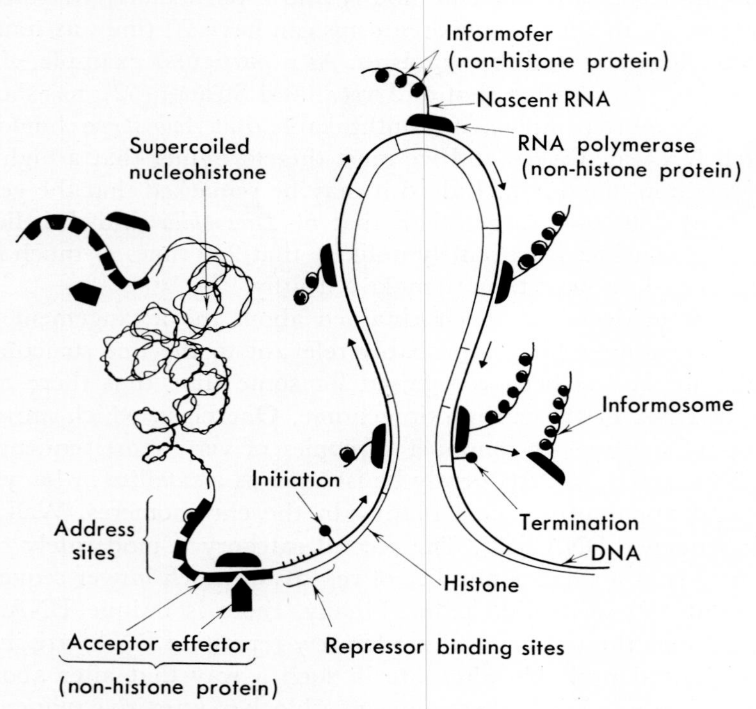

FIG. 3. A hypothetical model for the basic eukaryotic transcriptional unit. It is postulated that nucleohistone is supercoiled in such a way that large molecules like RNA polymerase cannot approach DNA. Non-histone proteins, bound to 'address sites,' are postulated to extend the nucleohistone in promoter-like regions so as to permit the attachment of large molecules. In this version of the model 2 components are envisaged at each address site, a receptor molecule, already bound to DNA, and an effector which can modify the receptor-DNA complex in such a way as to permit RNA polymerase attachment. If polymerase attaches and successfully passes the repressor binding sites, it begins to polymerise nucleotides to make RNA. Nascent RNA molecules bind informofer proteins and these produce further unwinding of nucleohistone with the result that an extended RNA- and protein-rich region develops.

in the genome, it is not difficult to imagine how random mutation could lead to a drift towards a structure in which, with the removal of termination sequences, long transcripts could be obtained, only part of which might contain meaningful information.

One requirement of this model is difficult to explain and that is that it demands that there must be some kind of selective pressure which leads to the acquisition of a very large genome. The experimental facts seem to indicate that unquestionably some kind of selective pressure does exist to this effect although the nature of this is at present elusive. It may simply be that, in eukaryotes, the amount of DNA in the genome is by itself selectively neutral and that it can therefore drift up and down. Perhaps no more significance than this should be attributed to variation in gene size but it may be pointed out that a large genome could conceivably bestow indirect advantages in at least 2 ways. First, it could permit a rather fine adjustment of gene dosage through selection pressure. Secondly, by maintaining a pool of sequences in each transcriptional unit, ancestrally related to functional genes, the evolution of new proteins such as iso-enzymes, might be facilitated in organisms which, because of their long generation times might otherwise evolve very slowly.

Co-ordination of Transcription and Translation

These speculations about the transcriptional unit suggest that transcription might be regulated by both positive and negative controls as in bacteria, the main difference being that the positive controls may be of a kind unique to eukaryotes. A question which arises therefore is: are the characteristics of these control circuits, similar to those observed in bacteria? In particular, are these controls freely reversible? The question has to be asked because we have evidence on the one hand that transcriptional controls are important to cytodifferentiation and, on the other, that cytodifferentiation is rather stable (*6*, *9*, *20*). Two very different mechanisms might easily be envisaged. On the one hand, non-histone proteins, responsible for activating transcription in specific sites, might bind only when allosterically altered as in many bacterial systems. Control might therefore be mediated by a mechanism involving a dynamic equilibrium between an effector and a receptor (Fig. 3). On the other hand, it is also possible to conceive of a mechanism in which, during replication of the chromosome, the presence of an inducer could result in a heritably altered structure. Nuclear transplantation and cell hybridisation experiments do not help to distinguish between these 2 possibilities. It is necessary to find a model system in which differentiation can be induced at will and which provides sufficient materials for analysis. Such a system has recently become available as a result of the discovery by Friend and her colleagues (*11*, *12*) that a mouse erythroleukaemic line can be induced to synthesise haemoglobin on treatment with dimethylsulphoxide (DMSO). We have undertaken some detailed studies of transcriptional and translational controls in this system which have indicated that it has considerable promise. A basic technique in most of these studies has been the use of cDNA to measure globin mRNA sequences in RNA populations. In one series of studies, we have investigated the template behaviour of chromatin isolated on the one hand from un-

TABLE 1. Globin mRNA Content of Different RNAs from Induced and Uninduced Friend Cells

Treatment	Fraction of RNA as globin mRNA ($\times 10^5$)				
	Chromatin-primed	HnRNA	Total nuclei	Cytosol	Polysomal
Uninduced	0	2	14	4	3–5
Induced	2–4	12	14	15–18	50–70

Each sample was titrated against globin cDNA and chromatin-primed RNA was prepared as outlined in the legend to Fig. 1. All other RNAs were prepared by a standard RNA preparative procedure from cell fractions. These were made by disrupting cells and separating out nuclei, polysomes and polysomal supernatant (cytosol) by standard procedures. Heterogeneous nuclear RNA (HnRNA) was prepared by fractionating total nuclear RNA on a sucrose gradient and recovering the RNA larger than 35S.

induced cells and on the other from cells induced to synthesise haemoglobin, using the technique earlier described. These studies have revealed that when RNA is transcribed from chromatin from uninduced cells, this does not contain any globin mRNA sequences whereas RNA transcribed from chromatin from induced cells does (*28*) (Table 1). This finding demonstrates that a transcriptional control involving the masking of chromatin operates in these cells and indicates that this is likely to be a very useful experimental system.

In the same cells, we have measured the concentration of mRNA sequences in nuclear RNA, non-polysomal cytoplasmic RNA and polysomal RNA (Table 1). Large differences in globin mRNA content are found in the polysomes of induced as compared with uninduced cells, as might be expected. It is of very great interest, however, that very much smaller differences are found in the non-polysomal cytoplasmic RNA and very little difference in total nuclear RNA (*13, 16a*). These latter findings argue for a control at the translational level. The overall picture seems to be therefore that the induction of haemoglobin synthesis in these cells involves coordinated transcriptional and translational controls. Since both α and β globin genes, which are not genetically linked, are synthesised in the course of induction, one therefore suggests that a large number of transcriptional events are regulated simultaneously during the induction of erythroid differentiation. There are many ways in which this might occur but clearly it is an attractive idea that one particular gene product might activate many other genes. Several authors have proposed a hierarchical arrangement according to which 'integrated genes' might determine the expression of many other genes. However, I have pointed out that it is not formally necessary to propose a hierarchical network of this kind (*24*). It is possible to propose a system in which an integrator gene is part of a self-regulating complex. An example of a scheme of this kind is shown in Fig. 4. In this scheme it is envisaged that each transcriptional locus has binding sites for both positive and negative controlling elements. Capital letters represent binding sites for positive controlling elements and small letters binding sites for negative controlling elements. The principle is simply that all the genes within a co-ordinated complex have a common binding site for a positive controlling element ($+_K$) which is absent from all other genes. The controlling element is itself produced by one of these genes. The result is therefore a self-locking positive feedback circuit. One can also postulate the con-

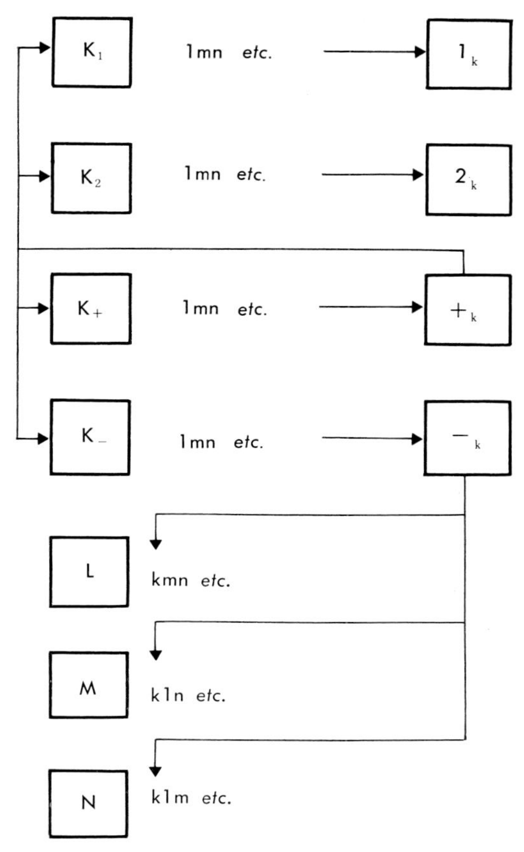

FIG. 4. A model of a simple self-stabilising circuit including positive and negative feedback loops such as might function in cellular differentiation. Capital letters (K, L, M, N) represent address sites and small letters (k, l, m, n) repressor binding sites. If an effector ($+_K$) is made by one gene having a K site it will then be made continuously and will switch on all other 'K' genes. Similarly if all genes, *except those with a 'K' address site*, have a 'k' repressor binding site and a repressor ($-_K$) is made by one 'K' gene, then all genes, except the 'K' genes, will be switched off.

verse, that is, binding sites for negative controlling elements which are absent from a similar complex but present in all other genes. If the negative controlling element is produced by a gene in the complex, this will of course result in shutting off all other genes but not genes within the complex. It seems quite likely that both kinds of control might operate and indeed experimental evidence would seem to support this suggestion.

Evidence for negative controls comes from cell hybridisation studies in which differentiated cells have been hybridised with other differentiated cells, usually from a different species. In most of these experiments, the differentiated function has been extinguished in the hybrid cell. This is not, however, an invariable rule and the opposite result has been obtained in some experiments. For example, in experiments conducted by Dr. Hickey and myself (*29*), hybrids were prepared between 2 different clones of the Friend cell. In one of these, haemoglobin synthesis could readily be induced by treatment with DMSO whereas the other was refractory to induction. The hybrid cells were inducible. In this system we therefore have evidence for co-ordinate induction of transcription of several genes, co-ordinate control of transcription and translation of the globin genes and for the existence of positive controlling elements.

ACKNOWLEDGMENT

Original work referred to in this communication was supported by grants from the Medical Research Council and Cancer Research Campaign of Great Britain.

REFERENCES

1. Axel, R., Cedar, H., and Felsenfeld, G. Synthesis of globin ribonucleic acid from duck-reticulocyte chromatin *in vitro*. Proc. Natl. Acad. Sci. U.S., *70*: 2029–2032, 1973.
2. Beerman, W. Chromosomes and genes. *In;* W. Beerman (ed.), Results and Problems in Cell Differentiation, vol. 4, pp. 1–33, Springer Verlag, Berlin/Heidelberg/New York, 1972.
3. Bishop, J. O., Pemberton, R., and Baglioni, C. Reiteration frequency of haemoglobin genes in the duck. Nature New Biol., *235*: 231, 1972.
4. Bishop, J. and Rosbash, M. Reiteration frequency of duck haemoglobin genes. Nature New Biol., *241*: 204–207, 1973.
5. Britten, R. J. and Kohne, D. E. Repeated sequences in DNA. Science, *161*: 529–540, 1968.
6. Cahn, R. D. and Cahn, M. B. Heritability of cellular differentiation: clonal growth and expression of differentiation in retinal pigment cells *in vitro*. Proc. Natl. Acad. Sci. U.S., *55*: 106–114, 1966.
7. Callan, H. G. The organisation of genetic units in chromosomes. J. Cell Sci., *2*: 1–7, 1967.
8. Chovnick, A. V., Finnerty, A., Schaler, A., and Duck, P. Studies on genetic organisation in higher organisms. I. Analysis of a complex gene in *Drosophila melanogaster*. Genetics, *62*: 145–160, 1969.
9. Coon, H. G. Clonal stability and phenotypic expression of chick cartilage cells *in vitro*. Proc. Natl. Acad. Sci. U.S., *55*: 66–73, 1966.
10. Davidson, E. H., Hough, B. R., Amenson, C. S., and Britten, R. J. General intersperstion of repetitive with non-repetitive sequence elements in the DNA of Xenopus. J. Mol. Biol., *77*: 1–24, 1973.
11. Friend, C., Patuleia, M. C., and de Harven, E. Erythrocyte maturation *in vitro* of murine (Friend) virus-induced leukemic cells. Natl. Cancer Inst. Monogr., *22*: 505–522, 1966.
12. Friend, C., Scher, W., Holland, J. G., and Sato, I. Haemoglobin synthesis in murine virus-induced leukemic cells *in vitro*: stimulation of erythroid differentiation by dimethyl sulfoxide. Proc. Natl. Acad. Sci. U.S., *68*: 378-382, 1971.
13. Gilmour, R. S., Harrison, P. R., Windass, J. D., Affara, N. A., and Paul, J. Globin messenger RNA synthesis and processing during haemoglobin induction in Friend cells. I. Evidence for transcriptional control in clone M2. Cell Differentiation, *3*: 9–22, 1974.

14. Gilmour, R. S. and Paul, J. Role of non-histone components in determining organ specificity of rabbit chromatin. FEBS Letters, *9*: 242–244, 1970.
15. Gilmour, R. S. and Paul, J. Tissue specific transcription of the globin gene in isolated chromatin. Proc. Natl. Acad. Sci. U.S., 1973, in press.
16. Harrison, P. R., Hell, A., Birnie, G. D., and Paul, J. Evidence for single copies of globin genes in the mouse genome. Nature, *239*: 219–221, 1972.
16a. Harrison, P. R., Gilmour, R. S., Affara, N. A., Conkie, D., and Paul, J. Globin messenger RNA synthesis and processing during haemoglobin induction in Friend cells. II. Evidence for post-transcriptional control in clone 707. Cell Differentiation, *3*: 23–30, 1974.
17. Imaizumi, T., Diggelman, H., and Scherrer, K. Demonstration of globin messenger sequences in giant nuclear precursors of messenger RNA of avian myeloblasts. Proc. Natl. Acad. Sci. U.S., *70*: 1122–1126, 1973.
18. Judd, B. H., Shen, M. W., and Kaufmann, Z. C. The anatomy and functions of a segment of the X chromosome of *Drosophila melanogaster.* Genetics, *71*: 139–152, 1972.
19. Kacian, D. L., Spiegelman, S., Bank, A., Terada, M., Metafora, S., Dow, L., and Marks, P. A. *In vitro* synthesis of DNA components of human genes for globins. Nature New Biol., *235*: 167, 1972.
20. Konigsberg, I. Clonal analysis of myogenesis. Science, *140*: 1273–1284, 1963.
21. Lodish, H. F., Firtel, R. A., Jacobson, A., Tuchman, J., Alton, T., Young, B., and Baxter, L. Transcription and structure of the genome of the eukaryotic cellular slime mould *Dictyostelium discoideum* Cold Spring Harbor Symp. Quant. Biol., *38*: 1973, in press.
22. O'Brien, S. J. On estimating functional gene number in eukaryotes. Nature New Biol., *242*: 52–54, 1973.
23. Paul, J. General theory of chromosome structure and gene activation in eukaryotes. Nature, *238*: 444–446, 1972.
24. Paul, J., Carroll, D., Gilmour, R. S., More, I. A. R., Threlfall, G., Wilkie, M., and Wilson, S. Functional studies on chromatin. 5th Karolinska Symposium, 1972.
25. Paul, J. and Gilmour, R. S. Template activity of DNA is restricted in chromatin. J. Mol. Biol., *16*: 242–244, 1966.
26. Paul, J. and Gilmour, R. S. Organ-specific restriction of transcription in mammalian chromatin. J. Mol. Biol., *34*: 305–316, 1968.
27. Paul, J. and More, I. A. R. Properties of reconstituted chromatin and nucleohistone complexes. Nature New Biol., *239*: 134–135, 1972.
28. Paul, J., Gilmour, R. S., Affara, N., Birnie, G. D., Harrison, P. R., Hell, A., Humphries, S., Windass, J., and Young, B. The globin gene: structure and expression. Cold Spring Harbor Symp. Quant. Biol., *38*: 885–890, 1973.
29. Paul, J. and Hickey, I. Molecular pathology of the cancer cell. J. Clin. Pathol., *7*: 4–10, 1973.
30. Ross, J., Aviv, H., Scolonik, E., and Leder, P. *In vitro* synthesis of DNA complementary to purified rabbit globin mRNA. Proc. Natl. Acad. Sci. U.S., *69*: 264–267, 1972.
31. Smith, K. D., Church, R. B., and McCarthy, B. J. Template specificity of isolated chromatin. Biochemistry, *8*: 4271, 1969.
32. Strauss, N. Comparative DNA renaturation kinetics in amphibians. Proc. Natl. Acad. Sci. U.S., *68*: 799–802, 1971.

33. Suzuki, Y., Gage, L. P., and Brown, D. D. Genes for silk fibroin in *Bombyx mori.* J. Mol. Biol., *70*: 637–649, 1972.
34. Verma, I. M., Temple, G. F., Fan, H., and Baltimore, D. *In vitro* synthesis of DNA complementary to rabbit reticulocyte 10S RNA. Nature New Biol., *235*: 163, 1972.
35. Walker, P. M. B. 'Repetitive' DNA in higher organisms. *In;* J. A. V. Balter and D. Noble (eds.), Progress in Biophysics and Molecular Biology, vol. 23, pp. 145–190, Pergamon Press, Oxford, New York, 1971.
36. Williamson, R., Drewienkiewicz, C., and Paul, J. Globin messenger sequences in high molecular weight RNA from embryonic mouse liver. Nature New Biol., *241*: 66–68, 1973.

Discussion of Paper of Drs. Paul et al.

DR. SUGIMURA: Is there any possibility of the presence of contaminating globin mRNA in your preparation of non-histone protein from mouse foetal liver?

DR. PAUL: We have not measured significant amounts of globin mRNA in these preparations.

DR. RUTTER: The conclusion of the "lack of" gene amplification depends upon a quantitative isolation of all DNA from the cell. Perhaps "episomal DNAs" could be less efficiently isolated than the bulk DNA. What kind of control experiments have you carried out, and in particular have you added cDNA to your sample before isolation to check on its recovery?

DR. PAUL: We prepared DNA for these experiments by the method of Britten *et al.* This involves dissociating the whole tissue in high salt and SDS, and then passing it over hydroxylapatite. In our experience this gives quantitative yields and added cDNA, which is of quite low molecular weight ($\sim$ 130,000), is recovered in the same proportion as total DNA.

DR. OSTERTAG: I wonder how good the estimate of the globin gene number really is using cDNA to hybridise with total cell DNA. There are now several papers mainly by Dr. Popp at Oak Ridge showing that there are 2 α type and 2 β type genes in the adult mouse. The differences between these chains are minor. We have just now found a third α chain in BalbC mice in the foetus which has only one or very few amino acid substitutions. In the embryo (yolk sac erythropoiesis) we find in addition 2 β type embryonic chains with each about 10–20 substitutions as compared to the normal β chain, and also one embryonic α type chain different from α by 30–40 substitutions. This adds up to at least 4 α chain genes and 4 β chain genes. How does this agree with your hybridisation data?

DR. PAUL: It would be more accurate for me to say that the globin genes have the same multiplicity as the "unique" fraction of the mouse genome. This could be as high as 5 copies per genome although it is probably less than this. We discuss this problem in a paper by Harrison *et al.* which is in press in J. Mol. Biol. There is not likely to be interaction between embryonic and adult messengers for α chains in the conditions of stringency we have used.

Dr. Ikawa: In your Friend leukemia's system, how many hours after the inducer treatment does the translational control occur?

Dr. Paul: About 2 days.

Dr. Thompson: Do you propose that the only function of non-histone protein is a negative one?

Dr. Paul: No.

Dr. Rutter: The question of the specificity of transcription depends upon whether you have endogenous haemoglobin mRNA. The finding of ^{32}P-labelled RNA in the hybrid means only that one has synthesis of labelled "mRNA" reacting with the cDNA, not that you have no endogenous mRNA. Thus it seems to me that one must be overly cautious about the simple interpretation of the results.

Dr. Paul: Dr. Gilmour has measured globin mRNA in chromatin-associated RNA in our preparations and finds very little. Moreover one can calculate, knowing the specific activities of ^{3}H-cDNA and ^{32}P-RNA in the hybrid, that at least 80% of the RNA in the hybrid is labelled with ^{32}P and, therefore, newly made. In our experiments the ratio of RNA made in the reactions to endogenous RNA in the chromatins is high, of the order of 50–100.

Dr. G. Schapira: Do you demonstrate the true biosynthesis of α and β chain.

Dr. Paul: Yes.

Dr. Hozumi: I am interested in your work on the mechanism of differentiation of leukemia cells by DMSO. Do you have any evidence for the modification of non-histone proteins by DMSO?

Dr. Paul: No. We have done no work on this as yet.

Dr. Nakahara: As a matter of curiosity, I just like to ask what Dr. Paul expects if you used chromatin isolated from malignant tumor cells in your transcription experiments.

Dr. Paul: Some years ago we compared RNA transcribed from untransformed and transformed cells, using the low C_0t hybridisation methods available at the time. We found no differences. We are now undertaking some similar experiments, using high C_0t methods to look for transcriptions of specific genes but we have no results we can report as yet.

Dr. Dube: Your evidence very clearly shows that non-histone proteins are important in unmasking DNA. Now, this would suggest a tissue specificity for non-histone proteins. Is there some evidence on this?

DR. PAUL: The experiments concerned with reconstituting chromatin themselves argue strongly for this. In fact, using a 2 dimensional method (developed by Dr. MacGillivray and Dr. Rickwood) to separate ^{32}P-phosphorylated chromosomal non-histone proteins one can easily demonstrate tissue differences. However, I think such experiments have to be interpreted with great caution in relation to chromosomal function. Now that we have a technique which promises to allow us to study the functions of individual non-histone proteins. It is possible that we will be able to obtain definitive answers to this kind of question.

DR. TOMKINS: Do you think that all mutants in eukaryotic cells are polar? If so I suppose this could explain the Judd finding in *Drosophila melanogaster* that one band corresponds to one complementation group.

DR. PAUL: I would not like to maintain that all are polar but I believe there is very good evidence that some are.

Ribosomes, Initiation Factors, and the Regulation of Gene Expression

Georges Schapira, Jean Delaunay, Daria Bogdanovsky, Louise Reibel, Catherine Vaquero, and Francine Creusot

Institut de Pathologie Moléculaire, Rue du Faubourg-Saint-Jacques, Paris, France*

Abstract: For many years, it seemed that ribosomes were not involved in the modulation of genetic information. Soon after the discovery of mRNA by Jacob and Monod, we showed that guinea-pig reticulocyte ribosomes accepted a message from rabbit reticulocytes contained in the pH 5 fraction of reticulocytes by first isolating labeled hemoglobin, then by isolating specific tyrosine-containing peptides. As a result, the first translation of mRNA from a rabbit allelic for one hemoglobin (α T IV Leu) on ribosomes from another one (α T IV Val) was established.

Subsequently, experiments were carried out with various crossed cell-free systems from mammals and birds. Going further down the phylogenic tree, we demonstrated that fish and plants are able to accept mRNA from rabbit reticulocytes.

On the other hand, there is proof of ribosomal specificity.

1) There is a barrier between prokaryotes and eukaryotes as far as biosynthesis is concerned.
2) Ribosomes from duck reticulocytes preferentially accepted mRNA from rabbit rather than from duck.
3) Ribosomes accept nucleus RNA from the same origin.
4) RNA from free polysomes can stimulate amino acid incorporation in a system with free polysomes. The same is true for bound polysomes, but the systems cannot be mixed.
5) The Mg dependence of ribosomes is different according to the source tissue.

In prokaryotes, one protein is necessary to control sensitivity to streptomycin and translational fidelity.

Ontogeny and oncogeny as investigated by 2-dimensional gel electrophoresis of ribosomal proteins showed no differences. Differences between ribosomal protein tissues of the same animal have been observed but were minimal. Coelectrophoresis of proteins from rabbit ribosomes mixed with the ribosomal proteins from other species showed no difference between mammals, birds, and reptiles.

Following further down the phylogenic tree, changes appeared and increased

* Groupe U 15 de l'Institut National de la Santé et de la Recherche Médicale, Laboratoire Associé au Centre National de la Recherche Scientifique.

proportionally as the species concerned became more distant from the rabbit. However there were still about 25 proteins with the same electrophoretic migration in species as distant as rabbit and plants. We suggest that a set of ribosomal proteins has been conserved during evolution, which would account for the possibility of translation on ribosomes from very distant species.

Initiation factors and interference factors are not always totally contained in ribosomes. We will emphasize 2 points: the exchangeability of initiation factors between prokaryotes and eukaryotes, and the role of a new RNA (M.W. 10,000) which links protein initiation factors and mRNA.

For many years, it was not certain whether or not ribosomes were involved in genetic regulation. We will first present evidence for the absence of genetic regulation, then some other evidence in favor of such regulation, which perhaps is not controlled directly by the ribosomes (but by initiation and interference factors). The role of ribosomal proteins will also be discussed.

Ribosomes and Translational Control

Soon after the discovery of mRNA by Jacob and Monod (*12*) we were able, with Kruh and Dreyfus (*14*), to demonstrate the nonspecificity of ribosomes among mammals: ribosomes from guinea-pig reticulocytes accepted the information contained in the pH 5 enzyme fraction from rabbit reticulocytes and *vice versa;* this was demonstrated by the labeling of hemoglobin (Hb) in cell-free systems.

The demonstration was more conclusive when we checked the labeling of specific tyrosine-containing peptides (*26*) instead of the labeling of Hb (Fig. 1). When we used allelic rabbit with either Hb containing α T IV Leu peptide or Hb containing α T IV Val, mRNA isolated from one kind of rabbit gave information to ribosomes from the other (*27, 28*) (Figs. 2 and 3). Afterwards, we (*30*) and some other teams (*18*) described cell-free crossing experiments with other mammals and between mammals and birds, synthesizing different Hb's and different proteins.

In Oxford, Gurdon (*10*) using another system, an *in vivo* system, translated mRNA from rabbit Hb into frog oocytes.

However, all the species involved in those experiments are closely related: they diverged only 300 million years ago. Furthermore, the proteins of their ribosomes display identical fingerprints after 2-dimensional polyacrylamide gel electrophoresis (*5, 7a*). We succeeded in synthesizing rabbit Hb on ribosomes from more distant species: trout liver and bean germ (*33*) (Figs. 4, 5, and 6). At the same time, 2 groups succeeded in translating Hb mRNA on wheat germ (*8, 24*). Fish and plant ribosomal proteins have 2-dimensional patterns which differ from those of mammals, reptiles and birds (*7a*).

It seems that one ribosome can replace any other (at least among eukaryotes).

On the other hand, *Escherichia coli* ribosomes are unable to accept information from rabbit Hb mRNA in a cell-free system (*9, 21*). This is in contradiction with the results previously obtained for cell-free systems by Laycock and Hunt (*16*).

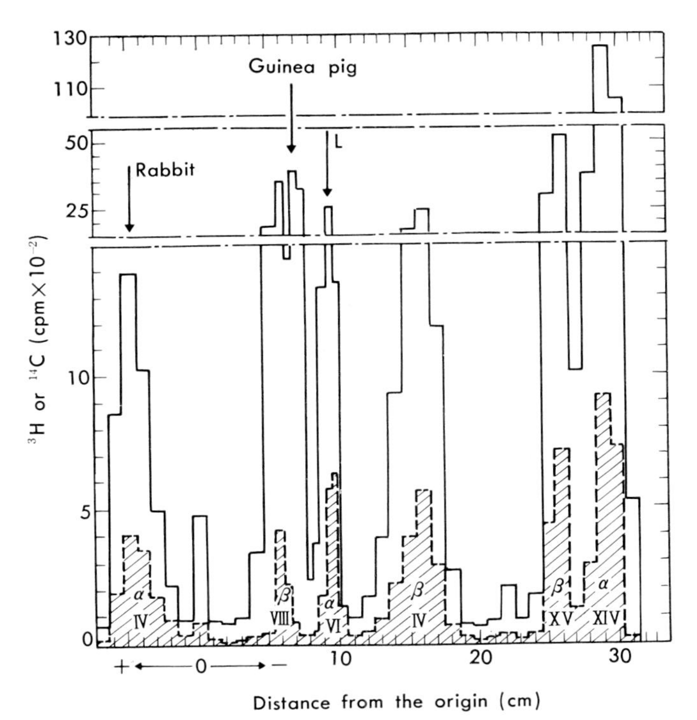

FIG. 1. Heterologous cell-free synthesis with guinea-pig ribosomes and rabbit "pH 5 enzyme" obtained after 2.5-hr ultracentrifugation at 48,000 rpm with ^{3}H-Tyr. *In vitro*-synthesized rabbit and guinea-pig hemoglobins were not separated. ^{14}C-labeled rabbit hemoglobin was used as a carrier. (1) Peptides characteristic of α T IV and α T VI incorporated ^{3}H-Tyr. They can be identified by the parallel elution curve of the ^{14}C-Tyr-labeled peptides. (2) The guinea-pig characteristic peptide placed at −7 cm can be identified by the contrast between the break in the ^{3}H curve, which is a control for its biosynthesis, and the absence of any peak corresponding to the ^{14}C curve (originating from the rabbit).

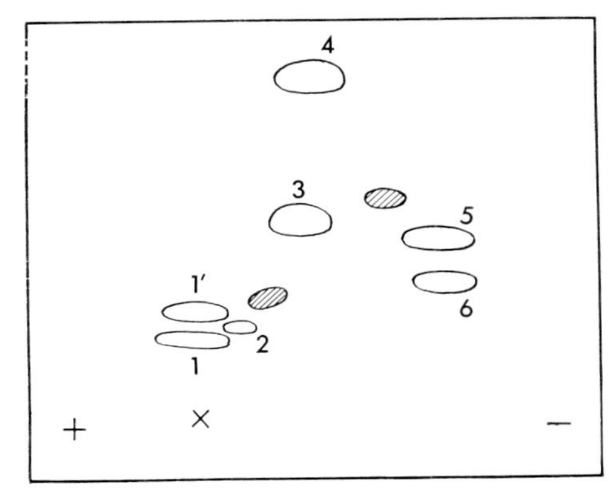

FIG. 2. Fingerprint of the tryptic hydrolyzate of α T IV Val and α T IV Leu rabbit globin. Tyrosine- and tryptophan-specific staining. (hatched) tryptophan peptide; (open) tyrosine peptide. 1, α T IV Val; 1′, α T IV Leu; 2, β T XIV; 3, α T VI; 4, β T IV; 5, β T XVI; 6, α T XV.

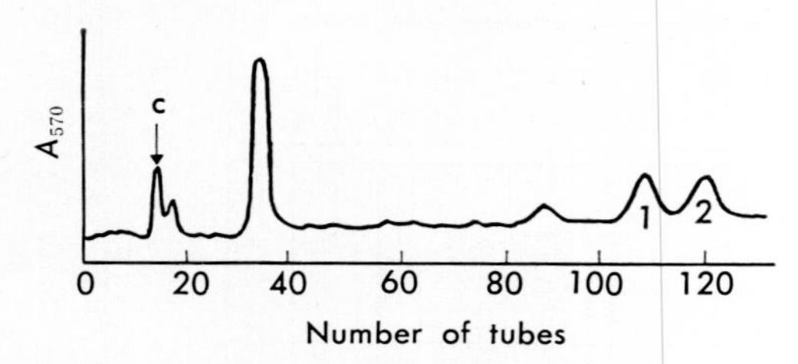

FIG. 3. Column chromatography of a tryptic hydrolyzate of the electrophoresis eluate (cf. Fig. 2). 1, α T IV Val peptide; 2, α T IV Leu peptide; c, cysteic acid.

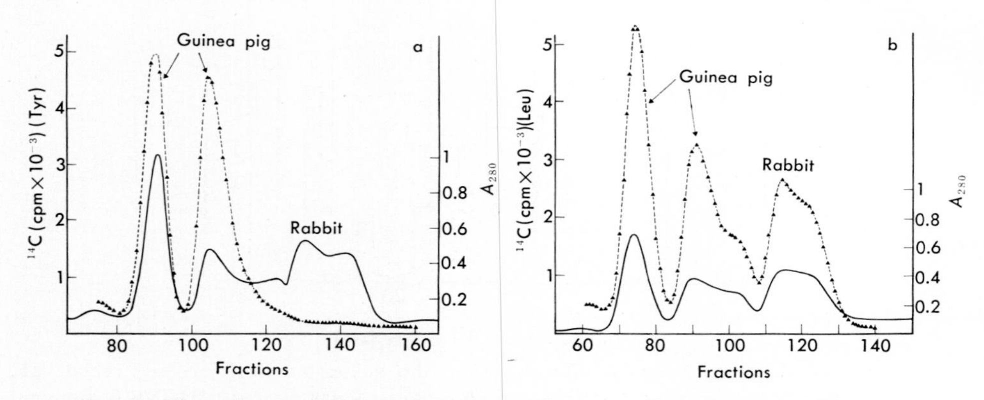

FIG. 4. Heterologous cell-free synthesis. Trout ribosomes were incubated with "pH 5 enzyme" and 0.5M KCl ribosome wash from guinea-pig reticulocytes. Chains were separated by chromatography and the fractions were counted in a liquid scintillation counter. In experiment (a) no rabbit Hb mRNA was added; in experiment (b) rabbit Hb mRNA was added. The continuous line (——) shows optical density at 280 nm (rabbit globin carrier) and the closed triangles (▲) show cts/min (*ni vitro*-synthesized rabbit globin).

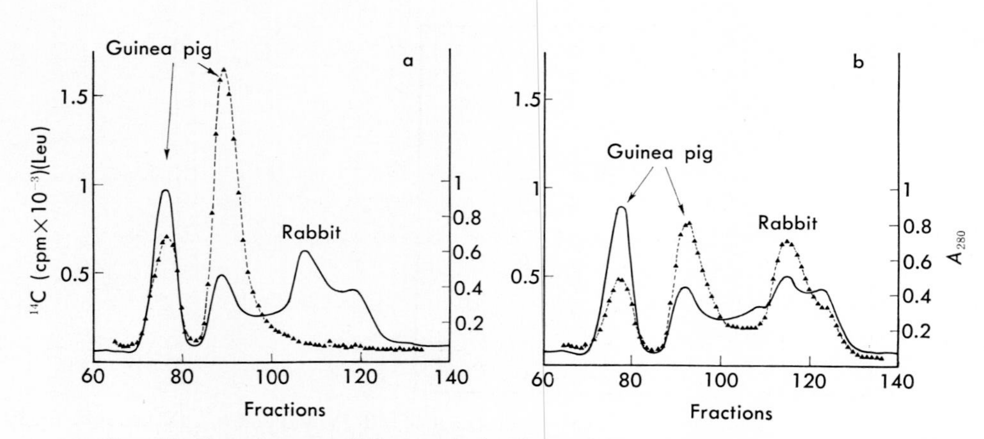

FIG. 5. Heterologous cell-free synthesis. Kidney-bean polysomes were incubated with "pH 5 enzyme" and 0.5M KCl wash from guinea-pig reticulocyte ribosomes. Details are given in the legend of Fig. 4.

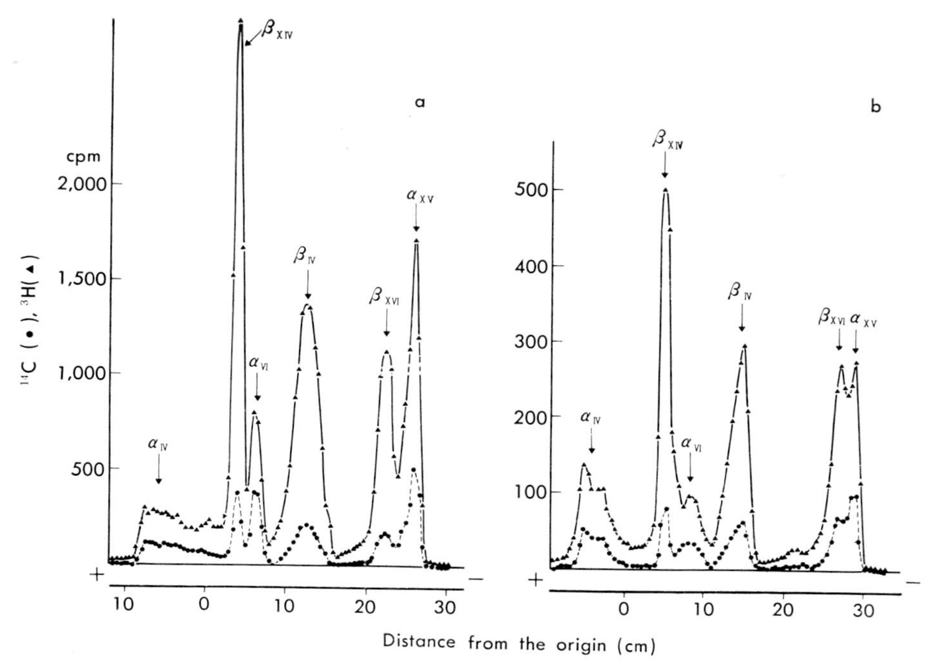

FIG. 6. Pattern of high-voltage electrophoresis of rabbit globin tyrosine-containing peptides synthesized in a cell-free system with either trout liver or kidney-bean root tip polysomes and rabbit mRNA. These incubations were carried out in presence of L-^{3}H-Tyr. L-^{14}C-Tyr rabbit Hb was then added as the carrier. The ^{3}H product comigrating with L-^{14}C rabbit Hb carrier on starch-block electrophoresis was digested with trypsin. (a) incubation with trout polysomes; (b) incubation with kidney-bean polysomes. (▲) refers to ^{14}C cts/min and (●) to ^{3}H cts/min.

Working in collaboration with Ben Hamida, we were unable to observe any translation of rabbit globin mRNA when we used plasmolyzed *E. coli* (*1*). There is a barrier, as far as biosynthesis is concerned, between prokaryotes and eukaryotes (Figs. 7 and 8).

Even if ribosomes from one species can replace ribosomes from any other species, *there are many differences between ribosomes which suggest their regulatory role.* Ribosomes have a specificity depending on their behaviour in relation to mRNA.

For instance, when ribosomes are taken from duck reticulocytes, the competition between duck and rabbit Hb mRNA is not in favor of the duck, as would be expected, but of the rabbit (*31*).

The interaction between polysomes and intracellular membranes has been proposed as being involved in the regulation of genetic expression. A model, termed by Pitot (*23*) as the "membron," is described as a functioning, regulatable, translating polyribosome complex, with a specific area of membrane. The altered membron biochemistry, inherent in most, if not all, neoplastic cells, changes the modulation of the membron, which does or does not select certain templates.

Experiments by Uenoyama and Ono (*32*) favor this theory: RNA from free polysomes can stimulate amino acid incorporation in a system with free polysomes, and to which KCl wash from free polysomes, not bound polysomes, is added. On

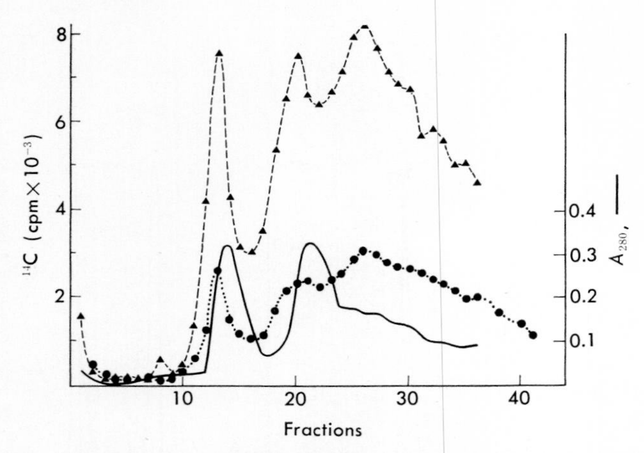

FIG. 7. Translational response of an *E. coli* "permeabilized cell system" to different exogenous mRNA's. After addition of α- and β-hemoglobin chains as carriers (full line), the radioactivity of the control and of the experiment was estimated. There are 2 peaks of radioactivity similar to the peak of the carrier.

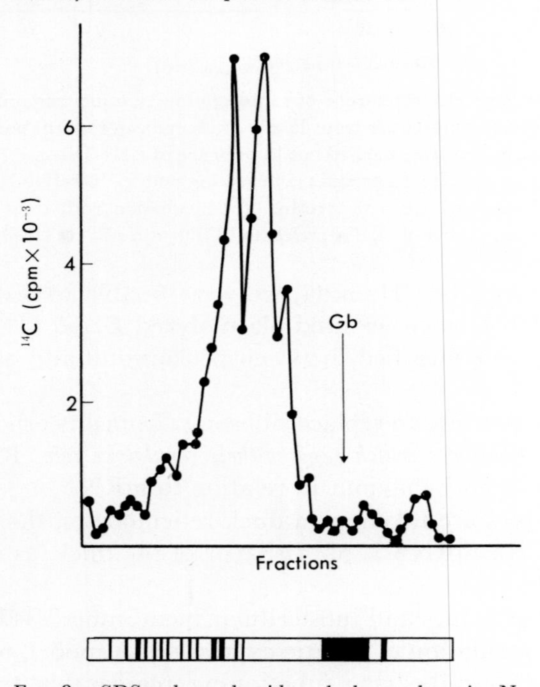

FIG. 8. SDS-polyacrylamide gel electrophoresis. No radioactivity of the globin was observed.

the other hand, RNA from bound polysomes can stimulate amino acid incorporation in a cell-free system with bound polysomes. These observations suggest that in the course of protein synthesis, a distinct specificity is present between mRNA and ribosomes.

The dependence of ribosomes on Mg^{2+} concentration is not the same for different organs of the same species: for instance, brain ribosomes need more than twice as much magnesium as kidney or liver ribosomes (*17*). The meaning of this is not clear but suggests that Mg^{2+} interaction with ribosomes has an important role in translation.

Working with nuclear RNA (nRNA) instead of mRNA, Naora and Kodaira (*20*) asserted that there are specific ribosomes which permit selective recognition of homologous nRNA. Hepatoma ribosomes retain the ability to recognize liver nRNA and are capable of binding liver nRNA preferentially to their own hepatoma nRNA or to phage f_2 RNA.

A more sophisticated form of genetic regulation is probably the change which occurs in the 30S subunit of *E. coli* ribosomes. It controls the streptomycin sensitivity and the translational fidelity. The protein involved is well-known and we will now discuss the role of ribosomal protein and genetic regulation.

Ribosomal Proteins

In some special cases among prokaryotes, mRNA translation is regulated by ribosomal proteins.

The first instance was observed with streptomycin and protein P 10 (S 12 in the new nomenclature (*34*)) of the 30 S subunit (*22*).

Recently, the requirement for a specific ribosomal protein for the biosynthesis of β-galactosidase in *E. coli* was demonstrated (*15*), but the translation of other mRNA's from *E. coli* has not been checked.

The situation is not clear with ribosomal proteins from eukaryotes. Typical 2-dimensional fingerprints of eukaryotic ribosomal proteins are given in Fig. 9.

Ribosomal protein patterns of different tissues show some differences as observed by 2-dimensional polyacrylamide gel electrophoresis according to Kaltschmidt and Wittmann (*13*), as subsequently modified (*5*, *11*). For instance, in appendix ribosomes, there are 60 spots. From a well-defined cluster of 3 spots, 2 spots are absent in the liver and all three in the kidney (*5*).

These differences are very small. *In vitro* phosphorylation of ribosomes from rabbit liver and reticulocytes is qualitatively and quantitatively the same (*7*). Ontogenic differences were absent as far as embryonic and adult liver are concerned. Oncogenic differences in cancerous and normal livers were not found by us (*6*) or by Bielka (*3*). These observations however are controversial.

The ribosomal protein patterns of 7 widely different mouse cell types have been investigated by Rodgers (*25*) in an attempt to assess the possible role of these proteins in control mechanisms in the cell. The results show that although there are small differences between mouse tissues, the largest differences were between rapidly growing cells (Ehrlich ascites carcinoma, A9, B82) and nonproliferating tissues (brain, kidney, liver, gut). Rapidly growing cells contained an acidic protein (M.W. 58,000) not found in tissues and not previously described.

As far as phylogeny of ribosomal proteins is concerned, however, the situation is different.

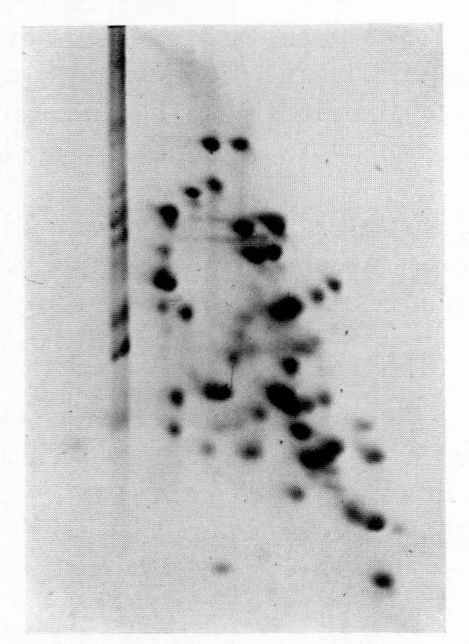

FIG. 9. Typical fingerprints of ribosomal proteins obtained by 2-dimensional polyacrylamide gel electrophoresis.

We studied 80 S ribosomal proteins of different species ranging from man to plants (*7a*) by 2-dimensional polyacrylamide gel electrophoresis according to the technique of Kaltschmidt and Wittmann (*13*) (Table 1).

For each species examined separately, there are about 60 spots. Some spots, however, result from the overlapping of several proteins, so that the number of proteins per ribosome is actually higher (between 70 and 80). In order to estimate the proximity between 2 species in terms of ribosomal proteins, we coelectrophoresed the ribosomal proteins from the 2 species. The higher the number, *n*, of proteins re-

TABLE 1. Species Investigated

Mammalia	Man	Placenta
	Calf	Liver
	Sheep	,,
	Rabbit	,,
	Guinea pig	,,
	Rat	,,
	Mouse	,,
Birds	Chicken	,,
Reptiles	Lizard	,,
Amphibians	Frog	,,
Fish	Perch	,,
	Herring	,,
	Mackerel	,,
Crustaceas	Crab	Hepatopancreas
Molluscaes	Scallop	,,
Plants	Kidney bean	Root tips
	Lentil	,,
	Maize	,,

solved out, the lower the proximity between the 2 species. We expressed the "degree of proximity," p, of 2 species as: $p=(120-n)/60$ (Fig. 10).

Figure 11 gives the degrees of proximity of various species with respect to the rabbit. As could be expected, the degree of proximity decreases proportionally to the taxonomic distance. The following points should be emphasized:

A. The degree of proximity remains equal to unity for all species between man and reptiles. Since the branch point of mammals and birds, *i.e.*, 300 million years ago, the evolution of ribosomal proteins has been restricted to amino acid substitutions resulting in no change of the electric charge. That such an evolution was so slow becomes understandable if one takes into account the numerous interactions of

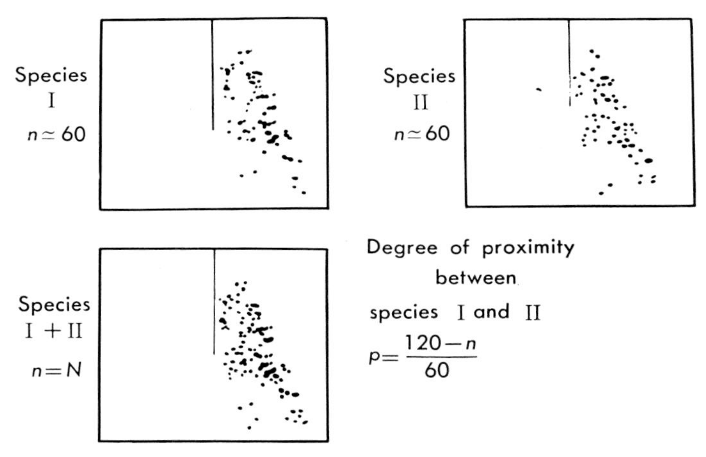

FIG. 10. Definition of "degree of proximity."

Branch point	Species		N	120−N	P
	Mammalia	Man	60	60	1
240M	Birds	Chicken	60	60	1
300M	Reptiles	Lizard	60	60	1
350–400M	Amphibians	Frog	65	55	0.92
440–500M	Fish	Perch	65	55	0.92
		Herring	67	53	0.88
		Mackerel	64	56	0.94
?	Crustaceans	Crab	80	40	0.66
?	Molluscs	Scallop	84	36	0.60
?	Plants	Kidney bean	93	27	0.45
		Maize	96	24	0.40
?	Bacteria	*E. coli*	109	4	0.07

FIG. 11. Correlation between the degree of proximity and the classical evolutionary tree.

ribosomal proteins with each other (and with ribosomal RNA's): the probability that a mutation might be acceptable is thus very low.
B. The degree of proximity starts to decrease with amphibians and fish, although very slowly. For more remote species, the decrease becomes evident. However, the degree of proximity is still equal to 0.40 for such remote species as rabbit and maize. In these 2 species, there are still about 25 proteins with the same migration. Although identical migration is only a presumption for the identity of primary structure, one may suggest that some ribosomal proteins have remained invariant throughout evolution.
C. Ribosomal proteins from *E. coli* and from mammals only have 10 spots which overlaps (to be published).

Interchangeability of Factors in Bacterial and Eukaryotic Translation Initiation Systems (*2*)

The question of the interchangeability of initiation factor (*29*) is open to discussion; we have seen that there is a barrier between ribosomes from prokaryotes and mRNA from eukaryotes.

A question of obvious relevance to the evolution of the cell machinery is whether elements from prokaryotic translation systems can be interchanged with corresponding elements of eukaryotic origin, to promote efficient protein synthesis.

We have reinvestigated the functionality of crossed initiation systems by systematically interchanging elements from *E. coli* with those from rabbit reticulocytes. The reaction under study was the poly AUG-dependent initiator tRNA binding to ribosomes and its transfer to puromycin.

The results of this work shed light on the functional correspondence between certain bacterial and eukaryotic factors and clearly emphasize the role of ribosomes in the "selection" of initiator tRNA with respect to its state of formylation and biological origin.

1. Interchangeability of tRNA's

Reticulocyte or sheep mammary gland initiator tRNA's, once formylated, can fully substitute for *E. coli* fMet-tRNA in an *E. coli* translation system (Tables 2 and 3). The reverse is not true, since poly AUG-dependent binding of *E. coli* Met-$tRNA_f$ to a reticulocyte system does not lead to formation of a "true" initiation complex (no transfer of the methionyl residue to added puromycin). Thus, in some respects, eukaryotic translation systems seem to impose, at a first glance, a more severe restriction with respect to the biological origin of initiator tRNA's.

2. Interchangeability of initiation factors (*IF*)

When acting on eukaryotic ribosomes, *E. coli* initiation factors do not discriminate between formylated and unformylated *E. coli* initiator tRNA's, at least in the binding reaction.

E. coli factors that "recognize" reticulocyte initiators on *E. coli* ribosomes fail to do so when acting on reticulocyte ribosomes.

Reticulocyte IF were tested on *E. coli* ribosomes. The kind of response differed

TABLE 2. Binding of Reticulocyte fMet-tRNA in an *E. coli* Initiation System

Poly AUG-dependent binding	fMet-tRNA (pmoles)	Met-tRNA (pmoles)
Complete	2.50	0.70
− IF	<0.01	<0.01
− Poly AUG	0.20	<0.01
− GTP	0.20	<0.01

The complete system contained: 1 A_{260} unit of 70 S; 0.12 A_{260} unit of poly AUG; 120 μg crude IF and excess tRNA (15 pmoles).

TABLE 3. Transfer of Reticulocyte fMet-tRNA to Puromycin in an *E. coli* Initiation System

Poly AUG-dependent transfer	fMet-tRNA (pmoles)	Met-tRNA (pmoles)
Complete	2.40	<0.01
− IF	<0.01	<0.01
− Poly AUG	0.20	<0.01

Incubation conditions were as given in Table 2 except that the mixture contained, in addition to the previous effectors, 1-nM puromycin, after incubation for 12 min at 25°C, 1-ml potassium phosphate, 0.1 M, pH 8.2, was added. Methionyl and formylmethionyl puromycin was extracted with 1.5-ml ethylacetate. The radioactivity contained in 1 ml was counted in 10-ml Bray liquid scintillation system.

with the type of preparation in use. The only preparation which showed stimulatory activity on heterologous ribosomes was $IF_{1/2\,hr}$, which catalyzed a GTP-independent poly AUG-dependent binding of *E. coli* fMet-tRNA to 70 S ribosomes (Table 4); the IF preparation obtained by longer KCl extraction ($IF_{2\,hr}$) was unable to function with *E. coli* ribosomes.

Since $IF_{1/2\,hr}$, when assayed with homologous ribosomes, exhibited some of the characteristic features previously ascribed to IF-M_1, namely the ability to catalyze the binding of reticulocyte fMet, but not Met-tRNA, the stimulation observed with $IF_{1/2\,hr}$ on *E. coli* ribosomes could, in principle, be due to the action of IF-M_1 proper. Hence IF-M_1 was further purified from crude $IF_{1/2\,hr}$ by DEAE-cellulose chromatography and its activity assayed with reticulocyte or *E. coli* ribosomes. As shown in Table 5, IF-M_1 allows a poly AUG-dependent but GTP-independent binding of *E. coli* fMet-tRNA to 40 S reticulocyte subunits, a result in agreement with previous reports. The fMet-tRNA binding to *E. coli* 70 S ribosomes is also stimulated but to a smaller extent than when reticulocyte subunits are used.

TABLE 4. Binding of *E. coli* fMet-tRNA to 70 S Ribosomes with Reticulocyte "$IF_{1/2\,hr}$"

Component omitted	*E. coli* fMet-tRNA (pmoles)
Complete	0.40
− IF	0.20
− Poly AUG	<0.01
− GTP	0.30

The complete system contained: 1 A_{260} unit of 70 S, 0.12 A_{260} unit of poly AUG, 150 μg crude $IF_{1/2\,hr}$ and excess tRNA (>15 pmoles). Other additions as usual.

TABLE 5. IF-M_1-dependent Binding of fMet-tRNA to Ribosomes

Component omitted	Retic. 40 S ribosomes		*E. coli* 70 S ribosomes	
	fMet-$tRNA_{coli}$ (pmoles)	fMet-$tRNA_{Retic.}$ (pmoles)	fMet-$tRNA_{coli}$ (pmoles)	fMet-$tRNA_{Retic.}$ (pmoles)
Complete	2.10	1.50	0.60	0.40
– IF-M_1	<0.01	0.01	<0.01	0.15
– Poly AUG	0.20	—	<0.01	0.04
– GTP	3.00	—	0.60	0.40

The complete system included: 0.6 A_{260} unit of reticulocyte 40 S (or 1 A_{260} unit of *E. coli* 70 S), 60 μg reticulocyte IF-M_1, 0.12 A_{260} unit of poly AUG and excess tRNA (>15 pmoles). Other effectors were present at the concentrations described in the legend of Table 4.

This shows that at least one of the reticulocyte initiation factors, *i.e.*, IF-M_1 is capable of interacting with bacterial ribosomes. In view of a possible analogy between *E. coli* IF_2 and reticulocyte IF-M_1 with respect to fMet-tRNA recognition, we have investigated whether *E. coli* IF_1 could complement IF-M_1 either with homospecific or with heterospecific ribosomes.

As shown in Fig. 12, somewhat unexpectedly, a clear complementation effect could be observed when reticulocyte 40 S subunits were used, but this was not seen with 70 S ribosomes.

Two factors among prokaryotes and eukaryotes can be crossed, but only from a eukaryotic into a prokaryotic system: tRNA from reticulocytes when formylated and IF-M_1 (into cell-free systems of *E. coli*).

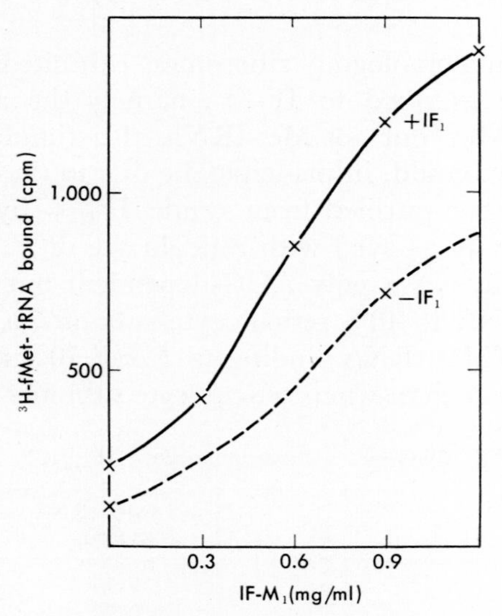

FIG. 12. Complementation between reticulocyte IF-M_1 and *E. coli* IF_1 for the binding of *E. coli* fMet-tRNA to reticulocyte 40 S subunits. The incubation mixture contained 0.5 A_{260} units 40 S, 0.12 A_{260} units of poly AUG; 15 pmoles *E. coli* fMet-tRNA; 1.3 μg *E. coli* IF_1 and reticulocyte IF-M_1 at the indicated concentrations (——). The background without IF_1 is also plotted (– – –).

The exact nature of eukaryotic IF (initiation factors) will be discussed on the basis of these last observations.

Miller and Schweet (*19*) first described the effect of 0.5M KCl ribosomal wash fractions on *in vitro* hemoglobin synthesis in a cell-free system which has been recognized as an IF of protein synthesis.

We observed that the 0.5M KCl wash fraction of rabbit reticulocyte ribosomes lost much of its activity when dialyzed and that the dialyzate had the UV spectrum of an RNA (Fig. 13). The dialyzate was concentrated, analyzed and isolated; it is

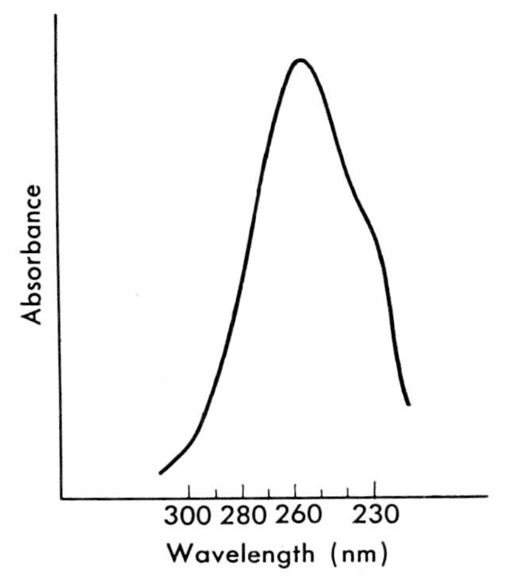

FIG. 13. UV spectrum of KCl wash RNA.

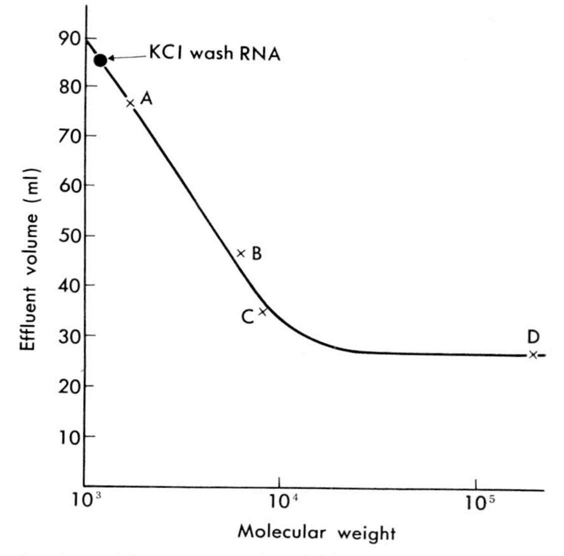

FIG. 14. Molecular weight of KCl wash RNA determined according to Andrews (*4*). The reference curve includes: A, myoglobin; B, hemoglobin; C, creatine kinase; and D, blue dextran.

FIG. 15. Electrophoresis of KCl wash RNA in polyacrylamide gel. Electrophoresis was carried out on 5% polyacrylamide gel at 5 mA/tube for 90 min. The large arrow shows the location of 4 S, which is not present in this electrophoresis, and the single band is the KCl wash RNA.

a single-stranded RNA with low molecular weight (M.W. 11,000) (Figs. 14 and 15). We were able to reconstitute to a large extent the activity of the dialyzed 0.5M KCl fraction by adding the lyophilized dialyzate (Tables 6 and 7, Fig. 16).

Sparsomycin is an antibiotic which inhibits chain elongation, but has little effect on initiation; its presence results in an accumulation of short initiation peptides during protein synthesis *in vitro*.

This reaction enables us to determine whether the RNA acts during initiation or elongation. In the case of complementation of the dialyzed 0.5M KCl wash by the RNA fragment, we obtained 3 times more dipeptides than when we used 0.5M KCl wash dialyzate alone (Fig. 17).

We concluded that the RNA fragment has some role in initiation. It appears that the initiation factor, probably IF_3, is a protein which requires an RNA fraction for activity. We can suppose that this RNA is complementary to the portion of the mRNA before the 5′-terminal segment.

TABLE 6. Activity of the KCl Wash RNA after Phenol Extraction

Factors	Activity pmoles of ^{14}C leucine
0.5 M KCl wash	326
0.5 M KCl wash dialyzate	39
0.5 M KCl wash dialyzate+RNA	187
0.5 M KCl wash dialyzate+RNA extracted with phenol	214

Activity was expressed in pmoles of ^{14}C-leucine incorporated in 130 μl containing 0.1 mg of washed ribosomes, 500 μg of protein of the 0.5 M KCl wash dialyzate or not, 0.5 μg of RNA, and 0.5 mg of pH 5 enzyme.

TABLE 7. Effect of KCl Wash RNA on the Synthesis of α and β Chains from Globin

Factors	Specific activity (cpm/$A_{280\ nm}$)	
	α chains	β chains
0.5 M KCl wash dialyzate (5 mg)	0.625	0.630
0.5 M KCl wash dialyzate (5 mg)+RNA (5 μg)	7.75	4.55

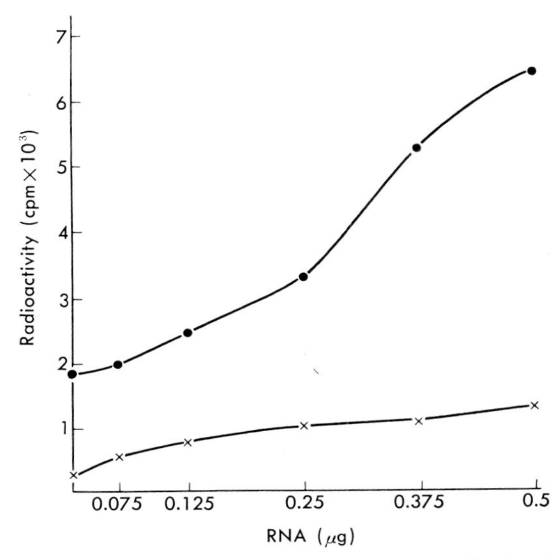

FIG. 16. Study of the biological activity of the KCl wash RNA as measured by the biosynthesis of hemoglobins. ● 0.5 M KCl wash dialyzate + RNA; × RNA alone.

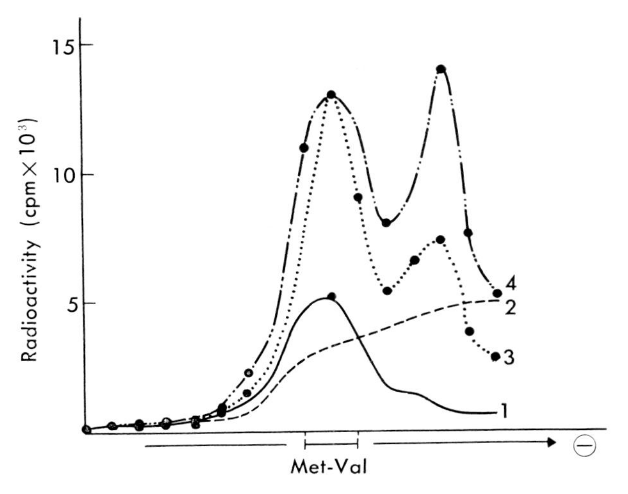

FIG. 17. Electrophoresis of sparsomycin peptides. The dipeptide Met-Val was a marker for initiation. 1, 0.5 M KCl wash dialyzate; 2, RNA alone; 3, 0.5 M KCl wash dialyzate + RNA; 4, 0.5 M KCl wash dialyzate.

REFERENCES

1. Ben Hamida, F., Vaquero, C., Reibel, L., and Schapira, G. Translational response of an *E. coli* "permeabilized cell system" (PCS) to different exogenous messenger RNA's. Unpublished.
2. Berthelot, F., Bogdanovsky, D., Schapira, G., and Gros, F. Interchangeability of factors and tRNA's in bacterial and eukaryotic translation initiation systems. Mol. Cell. Biochem., *1*: 63–72, 1973.
3. Bielka, H., Stahl, J., and Welfle, H. Studies on proteins of animal ribosomes. IX. Proteins of ribosomal subunits of some tumors characterized by two-dimensional polyacrylamide gel electrophoresis. Arch. Geschwulstforsch., *38*: 109–112, 1971.
4. Bogdanovsky, D., Hermann, W., and Schapira, G. Presence of a new RNA species among the initiation protein factors active in eukaryotes translation. Biochem. Biophys. Res. Commun., *51*: 25–32, 1973.
5. Delaunay, J., Mathieu, C., and Schapira, G. Eukaryotic ribosomal proteins: interspecific and intraspecific comparisons by two-dimensional polyacrylamide gel electrophoresis. Eur. J. Biochem., *31*: 561–564, 1972.
6. Delaunay, J. and Schapira, G. Rat liver and hepatoma ribosomal proteins. Two-dimensional polyacrylamide gel electrophoresis. Biochim. Biophys. Acta, *259*: 243–246, 1972.
7. Delaunay, J., Loeb, J. E., Pierre, M., and Schapira, G. Mammalian ribosomal proteins: studies on the *in vitro* phosphorylation patterns of ribosomal proteins from rabbit liver and reticulocytes. Biochim. Biophys. Acta, *312*: 147–151, 1973.

7a. Delaunay, J., Creusot, F., and Schapira, G. Evolution of ribosomal proteins. Eur. J. Biochem., *39*: 305–312, 1973.

8. Efron, D. and Marcus, A. Efficient synthesis of rabbit globin in a cell-free system from wheat embryo. FEBS Letters, *33*: 23–27, 1973.
9. Gielkens, A. L. J., Salden, M. H. L., Bloemendal, H., and Konigs, R. N. H. Translation of oncogenic viral RNA and eukaryotic messenger RNA in the *E. coli* cell-free system. FEBS Letters, *28*: 348–352, 1972.
10. Gurdon, J. B., Lane, C. D., Woodland, H. R. and Marbaix, G. Use of frog eggs and oocytes for the study of messenger RNA and its translation in living cells. Nature, *233*: 177–182, 1971.
11. Huynh-van-Tan, Delaunay, J., and Schapira, G. Eukaryotic ribosomal proteins. Two-dimensional electrophoretic studies. FEBS Letters. *17*: 163–167, 1971.
12. Jacob, F. and Monod, J. Genetic regulatory mechanisms in the synthesis of proteins. J. Mol. Biol., *3*: 318–356, 1961.
13. Kaltschmidt, E. and Wittmann, H. G. Ribosomal proteins. VI. Preparative polyacrylamide gel electrophoresis as applied to the isolation of ribosomal proteins. Anal. Biochem., *30*: 132–141, 1969.
14. Kruh, H., Rosa, J., Dreyfus, J. C., and Schapira, G. Synthèse d'hémoglobines par des systèmes acellulaires de réticulocytes. Biochim. Biophys. Acta, *49*: 509–519, 1961.
15. Kung, H., Fox, J. E., Spears, C., Brot, N., and Weissbach, H. Studies on the role of ribosomal proteins L_7 and L_{12} in the *in vitro* synthesis of β-galactosidase. J. Biol. Chem., *248*: 5012–5015, 1973.
16. Laycock, D. G. and Hunt, J. A. Synthesis of rabbit globin by a bacterial cell-free system. Nature, *221*: 1118–1122, 1969.

17. MacInnes, J. W. Mammalian brain ribosomes are behaviourally and structurally heterogenous. Nature New Biol., *241*: 244–246, 1973.
18. Mathews, M. B., Osborn, M., Berns, A. J. M., and Bloemendal, H. Translation of two messenger RNA's from lens in a cell-free system from Krebs II ascites cells. Nature New Biol., *236*: 5–7, 1972.
19. Miller, R. L. and Schweet, R. Isolation of a protein fraction from reticulocyte ribosomes required for *de novo* synthesis of hemoglobin. Arch. Biochem. Biophys., *125*: 632–646, 1968.
20. Naora, H. and Kodaira, K. Interaction of informational macromolecules with ribosomes. III. Binding of nuclear RNA by normal liver and hepatoma ribosomes. Biochim. Biophys. Acta, *224*: 498–506, 1970.
21. Noll, M., Noll, H., and Lingrel, J. Initiation factor IF-3-dependent binding of *Escherichia coli* ribosomes and N-formyl-methionine transfer-RNA to rabbit globin messenger. Proc. Natl. Acad. Sci. U.S., *69*: 1843–1847, 1972.
22. Nomura, M., Mizushima, S., Ozaki, M., Traub, P., and Lowry, C. V. Structure and function of ribosomes and their molecular components. Cold Spring Harbor Symp. Quant. Biol., *34*: 51–61, 1969.
23. Pitot, H. C. and Shires, T. K. Introductory remarks: membrane-polysome interactions. Fed. Proc. *32*: 76–79, 1973.
24. Roberts, B. E. and Bruce, M. P. Efficient translation of tobacco mosaic virus RNA and rabbit globin 9 S RNA in a cell-free system from commercial wheat germ. Proc. Natl. Acad. Sci. U.S., *70*: 2330–2334, 1973.
25. Rodgers, A. Ribosomal proteins in rapidly growing and nonproliferating mouse cells. Biochim. Biophys. Acta, *294*: 292–296, 1973.
26. Schapira, G., Padieu, P., Maleknia, N., Kruh, J., and Dreyfus, J. C. Information génétique portée par une fraction soluble de réticulocytes et traduite en protéine spécifique sur les ribosomes d'une autre espèce. J. Mol. Biol., *20*: 427–446, 1966.
27. Schapira, G., Dreyfus, J. C., and Maleknia, N. The ambiguities in the rabbit hemoglobin evidence for a messenger RNA translated specifically into hemoglobin. Biochem. Biophys. Res. Commun., *32*: 558–561, 1968.
28. Schapira, G., Benrubi, M., Maleknia, N., and Reibel, L. Hemoglobine de lapin: une variante résultant d'un allélomorphisme et non d'une ambiguité: Hb α 29 Val-Leu. Biochim. Biophys. Acta, *188*: 216–221, 1969.
29. Schapira, G., Reibel, L., and Cuault, F. Absence of species specificity of the KCl factor in the cell-free synthesis of hemoglobin. Biochimie, *54*: 465–469, 1972.
30. Schapira, G., Vaquero, C., and Reibel, L. Transfer of the genetic information carried by rabbit Hb mRNA onto guinea-pig and allelic rabbit reticulocyte ribosomes. Biochimie, *55*: 183–187, 1973.
31. Stewart, A. G., Gander, E. S., Morel, C., Luppis, B., and Scherrer, K. Differential translation of duck and rabbit globin messenger RNA's in reticulocyte-lysate systems. Eur. J. Biochem., *34*: 205–212, 1973.
32. Uenoyama, K. and Ono, T. Specificities in messenger RNA and ribosomes from free and bound polyribosomes. Biochem. Biophys. Res. Commun., *49*: 713–719, 1972.
33. Vaquero, C., Reibel, L., Delaunay, J., and Schapira, G. Translation of globin mRNA among eukaryotes. Biochem. Biophys. Res. Commun., *54*: 1171–1177, 1973.
34. Wittmann, H. G., Stoffler, H. G., Hindennach, I., Kurland, C. G., Randall-Hazerlbauer, M., Birge, E. A., Nomura, M., Kaltschmidt, E., Mizushima, S., Traut, R. R., and Bickle, T. A. Correlation of 30 S ribosomal proteins of *Escherichia coli* isolated in different laboratories. Mol. Gen. Genet., *111*: 327–333, 1971.

Discussion of Paper by Drs. G. Schapira et al.

DR. DUBE: It is rather surprising that duck mRNA binds better to rabbit ribosomes than to duck ribosomes. How much better is the binding?

DR. G. SCHAPIRA: Under the conditions used by Dr. Scherrer, there was no biosynthesis of duck hemoglobin at all. The initiation factors in this instance seems to be the same for duck and rabbit reticulocyte cell-free systems.

DR. DUBE: Initiation factors do play an important role in mRNA translation and the regulation of translation. A good example exists in bacterial systems, where this has been shown by Dube and Rudland (Nature, 1970).

DR. MURAMATSU: The work referred to by Dr. Schapira has been done by Dr. Scherrer and is published in Eur. J. Biochem. How many molecules of your 11,000 molecular weight RNA are there per one ribosome?

DR. G. SCHAPIRA: I cannot give the answer now, but certainly we will know in the near future.

DR. WEBER: Is your conclusion then, that the ribosomes do not play a role in regulation? Or is it possible that your conclusion may be restricted to hemoglobin processing?

DR. G. SCHAPIRA: I don't suppose that my conclusions on the relationship between ribosomes and hemoglobin are restricted to hemoglobin mRNA. Nevertheless, in order to be sure, I am trying the same experiments with fibroin (from silk worm) mRNA. Nevertheless, some specificity of ribosomes (strict or by the mediation of initiation and interference factors) cannot be ignored.

RNA Polymerases in Transcription and Replication in Eucaryotic Organisms

William J. Rutter, Frank Masiarz, and Paul W. Morris*
Department of Biochemistry and Biophysics, University of California, San Francisco, California, U.S.A.

Abstract: In contrast to procaryotes, eucaryotic organisms contain multiple RNA polymerases. The properties and subunit structure of these enzymes are generally similar to that of the procaryotic polymerase. We infer therefore that the general mechanism of action and regulation are also similar. Available evidence suggests that there are several different transcriptive systems (an RNA polymerase, the array of genes transcribed and regulatory elements) in eucaryotic organisms. Ribosomal RNA's and 4–5S RNA's are produced by polymerase I; most mRNA's and heterogeneous RNA's are produced by polymerase II. Organelles (*e.g.*, mitochondria) also have a specific polymerase. Proteins stimulating both chain elongation and initiation have been described. Factors that are required for transcription by polymerases I and II on double-stranded but not on single-stranded templates have been discovered in this laboratory. These may be analogous to the procaryotic "sigma" factors. DNA synthesis in eucaryotes as well as in procaryotes is initiated by RNA synthesis. Initiation may be facilitated by the usual polymerases associated with transcription at their initiating sites (replicons=transcriptons). However, evidence obtained from our studies on fertilized sea urchin eggs suggests that a specific polymerase (III) may be involved. Thus, DNA synthesis may be regulated *via* this specific RNA polymerase.

The major biological functions of the genome—gene expression and gene replication—may be controlled by the process of transcription. DNA-dependent RNA polymerases are responsible for the synthesis of the polyribonucleotides associated with protein synthesis: the ribosomal RNA's and mRNA's. It has recently become evident that the initiation of gene replication is also mediated by transcriptive events (*10*, *27*, *30*, *37*, *45*). This increase in our knowledge of the functions of the RNA

* Present address: Department of Biological Chemistry, University of Illinois at Medical Center, Chicago, Illinois, U.S.A.

polymerases raises significant questions concerning the specificity and regulation of the activity of these enzymes.

In procaryotic organisms, all of the transcriptive functions are carried out by a single enzyme whose activity is apparently blocked by rifamycin. In *Escherichia coli*, DNA synthesis of the bacteriophage M13 is inhibited by rifamycin (*6*), whereas DNA synthesis of bacteriophage ϕX174 is unexpectedly immune to the inhibitor (*38*). The lack of inhibition may be due to the presence of an as yet uncharacterized enzyme involved in DNA synthesis, the presence of a derivative enzyme no longer subject to the inhibition of rifamycin or the presence of an "initiator" RNA species.

In eucaryotic organisms, there are multiple forms of RNA polymerase with apparently different transcriptive functions. There appear to be 3 major classes of the enzymes in the nucleus; chromatographic heterogeneity has been observed in two of the classes (*34*). A distinct RNA polymerase has been reported in mitochondria (*24, 49*) and such a species may occur in chloroplasts. Thus a number of "transcriptive systems" composed of a particular RNA polymerase, the genes it transcribes and relevant regulatory molecules are present in the highly compartmentalized eucaryotic cell (*36*).

In this paper, we summarize the current state of knowledge concerning the structure, transcriptive role, and regulation of the activity of the various nuclear RNA polymerases. We present evidence that the activity of these enzymes on double-stranded templates can be influenced by specific factors and consider a possible role of these enzymes in replication.

Structure and Function of Eucaryotic Nuclear Polymerases

The subunit composition of the purified eucaryotic polymerases has been determined by SDS gel electrophoresis (see Table 1). The data indicate that the species are distinct but are nevertheless similar in general structure to procaryotic enzymes (*46*). The polymerases are comprised of 2 large subunits and at least 2 small subunits; the number of smaller subunits in the nuclear polymerases is not yet defined. The conspicuous difference in the mass of the several subunits emphasizes the asymmetry of the molecules. The apparent structural homology suggests that the mechanisms of transcription in procaryotes and eucaryotes are similar. However, there

TABLE 1. Structural Homologies between Procaryotic and Eucaryotic Nuclear Polymerases

Subunit	Procaryotic	Eucaryotic			
		I	II_A	II_B	III
β'	155,000	200,000	190,000	170,000	(200,000)
β	145,000	135,000	150,000	150,000	(135,000)
α	$(40,000)_2$	45,000	35,000	35,000	—
		25,000	25,000	25,000	—
		17,000	16,000	16,000	—
σ	90,000 *E. coli* 60,000 *B. subtilis*	?	?	?	?
Structure	$\alpha_2\beta\beta'\sigma$	$(\alpha)_n\beta\beta'$			

TABLE 2. Transcriptive Functions of Nuclear RNA Polymerases

Nuclear RNA polymerase	I	III	II
Localization	Nucleolus	Nucleoplasm	Nucleoplasm
Transcriptive range (% of total)	<10		>90
Specific products	rRNA	tRNA? 5 S RNA?	mRNA HnRNA
DNA synthesis	?	?	?

rRNA, ribosomal RNA; tRNA, transfer RNA; mRNA, messenger RNA; HnRNA, heterogeneous nuclear RNA.

are significant structural differences in the nuclear polymerases which are found consistently throughout the eucaryotic domain. These differences may be functionally relevant since they are evolutionarily conserved.

There is no correlation between the number of genes in a transcriptive system and its polymerase concentration. RNA polymerase concentrations are lower in eucaryotic than in procaryotic cells. In actively growing *E. coli*, there is an average of one polymerase molecule for every 10^4 nucleotides of template; in mammalian cells, this concentration is lower by at least an order of magnitude. However, this discrepancy may be more apparent than real since the eucaryotic genome is not transcribed as completely as the procaryotic one. For example, hybridization of total cellular RNA with DNA suggests that only 10–20% of the genome is actively transcribed in any particular cell (for example, see Ref. *13*). The enzyme activity correlates most directly with RNA output rather than the genetic complexity of a particular system.

Evidence summarized in Table 2 indicates that polymerase I catalyzes the synthesis of ribosomal RNA and that polymerase II makes most, if not all, of the heterogeneous and mRNA's. Transfer RNA's and 5S RNA are not produced by polymerase II; polymerase I or III must perform that function. Because ribosomal RNA's, tRNA's, and 5S RNA all function in the translation process, it is logical that a single polymerase (polymerase I) should synthesize all of these species. However, the activity of polymerase on each of these genes is controlled independently, since the production of these RNA's is not regulated coordinately. The results of hybridization experiments confirm that the heterogeneity of transcripts of polymerase II is much greater than that of polymerase I or III (*3*, *33*). Polymerase I probably transcribes less than 1% while polymerase II is responsible for transcription of the majority of the genome. In rapidly growing cells, the concentration of polymerase I is frequently as high as that of polymerase II and about 50% of the total RNA produced is ribosomal RNA. One of the physiological advantages of polymerase multiplicity is that the concentration of each enzyme can be maintained at a level appropriate for its transcriptive load.

Regulation of Polymerase Activity by Specific Factors

Although the diversity of polymerases remains unknown, it is clear that there cannot be a one gene-one enzyme relationship; this is obviously true for polymerase

II. The process of selective transcription of a gene requires the recognition of a specific DNA sequence for initiation, the indiscriminant transcription of the DNA and termination at a specific sequence. This modulation of specificity could be accomplished through specific allosteric transitions producing different functional states of the enzyme, or interactions with specific effectors. Early studies in procaryotic systems suggested that the specificity of core polymerases was regulated by "sigma" factors (determinants of initiation) and the "rho" factor (a determinant of termination). Studies by Chamberlin and colleagues (*8*, *15–18*) suggest that sigma factor may play a general role in initiation at appropriate sites, but there is no evidence for multiple factors that prescribe the transcription of particular genes.

The general structural resemblance between procaryotic and eucaryotic enzymes suggests that an analog of the sigma factor may be present in eucaryotic systems. There have been vigorous efforts to detect such components, presumably based on the assumptions that they would be determinants of specificity and involved in differentiative transitions. These studies have been hampered by the lack of an adequate assay. The sigma factor was discovered because it is required for the efficient transcription of the T4 bacteriophage genome; no such effect was evident on "non-physiological" templates such as bulk calf thymus DNA (*7*). Studies in eucaryotic systems have involved a search for a stimulation of the activity of polymerases on non-physiological templates. A number of such factors have been discovered, but some are non-specific. For example, deoxyribonucleases increase transcriptive activity presumably by providing additional initiation sites. Bovine serum albumin preferentially stimulates polymerase II activity; the activity on double-stranded templates is stimulated more than acitivity on single-stranded ones. This effect is probably due to a stabilization of the polymerases (*36*).

Some of the factors reported may simply be artifacts resulting from the assay method, while others appear to play physiologically significant roles. Stein and Hausen first reported a factor which selectively stimulates polymerase II (*42*). Most of the factors have only a modest effect on overall transcription, although in several cases (*26*, *39*, *43*), effects of at least an order of magnitude have been reported. Recent experiments by Lee and Dahmus show that both heat-stable and heat-labile factors are present in extracts of Novikoff ascites cells. One of these factors stimulates initiation while another affects chain elongation (*25*). Experiments in our laboratory have demonstrated the presence of heat-labile and heat-stable factors in calf thymus that are effective even at high concentrations of bovine serum albumin.

While attempting to purify polymerase I, Drs. M. Goldberg and J.C. Perriard in our laboratory observed that passage of a highly purified preparation of the enzyme over a ribosomal RNA-Sepharose column resulted in the loss of essentially all of its activity on native DNA. Subsequent experiments showed that a protein factor, which could be eluted from such ribosomal RNA columns, was capable of restoring the activity of the polymerase preparations on native DNA. These results are comparable to the basic observations concerning the sigma factor. Since this activity was obtained from highly purified enzyme preparations, the responsible factor either co-purifies with or is specifically bound to the polymerase. The factor

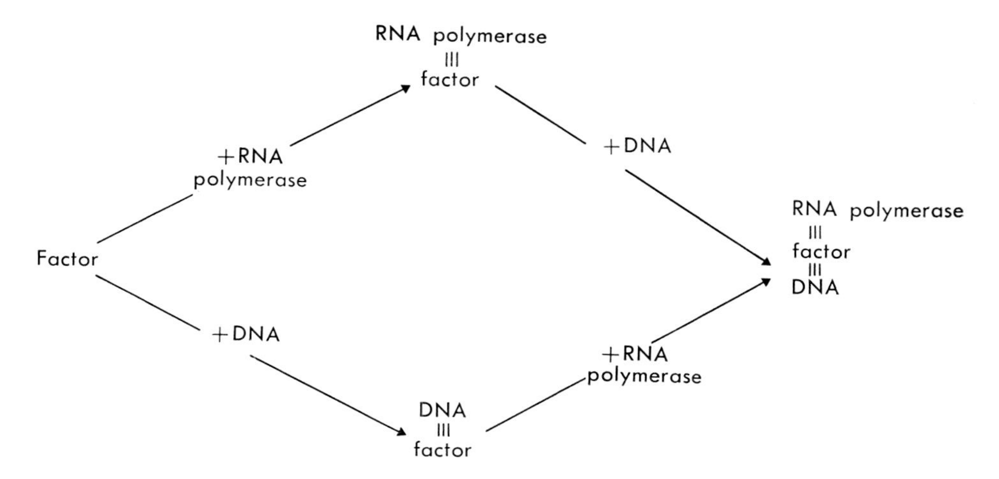

Fig. 1. Alternate pathways of formation of the ternary complex, DNA-regulatory factor-RNA polymerase.

is required for activity on native templates and restricts activity somewhat on poly-dC (*11*).

Although these factors may play significant roles in the transcriptive process, we do not believe that they are determinants of specificity. Our position is partially based on the biological utility of hypothetical polymerase-DNA-specificity factor complexes. Figure 1 shows two thermodynamically equivalent pathways for the formation of such a ternary complex. A DNA-factor complex appears to possess strong biological advantage while combination of a factor with the polymerase has 2 unfavorable consequences. In the latter case, each transcriptive cycle would require 2 bimolecular reactions, making the process kinetically cumbersome. A polymerase-factor complex restricts the transcriptive options of the enzyme and necessitates random search for a particular initiation site. In contrast, a complex of a factor with DNA could simplify the transcriptive process. Only a single bimolecular reaction need be involved in every cycle and each polymerase molecule could interact randomly with any initiation site. Such an argument seems valid for determinants of specificity, but it is not relevant for a factor generally required for initiation by a particular enzyme (*35*).

Transcription and Replication

Genome replication involves initiation at particular sites (replicons) on the DNA, bidirectional elongation *via* the formation of Okazaki fragments, and their ligation to form complete daughter DNA duplexes. DNA polymerases are apparently incapable of catalyzing the initiation of DNA synthesis; RNA polymerases may perform this function. Isolated Okazaki fragments contain short segments of RNA at the 5′-terminus which are eliminated prior to ligation (*30*). However, in mitochondria

TABLE 3. Sea Urchin Eggs Preloaded for DNA Synthesis

Parameter	pmoles ATP incorporated/min	
	Egg	2,000-cell embryo
DNA polymerase (total)	1	1
RNA polymerase (total)	0.04	0.04
Terminal ribosyl-transferase	18	18
Cyclic AMP[a)]	10^{-7} M	$>10^{-6}$ M
Histone mRNA[b)]	+	+

[a)] Nath and Rebhun (*29*). [b)] Skoultchi and Gross (*41*).

there appear to be several ribonucleotides which remain in the circular DNA genome (*12*). The initiation of discontinuous replication at various sites on the chromosome may be determined by particular sequences. In *Drosophila*, Hogness and colleagues have shown that the rates of DNA synthesis are determined largely by the number of initiation sites rather than alterations in the elongation rate itself (*4*, *23*). On the other hand, during chain elongation, the formation of Okazaki fragments may not be initiated by specific base sequences because Okazaki fragments are small compared to gene size. This would require that specific initiating sequences be interspersed among the ordered sequences of a gene.

We have selected the developing sea urchin embryo to study the role of transcription in replication. The urchin embryo is a useful eucaryotic model because the DNA synthetic rate is high, the cycles of synthesis are synchronous in the early phases of development and the cells can be obtained in reasonably large quantities. We have found that the eggs of the sea urchin are pre-loaded with sufficient amounts of RNA polymerase, DNA polymerase, and terminal ribosyltransferase activity for the completion of the early stages of development (Table 3). No increments in these enzyme activities occur before the late blastula stage (at least 2,000 cells). Other investigators have shown that histone mRNA is present in the unfertilized egg (*41*) and cyclic AMP levels are quite low (*29*). The initiation of rapid cell division after fertilization does not therefore appear to involve alterations in cyclic AMP concentration.

Dr. P. Morris in our laboratory discovered that more than 95% of the transcriptive activity of unfertilized sea urchin eggs resides in polymerase III (*36*). Studies of *Xenopus* eggs by Dr. R. Roeder have shown the presence of high levels of polymerase III, although polymerase I and a modified form of polymerase II are also present (*32*). Following the fertilization of sea urchin eggs, there is a rapid increase

TABLE 4. Changes in RNA Polymerase Activities after Fertilization of Sea Urchin Eggs

Polymerase	% total activity		
	Egg	16-cell embryo	Blastula (~2,000 cells)
I	0	0	15
II	<5	36	42
III	95	64	43

in the level of polymerase II; this is presumably related to its function in the specific synthesis of messenger and heterogeneous RNA's. Polymerase I activity increases somewhat later in the developmental schedule; this may coincide with a requirement for the synthesis of large amounts of ribosomal RNA (Table 4). The egg is also pre-loaded for translation and contains all of the ribosomes required for the first several thousand cells. There is no major synthesis of ribosomal RNA until after the blastula stage in either the sea urchin or *Xenopus*. The high level of polymerase III is consistent with a possible role of the enzyme in DNA replication, but it may also be a storage form or precursor of polymerase I and/or II.

TABLE 5. RNA Polymerases and Initiation of DNA Synthesis

1. Transcriptive specificity = replicative specificity (II ≫ I, III)
2. Non-specificity (II ~ I)
3. Specific enzyme involved (III?)

Which of the nuclear RNA polymerases are involved in the replicative process? Table 5 presents 3 possibilities for the involvement of the enzymes: (1) lack of specificity—any polymerase may initiate DNA synthesis at any site; (2) partial specificity—the most interesting case is that transcription and replication are affected by the same enzyme; (3) complete specificity—only a single species is involved in genome replication. Dr. F. Masiarz has recently attempted to distinguish between these 3 alternatives for the involvement of the enzymes in the synthesis of sea urchin DNA. Preliminary experiments measuring the effects of the specific polymerase II inhibitor, α-amanitin, on DNA synthesis in isolated blastula nuclei (where the RNA polymerases are present at similar activity levels) indicate that polymerase II is not involved. If the polymerases participated randomly according to their relative activity, then α-amanitin should inhibit the rate of DNA synthesis by about 40% (assuming the rate is dependent on reinitiation). On the other hand, one would predict a complete or nearly complete inhibition if polymerase II were required or if it were involved in the initiation of synthesis of the genes it transcribes. We have observed that α-amanitin has no detectable effect on the rate of DNA synthesis in these experiments (Table 6).

TABLE 6. Effect of Inhibition of Polymerase by α-Amanitin on DNA Synthesis

Parameter	% inhibition	
	Nuclei	M-bands
RNA synthesis	48	95
DNA synthesis	4	0

Studies of replication in procaryotic organisms indicate that the process may involve the specific attachment of initiation sites to membranous structures (*22*). In eucaryotic cells, recent evidence suggests that initiation does not involve the nuclear envelope (*9, 19, 47, 48*). However, in both types of system, newly synthesized DNA fragments have been found associated with membrane elements which selectively bind to magnesium N-lauryl sarcosinate crystals, forming a complex known as "M-bands" (*44*). We have isolated M-bands containing most of the newly synthesized DNA of the sea urchin embryo, as described by Infante *et al.* (*20*). They contain substantial amounts of RNA polymerase activity and low levels of DNA polymerase activity. α-Amanitin had no effect on the rate of DNA synthesis in these subnuclear preparations, although 95% of the RNA polymerase activity present was inhibited by this compound (Table 6). Our results suggest that α-amanitin resistant species may be involved in DNA replication. Since polymerase III activity is greatest during early post-fertilization events, it seems likely that the enzyme may serve to initiate DNA synthesis. Polymerase III is readily detected in sea urchin, *Xenopus* and yeast and occasionally detected as a labile activity in other systems. Sergeant and Krsmanovic have recently shown polymerase III to be thermolabile in KB cells (*40*). Perhaps the lack of routine detection of polymerase III activity is due to its inherent instability or its tight regulation during the cell cycle.

Replication of the genome is obviously under cyclic control and the initiation of DNA synthesis involves several gene products. Although regulation of DNA synthesis may be accomplished by the control of any of its components, control of initiation rates by the modulation of RNA polymerase activity is an attractive mechanism. RNA polymerase activity could be controlled by phosphorylation and dephosphorylation. Dr. M. Paule in our laboratory has obtained evidence for the phosphorylation of one of the small subunits of polymerase II from rat liver (*36*). Phosphorylation of the other polymerases may also occur. We have no evidence to indicate that phosphorylation affects the rate of RNA or DNA synthesis. Martelo (*28*) and Jungmann and colleagues (*21*) have reported that phosphorylation of proteins in crude extracts with protein kinases increases the rate of RNA synthesis in those extracts. The phosphorylation of an RNA polymerase or some other necessary component may regulate transcription.

Cyclic AMP appears to be a negative effector of DNA synthesis. In a variety of cells, both rates of mitosis and DNA synthesis are inversely related to cyclic AMP concentrations (*1*, *2*). In contrast, cyclic GMP appears to be a positive effector in some systems (*14*). The effects of cyclic AMP and cyclic GMP may be mediated through the phosphorylation of proteins by specific protein kinases. The phosphorylation of histones during the replication cycle has already been documented (*5*, *31*). The phosphorylation and concomitant regulation of other components of the DNA replication machinery may occur as well. It is also possible that the cyclic nucleotides are direct effectors of the catalytic activity of proteins involved in genome replication.

CONCLUDING REMARKS

RNA polymerases are involved in both transcription and replication of the genome. Control of both processes by the independent regulation of the multiple species of RNA polymerase represents a division of labor among enzymes of identical catalytic activity. It would also provide a single, unifying mechanism for the control of the major biological functions of the genome.

REFERENCES

1. Abell, C. W. and Monahan, C. W. The role of adenosine 3',5'-cyclic monophosphate in the regulation of mammalian cell division. J. Cell Biol., *59*: 549–558, 1973.
2. Anderson, W. B., Russell, T. R., Carchman, R. A., and Pastan, I. Interrelationship between adenylate cyclase activity, adenosine 3':5' cyclic monophosphate phosphodiesterase activity, adenosine 3':5' cyclic monophosphate levels, and growth of cells in culture. Proc. Natl. Acad. Sci. U.S., *70*: 3802–3805, 1973.
3. Blatti, S.P., Ingles, C. J., Lindell, T. J., Morris, P. W., Weaver, R. F., Weinberg, F., and Rutter, W. J. Structure and regulatory properties of eucaryotic RNA polymerase. Cold Spring Harbor Symp. Quant. Biol., *35*: 649–657, 1970.
4. Blumenthal, A. B., Kriegstein, H. J., and Hogness, D. S. The units of DNA replication in *Drosophila melanogaster* chromosomes. Cold Spring Harbor Symp. Quant. Biol., *38*: 205–223, 1973.
5. Bradbury, E. M., Inglis, R. J., and Matthews, H. R. Control of cell division by very lysine rich histone (F1) phosphorylation. Nature, *247*: 257–261, 1974.
6. Brutlag, D., Schekman, R., and Kornberg, A. A possible role for RNA polymerase in the initiation of M13 DNA synthesis. Proc. Natl. Acad. Sci. U.S., *68*: 2826–2829, 1971.
7. Burgess, R. R., Travers, A. A., Dunn, J. J., and Bautz, E. K. F. Factor stimulating transcription by RNA polymerase. Nature, *221*: 43–46, 1969.
8. Chamberlin, M. J. and Ring, J. Studies of binding of *Escherichia coli* RNA polymerase to DNA. V. T7 RNA chain initiation by enzyme-DNA complexes. J. Mol. Biol., *70*: 221–237, 1972.
9. Fakan, S., Turner, G. N., Pagano, J. S., and Hancock, R. Sites of replication of chromosomal DNA in a eukaryotic cell. Proc. Natl. Acad. Sci. U.S., *69*: 2300–2305, 1972.
10. Fox, R. M., Mendelsohn, J., Barbosa, E., and Goulian, M. RNA in nascent DNA from cultured human lymphocytes. Nature New Biol., *245*: 234–237, 1973.
11. Goldberg, M. I., Perriard, J. C., Hager, G., Hallick, R. B., and Rutter, W. J. Transcriptional systems in eucaryotic cells. *In;* B. B. Biswas, R. K. Mandal, A. Stevens, and W. E. Cohn (eds.), Control of Transcription (Proc. Int. Symp. on Control of Transcription, 1973, Calcutta, India), pp. 241–256, Plenum Press, New York, 1974.
12. Grossman, L. I., Watson, R., and Vinograd, J. The presence of ribonucleotides in mature closed-circular mitochondrial DNA. Proc. Natl. Acad. Sci. U.S., *70*: 3339–3343, 1973.
13. Grouse, L., Chilton, M. D., and McCarthy, B. J. Hybridization of ribonucleic acid with unique sequences of mouse deoxyribonucleic acid. Biochemistry, *11*: 798–805, 1972.
14. Hadden, J. W., Hadden, E. M., Haddox, M. K., and Goldberg, N. D. Guanosine 3':5'-cyclic monophosphate: A possible intracellular mediator of mitogenic influences in lymphocytes. Proc. Natl. Acad. Sci. U.S., *69*: 3024–3027, 1972.
15. Hinkle, D. C. and Chamberlin, M. J. Studies of the binding of *Escherichia coli* RNA polymerase to DNA. I. The role of sigma subunit in site selection. J. Mol. Biol., *70*: 157–185, 1972.
16. Hinkle, D. C. and Chamberlin, M. J. Studies of the binding of *Escherichia coli* RNA polymerase to DNA. II. The kinetics of the binding reaction. J. Mol. Biol., *70*: 187–195, 1972.

17. Hinkle, D. C., Mangel, W. F., and Chamberlin, M. J. Studies of the binding of *Escherichia coli* RNA polymerase to DNA. IV. The effect of rifampicin on binding and on RNA chain initiation. J. Mol. Biol., *70*: 209–220, 1972.
18. Hinkle, D. C., Ring, J., and Chamberlin, M. J. Studies of the binding of *Escherichia coli* RNA polymerase to DNA. III. Tight binding of RNA polymerase holoenzyme to single-strand breaks in T7 DNA. J. Mol. Biol., *70*: 197–207, 1972.
19. Huberman, J. A., Tsai, A., and Deich, R. A. DNA replication sites within nuclei of mammalian cells. Nature, *241*: 32–36, 1973.
20. Infante, A. A., Nauta, R., Gilbert, S., Hobart, P., and Firshein, W. DNA synthesis in developing sea urchins: role of a DNA-nuclear membrane complex. Nature New Biol., *242*: 5–8, 1973.
21. Jungmann, R. A., Hiestand, P. C., and Schweppe, J. S. Adenosine 3′,5′-monophosphate-dependent protein kinase and the stimulation of ovarian nuclear ribonucleic acid polymerase activities. J. Biol. Chem., *249*: 5444–5451, 1974.
22. Klein, A. and Bonhoeffer, F. DNA replication. Ann. Rev. Biochem., *41*: 301–332, 1972.
23. Kriegstein, H. J. and Hogness, D. S. The mechanism of DNA replication in *Drosophila melanogaster*: structure of replication forks and evidence of bidirectionality. Proc. Natl. Acad. Sci. U.S., *71*: 135–139, 1974.
24. Küntzel, H. and Schäfer, K. P. Mitochondrial RNA polymerase from *Neurospora crassa*. Nature New Biol., *231*: 265–269, 1971.
25. Lee, S. C. and Dahmus, M. E. Stimulation of eucaryotic DNA-dependent RNA polymerase by protein factors. Proc. Natl. Acad. Sci. U.S., *70*: 1383–1387, 1973.
26. Lentfer, D. and Lezius, A. G. Mouse myeloma polymerase B: Template specificities and the role of a transcription-stimulating factor. Eur. J. Biochem., *30*: 278–284, 1972.
27. Magnusson, G., Pigiet, V., Winnacker, E. L., Abrams, R., and Reichard, P. RNA-linked short DNA fragments during polyoma replication. Proc. Natl. Acad. Sci. U.S., *70*: 412–415, 1973.
28. Martelo, O. J. Phosphorylation of RNA polymerase in *E. coli* and rat liver. *In;* F. Huijing and E. Y. C. Lee (eds.), Protein Phosphorylation in Control Mechanisms (Miami Winter Symposia), vol. 5, pp. 199–216, Academic Press, New York, 1973.
29. Nath, J. and Rebhun, L. I. Studies on cyclic AMP levels and phosphodiesterase activity in developing sea urchin eggs: effects of puromycin, 6-dimethylamino purine and aminophylline. Exp. Cell Res., *77*: 319–322, 1973.
30. Okazaki, R., Sugino, A., Hirose, S., Okazaki, T., Imae, Y., Kainuma-Kuroda, R., Ogawa, T., Arisawa, M., and Kurosawa, Y. The discontinuous replication of DNA. *In;* R. D. Wells and R. B. Inman (eds.), DNA Synthesis *In Vitro*, pp. 83–106, University Park Press, Baltimore, 1973.
31. Oliver, D., Balhorn, R., Granner, D., and Chalkley, R. Molecular nature of F1 histone phosphorylation in cultured hepatoma cells. Biochemistry, *11*: 3921–3925, 1972.
32. Roeder, R. G. Multiple forms of deoxyribonucleic acid-dependent ribonucleic acid polymerase in *Xenopus laevis*. Levels of activity during oocyte and embryonic development. J. Biol. Chem., *249*: 249–256, 1973.
33. Roeder, R. G., Reeder, R. H., and Brown, D. D. Multiple forms of RNA polymerase in *Xenopus laevis*: Their relationship to RNA synthesis *in vivo* and their fidelity of transcription *in vitro*. Cold Spring Harbor Symp. Quant. Biol., *35*: 727–735, 1970.

34. Roeder, R. G. and Rutter, W. J. Multiple forms of DNA-dependent RNA polymerase in eucaryotic organisms. Nature, *224*: 234–237, 1969.
35. Rutter, W. J., Goldberg, M. I., and Perriard, J. C. RNA polymerases and transcriptional regulation in physiological transitions. *In;* J. Paul (ed.), Biochemistry of Differentiation and Development, Medical and Technical Publishing Co., London, 1974, in press.
36. Rutter, W. J., Morris, P. W., Goldberg, M., Paule, M., and Morris, R. W. RNA polymerases and transcriptive specificity in eucaryotic organisms. *In;* J. K. Pollak and J. W. Lee (eds.), The Biochemistry of Gene Expression in Higher Organisms, pp. 89–104, Australia and New Zealand Book Co., Sydney, 1973.
37. Sato, S., Ariake, S., Saito, M., and Sugimura, T. RNA bound to nascent DNA in Ehrlich ascites tumor cells. Biochem. Biophys. Res. Commun., *49*: 827–834, 1972.
38. Schekman, R., Wickner, W., Westergaard, O., Brutlag, D., Geider, K., Bertsch, L. L., and Kornberg, A. A. Initiation of DNA synthesis: Synthesis of ϕX174 replicative form requires RNA synthesis resistant to rifampicin. Proc. Natl. Acad. Sci. U.S., *69*: 2691–2695, 1972.
39. Seifart, K. H., Juhasz, P. P., and Benecke, B. J. A protein factor from rat liver tissue enhancing the transcription of native templates by homologous RNA polymerase B. Eur. J. Biochem., *33*: 181–191, 1973.
40. Sergeant, A. and Krsmanovic, V. KB cell RNA polymerases: Occurrence of nucleoplasmic enzyme III. FEBS Letters, *35*: 331–335, 1973.
41. Skoultchi, A. and Gross, P. R. Maternal histone messenger RNA: detection by molecular hybridization. Proc. Natl. Acad. Sci. U.S., *70*: 2840–2844, 1973.
42. Stein, H. and Hausen, P. Factors influencing the activity of mammalian RNA polymerase. Cold Spring Harbor Symp. Quant. Biol., *35*: 709–717, 1970.
43. Sugden, B. and Sambrook, J. RNA polymerase from HeLa cells. Cold Spring Harbor Symp. Quant. Biol., *35*: 663–669, 1970.
44. Tremblay, G. Y., Daniels, M. J., and Schaechter, M. Isolation of a cell membrane-DNA-nascent RNA complex from bacteria. J. Mol. Biol., *40*: 65–76, 1969.
45. Waqar, M. A. and Huberman, J. A. Evidence for the attachment of RNA to pulse-labeled DNA in the slime mold, *Physarum polycephalum*. Biochem. Biophys. Res. Commun., *51*: 174–180, 1973.
46. Weaver, R. F., Blatti, S. P., and Rutter, W. J. Molecular structures of DNA-dependent RNA polymerase (II) from calf thymus and rat liver. Proc. Natl. Acad. Sci. U.S., *68*: 2994–2999, 1971.
47. Williams, C. A. and Ockey, C. H. Distribution of DNA replicator sites in mammalian nuclei after different methods of cell synchronization. Exp. Cell Res., *63*: 365–372, 1970.
48. Wise, G. and Prescott, D. M. Initiation and continuation of DNA replication are not associated with the nuclear envelope in mammalian cells. Proc. Natl. Acad. Sci. U.S., *70*: 714–717, 1973.
49. Wu, G. J. and Dawid, I. B. Purification and properties of mitochondrial deoxyribonucleic acid-dependent ribonucleic acid polymerase from ovaries of *Xenopus laevis*. Biochemistry, *11*: 3589–3595, 1972.

Discussion of Paper of Drs. Rutter et al.

Dr. Shall: Is your data consistent with the possibility that polymerase III is a (chemically) modified form of polymerase I?

Dr. Rutter: There is no data available which refutes this possibility. However, attempts to interconvert the two have, in our hands, failed.

Dr. Sugimura: Is there any possible relation between RNA polymerase III and RNA which is attached to the nascent and short fragment DNA (Okazaki fragment)?

Dr. Rutter: If polymerase III were involved in DNA synthesis as proposed in this study, presumably it would be involved in the formation of such RNA sequences. The polymerase may also produce an initiator RNA.

Dr. Ostertag: Did you find any cytoplasmic (non-mitochondrial) RNA polymerase?

Dr. Rutter: We have not searched extensively for such a polymerase in recent years. But in our earlier studies, we had felt that the small amounts detected in the cytoplasm could be accounted for by nuclear leakage.

Dr. Holtzer: Is there any evidence suggesting that polymerase III is rate limiting in a system that is programmed to undergo a limited number of cell divisions? For example, in embryonic red blood cells or rotifera?

Dr. Rutter: No evidence as yet. The experiments to date have been largely hampered by the "lability" of the enzyme. More recent studies have shown that this enzyme has unique properties. This information should allow experiments of the sort you suggest.

Dr. Dube: I was very interested in your results showing fractionation of nuclear RNA polymerase into activities of different specificities. It is clear from your experiments that RNA polymerase II is specific for transcribing heterogeneous nuclear RNA (HnRNA) and mRNA. Now do you observe any tissue specificity for polymerase II, or perhaps is there a factor involved which might provide polymerase II with further specificity for transcribing different classes of messengers?

DR. RUTTER: This is a very significant question. With the rather crude methods for testing specificity which have been applied, no differences between polymerase II from thymus and liver are apparant. With the availability of probes for specific transcripts, questions like this will be answered.

DR. PRASAD: Do you find any change in the level of polymerase III in malignant cells.

DR. RUTTER: Such experiments have not been carried out. Most individuals studying polymerases in tumor cells have found only I and II. We believe this is due to the lability of III.

DR. PAUL: Chambon has recently reported that his RNA polymerase will not initiate on double-stranded DNA. Could you please comment?

DR. RUTTER: It has been known for some time that polymerase activity is lower in high molecular weight DNA than lower molecular weight DNA. Presumably this is due to more efficient initiation (non-specifically) at ends. If Chambon's preparations were missing II factors, then this could explain his result.

Properties of the Chromosomal Enzyme, Poly(ADP-Ribose) Polymerase in Cell Growth and Replication in Normal and Cancer Cells

Sydney SHALL, M. K. O'FARRELL, P. R. STONE, and W. J. D. WHISH
Biochemistry Laboratory, University of Sussex, Brighton, Sussex, U. K.

Abstract: The properties of a novel chromosomal enzyme system have been investigated in a number of different physiological states in both normal and tumour cells. The specific activities of both NAD pyrophosphorylase, which makes NAD, and poly(ADP-ribose) polymerase, which uses NAD as a substrate to make a chromosomal polymer, show a temporal correlation with DNA synthesis.

The specific activity of the polymerase fluctuates markedly in lymphoid cells during the transition from the non-growing to the growing state. Also, it varies during the growth cycle in tissue culture cells. Most cells have enzymes that can synthesise and degrade poly(ADP-ribose). The physiological function of this chromosomal enzyme system is unknown at present; we have as a working hypothesis the view that these enzymes of nuclear NAD metabolism directly integrate NAD and cellular energy metabolism with DNA biosynthesis both in progression through a single cell generation and in modulation of growth rate. In particular, we suggest that poly(ADP-ribose) may function by regulating the initiation of succeeding sets of replicons.

The essential features of differentiation and of malignancy may reasonably be looked for in the physiology of the chromosomes. An altered pattern of gene expression may emerge from the evolution of a new pattern of transcription. Control of malignancy may result from a new pattern of gene expression. On the other hand, it may be generated by a change in the gene complement. In either event, the result is that initiation of DNA and cell replication is altered either quantitatively or qualitatively. Initiation of cell replication occurs more frequently than previously, thus overriding the frequency of cell death by a significant amount; or, the mechanisms that initiate cell replication become insensitive or less sensitive to the extracellular forces which would usually restrain its rate of cell replication.

These general considerations direct our attention to the physiology of the chromosomes and to the control of the replication of the chromosomes and the cell. Cell cycle studies lead to the general conclusion that the control of cell replication subsists in the control of the initiation of DNA synthesis. Once the decision has been made to initiate DNA replication it usually is the case that chromosome replication is completed, followed by a G2 phase and then mitosis.Cells do not normally come to rest during these processes; this suggests that they are temporally related without the need for decisions. Once the original decision is achieved the cell growth processes will achieve their consummation in cell division.

NAD Metabolism

Accordingly, we direct our attention at the molecular physiology of chromosomal replication and its integration with general cell growth. For many years there has been an interest in the relationship between cell energy metabolism and cell replication in both normal and tumour cells. We have made our starting point the observation that there are two enzymes of NAD metabolism confined to the chromosomes in nucleated cells. This morphological fact is intriguing. At this time, the reason for this association is quite unknown.

The two enzymes are NAD pyrophosphorylase (NAD: NMN adenylyl-transferase) which catalyzes reaction (1)
Reaction 1: $ATP + NMN^{+} \rightleftharpoons NAD^{+} + PP_{i}$;
and poly(ADP-ribose) polymerase which catalyses reaction (2).
Reaction 2: $NAD^{+} + (ADP\text{-}ribose)_{n} \longrightarrow (ADP\text{-}ribose)_{n+1} + nicotinamide + H^{+}$

Synthesis of Poly(ADP-ribose)

The second reaction was first observed in the Laboratory of Chambon and Mandel in Strasbourg (*5, 9*). The structure of the polymer was independently established in Strasbourg and by Hayaishi (*27, 31*) in Kyoto and by Sugimura (*11, 12, 15, 40*) in Tokyo. The information on this novel nuclear polymer up to the end of 1971 has been comprehensively reviewed by Sugimura (*39*). More recent developments have been briefly reviewed by Shall (*34*).

The structure of the polymer is shown in Fig. 1. The polymer is probably linked covalently to chromosomal protein (*28, 41*). The enzyme responsible for the synthesis of this polymer is confined to the chromatin (*28, 41*), it is probably absent from nucleoli (*16*). It has been found in all nucleated cells in which it has been sought. Several different laboratories have shown that the synthesis of the polymer is not reversible, which is taken as evidence that the polymer is not merely a reservoir of NAD. The enzyme is specific, neither NADH nor NADP are substrates. The polymerase is strongly inhibited by nicotinamide and by thymidine.

Both DNA and histone will stimulate activity in partially purified preparations (*42*) and DNase will inhibit the activity (*5, 27*). There is as yet no evidence that the DNA has a template character in the synthesis of the polymer. The nucleoprotein may function to stabilize an enzymically active configuration of the enzyme.

The substrate for the synthesis of this polymer is NAD^{+}, which is synthesised

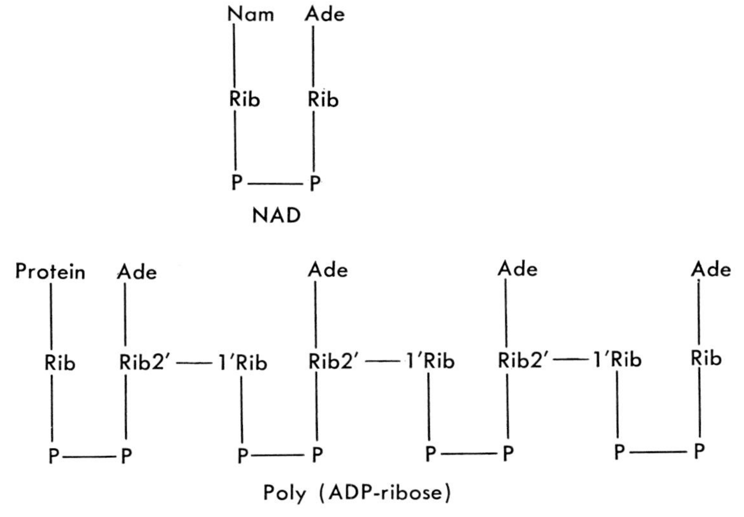

FIG. 1. Structure of NAD and of poly(ADP-ribose). Rib, ribose; Ade, adenine; Nam, nicotinamide.

by the other chromosomal enzyme (Reaction 1). A correlation between NAD pyrophosphorylase activity and mitosis has been reported (*26*). The activity of this enzyme has also been roughly correlated with DNA synthesis in various classes of rat liver nuclei (*14*). Finally, the activity of this enzyme increases along with the increase of DNA in regenerating liver (*10*).

NAD Pyrophosphorylase in Physarum polycephalum

To clarify a possible relationship between NAD pyrophosphorylase activity and DNA and RNA biosynthesis we have examined the specific activity of this enzyme in the nuclei of *Physarum polycephalum* during the cell generation. This organism may be grown as a multi-nucleate plasmodium in which there is complete, spontaneous synchrony of nuclear events. We have observed that the activity of this enzyme oscillates in time with DNA synthesis in this organism, but does not bear an obvious relationship with RNA biosynthesis.

We estimated the specific activity of NAD pyrophosphorylase in isolated nuclei of *P. polycephalum* throughout the cell cycle (Fig. 2). The activity of this enzyme showed a low basal level during G2 and about 1 hr before mitosis the activity began to increase and reached a peak at mitosis. The activity remained high throughout the S phase, which follows mitosis directly in this organism and then declined to the stable, basal, G2 level. There is clearly a positive temporal correlation with DNA biosynthesis. This temporal association suggests that there is a relationship between some aspect of NAD metabolism and DNA biosynthesis. This suggestion is strengthened by considering the behaviour of other enzymes in *P. polycephalum*. The specific activity of thymidine kinase fluctuates (*33*) like that of NAD pyrophosphorylase. On the other hand, glucose-6-phosphate dehydrogenase remains almost constant in

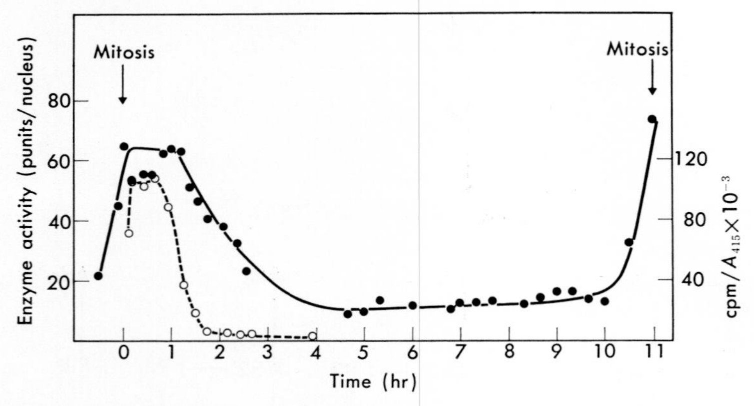

FIG. 2. NAD pyrophosphorylase activity in isolated nuclei of *P. polycephalum* during the cell generation. Horizontal axis: time (hr). Left hand vertical axis: ● specific activity of NAD pyrophosphorylase (p units/nucleus). Right hand vertical axis: ○ DNA synthesis (cpm/$A_{415} \times 10^{-3}$). (Reprinted with permission from Solao and Shall, Exp. Cell Res., *69*: 295, 1971.)

specific activity throughout the cell generation (*18, 33*). Phosphodiesterase, isocitrate dehydrogenase, acid phosphatase, β-glucosidase and histidase increase steadily in activity throughout the cell generation (*18*). Ribonuclease I (*3*) and glutamate dehydrogenase (*18*) increase stepwise between 3–4 hr and 4–5 hr after mitosis, respectively. NAD pyrophosphorylase does not show any of the three patterns shown by these metabolic enzymes, but on the contrary it shows a peak in time with DNA synthesis and thymidine kinase.

The variation in the rate of RNA synthesis in *Physarum* has a distinctive pattern (*23*). In each generation there are 2 maxima and 2 minima in the rates of RNA synthesis. The minima occur at mitosis and in early G2; the maxima occur in late S phase and in mid-G2 phase. It is observed that this particular temporal pattern is not reflected in the specific activity of NAD pyrophosphorylase (Fig. 2). This might suggest that there is not a close connection between these enzymes and RNA biosynthesis. However, it is possible that a subfraction of the total RNA synthesis, say mRNA synthesis, or the activity of one of the RNA polymerases, say the α-amanitin sensitive one, would by itself show a pattern more like those shown in Fig. 2 or 4.

The observed pattern of NAD pyrophosphorylase activity implies variations in the metabolism of NAD during the cell cycle. Sachsenmaier *et al.* (*32*) have measured the concentrations of NAD, NADH, and NADP at three points during the cell cycle (during G2, at mitosis and during early S phase) in *P. polycephalum*. The concentration of NAD was about half that of rat liver, but there were no significant variations in NAD concentration between the different points in the cell cycle. The pool of ATP in *Physarum* decreases from a peak level at prophase to a minimum just after mitosis and increases again in the post mitotic period (*7*). Thus the ATP pool is growing during the S period when NAD pyrophosphorylase activity is high.

NAD in Mouse Fibroblast

We have estimated the NAD+ content of mouse fibroblast (LS) cells both during the growth cycle and in synchronous cultures during a single cell generation. These mouse fibroblast cells, kindly provided by Dr. J. Paul, Glasgow, are grown in suspension culture. NAD+ was estimated by an enzyme recycling method (*22*). The NAD+ content found in the LS cells (about 10^{-15} moles/cell) was of the same order as that found in rat liver (0.23×10^{-15} moles/cell), and in HeLa cells grown in culture (0.85×10^{-15} moles/cell). The NAD+ content of the cells fluctuated during the growth cycle (Fig. 9). During the first 20–30 hr after inoculation the cell NAD+ level dropped; it rose consistently during the later logarithmic phase to a maximum coincident with the onset of the stationary phase. The level fell as the cells passed further into the stationary phase. The onset of the stationary phase varied in time with the cell density of inoculation; however, the maximum cell NAD+ content always coincided with the onset of the stationary phase.

In order to examine the NAD+ level during the cell cycle, a synchronous population was selected by a velocity gradient selection method (*35*), the degree of synchrony achieved was moderately good. There were significant fluctuations of NAD+ content through the cell cycle (Fig. 3). As in the previous experiment, the NAD+ level first dropped; but now it reached a minimum after about 2–3 hr and then rose to a maximum at about 15 hr, after which it declined again. The control,

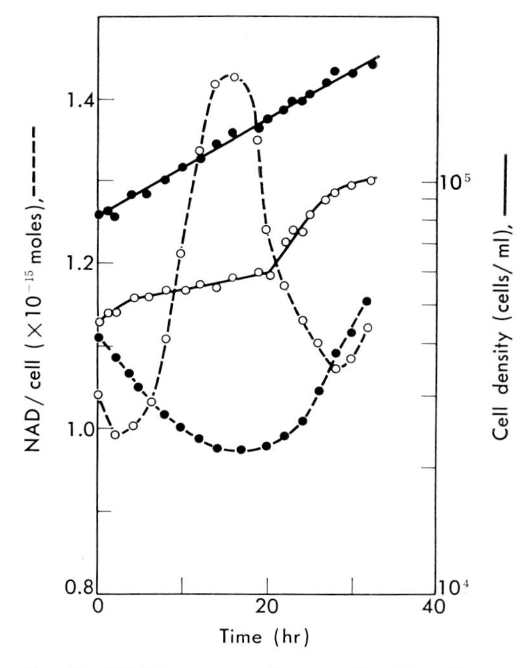

FIG. 3. NAD+ content of mouse fibroblast (LS) cells during the cell cycle. Cell density (cells/ml): ○—○ synchronous; ●—● control cells. NAD+ content per cell (10^{-15} moles/cell): ○---○ synchronous cells; ●---● control cells.

asynchronous culture displayed the same pattern as the previous experiment (Fig. 9). The peak in NAD^+ level correlates with the S period as measured by the incorporation of thymidine.

Tissue NAD Level and Growth Rate

There is available other indirect evidence suggesting a relationship between NAD biosynthesis, cell growth and DNA replication. However, this evidence suggests an inverse correlation between tissue NAD level and growth rate (*25*, *26*). Rapidly growing tissues like mouse mammary carcinoma (*19*), dye induced hepatoma and Krebs ascites tumour in the rat (*13*), and foetal rat liver (*8*), all have lower concentrations of NAD than the comparable tissues in the adult animal. For example, Briggs (*4*) found 0.54 mmoles NAD/kg wet weight in normal liver and 0.25 mmoles NAD/kg wet weight in dye induced hepatoma. The lower levels of NAD are associated with a lower activity of NAD pyrophosphorylase (Table 1), although biochemical evidence suggests that this enzyme is not the rate limiting step in NAD biosynthesis. At the present, we cannot resolve the apparent contradiction between this inverse correlation and our observation that during DNA synthesis there is an increased NAD pyrophosphorylase activity.

TABLE 1. NAD-pyrophosphorylase Activity in Tissues Growing at Different Rates

Tissues	p units/nucleus	Reference
C_3H mice		*2*
Normal lactating mammary gland	221±2.1	
Normal non-lactating gland	117±9.5	
Spontaneous mammary gland carcinoma (% activity=20)	46±2.5	
Strain A mice		*25*
Normal liver	700	
(Amino azotoluene) hepatoma (% activity=20)	150	
C_3H mouse liver		*2*
Foetal	34	
Young mice (7 and 17 days old)	200	
Adult	700	
P. polycephalum		*36*
Maximum activity during DNA synthesis	70	
Average activity during G 2 (% activity=about 15%)	About 10	

Poly(ADP-ribose) Polymerase in Physarum

We return now to a study of the second NAD metabolizing enzyme; poly (ADP-ribose) polymerase. We estimated the specific activity of poly(ADP-ribose) polymerase during the cell generation in *Physarum*. This enzyme showed a stable specific activity during the G2 period and a minimum in activity during the S phase (Fig. 4). The ratio of the peak to the basal level was about 8 for the NAD pyrophosphorylase; it was about 2 or 3 for the poly(ADP-ribose) polymerase. This observation of a minimum during S agrees with observations of Haines *et al.* (*14*) on iso-

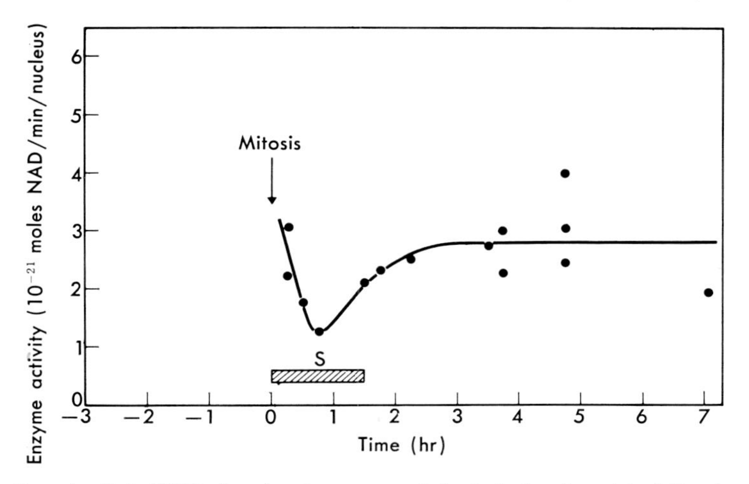

FIG. 4. Poly(ADP-ribose) polymerase activity in isolated nuclei of *P. polycephalum* during the cell generation. Mitosis occurred at 0 hr shown by the arrow. The hatched bar, labelled S, shows the period of DNA synthesis. (Brightwell, Whish, and Shall, manuscript in preparation.)

TABLE 2. Variation in the Activity of Poly(ADP-ribose) Polymerase in Isolated Nuclei during the Cell Cycle of *P. polycephalum*

Time in cell cycle	Activity of poly(ADP-ribose) polymerase (initial rate) (10^{-21} moles NAD/min/nucleus)
15 min before mitosis 2 (late G 2)	75
30 min after mitosis 2 (early S)	39
5 hr after mitosis 2 (early G 2)	69

Mitosis 2 is the second synchronous mitosis after the commencement of synchronous growth.

lated nuclei and with similar work by Miwa *et al.* (*24*) in transformed hamster lung cells.

The above experiment in *Physarum* was performed by measuring the product formed after a fixed period of incubation. This method may not take adequate account of the well-established observation that the kinetics of incorporation of NAD are not linear. This problem was overcome by Dr. M. K. O'Farrell who measured the initial rates of the enzyme reaction by an extrapolation method (*30*). The results (Table 2) clearly confirm the approximate 2-fold variation in enzyme activity in isolated *Physarum* nuclei during the cell cycle.

Poly(ADP-ribose) Polymerase in Different Growth Conditions

We have examined the behaviour of this chromosomal enzyme, poly(ADP-ribose) polymerase in different growth conditions (*20*). The experimental situation that we have examined is the transition of pig lymphocytes from the resting, non-growing state to the growing and dividing state caused by stimulation with phytohaemagglutinin (PHA). These cells were compared with rapidly growing mouse lymphoblastic leukaemia cells (L5178Y). Nuclei from all three cell types incorporate NAD^+ into acid insoluble material (Fig. 5), and all three cells show evidence of ability to degrade poly(ADP-ribose).

The enzyme activity was higher in those cells with the higher growth rate. The

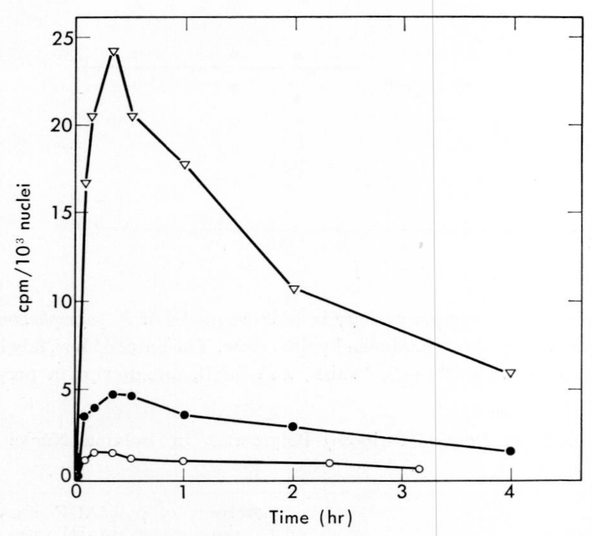

FIG. 5. Uptake of ^3H-NAD^+ into acid insoluble material by isolated nuclei from unstimulated (○), PHA-stimulated (●) pig lymphocytes, and from L5178Y cells (▽). Radioactivity of 1,000 cpm is equivalent to about 0.7 pmoles NAD^+. (Reprinted with permission from Lehmann, Kirk-Bell, Shall, and Whish, Exp. Cell Res., *83*: 63, 1974.)

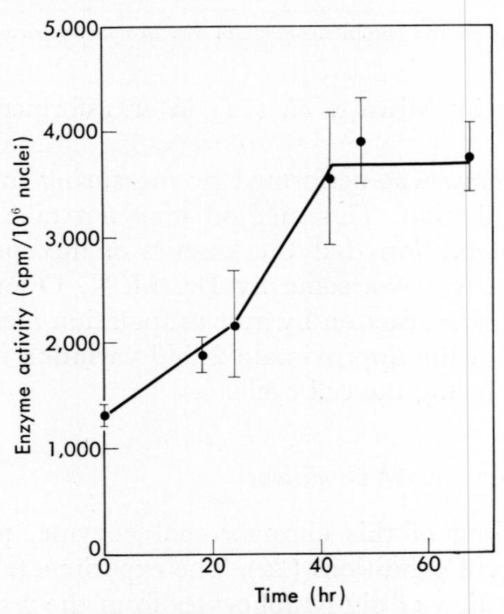

FIG. 6. Poly(ADP-ribose) polymerase activity in isolated nuclei of pig lymphocytes. PHA was added at time zero. The bars show the standard errors of the means of each time point. (Work of Lehmann, Kirk-Bell, and Shall.)

unstimulated, non-growing lymphocytes had a detectable, but very low activity. The PHA-stimulated cells at 48 hr when most cells were growing, making DNA and dividing, had a much higher specific activity. The leukaemic lymphoblasts (L5178Y) where all cells divide every 12 hr, were 5 to 20 times more active than the unstimulated lymphocytes.

Stimulation of pig lymphocytes by PHA led to a 3-fold increase in the specific activity of poly(ADP-ribose) polymerase in isolated nuclei (Fig. 6). The major part of the increase occurred between 24 and 48 hr, that is during DNA synthesis. However, there was a 40% increase over the resting level in the first 24 hr before any DNA synthesis occurred.

Poly(ADP-ribose) Polymerase Activity with Protein, RNA, and DNA synthesis

One may ask what relationship this increase in activity has to the synthesis of DNA, RNA, and protein. Following stimulation by PHA there is a large increase in the uptake of nucleosides, therefore uridine incorporation is not a satisfactory measure of RNA synthesis. The activities of nuclear RNA polymerases are reasonable estimates of RNA biosynthesis. The comparison of poly(ADP-ribose) polymerase activity with protein, RNA and DNA synthesis in PHA-stimulated lympho-

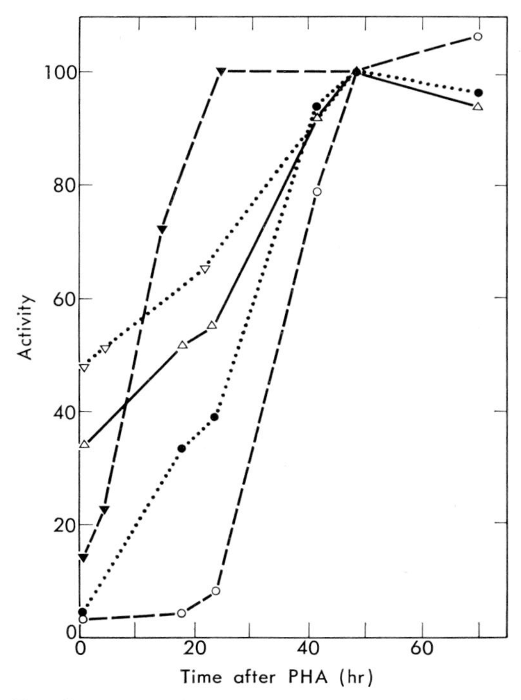

FIG. 7. Comparison of the synthesis of DNA, RNA, and protein with poly(ADP-ribose) polymerase in pig lymphocytes stimulated to grow by PHA. ○ DNA; ● protein; △ ^{3}H-NAD$^+$; ▼ amanitin-resistant RNA polymerase; ▽ amanitin-sensitive RNA polymerase. (Reprinted with permission from Lehmann, Kirk-Bell, Shall, and Whish, Exp. Cell Res., *83*: 63, 1974.)

cytes is shown in Fig. 7. DNA synthesis began about 20 hr after stimulation with PHA. The rate of protein synthesis started to increase soon after stimulation; incorporation was approximately linear up to 48 hr. The increase in the amanitin-resistant RNA polymerase activity, which is thought to be responsible for the nucleolar synthesis of the precursor of ribosomal RNA, occurred in the first 24 hr, after which the activity remained constant. On the other hand, the amanitin sensitive activity, presumably responsible for the synthesis of heterogeneous nuclear RNA and mRNA, revealed a different pattern of increase in activity, rather similar to the increase in rate of protein synthesis.

The kinetics of the increase in poly(ADP-ribose) polymerase do not correlate well with any of the other parameters measured. Perhaps the closest correlation is with the amanitin sensitive polymerase.

We may ask whether DNA synthesis is a prerequisite for the increase in poly (ADP-ribose) polymerase activity. This might be the case if the polymer is a structural component of the chromosome and the 2 syntheses are therefore tightly coupled.

The effect of inhibitors on the PHA-induced stimulation of poly(ADP-ribose) polymerase activity in pig lymphocytes reveal that DNA synthesis is not necessary for the increase in poly(ADP-ribose) polymerase activity. Not surprisingly, inhibition of protein synthesis gives a comparable inhibition of polymerase activity (Table 3).

In these experiments various inhibitors were added immediately after the addition of PHA to pig lymphocytes. After 42–48 hr, the poly(ADP-ribose) polymerase activity in the cells and the rates of DNA and protein synthesis were compared with uninhibited controls. 1 mM hydroxyurea completely prevented DNA synthesis; it partially inhibited the increase in the rate of protein synthesis and also partially inhibited the increase in poly(ADP-ribose) polymerase activity. Fluoro-deoxyuridine at concentrations between 1 and 5 μM also completely inhibited DNA synthesis, as measured by the incorporation of deoxycytidine into alkali-stable,

TABLE 3. Effects of Inhibitors on the PHA-induced Stimulation of Poly(ADP-ribose) Activity in Pig Lymphocytes

Inhibitor	Concentration	Poly (ADP-ribose) polymerase (% of control)	Increase in polymerase (% of control)	DNA synthesis (% of control)	Protein synthesis (% of control)
None (unstimulated)	—	44	0	3	4
None (stimulated)	—	100	100	100	100
Hydroxyurea	1 mM	60	29	2	36
Fluorodeoxyuridine	1 μM	84	71	0	86
Fluorodeoxyuridine	5 μM	83	70	0	82
Actinomycin D	5 ng/ml	65	38	27	34
Actinomycin D	25 ng/ml	31	−23	1	7

Lymphocytes were incubated after the addition of PHA in the presence of various inhibitors. After 42–48 hr, the poly(ADP-ribose) polymerase activity of the nuclei, and the rates of DNA and protein synthesis in the cells were measured. All activities are expressed as percentages of untreated PHA stimulated control.

acid-insoluble material. However, it seemed to be a more specific inhibitor than hydroxyurea; it produced only an 18% inhibition of protein synthesis at 48 hr as compared with controls. The inhibition of the increase in poly(ADP-ribose) polymerase activity was also about 18%.

Low doses of actinomycin D (5–50 ng/ml) in human lymphocytes prevent the onset of DNA synthesis and the appearance of DNA polymerase after stimulation by PHA. The increase in protein synthesis, poly(ADP-ribose) polymerase activity and DNA synthesis were all inhibited to similar degrees by actinomycin D. Thus we may conclude that the induction of poly(ADP-ribose) polymerase activity appears to require general protein synthesis but not to require concomitant DNA synthesis.

Chain Length of Poly(ADP-ribose)

The chain length of the polymer has been estimated in a number of experiments. The products of snake venom phosphodiesterase digestion, 5′-AMP and 2′-(5″-phosphoribosyl)-5′-AMP (PR-AMP), were separated in 3 different ways: (1) on Dowex 1×2 chromatography with a sodium chloride gradient (*29*), (2) on Dowex 1×2 formate, with stepwise elution by formic acid (*37, 38*), (3) on PEI-TLC using a lithium chloride acetic acid solvent (*37, 38*).

Using system 1 and *Physarum* nuclei after incubation with ^{3}H-NAD^{+} (3 μM) for 1 hr at 15°C, the average chain length of the polymer was 6.0. In the lymphoid cells the results obtained with systems 2 and 3 are shown in Table 4. Increase in the incorporation of NAD is associated with synthesis of more chains rather than with elongation of chains under these conditions.

TABLE 4. Chain Length of Poly(ADP-ribose) Synthesised by Isolated Nuclei

Cells	Activity relative to U	Chain length	Chain length relative to U
Unstimulated lymphocytes (U)	1.0	6.2	1.0
Stimulated lymphocytes (S)	4.0	8.3	1.4
Lymphoma (L)	5.8	5.3	0.9

Nuclei were incubated for 5 min with ^{3}H NAD^{+}. The acid-insoluble material was digested with snake venom phosphodiesterase and pronase and then subjected to cellulose polyethyleneimine thin layer chromatography. Chain length: total radioactivity/radioactivity in AMP.

Poly(ADP-ribose) Polymerase and Growth Cycle

We have also asked whether this nuclear enzyme is correlated with the growth state of cells grown in tissue culture. For these experiments we have estimated the specific activity of poly(ADP-ribose) polymerase in the isolated nuclei of mouse fibroblast (LS) cells grown in suspension culture (Fig. 8). The specific enzyme activity shows an initial drop and then an increase during the logarithmic phase. The increase in activity levels off before the stationary phase is reached. The cell number is still increasing logarithmically when the increase in enzyme activity

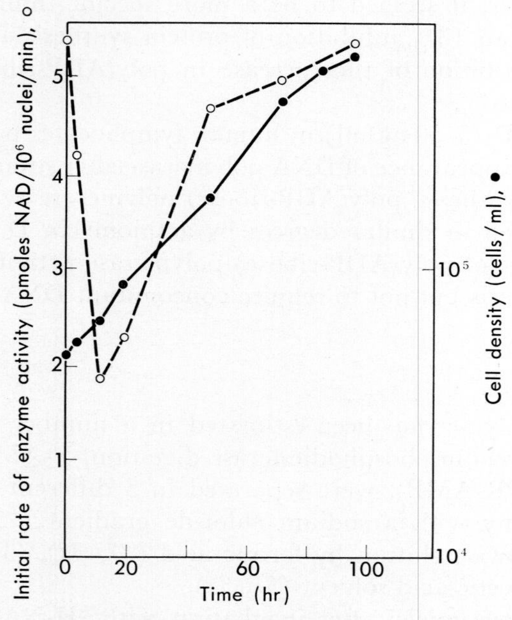

FIG. 8. Poly(ADP-ribose) polymerase specific activity during the growth cycle of LS cells grown in suspension culture. ● cell density (cells/ml); ○ initial rate of enzyme activity (10^{-18} moles NAD/min/nucleus).

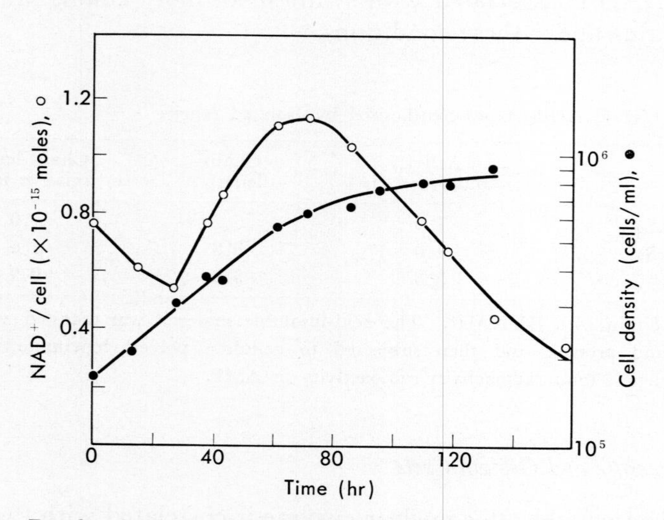

FIG. 9. Content of NAD^+ in LS cells during the growth cycle. ● cell density (cells/ml); ○ NAD^+ content per viable cell (10^{-15} moles/cell).

begins to level off; the change in enzyme activity is an early event in the approach to the stationary phase.

This pattern of enzyme activity is very similar to the variation in cellular NAD^+ levels (Fig. 9); again there is an initial drop and then an increase during the loga-

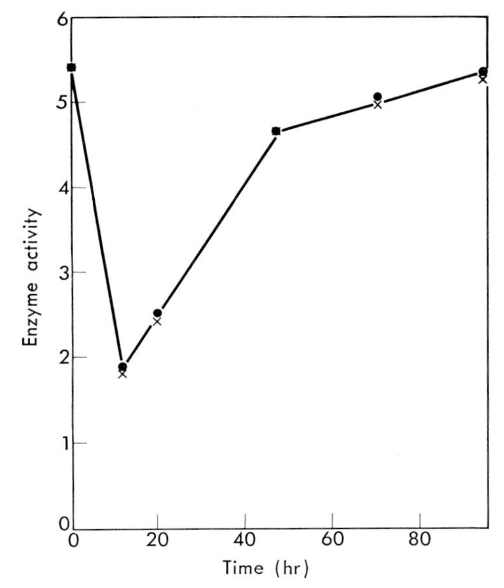

FIG. 10. Effect of the medium on the poly(ADP-ribose) polymerase specific activity during the growth cycle of LS cells in suspension culture. Enzyme activity: 10^{-18} moles NAD/min/nucleus. ● fresh medium; × 5 day conditioned medium.

rithmic phase, reaching a maximum co-incident with the onset of the stationary phase. NAD^+ is the specific substrate for poly(ADP-ribose) polymerase.

The time during the growth cycle of LS cells at which the poly(ADP-ribose) polymerase activity is at a minimum is approximately 12 hr (Fig. 8). The time at which the wave of DNA synthesis starts upon subculturing LS cells is also 12 hr (*1*). There is therefore a temporal correlation between the lowest value of poly(ADP-ribose) polymerase specific activity and the onset of DNA synthesis.

These changes may in the end have a trivial explanation; and these possibilities must clearly be systematically and carefully investigated. The observed variation in enzyme activity does not seem to be caused by the state of the medium in which the cells grow. The same pattern of poly(ADP-ribose) polymerase specific activity during the growth cycle is observed whether the cells were subcultured into fresh medium or into 5-day old conditioned medium (Fig. 10). The growth rates of the cells were exactly the same in both the fresh and conditioned medium.

The cell density may be relevant to the observed fluctuations in enzyme activity. The effect of varying the cell density upon the subsequent poly(ADP-ribose) polymerase activity is surprising (Table 5).

TABLE 5. Effect of Altering the Cell Density on Poly(ADP-ribose) Polymerase Activity

Cell culture	Cell density (at 47 hr) (cell/ml)	Initial rate of enzyme activity (10^{-18} moles NAD/min/nucleus)
Control	1.9×10^5	4.5
Concentrated	6.8×10^5	5.4
Not concentrated	1.9×10^5	4.4

TABLE 6. DNA Content of LS Cells during the Growth Cycle

Time (hr)	Cell density (cells/ml)	DNA content (pg/cell)
0	0.65×10^5	14.5
23	1.32×10^5	13.7
48	2.8×10^5	14.0
72	5.25×10^5	13.7
96	6.0×10^5	13.9

Cells were seeded from a density of 5.6×10^5 cells/ml to a density of about 0.5×10^5 cells/ml and allowed to grow for approximately 30 hr, when the density was 1.2×10^5 cells/ml. Part of the culture was then gently centrifuged ($100 \times g$, 5 min) and resuspended at a density of 5.4×10^5 cells/ml; a second part was centrifuged and resuspended at a density of 1.2×10^5 cells/ml; a third part was left undisturbed. Seventeen hours later the cells were harvested and the specific activity of poly (ADP-ribose) polymerase was estimated (Table 5). The high density culture has a higher specific activity; the 2 controls show that the high level is not the result of time in culture nor of the manipulations.

Hilz and Kittler (*17*) reported a correlation between poly(ADP-ribose) polymerase activity and DNA content in rat liver tissues of normal and malignant origin which differed widely in proliferation rates. To investigate this, the DNA content of the LS cells was determined at different times during the growth cycle of these cells (Table 6). The DNA content of LS cells is constant during the growth cycle at a level of about 14 pg/cell. Thus, the 2- to 3-fold variation in poly(ADP-ribose) polymerase specific activity during the growth cycle of LS cells is not correlated with the DNA content. We infer from this that the polymer is probably not concerned simply with maintaining chromatin structure.

On the other hand, there is a good inverse correlation between the rate of DNA synthesis and the specific activity of the poly(ADP-ribose) polymerase. The minimum activity occurs at about 12 hr and this correlates with a rise in supernatant DNA polymerase (*21*); the maximum activity occurs in late logarithmic phase when DNA synthesis decreases to about 0.1 % of the maximum rate which occurs at about 24 hr (*6*).

CONCLUDING REMARKS

In summary, we find that the specific activity of the chromosomal enzyme, poly(ADP-ribose) polymerase fluctuates with changes in the growth state of the culture. During a single cell generation the activity is at a minimum during the S phase. During a single growth cycle, the activity is low when DNA synthesis begins and high when DNA synthesis is terminating. However, the transition from a non-growing state to a growing state in lymphocytes is associated with a substantial increase in activity, most of which occurs during the first DNA synthetic period.

So, in conclusion we would observe that *a priori* poly(ADP-ribose) and its polymerase may be involved in the regulation of DNA synthesis, of RNA synthesis, in the structure of interphase, dispersed chromatin or in the structure of condensed chromosomes at cell division. We have previously suggested that enzymes of nuclear NAD metabolism directly integrate NAD and cellular energy metabolism with DNA biosynthesis both in progression through a single cell generation and in modulation of growth rates. In particular, we suggested that the poly(ADP-ribose) may function by regulating the initiation of succeeding sets of replicons (*34*, *36*). This is still a reasonable hypothesis.

ACKNOWLEDGMENTS

This work has been generously supported by the Science Research Council, the Medical Research Council, and the Cancer Research Campaign.

REFERENCES

1. Adams, R. L. P. The effect of endogenous pools of thymidylate on the apparent rate of DNA synthesis. Exp. Cell Res., *56*: 55–58, 1969.
2. Branster, M. J. and Morton, R. K. Comparative rates of synthesis of diphosphopyridine nucleotide by normal and tumour tissue from mouse mammary gland: studies with isolated nuclei. Biochem. J., *63*: 640–646, 1956.
3. Braun, R. and Behrens, K. A ribonuclease from Physarum: Biochemical properties and synthesis in the mitotic cycle. Biochim. Biophys. Acta, *195*: 87–98, 1969.
4. Briggs, M. H. Vitamin and enzyme content of hepatomas induced by Butter Yellow. Nature, *187*: 249–250, 1960.
5. Chambon, P., Weill, J. D., Doly, J., Strosser, M. T., and Mandel, P. On the formation of a novel adenylic compound by enzymatic extracts of liver nuclei. Biochem. Biophys. Res. Commun., *25*: 638–643, 1966.
6. Chang, L. M. S., Brown, Mc., and Bollum, F. J. Induction of DNA polymerase in mouse L cells. J. Mol. Biol., *74*: 1–8, 1973.
7. Chin, B. and Bernstein, I. A. Adenosine triphosphate and synchronous mitosis in *Physarum polycephalum*. J. Bacteriol., *96*: 330–339, 1968.
8. Dawkins, M. J. R. Respiratory enzymes in the liver of the newborn rat. Proc. Roy. Soc. *B.*, *150*: 284–298, 1959.
9. Doly, J. and Petek, F. Study of the structure of a polymer "poly ADP-ribose" synthesised in extracts of chicken liver. C. R. Acad. Sci., *D263*: 1341–1342, 1966.
10. Ferris, G. M. and Clark, J. B. Nicotinamide nucleotide synthesis in regenerating rat liver. Biochem. J., *121*: 655–662, 1971.
11. Fujimura, S., Hasegawa, S., Shimizu, Y., and Sugimura, T. Polymerization of the adenosine 5′-diphosphate-ribose moiety of NAD by nuclear enzyme. I. Enzymatic reactions. Biochim. Biophys. Acta, *145*: 247–259, 1967.
12. Fujimura, S., Hasegawa, S., and Sugimura, T. Nictotinamide mononucleotide-dependent incorporation of ATP into acid insoluble material in rat liver nuclei preparation. Biochim. Biophys. Acta, *134*: 496–499, 1967.
13. Glock, G. E. and McLean, P. Levels of oxidised and reduced DPN and TPN in tumours. Biochem. J., *65*: 413–416, 1957.
14. Haines, M. E., Johnston, I. R., Mathias, A. P., and Ridge, D. The synthesis of NAD and poly(ADP-ribose) in various classes of rat liver nuclei. Biochem. J., *115*: 881–887, 1969.
15. Hasegawa, S., Fujimura, S., Shimizu, Y., and Sugimura, T. The polymerization of adenosine 5′-diphosphate-ribose moiety of NAD by nuclear enzyme. II. Properties of the reaction product. Biochim. Biophys. Acta, *149*: 369–376, 1967.

16. Hilz, H. and Kittler, M. On the localisation of poly ADP-ribose synthetase in the nucleus. Hoppe-Seyler's Z. physiol. Chem., *349*: 1793–1796, 1968.
17. Hilz, H. and Kittler, M. Lack of correlation between poly ADP-ribose formation and DNA synthesis. Hoppe-Seyler's Z. physiol. Chem., *352*: 1693–1704, 1971.
18. Hüttermann, A., Porter, M. T., and Rusch, H. P. Activity of some enzymes in *Physarum polycephalum*. I. In the growing plasmodium. Arch. Mikrobiol., *74*: 90–100, 1970.
19. Jedeikin, L. A. and Weinhouse, S. Metabolism of neoplastic tissue. VI. Assay of oxidized and reduced DPN in normal and neoplastic tissues. J. Biol. Chem., *213*: 271–280, 1955.
20. Lehmann, A. R., Kirk-Bell, S., Shall, S., and Whish, W. J. D. The relationship between cell growth, macromolecular synthesis and Poly ADP-ribose polymerase in lymphoid cells. Exp. Cell Res., *83*: 63–72, 1974.
21. Lindsay, J. G., Berryman, S., and Adams, R. L. P. Characteristics of DNA polymerase activity in nuclear and supernatant fractions of cultured mouse cells. Biochem. J., *119*: 839–848, 1970.
22. Lowry, O. H., Passonneau, J. V., Schulz, D. W., and Rock, M. K. The measurement of pyridine nucleotides by enzymatic recycling. J. Biol. Chem., *236*: 2746–2755, 1961.
23. Mittermayer, C., Braun, R., and Rusch, H. P. RNA Synthesis in the mitotic cycle of *Physarum polycephalum*. Biochim. Biophys. Acta, *91*: 399–405, 1964.
24. Miwa, M., Sugimura, T., Inui, N., and Takayama, S. Poly (adenosine diphosphate ribose) synthesis during the cell cycle of transformed hamster lung cells. Cancer Res., *33*: 1306–1309, 1973.
25. Morton, R. K. Enzymic synthesis of coenzyme 1 in relation to chemical control of cell growth. Nature, *181*: 540–542, 1958.
26. Morton, R. K. New concepts of the biochemistry of the cell nucleus. Aust. J. Sci., *24*: 262–278, 1961.
27. Nishizuka, Y., Ueda, K., Nakazawa, K., and Hayaishi, O. Studies on the polymer of adenosine diphosphate ribose. I. Enzymic formation from NAD in mammalian nuclei. J. Biol. Chem., *242*: 3164–3171, 1967.
28. Nishizuka, Y., Ueda, K., Yoshihara, K., Yamamura, H., Takeda, M., and Hayaishi, O. Enzymic adenosine diphosphoribosylation of nuclear proteins. Cold Spring Harbor Symp. Quant. Biol., *34*: 781–786, 1969.
29. Nishizuka, Y., Ueda, K., and Hayaishi, O. Adenosine diphosphoribosyl transferase in chromatin. *In;* D. B. McCormick and L. D. Wright (eds.), Methods in Enzymology, vol. 18B, pp. 230–233, Academic Press, New York, 1971.
30. O'Farrell, M. K. Physical and enzymic properties of nuclei from *Physarum polycephalum*. D. Phil. Thesis, University of Sussex, England, 1972.
31. Reeder, R. H., Ueda, K., Honjo, T., Nishizuka, Y., and Hayaishi, O. Studies on the polymer of adenosine diphosphate ribose. II. Characterisation of the polymer. J. Biol. Chem., *242*: 3172–3179, 1967.
32. Sachsenmaier, W., Immich, H., Grunst, J., Scholz, R., and Bücker, Th., Free ribonucleotides of *Physarum polycephalum*. Eur. J. Biochem., *8*: 557–561, 1969.
33. Sachsenmaier, W. and Ives, D. H. Periodische Änderungen der Thymidinkinaseaktivität im synchronen Mitosecyclus von *Physarum polycephalum*. Biochem. Z., *343*: 399–406, 1965.
34. Shall, S. Poly(ADP-ribose). FEBS Letters, *24*: 1–6, 1972.

35. Shall, S. and McClelland, A. J. Synchronization of mouse fibroblast LS cells grown in suspension culture. Nature New Biol., *229*: 59–61, 1971.
36. Solao, P. B. and Shall, S. Control of DNA replication in *Physarum polycephalum*. Specific activity of NAD pyrophosphorylase in isolated nuclei during the cell cycle. Exp. Cell Res., *69*: 295–300, 1971.
37. Stone, P. R. and Shall, S. Poly(ADP-ribose) polymerase in mammalian nuclei. Characterisation of the activity in mouse fibroblast (LS cells). Eur. J. Biochem., *38*: 146–152, 1973.
38. Stone, P. R., Whish, W. J. D., and Shall, S. Poly(ADP-ribose) glycohydrolase in mouse fibroblast cells (LS cells). FEBS Letters, *36*: 334–338, 1973.
39. Sugimura, T. Poly(adenosine diphosphate ribose). Prog. Nuclei Acid Res. Mol. Biol., *13*: 127–151, 1973.
40. Sugimura, T., Fujimura, S., Hasegawa, S., and Kawamura, Y. Polymerization of the adenosine 5′-diphosphate ribose moiety of NAD by rat liver nuclear enzyme. Biochim. Biophys. Acta, *138*: 438–441, 1967.
41. Ueda, K., Reeder, R. H., Honjo, T., Nishizuka, Y., and Hayaishi, O. Poly(adenosine diphosphate ribose) synthesis associated with chromatin. Biochem. Biophys. Res. Commun., *31*: 379–385, 1968.
42. Yamada, M., Miwa, M., and Sugimura, T. Studies in poly(adenosine diphosphate ribose). X. Properties of a partially purified poly(adenosine diphosphate ribose) polymerase. Arch. Biochem. Biophys., *146*: 579–586, 1971.

Discussion of Paper by Drs. Shall et al.

Dr. Potter: Is the activity of the poly(ADP-ribose) polymerase affected by the NAD/NADH ratio? You spoke of integrating NAD and cellular energy metabolism with DNA biosynthesis. Do you have any information on the NAD/NADH ratio or the lactate/pyruvate ratio in *Physarum?* You speak of enzyme activity, probably enzyme amount at various times, but activity *in vivo* might depend on NAD/NADH ratio.

Dr. Shall: NADH is not a substrate for the poly(ADP-ribose) polymerase; only NAD^+ is a substrate. However, since changes in the NAD^+/NADH ratio are usually not accompanied by changes in the total concentrations of the 2 enzymes, it follows that changes in the redox ratio will lead to changes in the absolute concentration of the substrate, NAD^+ thus, the redox ratio may influence poly(ADP-ribose) synthesis. The NAD^+/NADH ratio in *Physarum* is similar to rat liver, but we have not yet estimated the ratio during the cell cycle. We shall do so. The lactate/pyruvate ratio in *Physarum* is not yet known. We measure enzyme activity, not enzyme amount. Yes, the *in vivo* activity will be modified by the NAD^+ concentration in the cell. We need to estimate the amount of poly(ADP-ribose) in intact cells. We are busy with these estimations.

Dr. Dube: No doubt your interesting work suggests that poly(ADP-ribose) is probably doing something important. What do you think is its function; is anything known?

Dr. Shall: We do not yet know the function of poly(ADP-ribose). As a working hypothesis we think that the chromosomal enzymes of NAD^+ metabolism integrate cellular growth and DNA replication both in transition from one growth state to another and in progression through the cell cycle. In particular, we suggest that poly(ADP-ribose) regulates the initiation of succeeding sets of replicons.

Dr. Dube: Does poly(ADP-ribose) exist in a helical form? If it does, it probably would be a left handed helix; perhaps you would comment on this.

Dr. Shall: Yes, you are quite right! Model building shows that poly(ADP-ribose) can form a double-stranded, anti-parallel, left-handed helix. We are, at this moment, attempting by physical measurements to determine whether poly (ADP-ribose) has secondary structure.

DR. THOMPSON: Is there not a paradox between the experiments which show an inverse correlation between the polymerase and DNA synthesis during the cell cycle and the growth experiments with L cells in with the enzyme falls sharply and then rises diring continuous log growth? Presumably DNA synthesis is constant during log phase.

DR. SHALL: No, there is no paradox here. When the poly(ADP-ribose) polymerase is low in early log phase, the activity of DNA polymerase is high; during late log phase the poly(ADP-ribose) polymerase activity is high and the DNA polymerase activity is low. There is an apparent contradiction between this inverse correlation and the increase in the activity soon after PHA stimulation of lymphocytes. In this case, however, we are simultaneously observing progression through a cell cycle and the transition from a non-growing to a growing state. Perhaps, the latter transition is associated with a very large inverse in the specific activity of poly(ADP-ribose) polymerase which overshadows a smaller decrease in specific activity at the time of DNA synthesis.

DR. WEINHOUSE: How specific is the polymerase with regard to the specificity toward the nucleoside triphosphate? Is the enzyme active toward nucleoside triphosphates other than ATP?

DR. SHALL: The enzyme which synthesises NAD has a low specificity; but poly (ADP-ribose) polymerase which synthesises the polymer, poly(ADP-ribose), is highly specific for NAD^+.

DR. SUGIMURA: There are 2 enzymes to degradate poly(ADP-ribose). One is phosphodiesterase, splitting pyrophosphate bond. The other is glycohydrase splitting ribose-ribose bonds. The former is abundant in the liver, while the latter is abundant in the thymus, fast growing hepatomas. The slowly growing Morris hepatoma contained principally the former enzyme. Apparently the patterns on the degradating enzymes of poly(ADP-ribose) fall in the case of class I of the molecular correlation concept called "malignancy linked discriminants" by Dr. Weber.

DR. WEBER: The results of Dr. Sugimura are of great interest. Would it be too early to ask whether these enzymes behave similarly to tumours or they rather resemble in pattern to that of the normal liver?

DR. SUGIMURA: Generally speaking, these enzymes do not behave similarly to tumours neither to the normal liver.

Cyclic AMP and Cellular Transformation

George S. Johnson

Laboratory of Molecular Biology, National Cancer Institute, National Institutes of Health, Bethesda, Maryland, U.S.A.

Abstract: Cyclic AMP levels are lower in transformed cells than in normal cells. This is due to decreased activity of the synthetic enzyme, adenylate cyclase. Addition of analogs of cyclic AMP or agents which elevate cyclic AMP levels tends to correct numerous abnormal properties. Thus, low levels of cyclic AMP are responsible for some but not all abnormal properties of transformed cells.

A common technique employed in cancer research is the growth of cells in tissue culture. Normal as well as transformed cells can be grown and comparisons between them can be made. Fibroblasts are commonly used for these comparisons, since primary cultures of fibroblasts are easily obtainable and established lines from several species are available. Also fibroblasts are easily transformed in culture by chemicals, viruses, X-rays, and spontaneous selection techniques.

What are some of the properties which distinquish normal from transformed cells? In general, transformed cells grow faster than normal cells and have a greatly increased saturation density; the transformed cells look different, usually they are less spindly and more refractile in appearance; the transformed cells are less adhesive to the substratum on which they are growing and will grow in soft agar; transformed cells have altered surface molecules, are readily agglutinable by plant lectins, often secrete abnormal amounts of micromolecules, and have increased glycolysis (see Ref. *16*).

The mechanism of transformation whereby these properties are altered is a crucial problem which at the present time is under extensive investigation. In formulating such a mechanism 2 important points must be considered. First of all numerous properties are altered. Secondly certain RNA and DNA viruses, which contain relatively few genes, readily transform cells. These points lead one to conclude that a central control locus in the cell may be involved in the initial event of transformation.

In a search for possible control loci, cyclic AMP emerges as a likely possibility.

Cyclic AMP is an ubiquitous nucleotide which has been shown to have regulatory functions in numerous cell types from relatively simple *Escherichia coli* to larger and more complex animal cells (*18*).

Cellular Properties Influenced by Cyclic AMP

The first experiments designed to show regulatory functions for cyclic AMP involved the addition of analogs of cyclic AMP to the growth media. These experiments demonstrated that cyclic AMP has numerous effects on cells. The most obvious change induced by cyclic AMP analogs is cell morphology (*6, 12*). The

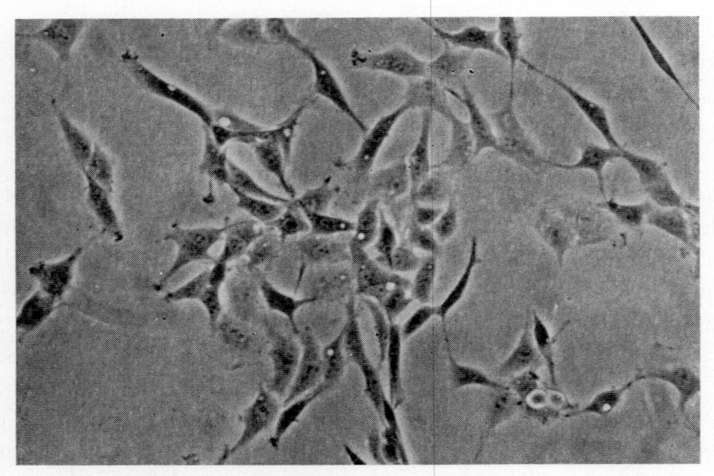

FIG. 1. Morphology of NRK cells infected with a temperature sensitive strain of Kirsten sarcoma virus (see Ref. *3*) and grown at the permissive temperature. Phase contrast, ×200.

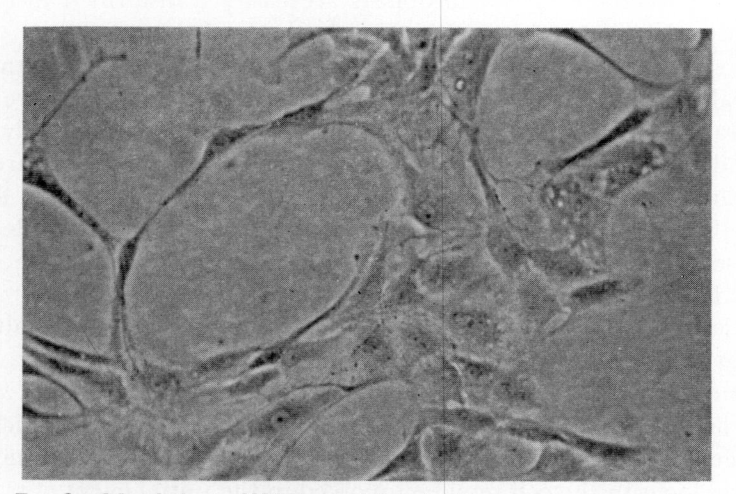

FIG. 2. Morphology of NRK cells infected with a temperature sensitive strain of Kirsten sarcoma virus and grown at the permissive temperature for 4 days in the presence of 1.5 mM Bt_2cAMP. Phase contrast, ×200.

TABLE 1. Detachment of L-929 Cells from the Substratum with Trypsin

Treatment	Time for 50% detachment (min)
Control	16
Bt_2cAMP+theophylline	40
Bt_2cAMP+theophylline for 3 days, then removed 30 min	12

Cells were detached 3 days after planting with 0.02% trypsin. Cells were grown in control medium or medium supplemented with 1 mM Bt_2cAMP plus 1 mM theophylline. Data taken from Ref. *10.*

TABLE 2. Migration of L-929 Cells

Treatment	Migration (μm/min)
Control	0.65±0.05
Bt_2cAMP	0.24±0.03
Bt_2cAMP for 4 days then removed	0.54±0.05

Migration was determined using time lapse cinematography for 150 min beginning 20 min after initiation of the treatment. The concentration of Bt_2cAMP was 1.2 mM. Data taken from Ref. *11.*

TABLE 3. Sulfated Acid Mucopolysaccharide Production by 3T3 and SV40 Transformed 3T3 Cells[a)]

Cell line		Sulfate incorporation ($cpm \times 10^{-3}$/min/protein)	Ratio (treated/control)
3T3:	Control	444	1.5
	Treated	660	
3T3 SV ClX:	Control	105	3.5
	Treated	371	

Treated cells were grown for days in 1 mM Bt_2cAMP plus 1 mM theophylline.
[a)] Data taken from Ref. *5.*

TABLE 4. Pigment Production in Melanoma Cells[a)]

Addition	Absorbance (420 mM)
Control	0.64
Bt_2cAMP (0.12 mM)	1.66

[a)] Data taken from Ref. *8.*

cells change from a rounded form to a more flat, elongated, and spindly form. The effect of Bt_2cAMP (N^6, 2′-0-dibutyryl cyclic AMP) on the morphology of transformed NRK (normal rat kidney) cells are shown in Figs. 1 and 2. Other properties of the cells are also affected. The cells are more difficult to remove from the substratum with trypsin (*10*) (Table 1), and using time lapse cinematography it was demonstrated that treated cells have a decreased motion across the substratum (*11*) (Table 2). Transformed 3T3 cells synthesize decreased amounts of sulfated acid mucopolysaccharides. Addition of Bt_2cAMP plus theophylline increases the synthesis by the transformed cells to about the same level as that of the parent 3T3 cells (*5*) (Table 3). In melanoma cells Bt_2cAMP induces flattening of the cells with extension of the cellular processes, and in addition, it increases the synthesis of pigment (*8*) (Table 4).

TABLE 5. Properties of Fibroblasts Influenced by Cyclic AMP Levels[a)]

1.	Morphology	8.	Collagen synthesis
2.	Adhesion to substratum	9.	Tumor growth
3.	Cell motion	10.	Virus production and viral transformation
4.	Logarithmic growth rate	11.	Transport
5.	Saturation density	12.	Cell membrane composition
6.	Growth in soft agar	13.	Agglutination by plant lectins
7.	Acid mucopolysaccharide synthesis	14.	Synthesis of cyclic AMP phosphodiesterase

a) These properties are reviewed in Ref. *16*.

Normal cells in culture divide logarithmically until a crucial cell density is reached. At this time growth ceases. This phenomenon has been termed "contact or density dependent inhibition of growth." Transformed cells divide logarithmically, usually at a faster rate than normal cells, however, at confluency, they do not cease growth but continue to divide to very high cell densities. Could cyclic AMP have a regulatory role in these growth properties? Bt_2cAMP decreases the rate of cell division (*2*, *9*, *19*), and there is an inverse relationship between the doubling time and the endogenous level of cyclic AMP; the faster the cells divide the lower the levels of cyclic AMP (*14*). Thus the cyclic AMP level is an important factor in the logarithmic growth rate. Bt_2cAMP alone or plus theophylline will not restore density dependent inhibition of growth to transformed cells (*9*). Is then cyclic AMP important in the cessation of growth at confluency? Several types of evidence suggest that it is. Cyclic AMP levels rise at confluency in normal cells but not in transformed cells (*13*, *14*). Agents which stimulate growth in confluent cultures of normal cells rapidly lower cyclic AMP levels (*13*, *22*) this stimulation is prevented by Bt_2cAMP (*4*, *23*). And Bt_2cAMP decreases the saturation density of 3T3 cells, a highly contact inhibited cell line (*9*).

Other cellular properties are affected by cyclic AMP. These properties are summarized in Table 5 and discussed in Ref. *16*.

Specificity

How specific are these effects for cyclic AMP and its analogs? Table 6 lists the compounds which are active in inducing the morphological changes in fibroblasts. Table 7 lists those which are ineffective. As can be seen, the effects are relatively

TABLE 6. Compounds Active in Inducing Morphological Responses in Fibroblasts

I. Cyclic AMP derivatives:	Bt_2cAMP
	N^6-monobutyryl cyclic AMP
	8-Br-cyclic AMP
II. Phosphodiesterase inhibitors:	Theophylline
	1-Methyl-3-isobutyl xanthine
	Papaverine
III. Adenylate cyclase activators:	Prostaglandin E_1
	Prostaglandin E_2
IV. Other compounds:	N^6-substituted adenines

TABLE 7. Compounds Not Effective in Inducing the Morphological Response in Fibroblasts

I. Adenine derivatives:	II. Guanine derivatives:	III. Other compounds:
Adenine	Cyclic GMP	BrdU
Adenosine	8-Br-cGMP	Prostaglandin B_1
5′ AMP	Bt_2cGMP	Indomethicin
ADP		Acetylsalicylic acid
ATP		Epinephrine
		Norepinephrine
		Serotonin
		Thymidine
		Sodium butyrate

TABLE 8. Inhibition of Cyclic AMP Phosphodiesterase in 3T3 SV40 ClX Homogenates[a]

Compounds	K_i (M)
Theophylline	3×10^{-4}
1-Methyl-3-isobutyl xanthine	2×10^{-5}
Adenine	1.8×10^{-3}
Zeatin	4×10^{-4}
Anilinopurine	3×10^{-4}

[a] Data taken from Ref. *7*.

TABLE 9. Cyclic AMP Levels in 3T3 SV40 ClX Cells[a]

Treatment		Cyclic AMP (pmoles/mg nucleic acid)
Control		20
Theophylline:	0.5 mM	30
	1.0	73
1-Methyl-3-isobutyl xanthine:	0.5	80
	1.0	150
Zeatin:	0.5	25
	1.0	35

[a] Data taken from Ref. *7*.

specific for cyclic AMP analogs and agents which elevate the levels of cyclic AMP. Interesting exceptions to this are the N^6-substituted derivatives of adenine. They quickly induce elongation of cellular processes, and also quickly decrease cell motion, and increase adherence to the substratum (*7*). The most obvious question to answer is if they elevate cyclic AMP levels. In cell homogenates they inhibit phosphodiesterase about as effectively as theophylline, a common phosphodiesterase inhibitor (Table 8). In the intact cell, however, they do not greatly affect cyclic AMP levels (Table 9). The mechanism of action of these compounds is not known but they should be useful in elucidating the actions of cyclic AMP on cells.

Cellular Properties Not Influenced by Cyclic AMP

Not all properties of cells are changed when Bt_2cAMP is added to the growth medium. The abnormal ganglioside pattern of L-929 cells and polyoma transformed

3T3 cells (PY-11) is not affected by Bt_2cAMP. (Johnson *et al.* unpublished results). Sakiyama and Robbins (*20*) found that Bt_2cAMP has no effect on the altered glycolipid pattern of Nil cells transformed by hamster sarcoma virus. Another function not affected by cyclic AMP is the ratio of $NAD^+/NADH$ (*21*). In logarithmic normal and transformed cells the ratio of total $NAD^+/NADH$ is approximately 2–3. As growth ceases at confluency in normal cells this ratio rises to about 10. In transformed cells the ratio does not rise at confluency. Bt_2cAMP or agents which affect cyclic AMP levels do not change this ratio in either normal or transformed cells.

Are Cyclic AMP Levels Low in Transformed Cells?

In mouse and rat cells the cyclic AMP level in logarithmically growing normal and transformed cells is related to the doubling time (see above). The cyclic AMP levels in transformed cells are lower in that transformed cells tend to grow faster than untransformed cells. The largest difference between normal and transformed cells is at confluency where cyclic AMP levels in normal cells rise, whereas the levels in transformed cells remain low or even fall slightly (*3*, *13*, *14*). This has been most conclusively demonstrated in NRK cells transformed by wild type or a temperature sensitive mutant of Kirsten sarcoma virus (*3*). The cyclic AMP levels rise in the parent NRK cells as the cells become confluent, whereas the cells transformed by the wild type virus remain low and independent of the cell density. The cells infected with the temperature sensitive virus do not elevate their cyclic AMP levels at confluency when grown at the permissive temperature but do when grown at the non-permissive temperature.

Cyclic AMP levels are lower in chick embryo cells transformed by wild type Rous sarcoma virus than in the untransformed parent cells. In cells infected with a temperature sensitive mutant of Rous sarcoma virus, the cyclic AMP levels fall when the cells are shifted form the non-permissive to the permissive temperature (*15*). The dependence of cyclic AMP on cell density has not yet been studied in detail in chick cells.

Effect of Transformation on Adenylate Cyclase Activity

Burk first measured adenylate cyclase in transformed fibroblasts (*2*). He found that one line of BHK cells transformed by polyoma virus has decreased adenylate cyclase activity. However BHK cells transformed by Rous sarcoma virus have increased activity. Later studies by Peery *et al.* (*17*) failed to show any consistent alterations in adenylate cyclase activity by transformation. More recent studies by Anderson *et al.* (*1*) in transformed chick cells showed that the K_m for ATP is altered by transformation and that a detailed kinetic analysis is required to show alterations by transformation.

No consistent differences in cyclic AMP phosphodiesterase activity with transformation have yet been reported.

CONCLUDING REMARKS

The initial event in transformation leads to many alterations in the cell, among them being a decrease in adenylate cyclase activity. As a result of this decrease, the cyclic AMP levels fall, and the cellular properties influenced by cyclic AMP are altered. The addition of cyclic AMP analogs tends to restore these properties to normal. Other abnormal properties are not corrected by cyclic AMP. They may be the result of the initial transformation event unrelated to the decrease in adenylate cyclase activity.

Using adenylate cyclase activity as an assay, the transformation factor or transformation event may be more easily studied. Also an analysis of abnormal properties not influenced by cyclic AMP may furnish a better understanding of the nature of the initial transformation event.

REFERENCES

1. Anderson, W. B., Johnson, G. S., and Pastan, I. Transformation of chick embryo cells by wild type and temperature sensitive Rous sarcoma virus alters adenylate cyclase activity. Proc. Natl. Acad. Sci. U.S., *70*: 1055–1059, 1973.
2. Burk, R. R. Reduced adenyl cyclase activity in a polyoma virus transformed cell line. Nature, *219*: 1272–1275, 1968.
3. Carchman, R. A., Johnson, G. S., Pastan, I., and Scolnick, E. M. Studies on the levels of cyclic AMP in cells transformed by wild type and temperature-sensitive kirsten sarcoma virus. Cell, *1*: 59–64, 1974.
4. Frank, W. Cyclic AMP and cell proliferation in cultures of embryonic rat cells. Exp. Cell Res., *71*: 238–241, 1972.
5. Goggins, J. F., Johnson, G. S., and Pastan, I. The effect of dibutyryl cyclic adenosine monophosphate on synthesis of sulfated acid mucopolysaccharides by transformed fibroblasts. J. Biol. Chem., *247*: 5759–5764, 1972.
6. Hsie, A. and Puck, T. T. Morphological transformation of Chinese hamster cells by dibutyryl adenosine cyclic 3′,5′-monophosphate and testosterone. Proc. Natl. Acad. Sci. U.S., *68*: 358–361, 1971.
7. Johnson, G. S., D'Armiento, M., and Carchman, R. A. N^6 substituted adenines induce cell elongation irrespective of the intracellular cyclic AMP levels. Exp. Cell Res., *85*: 47–56, 1974.
8. Johnson, G. S. and Pastan, I. N^6,$O^{2'}$-dibutyryl adenosine 3′,5′-monophosphate induces pigment production in melanoma cells. Nature New Biol., *237*: 267–268, 1972.
9. Johnson, G. S. and Pastan, I. Role of 3′,5′-adenosine monophosphate in regulation of morphology and growth of transformed and normal fibroblasts. J. Natl. Cancer Inst., *48*: 1377–1387, 1972.
10. Johnson, G. S. and Pastan, I. Cyclic AMP increases the adhesion of fibroblasts to substratum. Nature New Biol., *236*: 247–249, 1972.
11. Johnson, G. S., Morgan, W. D., and Pastan, I. Regulation of cell motility by cyclic AMP. Nature, *235*: 54–56, 1972.
12. Johnson, G. S., Friedman, R. M., and Pastan, I. Restoration of several morphological characteristics of normal fibroblasts in sarcoma cells treated with adenosine 3′,5′-

cyclic monophosphate and its derivatives. Proc. Natl. Acad. Sci. U.S., *68*: 425–429, 1971.
13. Otten, J., Johnson, G. S., and Pastan, I. Regulation of cell growth by cyclic adenosine 3′,5′-monophosphate. J. Biol. Chem., *247*: 7082–7087, 1972.
14. Otten, J., Johnson, G. S., and Pastan, I. Cyclic AMP levels in fibroblasts: relationship to growth rate and contact inhibition of growth. Biochem. Biophys. Res. Commun., *44*: 1192–1198, 1971.
15. Otten, J., Bader, J., Johnson, G. S., and Pastan, I. A mutation in a Rous sarcoma virus gene that controls adenosine 3′,5′-monophosphate levels and transformation. J. Biol. Chem., *247*: 1632–1633, 1972.
16. Pastan, I. and Johnson, G. S. Cyclic AMP and the transformation of fibroblasts. Adv. Cancer Res., *19*: 303–329, 1974.
17. Peery, C. V., Johnson, G. S., and Pastan, I. Adenyl cyclase in normal and transformed fibroblasts in tissue culture. J. Biol. Chem., *246*: 5785–5790, 1971.
18. Robison, G. A., Butcher, R. W., and Sutherland, E. W. Cyclic AMP. Academic Press, New York, 1971.
19. Ryan, W. L. and Heidrick, M. L. Inhibition of cell growth *in vitro* by adenosine 3′,5′-monophosphate. Science, *162*: 1484–1485, 1968.
20. Sakiyama, H. and Robbins, P. W. The effect of dibutyryl adenosine 3′,5′-cyclic monophosphate on the synthesis of glycolipids by normal and transformed nil cells. Arch. Biochem. Biophys., *154*: 407–414, 1973.
21. Schwartz, J. P., Johnson, G. S., Pastan, I., and Passonneau, J. V. The effect of growth conditions on NAD^+ and NADH levels and the NAD^+/NADH ratio in normal and transformed fibroblasts. J. Biol. Chem., *249*: 4138–4143, 1974.
22. Sheppard, J. R. Difference in the cyclic adenosine 3′,5′-monophosphate levels in normal and transformed cell. Nature New Biol., *236*: 14–16, 1972.
23. Willingham, M. C., Johnson, G. S., and Pastan, I. Control of DNA synthesis and mitosis in 3T3 cells by cyclic AMP. Biochem. Biophys. Res. Commun., *48*: 743–748, 1972.

Discussion of Paper of Dr. Johnson

DR. IKAWA: What is the correlation between the endogenous level of cyclic AMP and the exogenous level of cyclic AMP analogues?

DR. JOHNSON: That is a very difficult question to answer. The concentrations of cyclic AMP analogues added to the growth medium (1 mM) is much higher than normal intracellular cyclic AMP levels (1–10 μM). Thus a small percentage contamination of cyclic AMP in the analogue solution could influence the analysis. It should be mentioned that the mechanism of action of these analogues is not clear. They may act directly or they may act indirectly by inhibiting the degradation of cyclic AMP thereby elevating the cyclic AMP levels.

DR. IKAWA: You referred that cyclic AMP affects tumorigenicity of the transformed cells. At the same time, you mentioned that dibutyryl cyclic AMP will go away within 20 min after removal. How did the scientist maintain the cyclic AMP level in the transplantation study?

DR. JOHNSON: These experiments were done by growing the cells in dibutyryl cyclic AMP and then injecting these cells into the animals. No effort was made to maintain cyclic AMP levels at a high level. Apparently this effect is not as rapidly reversible as some of the effects I have described.

DR. POTTER: Regarding cell detachment by trypsin, you showed the effect of cyclic AMP+theophylline. Have you tried theophylline alone, or glucagon alone, or measured endogenous cyclic AMP after these agents?

DR. JOHNSON: Theophylline at 1 mM concentration will elevate the cyclic AMP levels 3–4 fold after about 5 hr. 1-Methyl-3-isobutyl xanthine, a more potent phosphodiesterase inhibitor (about 20 times more effective) will elevate the cyclic AMP levels 7–8 fold. Theophylline is only marginally effective in inducing these effects I have mentioned but it will potentiate the effects of Bt_2cAMP. 1-Methyl-3-isobutyl xanthine is potent by itself at inducing these effects. Glucagon has no effects in our cells.

DR. SILAGI: You said you were unable to restore "contact inhibition" (density dependent inhibition of growth) with dibutyryl cyclic AMP using transformed cells, but were able to do this with L cells, *i.e.*, change growth pattern from piled foci to oriented growth. How do you reconcile these divergent results?

DR. JOHNSON: We were unable to restore normal growth characteristics to L cells. Indeed the orientation and morphology are somewhat characteristic of confluent normal cells, however when cell counts were determined, it was clear that the cells continued growth to very high levels.

DR. SILAGI: With L cells, you found a decrease of growth rate in the presence of dibutyryl cyclic AMP. Could this not be interpreted as a reaction to a toxic effect of the compound rather than a return to a more normal growth rate?

DR. JOHNSON: Since the addition of dibutyryl cyclic AMP inhibits growth, the problem of cell death is a distinct problem. However, we feel that the inhibition of growth does represent a specific regulating function for cyclic AMP for a number of reasons. The effects are highly specific for cyclic AMP analogues and agents which elevate the intracellular cyclic AMP levels. This is most evident with the use of prostaglandin. Prostaglandin E_1 elevates the cyclic AMP levels in L cells; prostaglandin B does not. Prostaglandin E_1 inhibits growth, prostaglandin B does not. In different cell lines, prostaglandin E_1 cannot elevate cyclic AMP. In these cells prostaglandin E_1 cannot inhibit growth. The specific regulating function for cyclic AMP in inhibiting growth is further supported by the correlation of growth rates and cyclic AMP levels.

DR. SILAGI: Did you compare plating efficiencies of cells (L cells or others) grown with and without dibutyryl cyclic AMP as a check on toxicity?

DR. JOHNSON: We have not done detailed plating efficiency studies.

DR. RUTTER: Does cyclic GMP exert a competitive effect on cyclic AMP?

DR. JOHNSON: Neither cyclic GMP nor its dibutyryl or 8-bromo derivatives have any effects on our cells. We have not tested these compounds in the presence of dibutyryl cyclic AMP to test for competitive effects.

DR. RUTTER: Have you tested for cyclic GMP levels in your studies? Is it inversely related to cyclic AMP? As you know Dr. Goldberg and collaborators postulate a reciprocal relationship between cyclic AMP and cyclic GMP.

DR. JOHNSON: We have not measured cyclic GMP levels.

Hemoglobin-synthesizing Mouse and Human Erythroleukemic Cell Lines as Model Systems for the Study of Differentiation and Control of Gene Expression

S. K. Dube,*[1] G. Gaedicke,*[2] N. Kluge,*[1] B. J. Weimann,*[1] H. Melderis,*[1] G. Steinheider,*[1] T. Crozier,*[1] H. Beckmann,*[2] and W. Ostertag*[1]

Max-Planck-Institut für Experimentelle Medizin, Göttingen, West Germany[1] and Molekularbiologisch-Hämatologische Arbeitsgruppe, Universitäts-Kinderklinik, Hamburg-Eppendorf, West Germany*[2]*

Abstract: A small fraction of Friend or Stansly virus (SFFV) transformed erythroleukemic cells of spleen, liver, or peripheral blood, if maintained under tissue culture conditions, continuously differentiate into erythroblasts. Addition of dimethyl sulfoxide (DMSO), cyclic AMP, or etiocholanolone to these cells in culture, stimulates their differentiation along the erythrocytic line much further. Under favorable conditions the cells are able to synthesize up to 50% hemoglobin which is composed of adult α- and β-globin chains.

Induced hemoglobin synthesis is a good marker to study the processes involved in cellular differentiation. The differentiation, as evidenced by a 40–100 fold increase in globin synthesis, can be due to (a) increased transcription of the mRNA for proteins characteristic for the differentiated state of the cell; (b) processes occurring after transcription and leading to translatable globin mRNA (transport and processing of globin mRNA); (c) increased translation.

The results show that there is an increase in available globin mRNA after induction. This increase seems to be due to increased transcription. Furthermore there appears also a control at the level of translation.

Virus transformed erythroleukemic cells in culture offer a very useful experimental system for studying the mechanisms of cellular differentiation and transformation. We began our studies with Friend virus (FV)-transformed (*23*) mouse erythroleukemic cells (*24*, *26*, *33*, *63*, *69*, *80*, *81*, *89*, *92*, *97*). However, recently we have been able to establish human cell lines from the bone marrow of polycythemia vera patients (*102*) or patients with erythroleukemia. These cell lines are especially interesting because they provide for the first time a tissue-culture system synthesizing human hemoglobin. In the first part of this chapter we will discuss various aspects of the FV system and in the second we will summarize our preliminary results with the human cell lines. Another chapter (*66*) deals with those aspects of the transformation and differentiation problem which are related to the virus itself.

FV System

FV is an RNA tumor virus which transforms erythroid precursor cells in mouse and causes erythroblastic leukemia (*23*). The disease is characterized by a rapid proliferation of basophilic erythroblasts (*97*). The transformed cells, which are most likely at the proerythroblast stage in erythroid maturation (*57*), can be easily established in tissue culture and provide a very useful experimental system for the study of mechanisms involved in cellular differentiation and virus-mediated cell transformation (*24, 26, 33, 63, 69, 80, 81, 89, 92, 97*). This is so because in this system it is possible to switch the differentiation on or off by simple manipulations. Addition of dimethyl sulfoxide (DMSO) (*25*), etiocholanolone or cyclic AMP (*27*) to

TABLE 1. Mouse and Human Cell Lines

	Cell line	Origin		Origin of cell	Basal level globin synthesis (%)
a)	FSD-1	*Mouse*:	DBA-2	Spleen → solid tumor	Low
	FSD-2		Balb C	Spleen	High
	FLD-2		"	Liver	High
	FPB-1		"	Peripheral blood	Medium
	FSD-3		DBA-2	Spleen	Low
	FLD-3		"	Liver	Low
	FSD-4		"	Spleen	High
	FLD-4		"	Liver	High
	FSD-5		"	Spleen	Medium
	FLD-5		"	Liver	"
	FTD-5		"	Spleen → solid tumor	"
	F4		"	Clone of FSD-1	0.5–1
	F13		"	"	0.4
	F23		"	"	n.t.
	F4AO		"	Subline of F4	0.5–1
b)	F4S		DBA-2	Subline of F4	0.5–1
c)	F4D		DBA-2	Subline of F4	0.5–1
	F4AG		"	Subline of F4	0.5–1
	FA1		"	Clone of F4AG	1
	F4B		"	Subline of F4	0.5–1
	B1		"	Clone of F4B	Low
	B8		"	Clone of F4B	2%
	B8/4		"	Clone of B8	1–2
	B16		"	Clone of F4B	1
	F4TdR		"	Subline of F4	0.5–1
d)	PHD	*Human*:	Polycythemia	Bone marrow	0.1–0.3
	PBW		"	Bone marrow	0.1–0.3
	ENG		Erythroleukemia	Leukemic infiltrate in pleura	1–5
	EHN		"	Bone marrow	0.1–0.3

a) Cell lines and clones established from FV-transformed mouse erythroleukemic cells. b) Cell line established from FV-transformed mouse erythroleukemic cells grown in serum-free medium. c) Cell lines and clones established from FV-transformed mouse erythroleukemic cells resistant to BrdU, azo-the growing cells induces differentiation. This can be easily followed because the induced cells synthesize hemoglobin and turn red. On the other hand, addition of bromodeoxyuridine (BrdU), a thymidine analogue, to the growing cells inhibits the induction of differentiation (*64, 85*) except in the case of thymidine kinase-deficient cells (*64*), where BrdU induces differentiation.

The inhibitory effect of BrdU in this system is comparable to the reported effect of BrdU in other systems synthesizing specialized proteins characteristic of differentiated cells (*94, 96, 101, 106*).

In this chapter we will present evidence related to the induction and inhibition of globin synthesis. We will examine the temporal sequence of events during induced differentiation of Friend erythroleukemic cells. Finally, we will show that

DMSO maximal induced rate (%)	β-Chain synthesis (%)	Karyotype (M=metacentric chromosomes)	Additional information
10–15	75	Diploid, 0–2 M	4 years tissue culture and frozen
20–25	57	n.t.	6 months tissue culture and frozen
20–25	52	”	6 months tissue culture and frozen
10–20	75	”	2 years tissue culture and frozen
10	50	Heteroploid	6 months tissue culture and frozen
10–15	n.t.	n.t.	6 ” ” ” ” ”
20–25	50	”	6 ” ” ” ” ”
20–25	n.t.	”	6 ” ” ” ” ”
10–15	”	”	2 ” ” ” ” ”
”	”	”	2 ” ” ” ” ”
”	”	”	2 ” ” ” ” ”
20–25	50	Diploid, 2 M	4 years tissue culture spleen clone
10	64	Hyper diploid, 2 M	1 year tissue spleen clones from 3 years old continuous culture of FSD-1
20	67	Diploid, 2–3 M	1 year tissue culture
1–2	50	Diploid, 2 M	Not inducible
4–7	50	Diploid, 2 M	Grown in serum-free medium 2 years
20–50	50	Diploid, 2 M	DMSO-resistant 4.5%
20–25	”	”	Azoguanine-resistant
20–30	”	”	Azoguanine-resistant
10–20	55–60	Close to diploid	200 μg/ml BrdU-resistant
10–15	80–90	Several M	” ”
25–30	50	”	” ”
20	”	”	TK⁻, cells grow attached
10–15	”	”	200 μg/ml BrdU-resistant
20	”	Diploid, 2 M	3 mg/ml thymidine-resistant
0.1–0.3	70	Diploid +2, F⁻	Polycythemia vera
0.1–0.3	70	Diploid	Polycythemia vera
1–5	90–100	Heteroploid	Paraproerythroblast leukemia
0.1–0.3	70	n.t.	Erythroleukemia

guanidine, and DMSO. d) Human cell lines established from polycythemia vera and erythroleukemia patients.

the control of globin synthesis in these cells involves mechanisms operating both at the level of transcription and translation.

1. Establishment of cell lines

a) Parent cell line and clones: We started with Balb C-adapted FV obtained from the United States National Institutes of Health. The virus was injected intraperitoneally into DBA-2 mice. After 3 weeks the mice had enlarged spleens and were leukemic. Spleen cells from these mice were injected subcutaneously into normal DBA-2 mice. This procedure resulted in the production of solid tumors in mice. The tumor cells were grown in tissue culture and this cell line is designated FSD-1 (*63*). Cell-free supernatant from a clone of the FSD-1 cell culture was injected intraperitoneally into Balb C mice. The peripheral nucleated blood cells from these mice were used to start the cell culture line FPB-1. Several clones such as FSD-1/clone F4 (Table 1) from the cell line FSD-1 were established (*63*) using the spleen cloning procedure (*99*). The details of spleen cloning are briefly as follows. The mice were irradiated with a total radiation dosage of 900 R (^{60}Co or X-ray). On the same day, cells from culture were injected into the lateral tail vein of mice. Different dilutions of cell suspension were used to obtain not more than 4 clones/spleen. The clones are easily visible after 9–12 days and were picked up to start a new clonal culture. Other clones were isolated by agar cloning (*51*).

b) Cell lines grown in serum-free medium: Cells growing permanently under serum-free conditions were obtained as follows: cells of clone FSD1/F4 were fed every second day with serum-free medium by diluting the original cell suspension, which contained 10% serum. In this way the serum concentration was reduced in step with cell increase. The cell division rate decreased continuously. After 6 days the surviving cells were returned to a medium containing 10% serum. After recovery the same adaptation procedure with serum-free medium was repeated. The survival time under serum-free conditions increased with each step. After 6 such cycles, the cells could be grown continuously without any additional constituents (=cell line FSD-1/F4S). The optimal density of these cells is between 5×10^5–2×10^6 cells/ml. Cell densities below these values lead to cell death. Cell division rate is reduced to one doubling per 35–40 hr as compared to 15–25 hr for the same cells in 10% serum-containing media (*45*).

c) Cell lines and clones resistant to BrdU: Resistance to BrdU was obtained as follows. We started with the cloned line FSD-1/F4. Cells were grown in the presence of 2.5 μg/ml BrdU for 4 weeks. This concentration of BrdU is slightly inhibitory to BrdU-sensitive cells. The cells were then shifted to 200–500 μg/ml BrdU. A great proportion of the cells were resistant to these high concentrations of BrdU. F4 cells resistant to BrdU, F4/B, are very heterogeneous in karyotype due to chromosome breakage and rearrangements. Clones of BrdU-resistant cells F4/B1-24 were isolated using the spleen-cloning procedure (*64*). The various clones differ in the degree of BrdU or TdR incorporation into their DNA. This correlates with the thymidine kinase (TK) activity of the cells (*102*). For example, clone B8 incorporates essentially no TdR into its DNA and is TK$^-$.

d) Karyotype of FV-transformed erythroleukemic cells: The karyotype of virus-

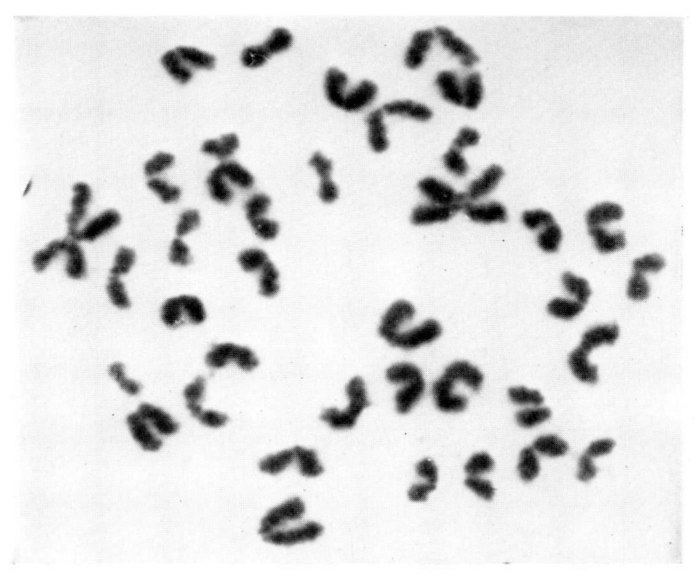

Fig. 1. Metaphase chromosomes of FSD-1/clone F4. Cells were grown in Eagle's medium with Earle's balanced salt solution, supplemented with twice the usual amount of essential and non-essential amino acids and vitamins; 5 times the usual amount of glutamine and 10–15% fetal calf serum. Chromosomes were stained with Giemsa. Thirty-eight chromosomes are shown with 2 metacentrics which look alike, especially in respect to length, arm ratio and morphology of the centromere regions. All the other chromosomes are indistinguishable from the normal mouse karyotype.

transformed cells is in general quite similar to the normal mouse karyotype. The cells are nearly always diploid with the chromosome number ranging between 36–41 in the case of FSD-1 and 38–39 in the case of FSD-1/clone F4. However, there is one feature of the karyotype of these leukemic cells which is striking; namely, the presence of 2 apparently homologous metacentric marker chromosomes (Fig. 1). The 2 metacentric chromosomes are present in only about 15% of the FSD-1 cells but in all of the cells of FSD-1/clone F4 (*63*). It has been shown that FSD-1 cells show unbalanced synthesis of globin chains while in the clone, F4, the synthesis is balanced, *i.e.*, the ratio of α to β chains is one (see also the section on induction by DMSO). Preliminary experiments (Schnedl, personal communication) suggest that one of the metacentrics has arisen by a translocation event involving chromosome number 4. It is interesting to note that this chromosome carries the FV-1 locus which controls the host range of LLV-F, the helper virus component of the FV complex (*82*, *83*).

2. Induction of hemoglobin synthesis by DMSO

In the uninduced state, little hemoglobin is synthesized by either FSD-1 cells or FSD-1/clone F4, the amount not exceeding 0.5% of the soluble cytoplasmic protein in the former case and being barely detectable in the latter. The FPB-1 cells do synthesize relatively large quantities of hemoglobin without DMSO. However, in all cases, addition of DMSO to the culture medium (final concentration, 1–2%)

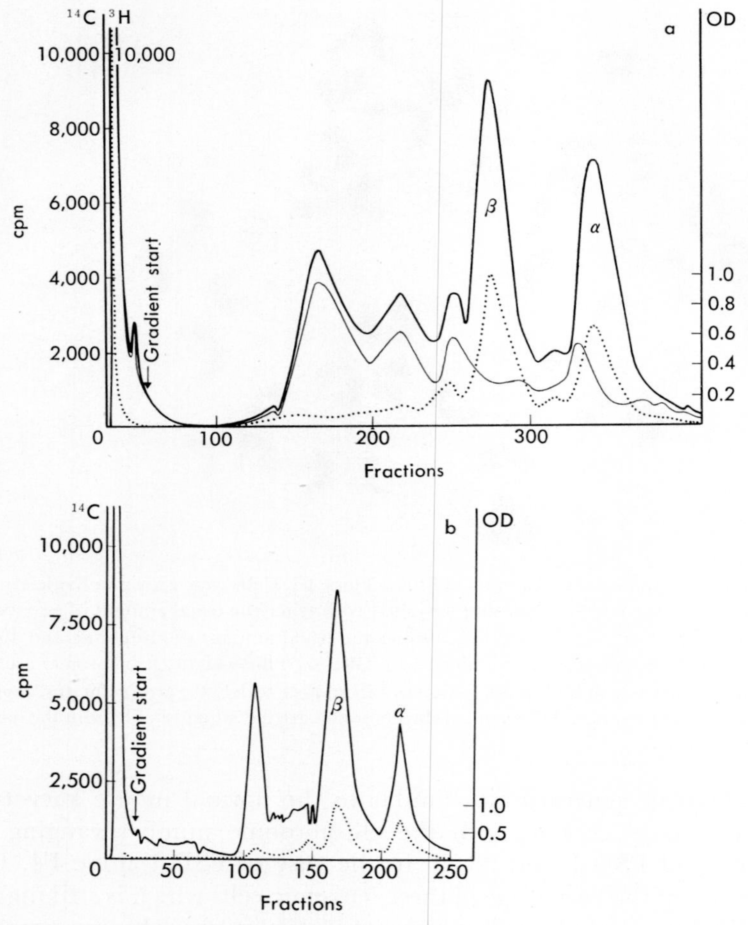

FIG. 2. Induction of hemoglobin synthesis by DMSO. Cells were grown as in Fig. 1. Cells were fed every 24 hr for 4–5 days with an increasing amount of DMSO in the medium (1–1.25–1.5–2–2%). During this induction schedule, the cells were diluted 1 : 1 at each feeding except on the last 2 days. (2% DMSO for longer than 3 days resulted in inhibition of cell division at day 4 and in cell death at day 6–7). Cells were labeled in leucine-free medium for 16–24 hr with either ^{14}C-leucine or ^{3}H-leucine. Cells were lysed for 10 min in 0.5% NP40 (Shell, Hamburg) in a pH 7.4 buffer containing 0.15M NaCl, 0.01M Tris-HCl, and 1.5×10^{-3}M $MgCl_2$ (*7*). The nuclei and particulate cytoplasm were removed by centrifugation at 18,000 rpm for 10 min. The supernatant was precipitated at −20°C in acetone-HCl (20 ml concentrated HCl in one liter of acetone). At this step, 10–20 mg DBA-2 mouse hemoglobin was added as a carrier. Globin chains were separated on CMC ("CM 52 Whatman") columns in 8M urea buffer containing mercaptoethanol (*15*). A linear gradient of 0.005M Na_2HPO_4, pH 6.5, to 0.045M Na_2HPO_4, pH 6.7, was used. The column eluates were monitored on an "LKB-Uvicord II" recorder and fractions counted in a scintillation counter. The ^{14}C-leucine and ^{3}H-leucine counts were summed and the ratio of the two was used to correct the scale. The ^{14}C- and ^{3}H-labeled peaks are in proportion to their relative contribution to the total leucine incorporation into soluble proteins. a) FSD-1clone F4 (genetically homogeneous) shows balanced β- and α-chain synthesis (50% of each) if corrected for background label and leucine residues per globin chain. ···carrier globin: the higher peak height of the β chain reflects the higher tryptophan content; —— uninduced control, label ^{3}H-leucine; —— DMSO-induced cells, label ^{14}C-leucine. b) cell culture line FSD-1/(genetically heterogeneous) shows unbalanced β- and α-chain synthesis (75% β, 25% α). ···carrier globin; —— DMSO-induced cells, label ^{14}C-leucine.

causes a dramatic increase in hemoglobin synthesis. Under favourable conditions, up to 50% of the cytoplasmic water-soluble, 10% TCA-precipitable labeled protein is hemoglobin. Chromatography of the *in vitro*-labeled proteins on carboxymethylcellulose (CMC) columns and 2-dimensional fingerprint analysis of tryptic peptides of the fractionated globin chains has shown that the hemoglobin is composed of adult α- and β-globin chains (*63*). A typical chromatographic fractionation of globin chains on CMC-urea columns is shown in Fig. 2. Data in Fig. 2 demonstrate that globin synthesis in FSD-1 cells is unbalanced, approximately 75% β and 25% α chains; whereas in FSD-1/clone F4, α and β chains are synthesized in equal proportions (see also the section on karyotype).

3. *Induction of hemoglobin synthesis by DMSO in cells grown in serum-free medium*

Cells adapted to serum-free growth, FSD-1/F4S, undergo induced differentiation if DMSO and iron are added to the medium. Erythropoietin is ineffective in promoting erythroid differentiation in these cells. Figures 3 and 4 show the results. The amount of hemoglobin synthesized in induced FSD-1/F4S cells corresponds to about 5% of the labeled soluble cytoplasmic protein.

These data clearly demonstrate that although serum enhances the capacity of the cells to differentiate, serum constituents or macromolecules are not required for induced differentiation of Friend erythroleukemic cells. Also, transferrin is not necessary for the transport of iron in these cells, since iron can be utilized in the absence of any macromolecular components in the medium. The fact that erythropoietin is without any effect on the induction of differentiation in these cells (see

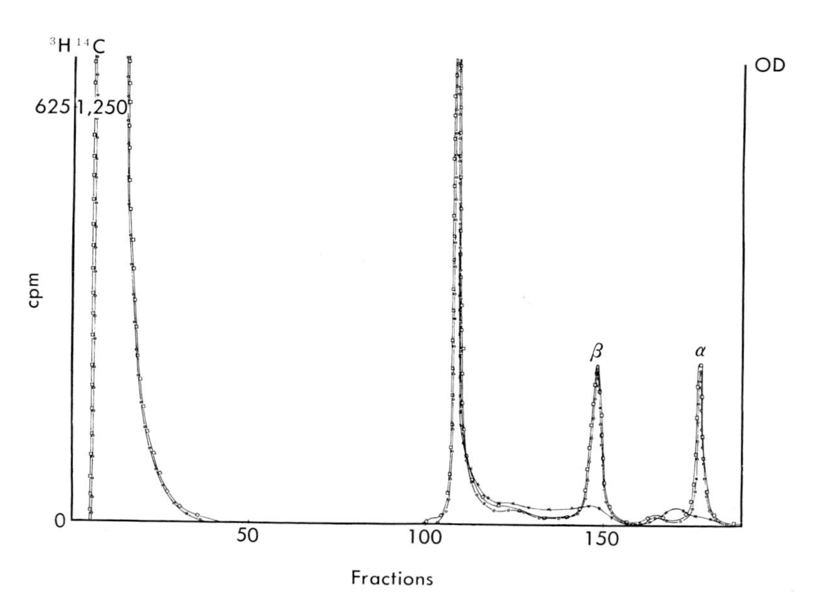

Fig. 3. Induction of hemoglobin synthesis in cells grown in serum-free medium. Cells were grown as described in text. Labeling with ^{3}H- or ^{14}C-leucine and fractionation of α- and β-globin chains on CMC-urea columns was as in Fig. 2. × control; ○ +DMSO; □ +DMSO +Fe^{2+}.

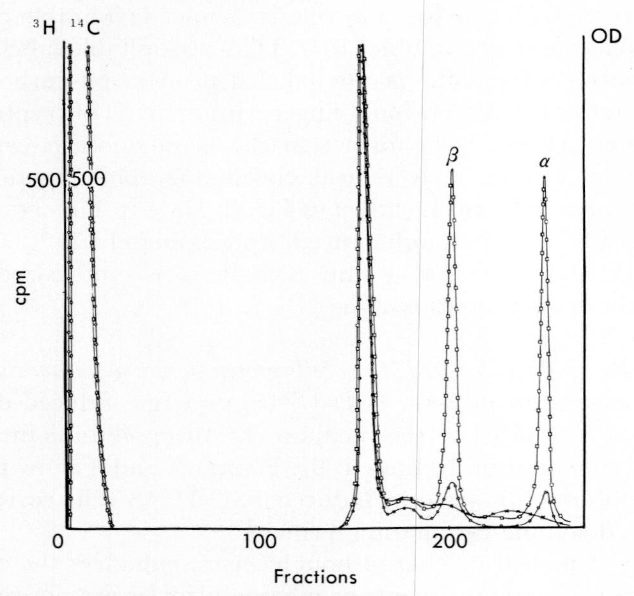

FIG. 4. Lack of responsiveness to erythropoietin (EPO). Details as in Fig. 3. × EPO+Fe^{2+}; ○ DMSO+Fe^{2+}; □ EPO+DMSO+Fe^{2+}.

also Ref. *57*) indicates that they are past the erythropoietin-sensitive stage in erythroid maturation. Since erthropoietin is necessary not for the commitment of the stem cell but for the induction of cell division and differentiation of committed stem cells (*3, 5, 8, 9, 12, 13, 22, 31, 38, 44, 86, 95*), our data suggest that FV-transformed cells are at the proerythroblast stage.

4. *Induction of hemoglobin synthesis by etiocholanolone and cyclic AMP*

We have found that both etiocholanolone and cyclic AMP are effective in inducing differentiation in Friend erythroleukemic cells (*27*). Etiocholanolone has been shown to enhance heme and hemoglobin synthesis in other systems (*28, 29, 42, 55, 58, 61, 62*). In our case the optimal concentration of etiocholanolone in the culture medium is 10^{-7} M. Higher concentrations are toxic to the cells. Under favorable conditions of induction by etiocholanolone, up to 7% of the water-soluble TCA-precipitable labeled protein is hemoglobin. Although this is considerably less than that synthesized by DMSO-induced cells, on a molar basis etiocholanolone is far more effective than DMSO. Figure 5 shows that the hemoglobin produced by the etiocholanolone-induced cells is composed of adult α and β chains. Similar results are obtained with cyclic AMP. Under optimal conditions, *i.e.*, 10^{-4} M cyclic AMP in the culture medium, up to 5% of the labeled cytoplasmic protein is hemoglobin (Fig. 6).

Cyclic AMP has been shown to induce differentiation in other systems (*6, 10, 11, 34, 41, 72–74, 79, 100*). It has also been shown to affect cell division rate (*40, 67, 105*) and membrane permeability (*46, 47, 76, 105*). The fact that it also induces differentiation in this system, albeit less effectively than DMSO, is very interesting.

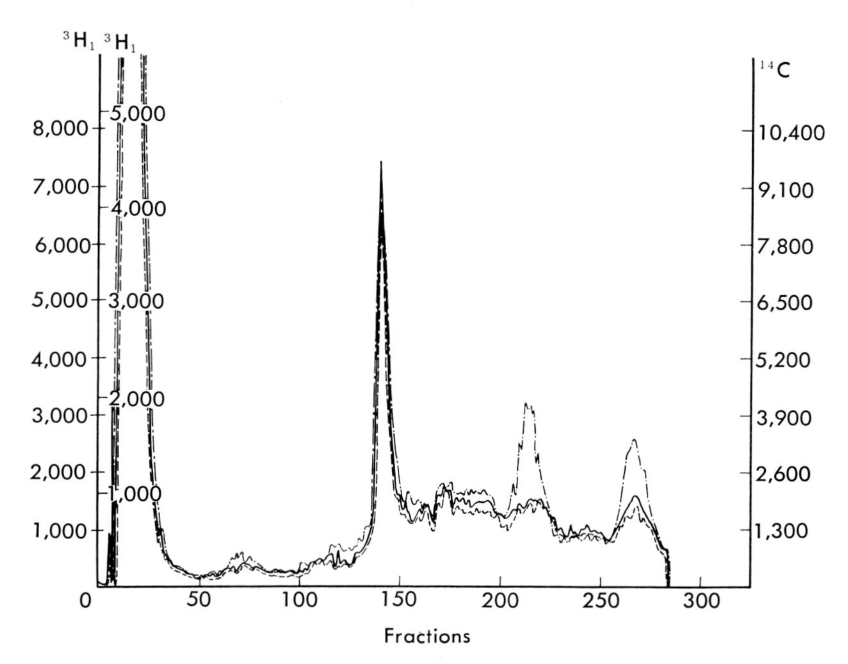

FIG. 5. Induction of hemoglobin synthesis by etiocholanolone. FSD-1/F4 cells were grown as in Fig. 1. Cells were fed with etiocholanolone in the medium for 4 days. Cells were labeled with ^{3}H-leucine and the globin chains fractionated on CMC-urea columns as in Fig. 2. Results with 2 different concentrations of etiocholanolone are shown in the figure. – – – 10^{-9} M etiocholanolone; —–— 10^{-7} M etiocholanolone; —— control ^{14}C-leucine-labeled globin chains.

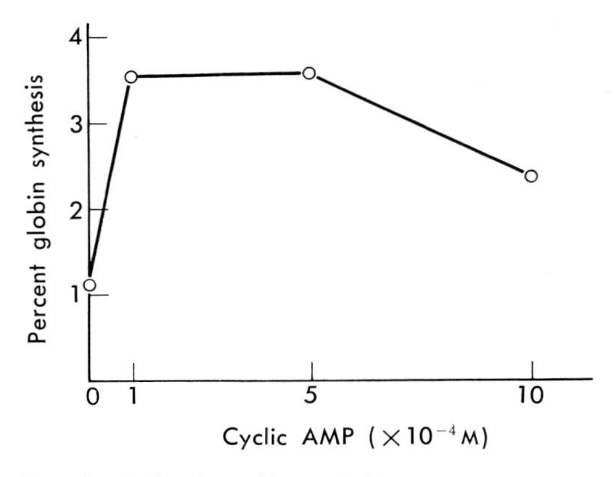

FIG. 6. Induction of hemoglobin synthesis by cyclic AMP. F4 cells were grown as in Fig. 1. Cells were fed with cyclic AMP for 86 hr. Details of labeling, identification, and quantitation of globin chains were as in Fig. 2. Results are plotted as percent of globin synthesis concentration *vs.* cyclic AMP in the medium.

The similarity of action of DMSO and cyclic AMP is further emphasized by the fact that during induced differentiation of erythroleukemic cells, membrane permeability and cell division rate are decreased (see the section on "Decrease in membrane permeability," p. 111).

5. Induction of hemoglobin synthesis by azidothymidine (AT) and in B8 cells also by BrdU

AT (*66*) and BrdU are thymidine analogues; AT has an N_3 group instead of 3′-OH in the deoxyribose moiety and in BrdU, the CH_3 group on C-5 is replaced by Br. Both BrdU and probably AT can be incorporated into DNA. However, because of the N_3 group in the 3′-OH position, AT presumably interferes with the chain elongation of DNA.

We have found that AT induces hemoglobin synthesis. Concentrations of AT up to 5×10^{-4} M are not toxic to the cells. Under optimal conditions about 4% of the labeled, water-soluble TCA-precipitable cytoplasmic protein is hemoglobin (Table 2). When AT is added together with DMSO, a synergistic effect on the induction of hemoglobin synthesis is observed.

BrdU normally inhibits induction of hemoglobin synthesis, as shown in the next section. However, in the case of BrdU-resistant clone B8 cells which are TK^- and do not incorporate BrdU or TdR into their DNA, an inducing effect of BrdU is observed. The data in Table 2 show that in the presence of 5×10^{-4} M/ml BrdU in the culture medium, hemoglobin synthesis in B8 cells is induced and corresponds to about 5% of the labeled, cytoplasmic protein.

This inducing effect of BrdU seems, at first, very surprising and paradoxical. However, hidden in this paradox may be an explanation for the mechanism of action of BrdU. Let us consider that BrdU has a dual mode of action, *i.e.*, it decreases membrane permeability and induces differentiation; it is incorporated into DNA and inhibits differentiation. The evidence for the latter is overwhelming (see the next section) and for the former, at present, circumstantial. The data in this chapter clearly demonstrate a correlation between induction of differentiation and decrease in membrane permeability. DMSO, etiocholanolone, and cyclic AMP which induce hemoglobin synthesis in Friend erythroleukemic cells decrease membrane permeability, and amphotericin B which is known to increase membrane permeability acts as an antagonist of DMSO. In view of these data, the induction of differentia-

TABLE 2. Effect of Various Substances on Globin Synthesis in FSD-1/F4 and B8 Cells

	Cells treatment	Globin synthesis (%)
B8	Untreated	2.6
	+BrdU 5×10^{-4} M	5.0
	+DMSO	23.5
	+BrdU, DMSO	30.4
F4	Untreated	1.1
	+Azido TdR 2×10^{-4} M	4.0
	+DMSO	22.5
	+Azido TdR, DMSO	26.0
	+TdR 1.2×10^{-4} M	0.9
	+TdR 4×10^{-4} M	0.6
	+Amphotericin B, DMSO	0.9

Globin synthesis was measured by labeling the cells and fractionating the globin chains on CMC-urea columns as in Fig. 2. The numbers represent percent of globin synthesis, *i.e.*, percent of cytoplasmic-labeled water-soluble TCA-precipitable counts under globin peaks on CMC-urea columns after accounting for the recovery. The numbers are directly comparable.

tion in B8 cells is best explained by postulating that BrdU decreases membrane permeability. Recent studies showing that BrdU decreases membrane permeability in 3T6 cells (*98*) provide further support for this view.

If the correlation between induction of differentiation and decrease in membrane permeability is not fortuitous, it may be very significant. This implies that BrdU acts primarily as an inducer of differentiation and the fact that it inhibits the expression of differentiated functions is a secondary effect, albeit a powerful one. It is easy to see that in a cell which is undergoing quantal mitosis (*101*) the inhibitory effect of BrdU by virtue of its incorporation into DNA (*64*) would overshadow the inducing effect and BrdU would appear to be a strong inhibitor of differentiation. However, in the case of a TK⁻ cell undergoing quantal mitosis BrdU cannot exert its inhibitory effect because it cannot be incorporated into DNA. Similarly, in a cell which does not require quantal mitosis for further differentiation, an inducing effect of BrdU would be expected. The inducing effect in neuroblastoma cells (*88*) is therefore easily explainable.

6. *Inhibition of induction by BrdU*

The thymidine analogue, BrdU, preferentially inhibits the synthesis of those proteins which are characteristic of the differentiated state of a cell (*94*, *96*, *101*, *106*). We have found (*64*) that BrdU, at 0.6–0.2 μg/ml, if added 1–2 days before DMSO application, interferes with the induction of globin synthesis (Fig. 7) and that BrdU-resistant cells, *e.g.*, B8, are able to synthesize globin in the presence of high doses (200 μg/ml) of BrdU (Fig. 8). In fact there is a slight stimulatory effect of BrdU on globin synthesis in clone B8. By analyzing the various BrdU-resistant clones we have

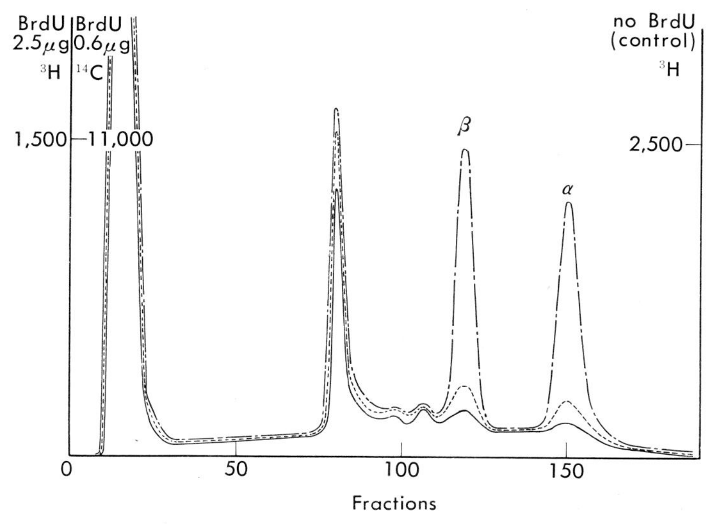

FIG. 7. Inhibition of induction of globin synthesis by BrdU. FSD-1/F4 was exposed to BrdU 2 days before induction with DMSO. Control was not treated with BrdU. Induction with 1.5 % DMSO was carried out for 4 days. Cells were labeled with ^{3}H- or ^{14}C-leucine and globin chains fractionated as in Fig. 2. - - - 0.6 μg BrdU/ml; —— 2.5 μg BrdU/ml, —-— control.

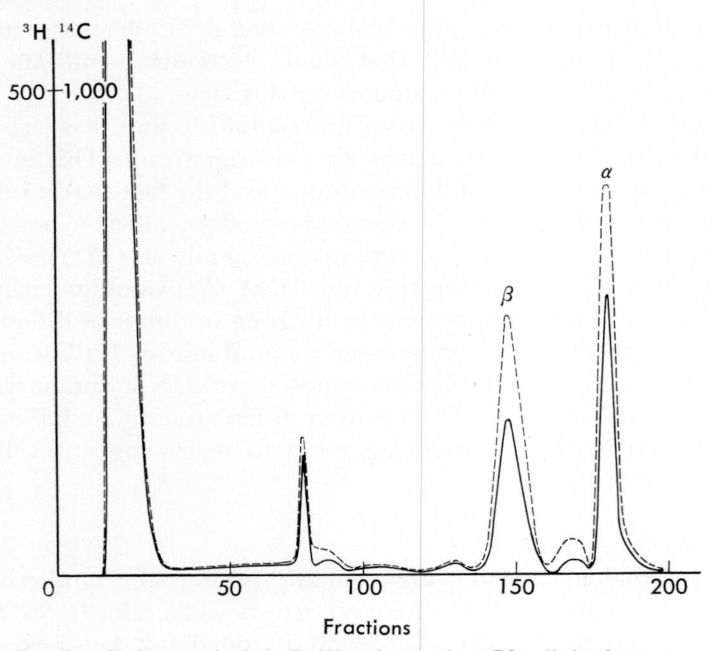

FIG. 8. Globin synthesis in BrdU-resistant clone B8 cells in the presence or absence of BrdU. Cells were induced with 1.5% DMSO in the presence (200 μg/ml) or absence of BrdU. Cells were labeled and globin chains fractionated as in Fig. 5. —— DMSO induction, no BrdU, label ^{3}H-leucine; - - - DMSO induction, BrdU 200 μg/ml 2 days before and during induction, label ^{14}C-leucine.

been able to show that the extent of inhibition of globin synthesis correlates with the degree of BrdU incorporation into DNA (*64*).

BrdU acts as an antagonist of DMSO (*64, 85*). If BrdU is added during induction of globin synthesis and not before, only a slight inhibition of globin synthesis is observed even at such high doses as 200 μg/ml BrdU. At all inhibitory doses of BrdU on globin synthesis during DMSO induction total protein synthesis remains virtually unaffected. Our data, therefore, suggest that BrdU does not act on globin mRNA translation on the polysomes but on processes leading to transcription, transport or processing of globin mRNA (*64*).

That BrdU interferes specifically with the transcription of globin mRNA is suggested very strongly by the following experiment. To a culture of FSD-1/F4, 2.5 μg/ml BrdU was added. Three days later 1.5% DMSO was added. Aliquots were harvested at various times. Cytoplasmic RNA was isolated and tested for globin template activity. The results showed that this RNA programmed hardly any globin synthesis in contrast to the RNA isolated from cells to which no BrdU was added before induction (unpublished observations).

7. *Temporal sequence of events during DMSO-, etiocholanolone-, or cyclic AMP-induced differentiation*

We believe that the temporal sequence of events during induced differentiation of erythroleukemic cells is as shown in Fig. 9. Within the first 20 hr after the addition of DMSO, etiocholanolone, or cyclic AMP, dramatic changes take place in the cell. The membrane permeability is decreased. This is followed by a decrease first in nuclear RNA synthesis and then in that of cytoplasmic RNA. In contrast to the general decline in RNA synthesis, however, the synthesis of globin mRNA is augmented, as shown by the appearance of newly labeled globin mRNA in the cytoplasm. Between day 1 and 2, the synthesis of globin mRNA reaches its peak; however, the synthesis of globin chains stays essentially at the background level. After day 3, very little label appears in globin mRNA. At this stage, induced globin synthesis is noticeable and it reaches a peak on day 5. About this time the DNA synthesis and cell division come to a halt. On day 6 some reticulocytes without nuclei appear.

In the following sections we will examine the evidence for the various effects summarized above. In all the experiments discussed below, FSD-1/F4 and BrdU-resistant F4/B8 cells were used. Although cyclic AMP and etiocholanolone gave qualitatively the same results as DMSO, the effects observed with DMSO were much more pronounced. For this reason only the results obtained with DMSO are presented in the following sections. The cells were maximally induced with 1.5% DMSO in the case of F4 and 2% DMSO in the case of B8 cells.

8. Decrease in membrane permeability

That the membrane permeability is affected during induction, is suggested by 2 sets of observations. Firstly, that the incorporation of ^{32}P-phosphate, ^{3}H-uridine,

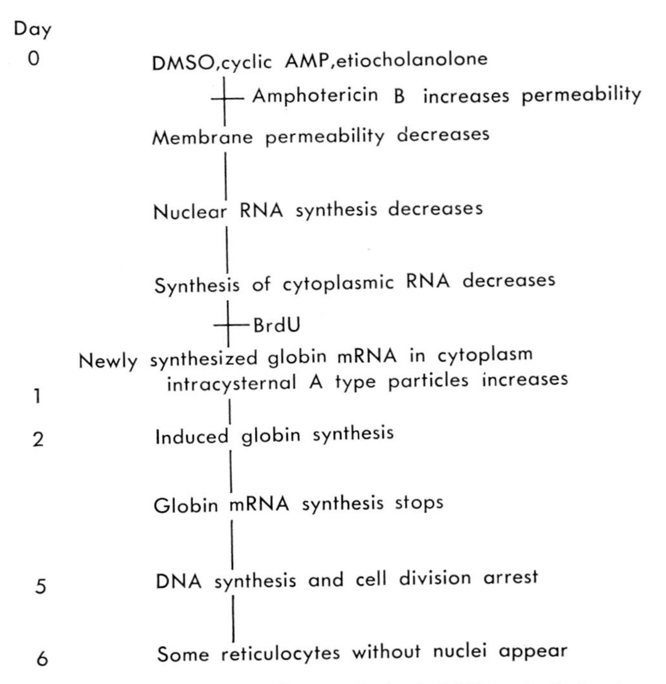

Fig. 9. Temporal sequence of events during DMSO-, etiocholanolone-, or cyclic AMP-induced differentiation of Friend erythroleukemic cells.

and ^{3}H-leucine in the cells decreases dramatically after induction; and secondly, amphotericin B which is known to increase membrane permeability (*1*, *14*) acts as an antagonist of DMSO.

a) *Decreased uptake of phosphate, uridine, and leucine:* Figure 10 shows that within 16 hr (approximately one generation time) of DMSO treatment, incorporation of ^{32}P in the cells decreases to about 30% and the synthesis of cytoplasmic RNA to about 15% of the normal levels. Similar results are obtained when the incorporation of ^{3}H-uridine in the cells is followed (Table 3). The amount of label in cytoplasmic RNA drops to 14%. In addition, uridine incorporation also shows a dramatic difference in the distribution of label between nucleus and cytoplasm. Whereas in uninduced cells the ratio of the label between nuclear and cytoplasmic RNA is close to one, in induced cells only about 10% of the label is found in the nucleus.

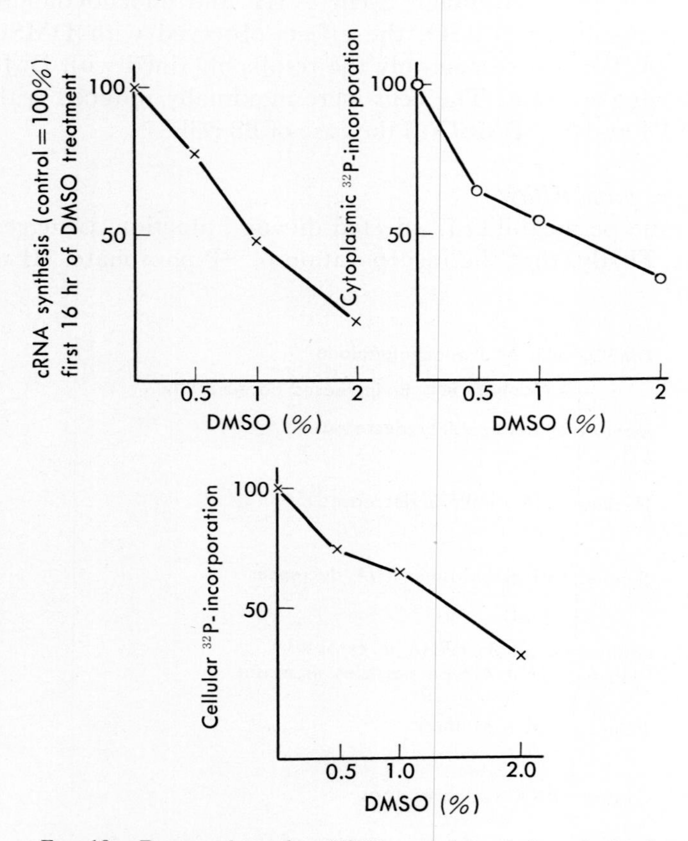

FIG. 10. Decreased uptake of $^{32}PO_4$ and decreased synthesis of cytoplasmic RNA in the first 16 hr of treatment with DMSO. FSD-1/F4 or B8 cells were grown as in Fig. 1. At zero time, DMSO and $^{32}PO_4$ were added together. In control cells only $^{32}PO_4$ was added. Cells were harvested 16 hr later. Cells were centrifuged and washed several times with cold medium. Cells were suspended in the medium and an aliquot counted for radioactivity. Cells were lysed with NP40 as in Fig. 2. One aliquot of the lysate was counted as such and the other after precipitation with TCA. The results are plotted as percent ^{32}P incorporation *vs.* DMSO concentration.

TABLE 3. Decreased Uptake of ^{3}H-uridine in the First 20 hr after DMSO Treatment of FSD-1/F4 Cells

Treatment	Cellular uptake in % of input	In % of control	RNA synthesis in % of input (TCA)	In % of control	In % of uptake	Synthesis cytoplasmic RNA in % of input (TCA)	In % of uptake	Synthesis nuclear RNA in % of total RNA
No DMSO	19.0	100	14.6	100	76	7.2	38	51
1.5% DMSO 20 hr	10.3	54	4.4	30	43	3.8	37	14
2% DMSO 20 hr	5.5	29	2.0	14	36	1.9	36	9

Cells were grown as in Fig. 2. At zero time, DMSO+^{3}H-uridine (50 Ci/mM, 200 μCi/20 ml culture medium) was added. Cells were harvested 20 hr later. Radioactivity determinations were done as follows: a) an aliquot was taken before pelleting the cells and divided in 2 parts. One part was counted as such, the other after precipitation with cold 10% TCA. b) Cells were lysed by suspending the pellet in 0.5% NP40 as in Fig. 2 (for 20ml culture 2 ml NP40 was used). An aliquot was taken before centrifugation, divided in 2 parts and counted as in (a). c) The nuclear pellet after NP40 treatment was suspended in a small volume of the medium. An aliquot was taken out, divided in 2 parts and counted as in (a). d) An aliquot from the supernatant after NP40 treatment was taken out and counted as in (a). Control: uninduced cells, labeled in the absence of DMSO. Experiments were done with 2 different concentrations of DMSO in the medium, *i.e.*, 1.5% and 2%. Experiments were also carried out with ^{3}H-uridine of low specific activity, with ^{14}C-uridine (50 mCi/mM) and with "cold" uridine (1 μg/ml) in the medium. The results were the same, *i.e.*, decreased uptake in induced cells.

This perhaps reflects a differential effect on the permeability of cytoplasmic and nuclear membranes, resulting also in increased transport from the nucleus. It is also conceivable that during induction little heterogeneous nuclear RNA is synthesized and that increased globin mRNA in the cytoplasm also arises by a mechanism involving RNA-dependent RNA polymerase, an enzyme recently demonstrated in reticulocytes (*19*). Qualitatively similar results are obtained when the incorporation of ^{3}H-leucine in the cells is measured. Table 4 shows that the cellular uptake of leucine drops to about 40%.

We have considered the possibility that the decreased uptake of phosphate, uridine, or leucine reflects their increased pool size after induction. However, it does not seem very likely, because in all the incorporation experiments we added "cold" phosphate, uridine, or leucine to offset the effects of pool size.

TABLE 4. Decreased Uptake of ^{3}H-leucine in the First 20 hr after DMSO Treatment of FSD-1/F4 Cells

Treatment	Cellular uptake in % of input	In % in control	Protein synthesis in % of input (TCA)	Protein synthesis of total uptake	Cytoplasmic protein synthesis in % of input (TCA)	Cytoplasmic protein in % of total uptake
No DMSO	12.5	100	8.7	61	5.9	47
1.5% DMSO 20 hr	5.2	42	3.1	60	2.7	52
2% DMSO 20 hr	5.3	43	3.0	57	2.2	42

Experimental details were as in Table 3 except that "hot" TCA precipitation was used, *i.e.*, after adding TCA, to a final concentration of 10%, the tubes containing the samples were heated in a boiling water bath for 10 min.

b) *Effect of amphotericin B:* Amphotericin B, a fungicide, has been shown to increase membrane permeability (*1*, *14*). We have found that if amphotericin B and DMSO are added to the cells together, the inducing effect of DMSO is totally wiped out, *i.e.*, no globin synthesis is observed. The results obtained with amphotericin B reinforce the interpretation of the incorporation data that the induced cells have decreased membrane permeability.

9. Decrease in the rate of cell division and in cytoplasmic RNA content after induction

The doubling time of the induced cells is considerably longer, about 35 hr, compared to that of the uninduced ones, about 20 hr. Table 5 shows that in 4 days, when the cell number of the uninduced cells has increased 25-fold, the induced cells show only a 6.4-fold increase. This corresponds to about 3 cell cycles. Comparing with normal erythropoiesis, where 5 cell cycles are believed to be required in going from stem cell to normoblast stage (*48*), it would appear that FV-transformed cells are at the proerythroblast stage. Another interesting feature of the data concerns the cytoplasmic RNA content. In uninduced cells, the increase in the amount of cytoplasmic RNA parallels the increase in cell number, *i.e.*, 25-fold. In the induced cells, on the other hand, the total cytoplasmic RNA content in the same time increases only by a factor of 1.3.

TABLE 5. Decrease in the Rate of Cell Division and in Cytoplasmic RNA Content after Induction

	Uninduced cells	DMSO-induced cells
Increase in cell number (4 days)	25	6.4
Cytoplasmic RNA content per cell in 10^{-11} g	2.0	0.4
Total cytoplasmic RNA of all cells in OD units day 0	9	9
Total cytoplasmic RNA of all cells day 4	225	11.5

FSD-1/F4 cells were induced with 1.5% DMSO for 4 days. Cells were fed by adding fresh medium, so no cells were lost. The number of cells was counted. For determining the amount of cytoplasmic RNA, cells were lysed with NP40 as in Fig. 2 and the supernatant deproteinized by phenol extraction as in Fig. 11. Optical density at 260 nm was measured. These measurements were done on day 0 and day 4 and also on uninduced cells. Note that the cytoplasmic RNA content as given in this table is probably lower than that in the cell because of losses during deproteinization. However, as a relative amount in induced and uninduced cells these figures are meaningful because the experiments were carried out under conditions as identical as possible.

10. Burst of synthesis of globin mRNA after induction

A few hours after the addition of DMSO to the medium, newly labeled globin mRNA appears in the cytoplasm, reaching a peak between 20 and 30 hr. After 48 hr, newly labeled globin mRNA is barely detectable. However, the content of globin mRNA relative to total RNA in the cell is maintained at a fairly constant level until day 5–6. In the experiments discussed below, we have determined the incorporation of ^{32}P-phosphate in cytoplasmic RNA and in fractionated globin mRNA at various times after induction.

The cultures of FSD-1/F4 and B8 cells were induced with DMSO for various lengths of time. The cells were shifted to a low-phosphate medium and after 3 hr

^{32}P-phosphate was added. The cells were harvested 17 hr later. For the first point, ^{32}P-phosphate and DMSO were added together. As a control, uninduced cells were labeled with ^{32}P under the same conditions. Cells were rapidly chilled before harvesting and the postmitochondrial supernatant was prepared as described in a previous section. RNA was isolated from the nucleoprotein particles which pelleted upon centrifugation of the postmitochondrial supernatant for 2 hr at $200{,}000 \times g$. Fractionation of RNA was carried out on polyacrylamide slab gels (*63*). Prior to loading the samples on the gel, RNA was heated in sealed capillary tubes at 100°C for 45 sec (*32*, *56*) to melt out any hydrogen-bonded aggregates. After electrophoresis, the gel was autoradiographed. All the bands and interbands were cut out and counted for radioactivity. The band corresponding to globin mRNA was identified by a separate experiment in which OD amounts of unlabeled and trace amounts of ^{32}P-labeled RNA were fractionated by electrophoresis on polyacrylamide gels and the RNA eluted from the individual bands was tested for globin template activity in *Xenopus* oocytes (*27*, *32*, *49*, *59*, *65*). The results are shown in Figs. 11 and 12.

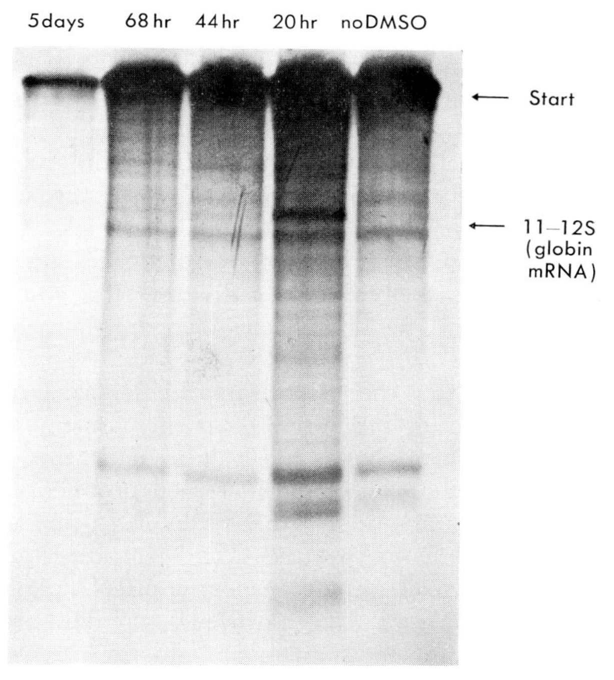

FIG. 11. Polyacrylamide gel electrophoresis of ^{32}P-labeled RNA isolated from B8 cells at different times after induction with DMSO. Cells were grown as in Fig. 1. Cells were induced with 2% DMSO and labeled with ^{32}P as described in text. Details for lysis with NP40, and preparation of postmitochondrial supernatant were as in Fig. 2. RNA from the nucleoprotein particles (see text) was extracted as described (*2*, *53*, *63*, 65). RNA was fractionated on 30 cm long 3, 4.25, and 5% polyacrylamide slab gels in Tris-borate buffer system (*18*). The gels were autoradiographed. The figure shows a typical fractionation pattern on a 3% gel. No DMSO: control uninduced cells; the other samples represent different times after induction as indicated.

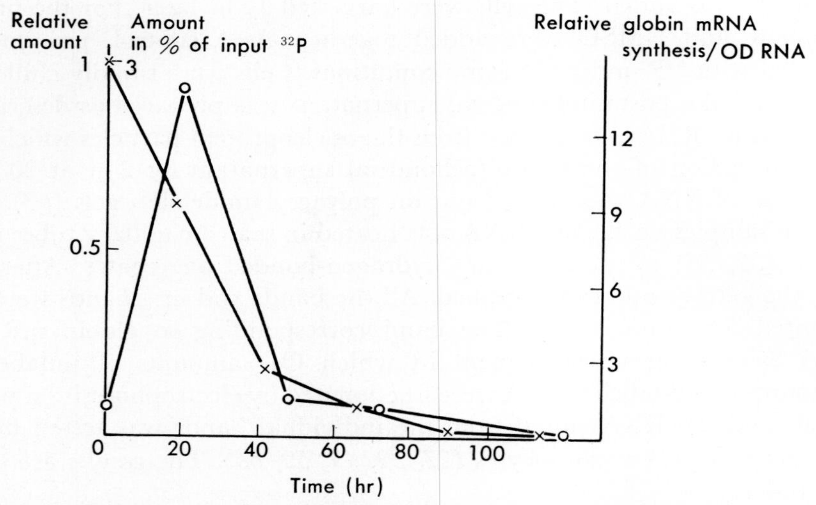

FIG. 12. ^{32}P incorporation into cytoplasmic RNA and globin mRNA during induction of B8 cells with 2% DMSO. Quantitative representation of data from Fig. 11. × cytoplasmic RNA; O globin mRNA.

Figure 11 shows the pattern obtained when the ^{32}P-labeled RNA isolated from B8 cells at different times after induction was fractionated on polyacrylamide gels. Most striking is the observation that a newly labeled, intense band in the globin mRNA region is present in the electrophoretogram of RNA from 20 hr induced cells but is barely detectable when the fractionated RNA is from cells induced for 2 or more days. These results are shown quantitatively in Fig. 12, where the relative amounts of newly labeled globin mRNA at various times after induction are plotted. The data for Fig. 12 were obtained by dividing the cpm in globin mRNA by the total cpm in the gel. The figure also shows the time course of ^{32}P incorporation in cytoplasmic RNA.

These data clearly demonstrate that there is a dramatic increase in the content of globin mRNA in the cytoplasm after induction. This increase must be due to newly transcribed RNA, because the mRNA is newly labeled. In this context it is interesting to consider that the mechanism of the increased production of globin mRNA may also involve the replication of globin mRNA by RNA-dependent RNA polymerase, an enzyme found in reticulocytes (*19*). Whether post-transcriptional control mechanisms involved in processing and transport of mRNA play an important role during induction is not at present clear. However, the data are probably best explained by postulating that the mechanism of induction involves switching on the transcription of globin genes. This would fit with the results on the hybridization of globin complementary DNA (cDNA) (*16, 33, 39, 70, 81*). Our data further show that the globin genes do not stay switched on, because label in globin mRNA is found only during the first 40 hr of induction, suggesting that the time interval between switching on and off of transcription is not more than 2 cell cycles. In fact it is more likely that this interval is only one cell cycle. This would support the idea of quantized cell cycles involved in differentiation (*35, 36, 101*).

11. Time lag between the transcription of globin genes and translation of globin mRNA

In the last section we have shown that the synthesis of globin mRNA reaches a peak approximately 20 hr after induction. However, as shown below, the synthesis of globin chains reaches a peak approximately 60 hr later.

Cultures of FSD-1/F4 cells were induced with DMSO. At various times after induction, the cells were shifted to a leucine-free medium and labeled with ^{3}H- or ^{14}C-leucine for 17 hr. As a control, uninduced cells were labeled with leucine under the same conditions. At the first time point after induction, DMSO and radioactive leucine were added together. The details of isolation, fractionation on CMC-urea columns and quantitative determination of globin chains were the same as described in a previous section. The radioactivity under globin peaks was determined as a percentage of total protein radioactivity on the column.

Figure 13 shows the time course of synthesis of globin chains and for comparison, also of globin mRNA. At the point when globin mRNA synthesis is at its peak, the synthesis of globin is barely above the uninduced level. Between days 2 and 3, when the globin mRNA synthesis is hardly detectable, the synthesis of globin is on its way up. Around day 5, when mRNA synthesis is undetectable, globin synthesis is at its peak and the labeled globin amounts to about 25% of the labeled cytoplasmic protein. The figure also shows that no label is found in globin chains if the cells are exposed to BrdU prior to addition of DMSO.

These data, showing that globin mRNA is transcribed early but translated late, argue in favour of a translational control mechanism. Messenger selection mechanisms operating at the level of polypeptide chain initiation have been documented in other systems (*20*, *30*, *54*, *84*, *93*) and it is quite likely that mechanisms of a similar nature control globin mRNA translation in this system. The fact that viral proteins are synthesized actively in the early phase of induction when globin synthesis is barely detectable (see Ref. *66*) provides further support for this notion. The data also throw some light on the stability of globin mRNA. Since globin synthesis occurs predominantly on day 4–6 and the peak of globin mRNA synthesis is on day 1–2, a crude estimate of the half-life of globin mRNA would be at least 3 days. The fact that globin mRNA can be pelleted in 2 hr at $200,000\times g$ indicates that it is in the form of ribonucleoprotein (RNP) particles. However, in the early phase of induction, very little label is detected in globin chains, indicating that at this stage the globin mRNA is not being translated. This makes it very unlikely that the RNP particles are polysomes. In fact, the data favor the possibility that they are informosome-like particles which have been shown to be present in differentiating systems (*68*, *75*, *90*).

12. Amount of globin mRNA in induced cells comparable to mouse reticulocytes

In the previous section we have considered control mechanisms operating at the level of translation. In this section we will show that most of the globin mRNA in induced cells on day 5 is on the polysomes. In uninduced cells, on the other hand, although there is very little globin mRNA, a major fraction is present in the 30–90S fraction. The amount of globin mRNA in induced cells is at least 80% of that

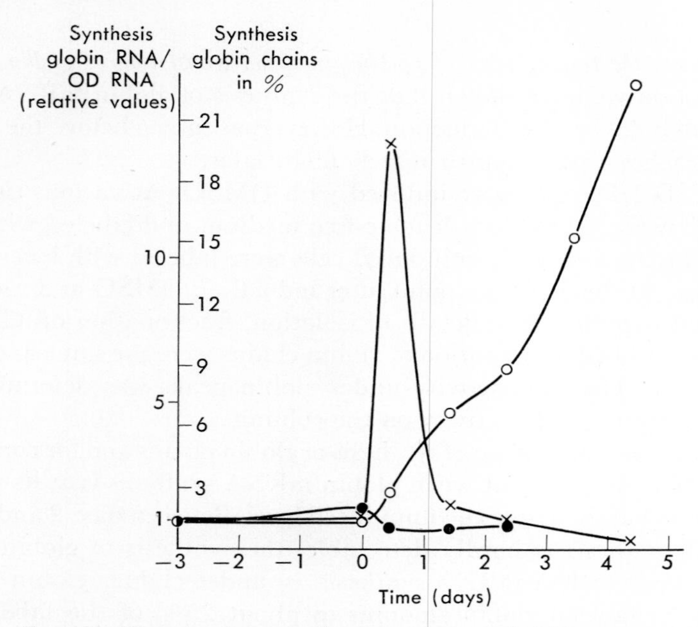

FIG. 13. Time lag between the transcription of globin genes and translation of globin mRNA. B8 cells were induced with 2% DMSO as in Fig. 11 and labeled with ^{3}H- or ^{14}C-leucine at various times after induction, as described in text. Globin chains were fractionated as in Fig. 2 and radioactivity under globin peaks as a percent of total protein radioactivity on the column was determined. The time course of synthesis of globin chains is shown and also, for comparison, that of globin mRNA. Also shown is the inhibitory effect of BrdU on globin synthesis in FSD-1/F4 cells. × synthesis globin mRNA (DMSO); ○ synthesis of globin (DMSO); ● synthesis of globin, BrdU 2.5 μg/ml (day 0), DMSO (day 3).

in mouse reticulocytes, indicating that these cells are in late stage of erythroid maturation.

A large culture of FSD-1/F4 cells was induced with DMSO for 5 days. Before harvesting, the cells were incubated with cycloheximide for 15 min to prevent run-off of monosomes (*27, 103*). The cells were collected and lysed with 0.5% NP40 (see legend to Fig. 2). The NP40 cytoplasmic suspension was layered on 12–32% sucrose gradients. The polysomes were separated from the monosomes and the postribosomal supernatant (Fig. 14). The indicated fractions were treated with EDTA and SDS at a high pH and low cation concentration (*53, 65*) and extracted with a mixture of phenol, cresol, chloroform, and isoamyl alcohol (*2, 63*). The RNA of each fraction was separated on 8–50% sucrose gradients. Figure 15 shows the separation of the polysomal RNA of induced leukemic cells and Fig. 16 that of the polysomal RNA of mouse reticulocytes as a control. It can be seen that the polysomal RNA extracted from reticulocytes and induced and uninduced erythroleukemic cells has a similar distribution profile in the sucrose gradient. However, much more material in the 12–16S region is present in the induced postribosomal fraction as compared with the same fraction extracted from uninduced cells (Figs. 17 and 18).

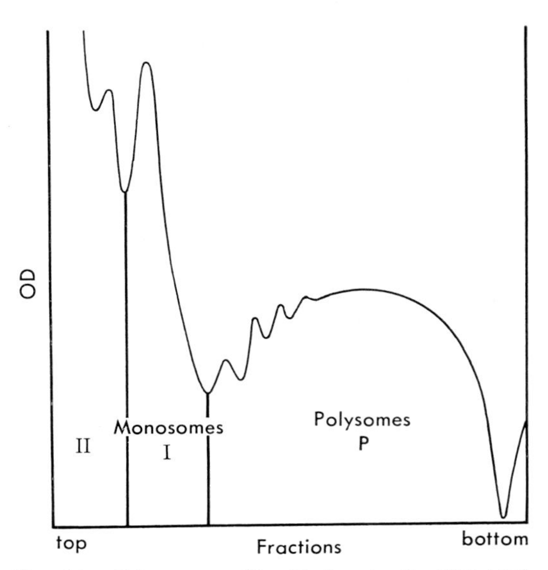

FIG. 14. Polysome profile of induced cells. FSD-1/F4 cells were grown as in Fig. 1. Cells were induced with 1.5% DMSO for 5 days. Uninduced and induced cells were exposed to 100 μg/ml cycloheximide for 15 min at 37°C. The cytoplasmic NP40-soluble supernatant was centrifuged in a linear 12–32% sucrose gradient containing 0.05M Tris-HCl, pH 7.4, 0.025M KCl, and 0.005M $MgCl_2$ in a Beckman SW27 rotor at 100,000 $\times g$ for 2 hr. The fractions indicated as P, I, II were collected.

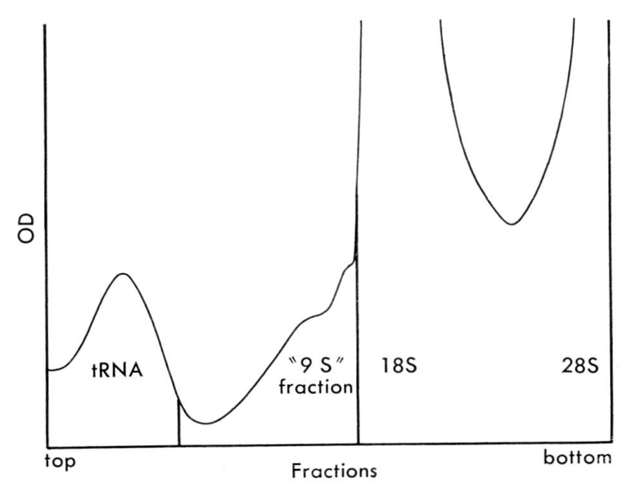

FIG. 15. Fractionation of polysomal RNA of induced cells on a sucrose gradient. Polysomal RNA (fraction P of Fig. 14) of induced cells was deproteinized as in Fig. 11. The RNA was applied to 8–50% sucrose gradients in 0.01M Tris-HCl, pH 7.4, 0.015M KCl (Beckman SW40 rotor, 190,000 $\times g$, 16 hr, 4°C). The 8–16S fraction was used for injection into *Xenopu* soocytes.

The recovery of the RNA during the isolation procedure was monitored at each step to allow an estimation of globin mRNA activity per total RNA of an average leukemic cell. The RNA indicated as "9S" RNA was collected and injected

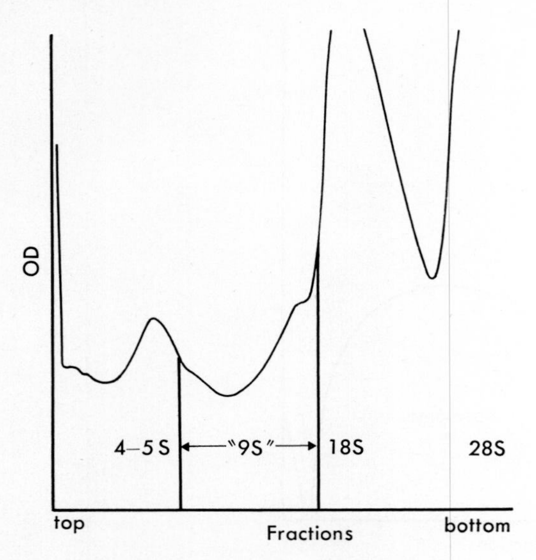

FIG. 16. Fractionation of polysomal RNA of mouse reticulocytes. Details as in Figs. 14 and 15.

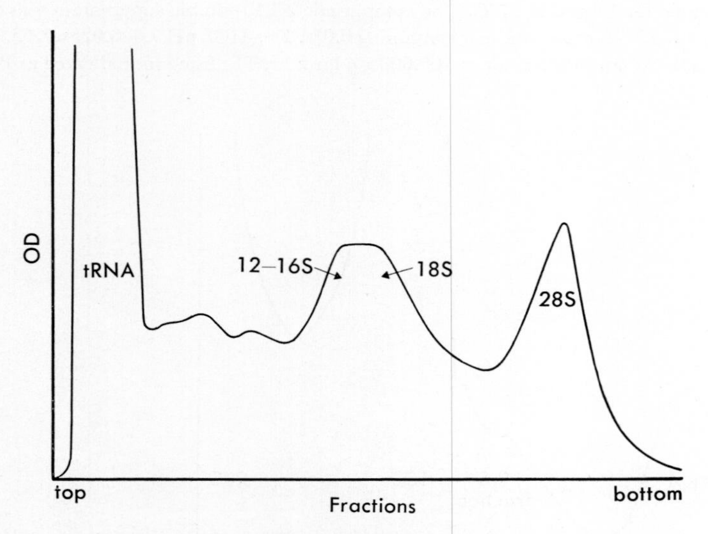

FIG. 17. Fractionation of postribosomal RNA fraction of induced cells. RNA of fraction II of Fig. 14 was separated on sucrose gradient as in Fig. 15. The fraction in 8–16S region was pooled and used for injection into *Xenopus* oocytes.

into frog oocytes (*16*). The proteins synthesized by the *Xenopus* oocytes were fractionated on CMC-urea columns (see legend to Fig. 2) using as carrier proteins either adult mouse globin or alternatively ^{14}C-labeled β and α chains. The latter were isolated from DMSO-induced erythroleukemic cells. Figure 19 shows the relative increase in globin synthesis in the oocytes after injection of equivalent amounts of polysomal RNA from induced and uninduced cells. Very little radioactivity in the

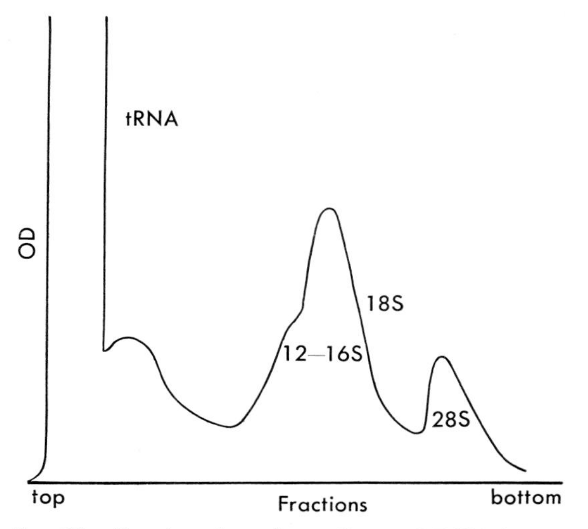

FIG. 18. Fractionation of postribosomal RNA fraction of uninduced cells. RNA extracted from the postribosomal fraction, corresponding to fraction II of Fig. 14, was separated on sucrose gradients as in Fig. 15. Again, the 8–16S region was pooled for globin template activity measurements in *Xenopus* oocytes.

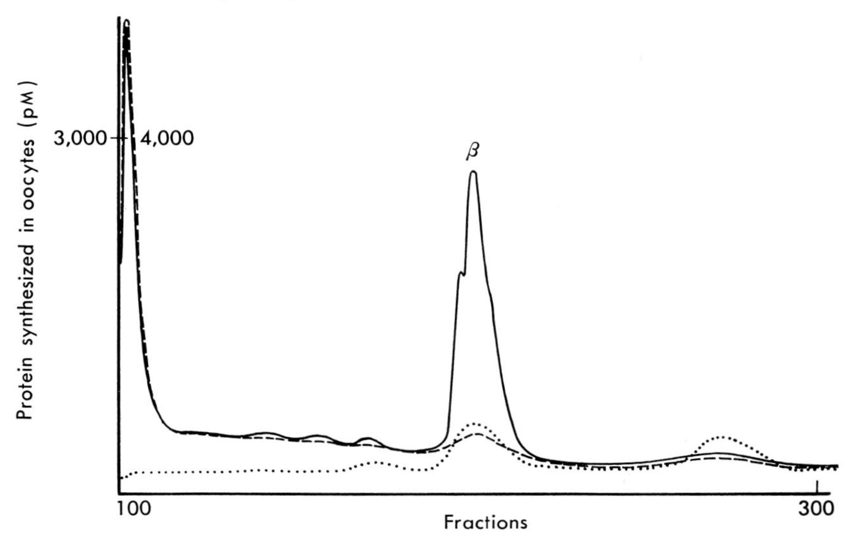

FIG. 19. Relative amounts of globin synthesized in frog oocytes in response to 8–16S polysomal RNA extracted from induced and uninduced cells. The experimental details were as described in text and in Ref. *65*. The fractionation of globin on a CMC-urea column (details as in Fig. 2) is shown. The concentration of RNA from induced and uninduced cells was adjusted so that the radioactivity under the globin peak is directly comparable. —— polysomal 8–15S RNA, stimulated cells; - - - polysomal 8–15S RNA, unstimulated cells; ······ carrier globin, OD.

α-globin chain region is seen in Fig. 19. This was consistently observed. Synthesis of the α chain is apparently quite sensitive to temperature. On low-temperature incubation of the oocytes, the yields are low. Also, the α chain precipitates out if no heme is present or when it is present at low concentrations. However, the recovery of β chain is quite reliable and therefore it is possible to correlate the amount of globin mRNA with the radioactivity under the β-chain peak. This was done to calculate the data shown in Table 6.

The fact that the RNA from uninduced cells programs very little globin synthesis in *Xenopus* oocytes indicates that this RNA contains very few globin sequences. This clearly establishes the point that the observed lack of hemoglobin in uninduced cells is due to a lack of globin transcripts and not due to a translational block in globin synthesis which is removed upon induction. Globin template activity measurements on the RNA from induced cells also argue in this direction. Figure 19 shows that this RNA is very active in globin synthesis in *Xenopus* oocytes indicating that it is rich in globin sequences. These data provide further support for the conclusion drawn in an earlier section that one of the major events during induced differentiation of erythroleukemic cells is switching on the transcription of globin genes.

Table 6 gives relative globin template activities of polysomal, monosomal and postribosomal "9S" RNA fractions from induced and uninduced cells. Also given is the globin mRNA activity in the mouse reticulocyte polysomal "9S" RNA fraction. The activities are given in terms of cpm in β chain. The amount of RNA was adjusted so that the values are directly comparable and reflect the relative

TABLE 6. Globin Template Activity of RNA Isolated from Induced and Uninduced Cells

RNA sources	Cellular globin synthesis (%)	Relative increase during stimulation	Relative globin mRNA activity corrected for OD	Relative increase of globin mRNA (oocyte test)
Uninduced				
Polysomal fraction	—	—	<225	1
Monosomal fraction	—	—	<1,100	1
Postribosomal fraction	—	—	<150	1
Total	<0.5	1	<1,500	1
Induced				
Polysomal fraction	—	—	25,600 (74%)	>100
Monosomal fraction	—	—	5,300 (16%)	>5
Postribosomal fraction	—	—	3,800 (11%)	>25
Total	20	>40	34,700	>23
Mouse reticulocyte polysomes	98	—	29,000	—

FSD-1/F4 cells were induced with 1.5% DMSO for 5 days. RNA from polysomal, monosomal, and postribosomal fractions was isolated as described in text. RNA was injected in *Xenopus* oocytes and the synthesis of globin chains measured as described in Ref. *65*. The table gives the relative globin mRNA activity of various fractions as cpm under the β-chain peak, incorporated in response to an equivalent OD amount of RNA injected. Also given in the table are the cpm under the β-chain peak when an equivalent amount of mouse reticulocyte polysomal RNA is injected into the oocytes. The numbers are directly comparable.

globin mRNA activity of the RNA injected. The data show that most of the globin template activity in 5-day induced cells is in the polysomal fraction. Furthermore, the amount of globin mRNA in induced cells is over 80% of that in mouse reticulocytes. On the other hand, experiments with another FV-transformed cell line have shown that the amount of globin mRNA in induced cells is only 30–40% of that in reticulocytes (*52*). In this context it is noteworthy that the number of benzidine-positive cells in this cell line is only 30–40% (*52*), as compared to 80–90% in our cell lines FSD-1/F4 or B8. The data show another interesting point. Although the uninduced cells are very low in globin mRNA content, the globin template activity is mostly in the monosomal fraction. This suggests that the mRNA in uninduced cells is not as effectively translated.

There is another feature of the data which deserves comment. Figures 17 and 18 show that much more 8–16S material is present in the postribosomal fraction of induced erythroleukemic cells as compared to the same fraction of uninduced cells. This RNA has very little globin template activity as compared to the polysomal RNA of 8–16S region. Probably during DMSO induction, some ribosomal RNA is degraded, and then appears in the postribosomal fraction. This agrees with observations on ribosomal RNA breakdown during normal reticulocyte maturation (*17, 50*). We have resolved some of this RNA on polyacrylamide gels and observed distinct bands in the 10–12S region (unpublished observations).

Human Cell Lines with Some Indication That They Synthesize Hemoglobin

1. Establishment of the cell cultures

The bone marrow was derived from untreated patients suffering from polycythemia vera (PV) (*60*) or erythroleukemia. It was obtained by iliac or sternal puncture using a heparinized syringe for aspiration. The marrow was immediately transferred to tissue-culture flasks containing complete medium at 4°C and transported to the laboratory. After incubation at 36°C in 4% CO_2 atmosphere for about 5 hr, the bone marrow fragments had become attached to the surface so that the erythrocytes could be removed by very gently turning the bottle and decanting the medium. After addition of fresh medium, the bottles were reincubated and fed every third day by decanting half of the medium and adding fresh medium. In the first week a strong proliferation of cells was observed. These were fibroblastoid as well as suspended nucleated cells. After 3 weeks the suspended cells died or had been lost by the feeding procedure. At this time one could observe a conglomeration of cells located in groups of 20–50 cells in the neighbourhood of the original bone marrow fragments. The cells were strongly attached to the surface but exhibited a circular, flat shape. These "nests" were surrounded by fibroblast-like cells and were sometimes partially overgrown by them. In the following weeks up to 50% of the "nests" disappeared by becoming fibroblastoid cells. After 6–9 weeks a "piling up" process in these "nests" could be observed. This process began with increased rounding up and poor division. These cells exhibited new root-like extrusions. Later on they formed clusters consisting of about 50 cells which loosened and floated in suspension. The permanent cell lines were derived from these clustered cells.

The above process could be observed in 4 bone marrows from patients with untreated PV and 2 with erythroleukemia. It could not be observed in cultures of at least 6 bone marrows from treated (^{32}P or chemotherapeutically) PV patients or one patient with secondary PV and one patient with erythroblastosis probably caused by ECHO-virus infection.

The permanent cell lines derived from untreated PV patients are designated as PHD and PBW, and those from patients with erythroleukemia as EHN, EHL, and ENG. The latter grow poorly in culture. The ENG line was derived from a pleural effusion caused by a leukemic infiltrate of a patient suffering from an atypical erythroleukemia. Growth of the ENG cells was immediate after transfer to tissue culture. The growth rate however is extremely slow. ENG cells grow attached and are morphologically epithelioid.

2. *Some characteristics of the cell lines established*

a) *Karyotype of cell lines:* The cell line PHD shows 48 chromosomes (*102*). This karyotype is observed in nearly 100% of the cells, indicating the clonal character of the culture. The 2 additional chromosomes seem to belong to the B and C group. There is also an indication of the presence of an F-group abnormality (deletion or inversion) as observed by others in disorders of the "red compartment" (*37, 43, 78*). PBW has 46 chromosomes. ENG cells are heteroploid.

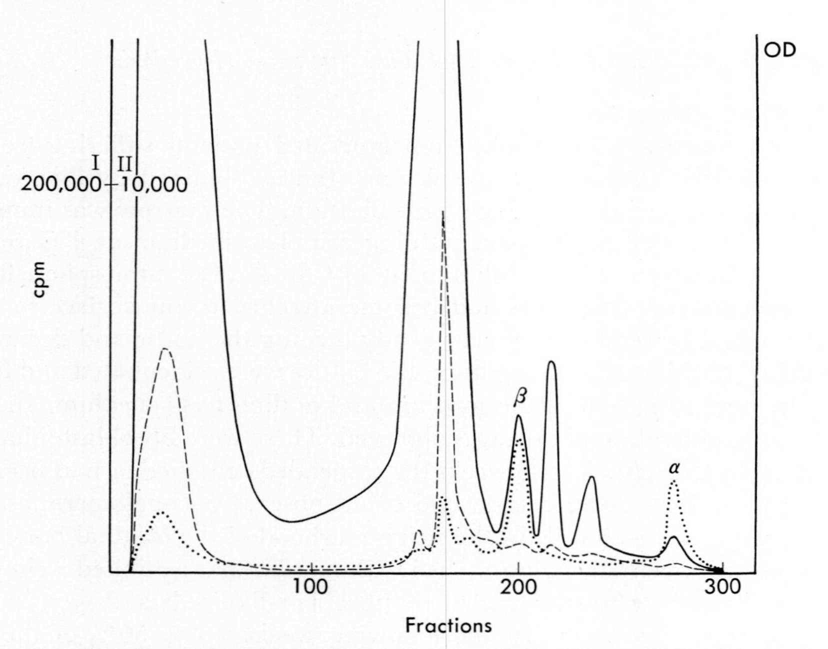

FIG. 20. Search for globin synthesis in human cell line PHD established from the bone marrow of a polycythemia vera patient. Details of the labeling of cells and fractionation of globin chain on CMC-urea columns are as in Fig. 2. Adult human hemoglobin was used as a carrier. A radioactive peak is seen coeluting with β chain. —— polycythemia vera, cell culture, ^{3}H labeled, DMSO stimulated (scale I); - - - polycythemia vera, cell culture, ^{14}C labeled, DMSO stimulated (scale II); ······ OD human globin (control).

b) *Search for globin synthesis:* Cytochemically no convincing results could be obtained to support the expectation that these cells were of erythroid nature. Induction with DMSO was not possible and the cells were found to be very sensitive to that agent.

To determine whether these cells synthesized any globin chains, we have carried out similar experiments, as described in an earlier section on hemoglobin synthesis in Friend erythroleukemic cells. PHD cells were labeled with ^{3}H- and ^{14}C-leucine, lysed with NP40 and the lysate precipitated with HCl-acetone after addition of carrier adult human hemoglobin. The radioactive material coeluting with the α and β chains on a CMC-urea column (Fig. 20) was digested with trypsin and fingerprinted. The added carrier material gave a good fingerprint whereas most of the radioactive material remained at the origin, indicating incomplete digestion. Nevertheless a leu-arg peptide very specific for the α chain could be identified.

We have also carried out column chromatographic analysis on PBW, EHN, and ENG cells. The elution patterns of PBW and EHN are similar to those obtained with PHD cells. ENG cells showed a prominent peak in the β region and no radioactive label in the α peak (Fig. 21). This imbalance in β and α synthesis is not unexpected since several reports have been published stating that in erythroid cells of patients with erythroleukemia, inclusion bodies of β_4 could be detected (*4*, *104*). This seems to be a general pattern in mouse and human erythroleukemia: excess synthesis of β chain or deficiency of α chain synthesis (Figs. 2, 20, and 21). More work is necessary to clearly establish the identity of coeluting radioactivity in PHD,

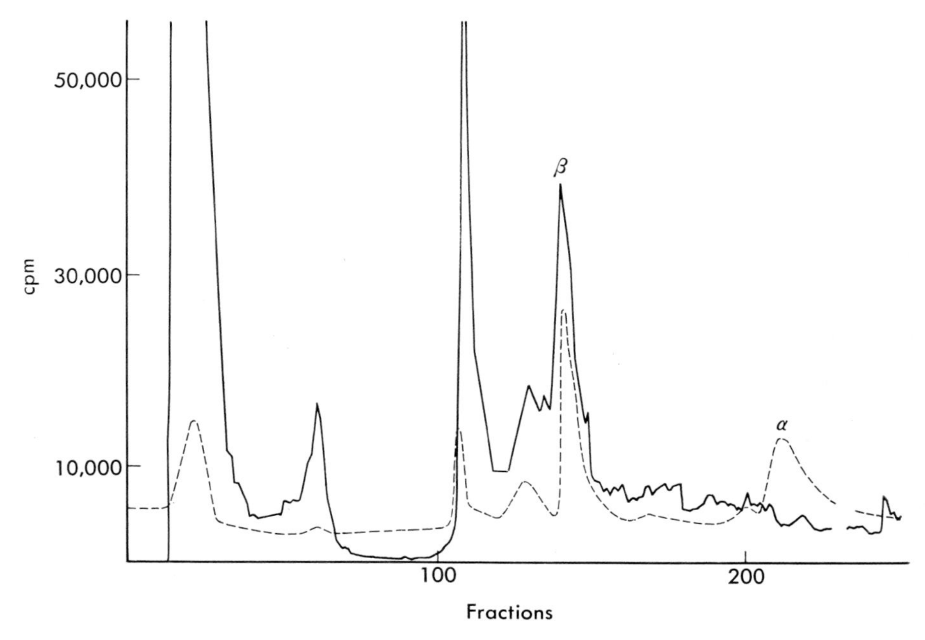

FIG. 21. Search for globin synthesis in human cell line ENG derived from patient suffering from erythroleukemia. Details as in Fig. 20. Human hemoglobin was used as a carrier. A prominent peak is seen in the β region. —— ^{3}H; - - - OD.

PBW, and EHN cells with the human globin chains. However, judging from the control experiments, *i.e.*, the CMC-urea column elution patterns of HeLa and other cell lines it appears that the cell lines we have established do indeed synthesize small amounts of globin chains; more β than α chains. Further work on the characterization and induction of hemoglobin synthesis is in progress.

REFERENCES

1. Amati, P. and Lago, C. Sensitivity to amphotericin B of *in vitro* established cell lines. Nature, *247*: 466–469, 1974.
2. Aviv, H. and Leder, P. Purification of biologically active globin messenger RNA by chromatography on oligothymidylic acid-cellulose. Proc. Natl. Acad. Sci. U.S., *69*: 1408–1412, 1972.
3. Barker, J. E., Last, J. E., Adams, S. L., Nienhuis, A. W., and Anderson, W. F. Hemoglobin switching in sheep and goats: Erythropoietin-dependent synthesis of hemoglobin C in goat bone marrow cultures. Proc. Natl. Acad. Sci. U.S., *70*: 1739–1743, 1973.
4. Beaven, G. H., Stevens, B. L., Dance, N., and White, J. C. Occurrence of haemoglobin H in leukaemia. Nature, *199*: 1297–1298, 1963.
5. Borsock, H., Ratner, K., Tattrie, B., Teigler, D., and Lajtha, L. G. Effect of erythropoietin *in vitro* which stimulated that of a massive dose *in vivo*. Nature, *217*: 1024–1026, 1968.
6. Borsock, H., Jiggins, S., and Wilson, R. T. Induction of erythroblast increase in rabbit marrow cells by extracts of leukocytes and certain metabolic compounds. IV. Studies on erythropoiesis. J. Cell Physiol., *79*: 277–282, 1972.
7. Borun, T. W., Scharff, M. D., and Robbins, E. Preparation of mammalian polyribosomes with the detergent nonidet P-40. Biochim. Biophys. Acta, *149*: 302–304, 1967.
8. Bottomley, S. S. and Smithee, G. A. Effect of erythropoietin on bone marrow Δ-aminolevulinic acid synthetase and heme synthetase. J. Lab. Clin. Med., *74*: 445–452, 1969.
9. Bruce, W. R. and McCulloch, E. A. The effect of erythropoietic stimulation on the hemopoietic colony-forming cells of mice. Blood, *23*: 216–232, 1964.
10. Butcher, R. W., Robinson, G. A., and Sutherland, E. W. Cyclic AMP and hormone action. *In;* G. Litwak (ed.), Biochemical Aspects of Hormones, pp. 21–54, Academic Press, New York, 1972.
11. Byron, J. W. Effect of steroids and dibutyryl cyclic AMP on the sensitivity of haemopoietic stem cells to ^{3}H-thymidine *in vitro*. Nature, *234*: 39–40, 1971.
12. Camiscoi, J. F. and Gordon, A. S. Bioassay and standardization of erythropoietin. *In;* A. S. Gordon (ed.), Regulation of Hematopoiesis, vol. 2, pp. 369–393, Appleton-Century-Crofts, New York, 1970.
13. Chui, D. H. U., Djaldetti, M., Marks, P. A., and Rifkind, R. A. Erythropoietin effects on fetal mouse embryoid cells. I. Cell population and hemoglobin synthesis. J. Cell Biol., *51*: 585–595, 1971.
14. Cirillo, V. P. Action on cell membrane structure. *In;* R. M. Hochster, M. Kates, and J. H. Quasil (eds.), Metabolic Inhibitors, vol. 3, pp. 64–68, Academic Press, New York and London, 1972.
15. Clegg, J. B., Naughton, M. A., and Weatherall, P. J. An improved method for the

characterization of human haemoglobin mutants: Identification of $\alpha_2\beta_2^{95glu}$, haemoglobin N (Baltimore). Nature, *207*: 945–947, 1965.
16. Conkie, D., Affara, N., Harrison, P. R., and Paul, J. 1974, in press.
17. De Jimenez, E. S. and Lotina, B. Degradation of ribosomal RNA during red cell differentiation. Biochem. Biophys. Res. Commun., *48*: 1323–1329, 1972.
18. Dingman, C. W. and Peacock, A. C. Analytical studies on nuclear ribonucleic acid using polyacrylamide gel electrophoresis. Biochemistry, *7*: 659–668, 1968.
19. Downey, K. M., Byrnes, J. J., Jurmark, B. S., and So, A. G. Reticulocyte RNA-dependent RNA polymerase. Proc. Natl. Acad. Sci. U.S., *70*: 3400–3404, 1973.
20. Dube, S. K. and Rudland, P. S. Control of translation by T4 phage: Altered binding of disfavoured messengers. Nature, *226*: 820–823, 1970.
21. Duesberg, P. H. Physical properties of Rous sarcoma virus RNA. Proc. Natl. Acad. Sci. U.S., *60*: 1511–1578, 1968.
22. Freshney, R. I., Paul, J., and Conkie, D. J. Effect of erythropoietin on haemoglobin synthesis and heme snythesizing enzymes of mouse foetal liver cells in culture. J. Embryol. Exp. Morphol., *27*: 525–532, 1972.
23. Friend, C. Cell-free transmission in adult Swiss mice of a disease having the character of a leukemia. J. Exp. Med., *105*: 307–318, 1957.
24. Friend, C., Patuleia, M. C., and DeHarven, E. Erythrocytic maturation *in vitro* of murine (Friend) virus-induced leukemic cells. Natl. Cancer Inst. Monogr., *22*: 505–520, 1966.
25. Friend, C., Scher, W., Holland, J. G., and Sato, T. Hemoglobin synthesis in murine virus-induced leukemic cells *in vitro*: Stimulation of erythroid differentiation by dimethyl sulfoxide. Proc. Natl. Acad. Sci. U.S., *68*: 378–382, 1971.
26. Furusawa, M., Ikawa, Y., and Sugano, H. Development of erythrocyte membrane-specific antigen(s) in clonal cultured cells of Friend virus-induced tumor. Proc. Japan Acad., *47*: 220–224, 1971.
27. Gaedicke, G., Abedin, U., Dube, S. K., Kluge, N., Neth, R., Steinheider, G., Weimann, B. J., and Ostertag, W. Control of globin synthesis during DMSO induced differentiation of mouse erythroleukemic cells in culture. *In;* R. Neth, R. Gallo, and F. Stohlman (eds.), Modern Trends in Human Leukemia-Biological, Biochemical and Virological Aspects, J. F. Lehmann's Verlag, Munich, 1973, in press.
28. Gordon, A. S., Zanjani, E. D., Levere, R. D., and Kappas, A. Stimulation of mammalian erythropoiesis by 5β-H steroid metabolites. Proc. Natl. Acad. Sci. U.S., *65*: 919–924, 1970.
29. Granick, S. and Kappas, A. Steroid induction of porphyrin synthesis in liver cell culture. I. Structural basis and possible physiological role in the control of heme formation. J. Biol. Chem., *242*: 4587–4593, 1967.
30. Groner, Y., Pollack, Y., Berissi, H., and Revel, M. Cistron specific translation control protein in *Escherichia coli*. Nature New Biol., *239*: 16–19, 1972.
31. Gross, M. and Goldwasser, G. On the mechanism of erythropoietin-induced differentiation. V. Characterization of the ribonucleic acid formed as a result of erythropoietin action. Biochemistry, *8*: 1795–1805, 1969.
32. Gurdon, J. B., Lane, C. D., Woodland, H. R., and Marbaix, G. Use of frog eggs and oocytes for the study of messenger RNA and its translation in living cells. Nature, *233*: 177–182, 1971.
33. Harrison, P. R., Conkie, D., and Paul. J. Localisation of cellular globin messenger RNA *in situ* hybridization to complementary DNA. FEBS Letters, *32*: 109–112, 1973.

34. Hier, D. B., Arnason, B. G. W., and Young, M. Studies on the mechanism of action of nerve growth factor. Proc. Natl. Acad. Sci. U.S., *69*: 2268–2272, 1972.
35. Holtzer, H., Weintraub, H., Mayne, R., and Mochan, B. The cell cycle, cell lineages, and cell differentiation. Current Topics Dev. Biol., *7*: 229–256, 1972.
36. Holtzer, H., Dienstman, S. Holtzer, S., and Biehl, J. Quantal cell cycles, normal cell lineages, and tumorigenesis. *In;* W. Nakahara *et al.* (eds.), Differentiation and Control of Malignancy of Tumor Cells, Univ. of Tokyo Press, Tokyo, pp. 389–400, 1974.
37. Hossfeld, D. K., Schmidt, C. G., and Sandberg, A. A. Die " F "-chromosomen Anomalie in Erkrankungen des erythropoetischen Systems. *In;* J. F. Bergmann (ed.), Verhandlungen der Deutschen Gesellschaft für Innere Medizin, vol. 78, pp. 126–129, München, 1972.
38. Hrinda, M. E. and Goldwasser, E. On the mechanism of erythropoietin-induced differentiation. VI. Induced accumulation of iron by marrow cells. Biochim. Biophys. Acta, *195*: 165–175, 1969.
39. Ikawa, Y., Ross, J., Gielen, J., Packman, S., Leder, P., Ebert, P., Hayashi, K., and Sugano, H. Erythrodifferentiation of cultured Friend leukemia cells. *In;* W. Nakahara *et al.* (eds.), Differentiation and Control of Malignancy of Tumor Cells, Univ. of Tokyo Press, Tokyo, pp. 515–546, 1974.
40. Johnson, G. S. and Pastan, I. Role of 3′,5′-adenosine monophosphate in regulation of morphology and growth of transformed and normal fibroblasts. J. Natl. Cancer Inst., *48*: 1377–1387, 1972.
41. Johnson, G. S. and Shimomura, O. Preparation and use of aequorin for rapid microdetermination of Ca^{2+} in biological systems. Nature New Biol., *237;* 287–288, 1972.
42. Kappas, A. and Granick, S. Steroid induction of porphyrin synthesis in liver cell culture. II. The effects of heme uridine diphosphate glucuronic acid, and inhibitors of nucleic acid and protein synthesis. J. Biol. Chem., *243*: 346–351, 1968.
43. Kay, H. E. M., Lawler, S. D., and Millard, R. E. The chromosomes in polycythaemia vera. Brit. J. Haematol., *12*: 507–527, 1966.
44. Keighly, G., Hammond, D., and Lowy, P. H. The sustained action of erythropoietin injected repeatedly into rats and mice. Blood, *23*: 99–107, 1964.
45. Kluge, N., Gaedicke, G., Steinheider, G., Dube, S. K., and Ostertag, W. Globin synthesis in Friend erythroleukemic mouse cells in protein- and lipid-free medium. Effects of dimethylsulfoxide, iron and erythropoietin. Exp. Cell Res., 1974, in press.
46. Kram, R., Mamont, P., and Tomkins, G. Pleiotypic control by adenosine 3′: 5′-cyclic monophosphate: A model for growth control in animal cells. Proc. Natl. Acad. Sci. U.S., *70*: 1432–1436, 1973.
47. Kram, R. and Tomkins, G. Pleiotypic control by cyclic AMP: Interaction with cyclic GMP and possible role of microtubules. Proc. Natl. Acad. Sci. U.S., *70*: 1659–1663, 1973.
48. Lajtha, L. G. Cytokinetics and regulation of progenitor cells. J. Cell Physiol., *67*: Suppl. 1, 133–148, 1966.
49. Lane, C. D., Marbaix, G., and Gurdon, J. B. Rabbit haemoglobin synthesis in frog cells: The translation of reticulocyte 9S RNA in frog oocytes. J. Mol. Biol., *61*: 73–91, 1971.
50. Lanyon, W. G., Paul, J., and Williamson, R. Studies on the heterogeneity of mouse globin messenger RNA. Eur. J. Biochem., *31*: 38–43, 1972.
51. Laskov, R. and Scharff, M. Synthesis, assembly, and secretion of gamma globulin by mouse myeloma cells. J. Exp. Med., *131*: 515–541, 1970.

52. Leder, P., Ross, J., Gielen, J., Packman, S., Ikawa, Y., Aviv, H., and Swan, D. Regulated expression of mammalian genes: Globin and immunoglobulin as model systems. Cold Spring Harbor Symp. Quant. Biol., *38*: 753–761, 1973.
53. Lee, S. Y., Mendecki, J., and Brawerman, G. A polynucleotide segment rich in adenylic acid in the rapidly-labeled polyribosomal RNA component of mouse sarcoma 180 ascites cells. Proc. Natl. Acad. Sci. U.S., *68*: 1331–1335, 1971.
54. Lee H. S. and Ochoa, S. Specific inhibitors of MS2 and late T4 RNA translation in *E. coli*. Biochem. Biophys. Res. Commun., *49*: 371–376, 1972.
55. Levere, R., Kappas, A., and Granick, S. Stimulation of hemoglobin synthesis in chick blastoderms by certain 5β androstane and 5β pregnane steroids. Proc. Natl. Acad. Sci. U.S., *58*: 985–990, 1967.
56. Maisel, J., Klement, V., Lai, M. M. C., Ostertag, W., and Duesberg, P. H. Ribonucleic acid components of murine sarcoma and leukemia viruses. Proc. Natl. Acad. Sci. U.S., *70*: 3536–3540, 1973.
57. Mirand, E. A. Nonerythropoietin-dependent erythropoieses. *In;* A. S. Gordon (ed.), Regulation of Hematopoiesis, vol. 1, pp. 635–647, Appleton-Century-Crofts, New York, 1970.
58. Mizoguchi, H. and Levere, R. D. Enhancement of heme and globin synthesis in cultured human marrow by certain 5β-H steroid metabolites. J. Exp. Med., *134*: 1501–1512, 1971.
59. Moar, V. A., Gurdon, J. B., Lane, C. D., and Marbaix, G. Translational capacity of living frog eggs and oocytes, as judged by messenger RNA injection. J. Mol. Biol., *61*: 93–103, 1971.
60. Modan, B. The Polycythemic Disorders, C. C. Thomas Publ., Springfield, 1971.
61. Necheles, T. F. and Rai, U. S. Studies on the control of hemoglobin synthesis: The *in vitro* stimulating effect of a 5β-H steroid metabolite on heme formation in human bone marrow cells. Blood, *34*: 380–384, 1969.
62. Necheles, T. F. Studies on the control of hemoglobin synthesis: A model of erythroid differentiation based upon the *in vitro* effect of rythropoietin and 5β-H steroids. *In;* H. Martin and L. Nowicki (eds.), Synthesis, Structure and Function of Hemoglobin, pp. 53–59, J. F. Lehmann's Verlag, Munich, 1972.
63. Ostertag, W., Melderis, H., Steinheider, G., Kluge, N., and Dube, S. K. Synthesis of mouse haemoglobin and globin mRNA in leukaemic cell cultures. Nature New Biol., *239*: 231–234, 1972.
64. Ostertag, W., Crozier, T., Kluge, N., Melderis, H., and Dube, S. K. Action of 5-bromodeoxyuridine on the induction of haemoglobin synthesis in mouse leukaemia cells resistant to 5-BUdR. Nature New Biol., *243*: 203–205, 1973.
65. Ostertag, W., Gaedicke, G., Kluge, N., Melderis, H., Weimann, B. J., and Dube, S. K. Globin messenger in mouse leukemic cells: Activity associated with RNA species in the region of 8 to 16S. FEBS Letters, *32*: 218–222, 1973.
66. Ostertag, W., Cole, T., Crozier, T., Gaedicke, G., Kind, J., Kluge, N., Krieg, J. C., Roesler, G., Steinheider, G., Weimann, B. J., and Dube, S. K. Viral involvement in differentiation of erythroleukemic mouse and human cells. *In;* W. Nakahara *et al.* (eds.), Differentiation and Control of Malignancy of Tumor Cells, Univ. of Tokyo Press, Tokyo, pp. 485–513, 1974.
67. Otten, J., Johnson, G. S., and Pastan, I. Cyclic AMP in fibroblasts: Relationship to growth rate and contact inhibition of growth. Biochem. Biophys. Res. Commun., *44*: 1192–1198, 1971.

68. Ovchinnikov, L. P. and Spirin, A. S. Ribonucleoprotein particles in cytoplasmic extracts of animal cells. Naturwissenschaften, *57*: 514–521, 1970.
69. Patuela, M. C. and Friend, C. Tissue culture studies on murine virus-induced leukemia cells: Isolation of single cell in agar-liquid medium. Cancer Res., *27*: 726–730, 1967.
70. Paul, J., Gilmour, R. S., Hickey, I., Harrison, P. R., Affara, N., and Windass, J. Chromosomal non-histone proteins and the transcriptional unit. *In;* W. Nakahara *et al.* (eds.), Differentiation and Control of Malignancy of Tumor Cells, Univ. of Tokyo Press, Tokyo, pp. 19–35, 1974.
71. Pope, J. H. The isolation of a mouse leukemia virus resembling Friend virus. Aust. J. Exp. Biol., Med., *40*: 263–269, 1962.
72. Prasad, K. N. and Hsie, A. W. Morphologic differentiation of mouse neuroblastoma cells induced *in vitro* by dibutyryl adenosine 3′: 5′-cyclic monophosphate. Nature New Biol., *233*: 141–142, 1971.
73. Prasad, K. N., Waymire, J. C., and Weiner, N. A further study on the morphology and biochemistry of X ray and dibutyryl cyclic AMP-induced differentiated neuroblastoma cells in culture. Exp. Cell Res., *74*: 110–114, 1972.
74. Prasad, K. N., Sahu, S. K., and Kumar, S. Relationship between cyclic AMP level and differentiation of neuroblastoma cells in culture. *In;* W. Nakahara *et al.* (eds.), Differentiation and Control of Malignancy of Tumor Cells, Univ. of Tokyo Press, Tokyo, pp. 287–309, 1974.
75. Preobrazhensky, A. A. and Ovchinnikov, L. P. RNA-binding protein from rabbit reticulocyte extract. FEBS Letters, 1974, in press.
76. Puck, T. T., Waldron, C. A., and Hsie, A. W. Membrane dynamics in the action of dibutyryl adenosine 3′: 5′-cyclic monophosphate and testosterone on mammalian cells. Proc. Natl. Acad. Sci. U.S., *69*: 1943–1947, 1972.
77. Rauscher, F. J. A virus-induced disease of mice characterized by erythrocytopoiesis and lymphoid leukemia. J. Natl. Cancer Inst., *29*: 515–543, 1962.
78. Reeves, B. R., Lobb, D. S., and Lawler, S. D. Identity of the abnormal F-group chromosome associated with polycythemia vera. Humangenetik, *14*: 159–161, 1972.
79. Riedel, V., Gerisch, G., Müller, E., and Beug, H. Defective cyclic adenosine-3′,5′-phosphate-phosphodiesterase regulation in morphogenetic mutants of *Dictyostelium discoideum*. J. Mol. Biol., *74*: 573–585, 1973.
80. Rossi, G. B. and Friend, C. Erythrocytic maturation of (Friend) virus-induced leukemic cells in spleen clones. Proc. Natl. Acad. Sci. U.S., *58*: 1373–1380, 1967.
81. Ross, J., Ikawa, Y., and Leder, P. Globin messenger-RNA induction during erythroid differentiation of cultured leukemia cells. Proc. Natl. Acad. Sci. U.S., *69*: 3620–3623, 1972.
82. Rowe, W. P. and Sato, H. Genetic mapping of the Fv-1 locus of the mouse. Science, *180*: 640–641, 1973.
83. Rowe, W. P., Humphrey, J. B., and Lilly, F. A major genetic locus affecting resistance to infection with murine leukemia viruses. J. Exp. Med., *137*: 850–853, 1973.
84. Rudland, P. S. Control of translation in cultured cells: Continued syntheses accumulation of messenger RNA in nondividing cultures. Proc. Natl. Acad. Sci. U.S., *71*: 750–754, 1974.
85. Scher, W., Preisler, H. D., and Friend, C. Hemoglobin synthesis in murine virus-induced leukemic cells *in vitro*. III. Effects of 5-bromo-2′-deoxyuridine, dimethylformamide and dimethylsulfoxide. J. Cell Physiol., *81*: 63–70, 1973.

86. Schooley, J. C. and Garcia, J. F. Some properties of serum obtained from rabbits immunized with human urinary erythropoietin. Blood, *25*: 204–217, 1965.
87. Schoolman, H. M., Spurrier, W., Schwartz, S. O., and Szants, P. B. Studies in leukemia. VII. The induction of leukemia in Swiss mice by means of cell-free filtrates of leukemic mouse brain. Blood, *12*: 694–700, 1957.
88. Schubert, D. and Jacob, F. 5-Bromodeoxyuridine-induced differentiation of a neuroblastoma. Proc. Natl. Acad. Sci. U.S., *67*: 247–254, 1970.
89. Soule, H. D., Albert, S., Wolf, P. L., and Stansly, P. G. Erythropoietic differentiation of stable cell lines derived from hematopoietic organs of mice with virus-induced leukemia. Exp. Cell Res., *42*: 380–383, 1966.
90. Spirin, A. S. Informosomes. Eur. J. Biochem., *10*: 20–35, 1969.
91. Stansly, P. G. and Soule, H. D. Transplantation and cell-free transmission of a reticulum-cell sarcoma in BALB/c mice. J. Natl. Cancer Inst., *29*: 1083–1105, 1962.
92. Steinheider, G., Melderis, H., and Ostertag, W. Mammalian embryonic hemoglobins *In;* H. Martin and L. Novicki (eds.), Synthesis, Structure and Function of Hemoglobin, pp. 225–235, J. F. Lehmann's Verlag, Munich, 1972.
93. Steitz, J. A., Dube, S. K., and Rudland, P. S. Control of translation by T4 phage: Altered ribosome binding at R17 initiation sites. Nature, *226*: 824–827, 1970.
94. Stellwagen, R. H. and Tomkins, G. M. Preferential inhibition by 5-bromodeoxyuridine of the synthesis of tyrosine aminotransferase in hepatoma cell cultures. J. Mol. Biol., *56*: 167–182, 1971.
95. Stepherson, J. R., Axelrad, A. A., McLeod, D. L., and Shreeve, M. M. Induction of colonies of hemoglobin-synthesizing cells by erythropoietin *in vitro*. Proc. Natl. Acad. Sci. U.S., *68*: 1542–1546, 1971.
96. Stockdale, F., Okazaki, K., Nameroff, M., and Holtzer, H. 5-Bromodeoxyuridine: Effect on myogenesis *in vitro*. Science, *146*: 533–535, 1964.
97. Tambourin, P. and Wending, F. Malignant transformation and erythroid differentiation by polycythemia-inducing Friend virus. Nature New Biol., *234*: 230–233, 1971.
98. Tsuboi, A. and Baserga, R. Effect of 5-bromo-2-deoxyuridine on transport of deoxyglucose and cycloleucine in 3T6 fibroblasts. Cancer Res., *33*: 1326–1330, 1973.
99. Till, J. E. and McCulloch, E. A. A direct measurement of the radiation sensitivity of normal mouse bone marrow cells. Radiat. Res., *14*: 213–222, 1961.
100. Waymire, J. C., Weiner, N., and Prasad, K. N. Regulation of tyrosine hydroxylase activity in cultured mouse neuroblastoma cells: Elevation induced by analogs of adenosine 3′: 5′-cyclic monophosphate. Proc. Natl. Acad. Sci. U.S., *69*: 2241–2245, 1972.
101. Weintraub, H., Campbell, G. L. M., and Holtzer, H. Identification of a developmental program using bromodeoxyuridine. J. Mol. Biol., *70*: 337–350, 1972.
102. Weimann, B. J., Kluge, N., Dube, S. K., von Ehrenstein, G., Gaedicke, G., Kind, J., Krieg, J. C., Melderis, H., and Ostertag, W. Differentiation and transformation of hematopoietic cell in culture. *In;* R. Neth, R. Gallo, and F. Stohlman (eds.), Modern Trends in Human Leukemia-Biological, Biochemical and Virological Aspects, J. F. Lehmann's Verlag, Munich, 1973, in press.
103. Wettstein, F. O., Noll, H., and Penman, S. Effect of cycloheximide on ribosomal aggregates engaged in protein synthesis *in vitro*. Biochim. Biophys. Acta, *87*: 525–528, 1964.
104. White, J. C., Ellis, M. J., Coleman, P. N., Beaven, G. H., Gratzer, W. B., Shooter,

E. M., and Skinner, E. R. An unstable haemoglobin associated with some cases of leukaemia. Brit. J. Haematol., *6*: 171–177, 1960.

105. Willingham, M. C., Johnson, G. S., and Pastan, I. Control of DNA synthesis and mitosis in 3T3 cells by cyclic AMP. Biochem. Biophys. Res. Commun., *48*: 743–748, 1972.

106. Wilt, F. H. and Anderson, M. The action of 5-bromodeoxyuridine on differentiation. Dev. Biol., *28*: 443–447, 1972.

Discussion of Paper by Drs. Dube et al.

DR. HOLTZER: Why just one burst of hemoglobin mRNA?

DR. DUBE: Of course, this would be expected on the basis of your theory of quantal mitosis, *i.e.*, the entire packaged program for further differentiation is established in a single mitosis. We have some evidence that in DMSO induced differentiation, a minor fraction of cells is refractory to DMSO and is not synthesizing any hemoglobin and globin mRNA whereas the major fraction synthesizes the amount of globin mRNA expected for normal differentiation, *i.e.*, similar in amount to that in mouse reticulocytes. This suggests an all or none phenomena in differentiation in one single cell cycle as was suggested by you recently on theoretical grounds.

DR. RUTTER: You mentioned that etiocholanolone induces differentiation. Do other steroids, for example, glucocorticoids exert a similar effect?

DR. DUBE: We have not checked other steroids but we would suspect that some other steroids are even better inducers of erythroid differentiation.

DR. RUTTER: Is it possible that the hemoglobin mRNA synthesis (a pulse in your graph) could come from heterogeneous mRNA containing hemoglobin mRNA sequence as described by Dr. Paul, this morning. Then, is it further possible that this effect is solely or largely post-transcriptional?

DR. DUBE: Yes, it is possible that post-transcriptional mechanisms are important but we have to make several assumptions to accomodate the absolute increase in globin mRNA solely by post-transcriptional events especially since we measure newly labeled mRNA.

DR. JOHNSON: Are the effects you observe specific for cyclic AMP or can you see the same effects with other adenine derivations?

DR. DUBE: We have not checked other adenine derivations.

DR. JOHNSON: Do you feel this represents a specific regulatory function for cyclic AMP or a general nucleotide effect?

DR. DUBE: Yes, our results would argue in favor of a specific regulatory function for cyclic AMP. I do not think that this is a general nucleotide effect.

DR. OHNO: Have you studied "heme" synthesis? The globin synthesis is said to be dependent upon the heme synthesis.

DR. DUBE: We have not studied heme synthesis directly but Dr. Friend has published some work that heme synthesis is elevated in noninduced cells and increases during induction. Our experiments with erythroleukemic cells adapted to serum free growth allow us to suggest some possible involvement of heme synthesis in induction. These cells can only be induced if both Fe and DMSO are present. Secondly if these cells are grown for long periods without added iron very little induction can be obtained even if Fe and DMSO are added suggesting that variants which are dificient in heme synthesis are favored under these selective (low Fe) conditions. We have some other evidence to support this conclusion.

DR. IKAWA: In both Friend's and the leukemia lines of your group the translational control of globin mRNA production was cited. However, in my Friend leukemia lines, which do not undergo erythroid differentiation in the routine passage, only 2.5 hr after actinomycin D treatment, the accumulation of globin mRNA stopped. Accordingly we interpreted the whole process under the transcriptional control.

DR. DUBE: Of course we think that transcriptional control mechanisms do indeed play an important role during induction of erythroid differentiation in these cells. However, we also think that in addition, the data argue in favor of translational control mechanisms. In the uninduced cell, although the amount of globin mRNA is relatively very small, a large proportion of it is present in nucleoprotein particles and not in polysomes. In induced cells we have to take into consideration 2 points: (1) the time lag between transcription and translation of globin mRNA and (2) the presence of globin mRNA in nucleoprotein particles during early phase of induction. As will be pointed out in the second talk of our group (*66*) there is also an induction of virus synthesis, approximately 10–100 fold above control level at the same time when globin mRNA appears. Viral mRNA is obviously being actively translated during this time, but not globin mRNA. This suggests that the lack of globin synthesis during this time is not due to a general translational block in the cell but represents a specific translational control.

DR. TOMKINS: Does erythropoietin have an effect on your system?

DR. DUBE: Erythropoietin has no effect on DMSO induced differentiation. Also in cells which are grown under serum free conditions erythropoietin by itself or in combination with Fe is unable to induce erythroid differentiation.

Dr. Tomkins: Since AT does not inhibit globin synthesis does it mean that globin induction is independent of DNA synthesis.

Dr. Dube: AT does not interfere markedly with cellular DNA synthesis. We presume that cellular repair systems take care of terminally incorporated AT. Virus release and possibly viral replication are inhibited by this substance.

Abnormal Gene Expression on Phenotype of Special Protease in Hepatomas

Nobuhiko KATUNUMA and Yoshiko BANNO

Department of Enzyme Chemistry, Institute for Enzyme Research, School of Medicine, Tokushima University, Tokushima, Japan

Abstract: The interrelated regulation between biosynthesis and degradation of enzyme protein is important for maintaining a state of dynamic equilibrium in every organ, but the mechanisms of intracellular enzyme degradation are unknown.

A proteolytic activity capable of inactivating several pyridoxal enzymes has been detected in rat liver and highly purified from heavy mitochondrial fraction of the liver. The liver protease possesses many properties common to the group-specific protease of other organs. However, the liver enzyme appears to differ from the proteases from other tissues in protein structure based on molecular weight, electrophoretic pattern, and ion-exchange column chromatographic pattern. It is well-known that many pyridoxal enzymes located in the liver respond quite sensitively to physiological conditions, such as dietary, hormonal, or nervous conditions. Regenerating liver and neonatal liver showed very low protease activity. These data suggest that there may be an important relation between the protease activity and cell-division cycle. Since it is possible to consider that the imbalance between biosynthesis and degradation of protein is one of the most important characteristics of cancer cells, it may be of significance to study the protease in various hepatomas of different growth rates. Judging from many physicochemical and enzyme chemical properties, the proteases in various hepatomas appear to be identical with that in normal liver. The reprogramming of gene expression in some hepatomas is manifested by a marked rise in the activity of the protease, but no organ-specific isozyme shift toward the muscle-type protein and enzymic properties was observed.

Except for the most rapidly growing hepatoma group, a reciprocal relation between ornithine transaminase (OTA) and protease activities was observed. A double-logarithmic plot of the two activities yields a straight line. Linearity of the double-logarithmic plots suggests the possibility that the level of OTA is controlled by the inactivating enzyme.

The level in normal liver is not on the line or on an extension of the line. It is possible to assume that the intracellular OTA level is not only simply controlled by the protease activity but also by other additional regulatory mechanisms.

It is important to note that normal liver differs completely from slow-growing minimum-deviated hepatomas in protease activity as a marker enzyme.

It has been proved that all intracellular enzyme proteins are subject to turnover, and exhaustive studies using combined immunologic and isotopic techniques have shown that all proteins of rat liver, including various enzymes, are continuously synthesized and degraded at rates that differ for each specific protein. The occurrence of intracellular protein degradation as an important feature of the turnover of proteins in mammals is known. Thus interrelated regulation between biosynthesis and degradation of the enzyme protein is important in maintaining a state of dynamic equilibrium in every organ. Intracellular enzyme levels are decided by the balance of biosynthesis and degradation rates.

It is possible to assume that a cancerous cell expresses an imbalance of protein catabolism and anabolism regarding certain groups of enzyme. Many examples of phenotype abnormality in hepatomas have been demonstrated in the pyridoxal enzyme group. In this connection, it is interesting to note that the special protease which acts on pyridoxal enzymes, as one of the representative examples of catabolic enzymes, shows abnormal phenotype expression in various hepatomas with different growth rate.

However, this important aspect of protein degradation at the molecular level has been neglected in comparison with the interest shown in protein synthesis. It is still not known what kinds of proteases participate in the degradation, what substrate forms are susceptible to the protease (the nature of the trigger reaction for enzyme protein degradation in physiological conditions), what is the process of enzyme protein degradation by the protease or what control mechanisms exist for the reaction.

Several new proteases which exhibit substrate specificity and respond to growth conditions were recently discovered in various organs of rat. We termed these

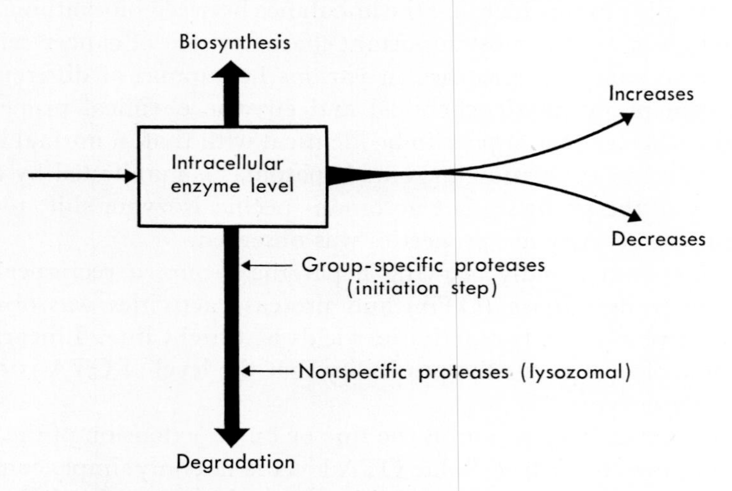

FIG. 1. Bioequilibrium of cellular enzyme level.

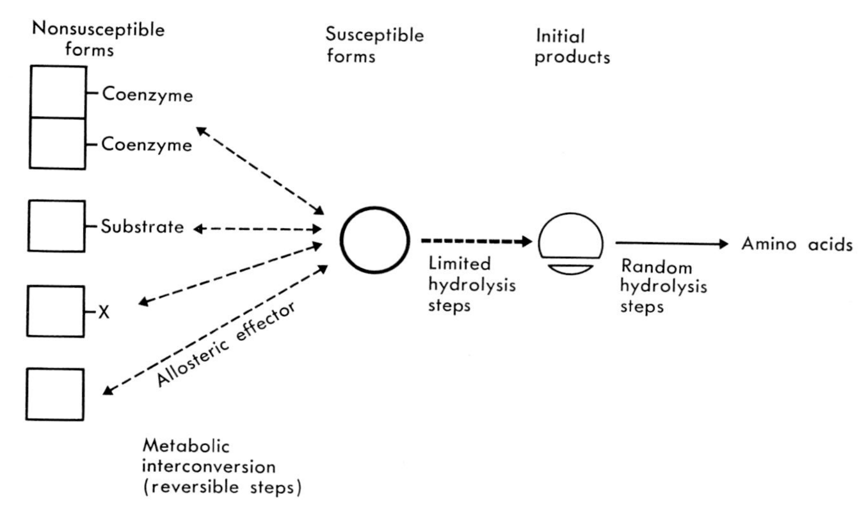

FIG. 2. Hypothetical initiating mechanisms of intracellular degradation of pyridoxal enzymes.

group-specific proteases (GSP). These are prime candidates for the initiating role in intracellular enzyme degradation (*2–4, 8*) (Figs. 1 and 2).

It was anticipated that similar proteases might be present in various hepatomas. The objects of this present investigation are to compare the proteases in various hepatomas and liver, and to clarify the relation between the growth rate of hepatomas and the activity of the protease, and also to confirm the inverse relation between the protease activity and intracellular ornithine transaminase (OTA) content as one of the substrates. The assay systems were used according to the method illustrated in Fig. 3.

1. Comparison of Group-specific Proteases for Pyridoxal Enzymes from Various Organs of Rat

There have been few previous reports of intracellular proteases which act in the alkaline pH range and no previous reports on proteases which show specificity for special groups of proteases. We have obtained evidence for such special groups of enzyme proteins, specifically for several such intracellular proteases in liver, skeletal muscle, muscle layer of the intestine, and mucosa layer of the intestine. All of these proteases were purified to homogeneity and the protease from the muscle layer of the intestine was crystallized. Several important properties are compared in Table 1. On the basis of these properties, the proteases located in different organs are regarded as a group of organ-specific isoenzymes. The present proteases have quite different properties from many intracellular proteases which have been reported under the term of lysozomal proteases (*1*).

TABLE 1. Comparison of Group-specific Proteases in Various Organs

Properties	Skeletal muscle	Intestine		Liver
		Muscle layer	Mucosa layer	
Common properties				
1. Substrate specificity for pyridoxal enzymes	High	High	High	High
2. Susceptibility of trypsin substrate (TAME)	0	0	0	0
3. Susceptibility of chymotrypsin substrate (ATEE)	Weak	Very strong	Weak	Weak
4. Inhibition by synthetic chymotrypsin inhibitor	No	No	No	No
5. Optimum pH (alkaline)	9.0	9.0	8.6	8.6
6. Protein nature inhibitor	Exists	Exists	Exists	Exists
7. Coenzyme protection	Exists	Exists	Exists	Exists
8. Influence by diet control	Exists	Exists	No	Exists
9. SH inhibitor	No effect	No effect	No effect	No effect
10. Histidine modification (photooxidation)		No activity		
11. Modification of seryl-OH by DFP	Inhibited	Inhibited	Inhibited	Inhibited
Different properties				
1. Molecular weight	12,000–14,000	25,000	30,000	16,000
2. Elution from DE-52 column	0.05 M	0.005 M	0.25 M	0.75 M (after protamine treatment 0.1 M)
3. Reaction with GSP antibody from intestine (muscle layer)	—	+	—	—
4. Susceptibility of ATEE[a]	30%	1,000%	25%	30%
5. Susceptibility of TEE[a]	0	100	0	0
6. Catalytic speed	Very fast	Very slow	Very fast	Fast
7. Effect by Ca^{2+}	Inhibited	No effect		No effect

DFP, diisopropylfluorophosphate. GSP, group-specific protease. [a] 100% with OTA as a substrate.

2. *Group-specific Protease in Normal Liver*

1) Properties

A proteolytic activity capable of inactivating several pyridoxal enzymes has been detected in rat liver and was highly purified from heavy mitochondrial fraction of the liver. The liver protease possesses many properties common to group-specific proteases from other organs: the optimum pH is on the alkaline side, it shows relative specificity for pyridoxal enzymes, *i.e.*, the apo-OTA (dimer) is susceptible to the protease, but both holo-OTA (tetramer) of the pyridoxal phosphate and pyridoxamine phosphate forms are nonsusceptible, and it is a kind of limited proteolytic enzyme. However, the liver enzyme appears to differ from the group-specific proteases from other tissues in protein structure, based on molecular weight, electrophoretic pattern, immunoanalysis and the elution pattern on ion-exchange column chromatography (see Table 1). It is important to identify the trigger mechanism starting intracellular enzyme degradation. In the case of OTA as a substrate, holo

TABLE 2. Effect of Coenzyme Derivatives and Substrate Derivatives on Conformation and Susceptibility of Apo-OTA to the Protease

Cofactor	Susceptibility (%)	Conformation
None	100	Dimer
PALP	6	Tetramer
PAMP	7	Tetramer
PINP	90	Dimer
PAL	88	Dimer
PAM	89	Dimer
Orn	86	Dimer
α-KG	86	Dimer
Gly	92	Dimer
Lys	97	Dimer

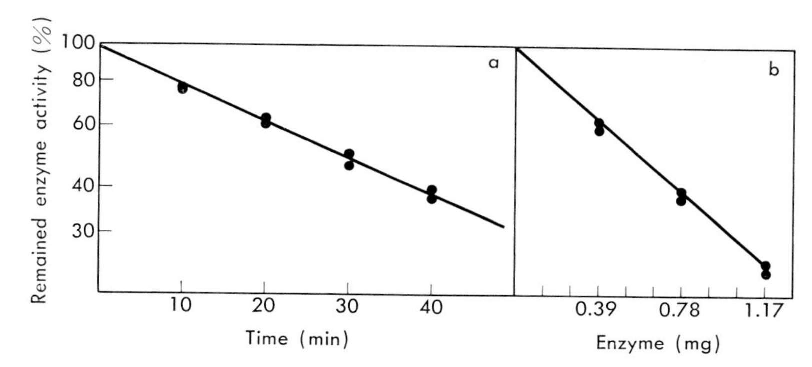

FIG. 3. Time course (a) and dose-response curve (b) for group-specific protease, and definition of the activity. The reaction mixture (final volume, 0.3 ml) contained 70 μmoles of KPB (potassium phosphate buffer) pH 8.5, 40–60 units (0.1 mg) of substrate enzymes, and a suitable amount of group-specific protease preparation. The enzyme was incubated at 37°C, and the reaction was stopped by 10-fold dilution with cold buffer. Then the remaining activity of the substrate enzyme was assayed, and the percentage inactivation was calculated. One unit of the enzyme is defined as the amount inactivating 50% of the substrate enzyme in 30 min under these conditions.

forms (PALP and PAMP forms) consist of a tetramer and are not susceptible to the protease, while the apo form is a dimer and is susceptible to the protease (Table 2). It might be important for the binding site for the relevant coenzymes to be vacant, and at the same time the susceptibility of the substrate may depend upon differences in the tertiary structure between apo and holo forms of the substrate protein (*4*). These forms are interconvertible under biological conditions. Thus the initiating protease which catalyzes limited proteolysis acts on the susceptible form of substrate enzymes. Our hypothesis for intracellular enzyme degradation is introduced in Fig. 2 (*5*).

2) Regulation of liver protease activity

The liver appears to play an important role in the turnover of pyridoxal enzymes. It is well known that many pyridoxal enzymes located in the liver respond

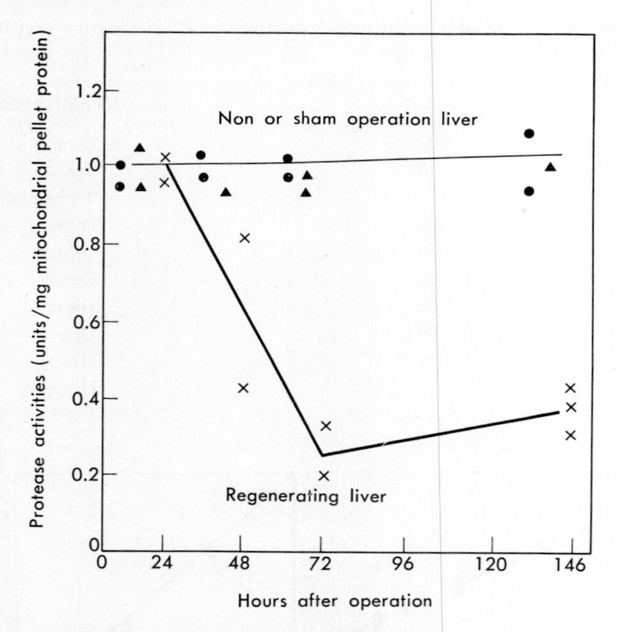

FIG. 4. Changes of the group-specific protease in regenerating liver. At the times indicated after operation, livers were collected. Ten percent mitochondrial suspensions in 0.05 M KPB (w/v) (pH 7.5) were sonicated and the precipitates were collected by centrifugation at 10,000 $\times g$. They were suspended in 0.5 M KPB (pH 8.5) to make a 10% suspension. The protease activities were assayed by the method described in Fig. 3.

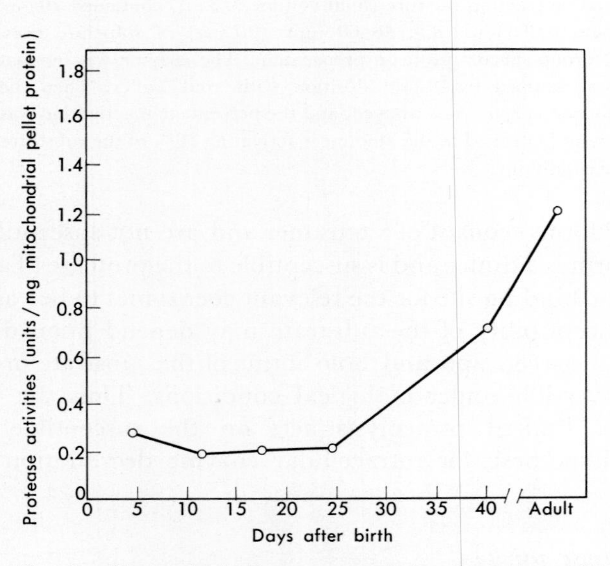

FIG. 5. Changes in the liver protease during development. Livers of neonatal rats were collected at various times after birth and divided into 2 groups. Precipitates of mitochondrial sonicate were suspended in 0.5 M KPB (pH 8.5) by the same method as in Fig. 4.

quite sensitively to physiological conditions, such as dietary, hormonal, or nervous conditions. Regenerating liver and neonatal liver showed very low protease activity (*4*). The changes of the liver protease in regenerating liver are shown in Fig. 4. Two days after partial hepatectomy, marked decreases of the protease activity were observed. Figure 5 shows the changes of the protease activity during development. The neonatal protease activity (within 3 weeks) is very low and then increases rapidly to the adult level. It can be speculated that there is a relation between the appearance of the protease activity and the cell-division cycle. The preparations illustrated in Figs. 4, 5, 6, and 7 for the assay of protease activity in liver and hepatomas were assayed on 0.5 M KPB extracts of centrifuged pellets after sonication of mitochondria.

3. *Group-specific Protease in Morris Hepatomas of Different Growth Rates*

1) Comparison of the properties of hepatoma proteases

Since it is possible to consider that the imbalance between biosynthesis and degradation of proteins, including enzymes, is one of the most important characteristics in cancer cells, it may be significant to study the relationship between the protease pattern and pyridoxal enzymes which are susceptible to the protease in various hepatomas of different growth rates. In the present investigation, varying activities of protease capable of inactivating OTA were observed in nearly all the Morris hepatomas tested. Judging from the physicochemical and enzyme chemical properties, the proteases in various hepatomas appear to be identical with that in normal liver. The reprogramming of gene expression in almost all of the hepatomas is manifested by a marked rise in the activity of the protease, but no organ-specific isozyme shift toward muscle-type protein and enzymic properties was observed (*6*).

TABLE 3. Purification of OTA-inactivating Enzyme from Hepatoma 8999

Enzyme	Total activity (units)	Total protein (mg)	Specific activity (units/mg)	Purity
Homogenate	123	15,000	0.008	1
Mitochondria	165	2,200	0.073	9
Pellet after sonication	1,125	278	4.041	493
0.6 M KPB extracts	526	190	2.765	337
Extracts from acetone 60% precipitate	138	72	3.307	403
0.75 M KCl eluates from DE-52	100	1.65	60.600	7,400
Sephadex G-75 fraction after protamine treatment	555	0.68	817.000	10,220

The purification of OTA-inactivating protease in hepatoma was carried out using the same method as for normal liver protease. Table 3 shows the applicability of the procedure for normal liver to the hepatomas. The purified protease from hepatoma 8999 was homogeneous judging from acrylamide gel disc electrophorasis.

In order to gain insight into the nature of the inactivating system, various properties of the enzyme were examined using the purified protease. The pH optimum was between 8.5 and 9.0. No effect on the activity of the liver enzyme was observed on addition of L-cystine, mercaptoethanol, or *p*-chloromercuribenzoate, but modification of seryl-OH by diisopropyl fluorophosphate caused complete loss of the activity. Thus seryl-OH is related to the activity, but thiol-SH is not. On the other hand, the active sites of almost all cathepsins involve thiol-SH as an active site (*1*). The inactivating enzyme attacked only apo-OTA, but not the holo form, so that, the addition of pyridoxal phosphate or pyridoxamine phosphate showed a protective effect against inactivation by the protease, as in the case of the enzyme from normal liver. Since in the inactivating enzyme activity was markedly higher in almost hepatomas than that in the liver, it was of interest to investigate whether the difference in gene expression, as revealed by measuring the enzyme activity, was also reflected in different properties of the enzyme in the neoplasms. For this reason, the mitochondrial proteases from liver and hepatomas were examined by Sephadex G-75 column chromatography. The results indicate that the OTA-inactivating protease in normal liver was eluted before cytochrome C. In all hepatomas examined including those of medium growth rate (8999), as well as rapidly growing tumors (7777 and 3924A), the proteases were eluted in the same position as the normal liver enzyme after addition of sufficient protamine sulfate. Thus, the molecular weight of the OTA-inactivating proteases of hepatomas is estimated as 14,000–16,000.

A synthetic substrate for trypsin, *p*-toluene sulfonyl-L-arginine methyl ester (TAME), is not attacked at all and a synthetic substrate for chymotrypsin, N-acetyl L-tyrosine ethyl ester (ATEE), is hydrolyzed 20 to 30% as rapidly by the hepatoma enzymes as compared with OTA inactivation (100%). These properties are the same as those of the enzyme from normal liver as itemized in Table 4. Much more work remains to be done in establishing the substrate specificity of the hepatoma OTA-

TABLE 4. Identical Nature of OTA-inactivating Enzymes in Liver and Hepatomas

1. Localization: membrane-bound in mitochondria	
2. Properties of the enzyme protein	
(a) Solubilization: 0.6 M KPB, pH 8.0	
(b) Elution pattern on DE-52 column chromatography: 0.75 M KCl in 0.005 M KPB, pH 7.5	
(c) Molecular weight judging from the pattern of Sephadex G-75 column chromatography	
i) Before protamine sulfate treatment: the same position as blue dextran (M.W.=2,000,000)	
ii) After protamine sulfate treatment: before cytochrome C (M.W.=12,500)	
(d) Both enzymes can be purified by the same methods.	
3. Properties of the enzyme	
(a) Substrate susceptibility	
Apo-OTA	100%
Holo-OTA	0
MDH	0
GDH	0
TAME	0
ATEE	
Before protamine treatment	50
After protamine treatment	20
(b) Optimum pH (alkaline), pH 8.6–9.2	

MDH, malate dehydrogenase; GDH, glutamate dehydrogenase.

inactivating protease, but the results indicate that it does not attack TAME, malate dehydrogenase, or glutamate dehydrogenase. If the OTA-inactivating protease in hepatomas is assumed to be identical with the liver protease, the liver protease has relative specificity for the pyridoxal enzyme group. It might be considered that the OTA-inactivating protease in hepatomas represents the same group-specificity for pyridoxal enzymes as that in normal liver.

The similarities of OTA-inactivating enzymes in liver and hepatomas are summarized in Table 4. Therefore, the existence of an isozyme shift toward muscle type OTA-inactivating protease seems unlikely (see Table 1).

2) *Relationship between OTA-inactivating protease and growth rates of hepatomas*

Thus, the reprogramming of gene expression in hepatomas is manifested by a marked increase in the activity of the OTA-inactivating protease. It is very interesting that slow- and medium-growth rate hepatomas contain extremely high activity of this protease, more than 100 times that of normal liver, and all of the very rapidly growing hepatomas have low activity of this enzyme. An inverse relation was observed between the protease activity and the growth rates of various hepatomas of different growth rates, as shown in Fig. 6.

It has been proved that normal regenerating liver, fetal liver, and neonetal liver show very low inactivating activity (*4*), and furthermore in the case of yeast, no activity was observed in the lag and logarithmic phases of growth but the enzyme

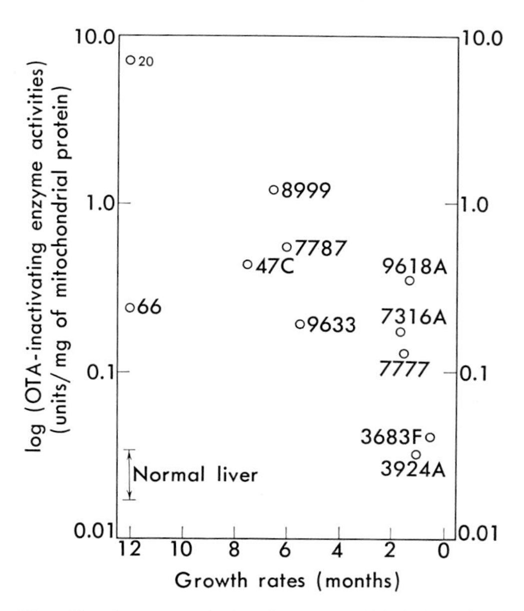

FIG. 6. Inverse relation between OTA-inactivating enzyme activities and growth rates of various hepatomas (units/mg mitochondrial protein). Partial purification of the inactivating enzymes was performed according to the method described in Fig. 4 to compare the inactivating enzyme levels in various hepatomas with different growth rates. The standard assay and the definition of enzyme activity are described in Fig. 3.

activity appeared in the stationary phase (*7*). Also, the inactivating enzyme activities in rapidly growing hepatomas were lower than in medium- and slow-growing hepatomas, as shown in Fig. 6. It may be speculated from these phenomena that the cells in the cell-division phases have low enzyme activity and the cells in resting phases show high activity.

Much current thinking regarding gene expression in liver neoplasia is concerned with the quantitative expression of structural genes and the relationship to growth rate. An important difference in gene expression between hepatomas and normal liver might be that the protease activities in almost of all hepatomas are much higher than in normal liver. It may be speculated that a certain repression is acting on the gene expression of the protease in normal liver, whereas the expression in hepatomas is free from such repression.

In all hepatomas examined up to date, irrespective of malignancy and growth rate, high activities of the OTA-inactivating protease was present. It is worth considering the possible use of this and other isozymes as marker enzymes for liver neoplasia and they may have possible diagnostic use in liver biopsies or in plasma samples, if the enzyme does appear in the blood (*10*, *11*).

4. *Relationship between OTA Activities and OTA-inactivating Protease Activities*

OTA activities in hepatomas of different growth rates were reported by us (*9*). The relationship between OTA activity and the inactivating enzyme activity in various hepatomas are illustrated in Fig. 7.

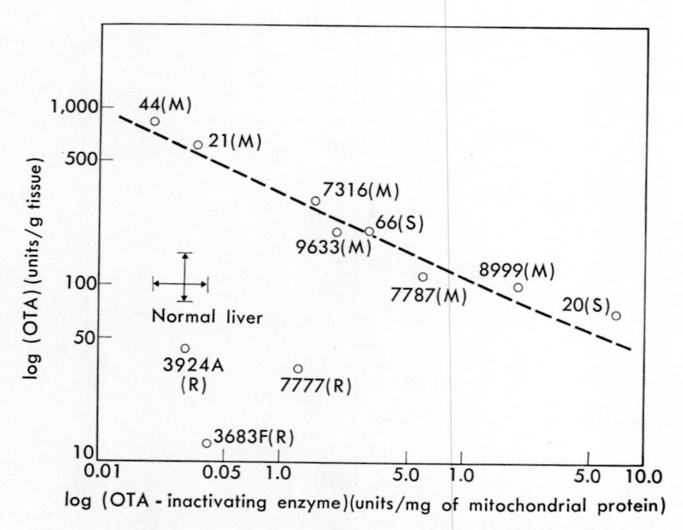

FIG. 7. Reciprocal relation between OTA activities and OTA-inactivating protease activities in various hepatomas of different growth rates. The data on OTA activities in hepatomas of different growth rates were taken from another paper by Tomino *et al.* (*9*). The data on the inactivating enzyme activities were taken from Fig. 6. Double-logarithmic plots were used to obtain the relation between OTA activities and -inactivating enzyme activities. R: rapidly growing hepatoma; M: medium-growing hepatoma; S: slowly growing hepatoma.

Except in the most rapidly growing hepatoma group, there seems to exist a reciprocal relation between both activities. A double-logarithmic plot of the two activities yields a straight line. The linearity of the double-logarithmic plots permits application of the following equation to express the relation between the both enzyme activities.

$$(\text{OTA}) = A/(\text{inactivating enzyme})^k$$

This equation suggests the possibility that the level of OTA is controlled by the inactivating protease in some groups of hepatomas.

Rapidly growing hepatomas show very weak activities of both enzymes, because both enzymes are located in mitochondria and rapidly growing hepatomas contain very few mitochondria in their cells.

The level in normal liver is not on the line or on an extension of the line. It is possible to assume that the intracellular OTA level is not simply controlled by the protease but also by other additional regulatory mechanisms.

The present results draw attention to a factor that may play an important role in regulating the level of gene products, the enzymes, by means of group-specific inactivating protease. It has been assumed that the biochemical phenotype reflects the abnormal operation of gene expression and its linking with malignancy. It is now also necessary to consider the role played by specific proteases that may control the biochemical phenotype, OTA level, by expression of their enzyme degradative activity in hepatomas. Since the steady-state enzyme level depends on both synthesis and degradation, further studies are required to establish the role of this dynamic equilibrium in the phenotypic expression of the biochemical pattern that characterizes hepatomas.

ACKNOWLEDGMENTS

A part of this work on the protease in hepatomas has been carried out jointly with G. Weber, H. P. Morris, and I. Tomino, in G. Weber's laboratory, Indiana Univ., School of Medicine, Indianapolis, U.S.A., and will be published elsewhere.

REFERENCES

1. Barret, A. J. *In;* J. T. Pingle and Honor B. Fell (eds.), Properties of Lysozomal Enzymes, Lysosomes in Biology and Pathology, vol. 2, pp. 245–312, North-Holland Publishing Co., Amsterdam and London, 1969.
2. Katunuma, N., Katsunuma, T., Kominami, E., Suzuki, K., Hamaguchi, Y., Chichibu, K., and Kobayashi, K. Regulation of intracellular enzyme levels by group-specific proteases in various organs. Adv. Enzyme Regulation, *11*: 37–51, 1973.
3. Katunuma, N. Enzyme degradation and its regulation by group-specific proteases in various organs of rats. *In;* B. L. Horecker and E. R. Stadtman (eds.), Current Topics in Cellular Regulation, vol. 7, pp. 175–203, Academic Press, New York and London, 1973.
4. Katunuma, N. New intracellular proteases in various organs and their regulation. *In;* E. H. Fischer (ed.), Metabolic Interconversion of Enzymes, pp. 313–324, Springer Verlag, New York and Berlin, 313–324, 1973.

5. Katunuma, N. Initial mechanisms of intracellular pyridoxal enzyme degradation and group-specific proteases. *In;* R. T. Schimke and N. Katunuma (eds.), Intracellular Protein Turnover (Japan-U.S.A. Scientific Cooperation Seminar), Academic Press, New York, 1974, in press.
6. Katunuma, N., Tomino, I., Morris, H. P., and Weber, G. Ornithine transaminase-inactivating system in hepatomas of different growth rates. Unpublished.
7. Katsunuma, T., Schott, E., Elsasser, S., and Holzer, H. Purification and properties of tryptophan synthetase-inactivating enzymes from yeast. Eur. J. Biochem., *27*: 520–526, 1972.
8. Kominami, E., Kobayashi, K., Kominami, S., and Katunuma, N. Properties of specific protease for pyridoxal enzymes and its biological role. J. Biol. Chem., *247*: 6848–6855, 1972.
9. Tomino, I., Katunuma, N., Morris, H. P., and Weber, G. Imbalance in ornithine metabolism in hepatomas of different growth rates as expressed in behavior of L-ornithine transaminase. Am. J. Cancer Res., *34*: 627–636, 1974.
10. Weber, G. Molecular correlation concept: ordered pattern of gene expression in neoplasia. GANN Monograph on Cancer Research, *13*: 57–77, 1972.
11. Weber, G. The molecular correlation concept: recent advances and implications. *In;* H. Busch (ed.), The Molecular Biology of Cancer, 1973, in press.

Discussion of Paper by Drs. Katunuma and Banno

DR. TOMKINS: Is the concentration of your enzyme constant when expressed in terms of units per mitochondrion? Or does the concentration vary? I ask this since you found low levels of enzyme in cells with few mitochondria.

DR. KATUNUMA: The protease activities are not the same per one mitochondrion. There are big differences from hepatoma to hepatoma per mitochondrion.

DR. DUBE: Your results are very interesting and could be taken to suggest a possible regulatory mechanism involving proteolytic enzymes under different nutritional conditions. This was the point that Dr. Tomkins was also getting at; is it possible that the different levels of proteolytic activity observed reflect varying amounts of inhibitor-enzyme complex? I think that the work of Tsai *et al.* (Eur. J. Biochem., 1973) would also support this notion. Also, is your inhibitor a protein?

DR. KATUNUMA: The muscle protease activity increases with a nonprotein diet, starvation, or denervation. In some purification steps, a marked increase of total units was observed, and therefore there are inhibitors in the relevant organs. We purified one of the inhibitors partially and it was a protein.

DR. THOMPSON: Do the levels of OTA and its inactivating enyme vary in regenerating liver?

DR. KATUNUMA: The level of inactivating enzyme decreases in regenerating liver, but the level of OTA is almost constant per gram weight, while a reciprocal tendency was observed in the case of B_6 deficiency. Since the holoenzyme is not susceptible but the apoenzyme is susceptible to the protease, the apo-holo ratio has more great significance for inactivation by the protease.

DR. WEINHOUSE: What is the nature of the proteolytic activity? Is the enzyme specific to certain peptide bonds; does the enzyme act on the native protein; and how is its activity regulated in the cell?

DR. KATUNUMA: The mode of inactivation is a limited proteolysis, we have much data on that, and the splitting position might be near the binding site of pyridoxal phosphate, as you said. Nonprotein diet, starvation, high protein diet, and denervation caused increases in the muscle protease activity, and fetal liver and regenerating liver showed very low activities.

Ordered and Specific Pattern of Gene Expression in Differentiating and in Neoplastic Cells

George WEBER

Department of Pharmacology, Indiana University School of Medicine, Indianapolis, Indiana, U.S.A.

Abstract: The objective of this paper is to analyze recent advances made by application of the molecular correlation concept as a conceptual and experimental method and to demonstrate the separation of the biochemical pattern of hepatic neoplasia from that of differentiating liver. The results provide evidence for the following conclusions.

1) Ordered pattern of gene expression in cancer cells: enzymatic and metabolic imbalance in cancer cells. In the model system of a spectrum of hepatomas of different growth rates, an ordered pattern in the reprogramming of gene expression is manifested in a progressive imbalance in the activities of opposing key enzymes and metabolic sequences in the synthetic and degradative pathways of carbohydrate, pyrimidine, purine, DNA, ornithine, and cyclic AMP metabolism.

2) Malignancy-linked biochemical imbalance: co-variance of imbalance with tumor growth rate (Class 1). The biochemical and enzymatic imbalance that is co-variant with hepatoma growth rate indicates a link in the metabolic and proliferative expression of the genome as manifested in the different degrees of neoplastic transformation. Enzymes and metabolic pathways that show co-variance with biological malignancy are grouped in Class 1 of the molecular correlation concept and they are called malignancy-linked discriminants.

3) Neoplastic transformation-linked biochemical imbalance: ubiquitous alterations occurring in all hepatomas (Class 2). The biochemical and enzymatic alterations that occur in all tumors are grouped by the molecular correlation concept in Class 2. These alterations indicate a link between the metabolic alterations and the neoplastic conversion of the cell that is characteristic of neoplasia *per se*. These changes then are characteristic of the reprogramming of gene expression that occurs in the malignant transformation and they are called neoplastic transformation-linked discriminants.

4) Isozyme shift. In malignancy the reprogramming of gene expression is also expressed in the shift of isozyme pattern where, as we recognized, there emerges a predominance of the low K_m isozymes. Concurrently, the high K_m, regulatory enzymes diminish. These alterations are the molecular basis, in part at least, for the decreased responsiveness of the cancer cells to regulatory signals.

5) *Specificity of metabolic imbalance to neoplasia.* The discriminating power of the biochemical pattern identified in this laboratory provides quantitative and qualitative discriminants that allow the recognition of differences at the molecular level between rapidly growing hepatomas and rapidly growing embryonic, differentiating or regenerating livers. These results indicate that the ordered pattern of gene expression as manifested in the metabolic and enzymatic imbalance discovered in the hepatoma spectrum is specific to neoplasia. There is no similar pattern of gene expression in fetal, differentiating, regenerating, or normal adult liver. The neoplastic pattern of gene expression is both quantitatively and qualitatively different from that of the various control tissues.

6) *Selective biological advantages of the imbalance in gene expression for cancer cells.* The biochemical imbalance confers selective biological advantages by the progressive predominance of the synthetic pathway over the declining catabolic one in pyrimidine, purine, and nucleic acid metabolism. Through a progressively emerging imbalance in ornithine metabolism the competition of the urea cycle for carbamyl phosphate and aspartate is gradually switched off, allowing an increase in channeling of these precursors to biosynthesis of nucleic acids and polyamines. The imbalance in the cyclic AMP system at the cell membrane favors neoplastic proliferation. The increase in the key enzymes and overall metabolic flow in the pentose phosphate pathway that occurs in all hepatomas should provide the pentose for nucleic acid biosynthesis.

A careful evaluation of the now extensive evidence leads to the conclusion that what is important about cancer is ordered; what is not, is the random element and the diversity.

The objective of this paper is to analyze recent advances made by application of the molecular correlation concept as an experimental and conceptual method in different strategic metabolic areas. Newly discovered aspects of the biochemical imbalance in neoplastic cells will be related to the earlier observations from these laboratories, and the linking of the imbalance with malignant transformation and with the degrees of expression of malignancy will be evaluated.

Attention will be given to the specificity to neoplasia of the ordered metabolic imbalance identified in the spectrum of hepatomas of different growth rates. The characteristic pattern of metabolic imbalance in neoplastic cells will be clearly distinguished from the biochemical pattern observed in differentiating, developing, and regenerating liver, as well as from that observed in normal rat liver.

The selective biological advantages conferred to cancer cells by the metabolic imbalance that we recognized will be outlined in the different areas of intermediary metabolism.

Conceptual Background

There has been splendid therapeutic progress in the clinical treatment of infectious diseases. This success is chiefly due to the advances achieved in the design

TABLE 1. Comparison of Problems in Strategy of Chemotherapy of Infectious and Neoplastic Diseases

Factors	Infectious diseases	Neoplastic diseases
Target cells	Foreign	Host
Immune response	Strong	Weak
Growth fraction	Very high	Quite variable
Total cell kill	Not necessary	Necessary
Selective chemotherapy	Available	Not available
Combination chemotherapy	Not desirable	Highly desirable

From Weber (*30*).

of selectively toxic anti-bacterial chemotherapy. As a result of the judicious use of anti-microbial drugs over the past 30 years the mortality due to infectious diseases has steadily diminished to the present very low figures. In dark contrast stands the steady rise in deaths due to neoplastic diseases. It is a fact that over the past 3 decades the standard treatment modalities, surgery, radiation, and chemotherapy, have so far failed to improve on the average cancer mortality of 50%. However, a comparison of the problems in the strategy of clinical chemotherapy of infectious and neoplastic diseases (Table 1) points out the complexity of the treatment of cancer in man.

Among the many special problems the clinical treatment of a cancer patient poses, the following are emphasized. In the neoplastic diseases we are seeking to selectively destroy the target cells of the host; the immune mechanisms appear to be weak as compared to the highly effective immune response in infectious diseases; the growth fraction is quite variable; and it seems that a total cell kill is necessary to cure neoplasia.

As a result of these and other considerations, for the design of more effective selectively toxic drugs it is vital to elucidate the biochemical pattern of the neoplastic cells, to identify the specificity to neoplasia of the metabolic imbalance, and to demonstrate the biological advantages the biochemical alterations confer to the cancer cells. Progress in this area should put us in a more advantageous position to achieve a rational design of drugs against key enzymes and against the overall pleiotropic cellular mechanism which underlies and maintains the neoplastic phenotype as revealed in the malignant behavior and its linked metabolic expression.

Molecular correlation concept: general and special theory

The molecular correlation concept was developed in my laboratories as a conceptual and experimental method for analyzing the control of gene expression at the molecular level (*17–20*, *22*, *25*, *28*). This approach is based on the assumption that the various cellular functions can be elucidated, the integrative controls analyzed and the phenotypic pattern understood in terms of molecular events. The approaches of the general theory were outlined elsewhere (*25*). The general theory was applied to molecular events linked with various normal cellular functions such as hormonal and nutritional adaptation mechanisms, differentiation and development, and normal cellular proliferative responses such as regeneration (*21*, *24*). As the general theory refers to the molecular events linked with the different normal cellular functions, the special theory refers to the application of the molecular correlation concept to a special case: neoplasia.

The special theory seeks to elucidate the pattern and mechanism of regulation of gene expression in normal and neoplastic growth. The conceptual and theoretical aspects of the special theory were outlined elsewhere (*25*). As a result of application of the molecular correlation concept of neoplasia, an ordered pattern of metabolic imbalance was discovered in carbohydrate, pyrimidine, DNA, purine, ornithine, and cyclic AMP metabolism (*3*, *17–20*, *22*, *25*, *28–30*). The molecular correlation concept was also useful in identifying the link between metabolic imbalance and the degrees in the expression of malignant transformation (*3*, *17–20*, *22*, *25*, *28–30*).

Experimental Background

The experimental testing ground of the molecular correlation concept was a spectrum of liver tumors of different malignancy (*3*, *17–20*, *22*, *25*, *28–30*). The spectrum contains the hepatomas of different growth rates produced by Morris (*12*), and the Novikoff (*13*) and Reuber tumors (*12*), and primary hepatomas in rat (*17*) and in man. This liver tumor system provides a spectrum of neoplasms exhibiting the same histological cell type, but discriminated by different degrees of malignancy.

Malignancy, for the purpose of this discussion, will be defined as "the ability of tumor cells to grow progressively and kill their host" (*10*). The malignancy was measured by biological, morphological, and biochemical criteria, as outlined elsewhere (*22*, *28*). Thus, the malignancy in the tumor spectrum can be determined by evaluating the growth rate of the neoplasms by several independent methods which provide an assurance of quantitation of malignancy. As a result the biochemical parameters can be correlated with the malignancy in the different tumor lines.

Choice of biochemical parameters: the concept of key enzymes

In analyzing the molecular events involved in the modulation and expression of various cell functions, our studies on starvation, differentiation (*21*), insulin (*24*), and steroid action (*27*), led to the recognition that the control of gene expression operates through regulation of certain key enzymes. Thus, metabolic control is achieved by regulation of the amount and activity of opposing and competing key enzymes and opposing and competing overall metabolic pathways (*21*, *24*, *27*). Characteristic features of key enzymes were identified elsewhere (*27*).

The application of the special theory of the molecular correlation concept led to the discovery that in neoplasia also the opposing and competing key enzymes and metabolic pathways are the ones that are linked with the neoplastic transformation *per se* and with the degrees of biological malignancy. The recognition of such correlations, however, can be missed and the main errors that may lead an investigator to overlook existing correlations were outlined elsewhere (*22*).

Ordered Pattern of Gene Expression in Cancer Cells: Enzymatic and Metabolic Imbalance

In the model system of a spectrum of hepatomas of different growth rates, an ordered pattern in the reprogramming of gene expression is manifested in 2 main classes of alterations. In Class 1 are grouped the biochemical parameters that altered in relation to the progressive expression of malignancy in the different tumor lines. In Class 2 are parameters that altered in the same direction in all or most tumors. Alterations in these Classes reveal a link between metabolic imbalance and the neoplastic transformation itself. In Class 3 are grouped the alterations that show no relation to growth rate; they are coincidental changes in gene expression.

In order to examine the ordered nature of the metabolic imbalance an analysis will be given of the reprogramming of gene expression in several metabolic pathways.

1. Carbohydrate metabolism: phenotypic evidence for reprogramming of gene expression in neoplasia

In Table 2 the 6 aspects of the reprogramming of gene expression are outlined. As discussed in detail in earlier publications, it was observed that in parallel with the increase in tumor growth rate there was an increase in the activities of the key glycolytic enzymes and a decrease in those of the key gluconeogenic ones. As a result there emerged a metabolic imbalance that favored glycolysis over gluconeogenesis. Furthermore, as we recognized earlier (*20*), there occurred an isozyme shift resulting in a decrease in the amount of isozymes susceptible to physiological regulatory signals and a decrease in the production of glycolytic isozymes that were not sensitive or much less susceptible to physiological control mechanisms. These alterations were co-variant with growth rate, indicating an imbalance in gene expression that was linked with the degrees in the expression of biological malignancy as measured by growth rate. These alterations in carbohydrate metabolism conferred selective advantages to the cancer cells.

The basic pattern of imbalance discovered in my laboratories in examining

TABLE 2. Carbohydrate Metabolism: Phenotypic Evidence for Reprogramming of Gene Expression in Neoplasia

1. Synthetic enzymes	Key gluconeogenic enzymes[a]	Decreased
2. Catabolic enzymes	Key glycolytic enzymes[b]	Increased
3. Metabolic imbalance	Ratios of key glycolytic/key gluconeogenic enzymes	Increased
4. Isozyme shift	High K_m isozymes[c]	Decreased
	Low K_m isozymes[d]	Increased
5. Relation to malignancy	Alterations are co-variant with growth rate	Malignancy-linked imbalance
6. Biological role	(a) Imbalance in glycolytic/gluconeogenic enzymes leads to increase in glycolysis (b) Isozyme shift leads to decreased responsiveness to physiological controls	Confers selective advantages to cancer cells

[a] Glucose-6-phosphatase (G-6-Pase), fructose-1,6-diphosphatase (FDPase), phosphoenolpyruvate carboxykinase, pyruvate carboxylase. [b] Hexokinase (HK), phosphofructokinase (PFK), pyruvate kinase. [c] Glucokinase, liver-type pyruvate kinase. [d] HK, muscle-type pyruvate kinase.

carbohydrate metabolism in hepatomas proved to be applicable to areas we subsequently investigated, namely, pyrimidine, purine, DNA, pentose phosphate, ornithine, and cyclic AMP metabolism. Some of the most recent results will be discussed as they are relevant to the present theme.

2. DNA metabolism: phenotypic evidence for reprogramming of gene expression in neoplasia

In Table 3 the 6 aspects of the reprogramming of gene expression in neoplasia in DNA metabolism are summarized. There is an imbalance in this metabolic area; however, it contrasts with that observed in carbohydrate metabolism, since here the synthetic enzymes are increased and the catabolic enzymes are decreased in parallel with tumor growth rate. This coordinated enzymatic imbalance results in the imbalance of the opposing overall metabolic pathways that channel thymidine into the synthetic and into the degradative pathways. These alterations

TABLE 3. DNA Metabolism: Phenotypic Evidence for Reprogramming of Gene Expression in Neoplasia

1. Synthetic enzymes	Key enzymes of UMP[a)], TTP[b)], and DNA[c)] synthesis	Increased
2. Degradative enzymes	Key enzymes of UMP and thymidine (TdR) catabolism[d)]	Decreased
3. Enzymatic imbalance	Ratios of synthetic/catabolic enzymes	Increased
4. Metabolic imbalance	Ratios of TdR to DNA/TdR to CO_2 pathways	Increased
5. Relation to malignancy	Alterations are co-variant with growth rate	Malignancy-linked imbalance
6. Biological role	(a) Imbalance in anabolic/catabolic enzymes of UMP metabolism leads to increased *de novo* DNA synthesis (b) Imbalance in anabolic/catabolic enzymes of TdR metabolism leads to increased salvage pathway to DNA synthesis	Confers selective advantages to cancer cells

[a)] Aspartate transcarbamylase, dihydroorotase. [b)] Ribonucleotide reductase, dCMP deaminase, dTMP synthase, TdR kinase, dTMP kinase. [c)] DNA polymerase. [d)] Dihydrouracil dehydrogenase (DH), dihydrothymine DH.

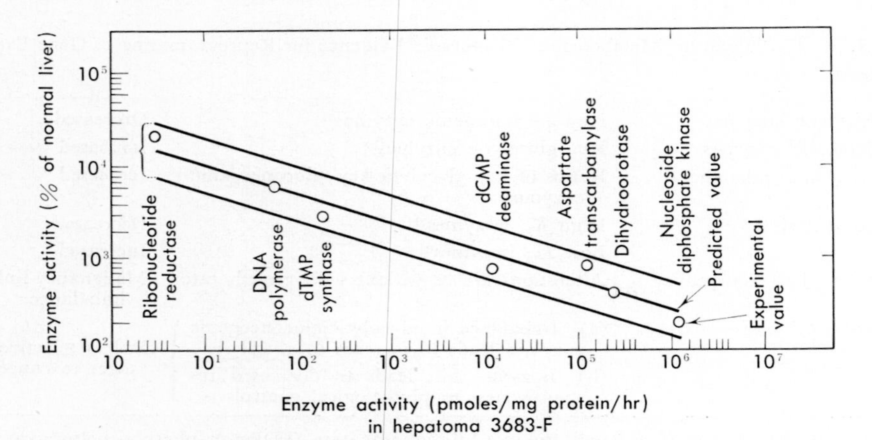

FIG. 1. Correlation of enzyme activity in normal liver with the extent of rise in the rapidly growing hepatoma. $y=ax^b$. a, 21,360; b, -0.31732. Correlation coefficient, -0.97.

are co-variant with tumor growth rate and they confer selective advantages to the cancer cell (*28, 30*).

In a previous publication it was pointed out that the lower the activity of an enzyme was in the resting adult rat liver, the higher it increased in the rapidly growing hepatomas (*21*). A mathematical expression of this statement and the plotting of the results from this earlier report (*21*) are in Fig. 1. The predicted value for the activity of nucleosidediphosphate kinase in the rapidly growing hepatoma 3683F was tested and the value observed (*21, 22*) fits well into the range mathematically predicted by the equation in Fig. 1. Further testing of this relationship is being carried out.

3. Cyclic AMP metabolism in the membrane: phenotypic evidence for reprogramming of gene expression in neoplasia

The sum of evidence for the possible role of the cyclic AMP metabolizing enzymes in liver neoplasia was recently analyzed and the conclusion was drawn that the major, biologically relevant alterations in such cancer cells may take place in the compartment where cyclic AMP is generated, in the plasma membrane (*26, 29*). This assumption is supported by the fact that in the liver cell nearly all the adenylate cyclase and a small, but appreciable fraction of the total cyclic AMP phosphodiesterase (PDE) activity are localized in the plasma membrane. Therefore, it was postulated that the pattern of imbalance in cyclic AMP metabolism in the

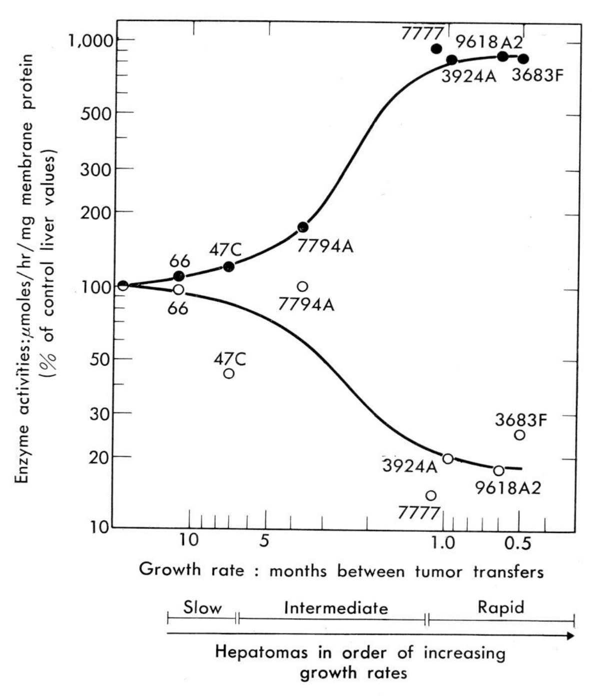

FIG. 2. Correlation of reciprocal behavior of membrane adenylate cyclase and cyclic AMP PDE activities with hepatoma growth rates. ● cyclic AMP PDE; ○ adenylate cyclase.

cancer cells may be detectable not by assaying the activities of total adenylate cyclase in the homogenate or the cyclic AMP PDE activity in the supernatant fluid, but by measuring the activities of the opposing synthetic and catabolic enzymes of cyclic AMP metabolism in the plasma membrane (*26*).

Studies were undertaken to test this hypothesis by determining the activities of adenylate cyclase and cyclic AMP PDE in the isolated plasma membrane from hepatomas of vastly different growth rates (*29*).

The behavior of the opposing enzymes of cyclic AMP synthesis and breakdown in different liver tumors is shown in Fig. 2. The membrane cyclic AMP PDE activity increased and concurrently the adenylate cyclase activity decreased in parallel with the increase in tumor growth rate. In the cell membrane of the rapidly growing hepatomas PDE activity was highly increased in presence of low levels of adenylate cyclase. The ratios of the activities of cyclic AMP PDE/adenylate cyclase yielded a good correlation with the growth rate of the hepatomas over 2 log scales (Fig. 3).

These observations reveal the presence of an imbalance in cyclic AMP metabolism that is manifested in the opposing behavior of the antagonistic enzymes of cyclic AMP metabolism in the tumor plasma membrane. The biochemical imbalance is linked with the expression of the increasing degrees of neoplastic transformation as manifested in a progressive increase in hepatoma growth rate. The reciprocal behavior of plasma membrane adenylate cyclase and cyclic AMP PDE activities is in agreement with the predictions of the molecular correlation concept. The behavior of these enzymes groups them in Class 1.

This malignancy-linked imbalance should result in a metabolic pattern unfavorable to the production of cyclic AMP or to maintenance of cyclic AMP con-

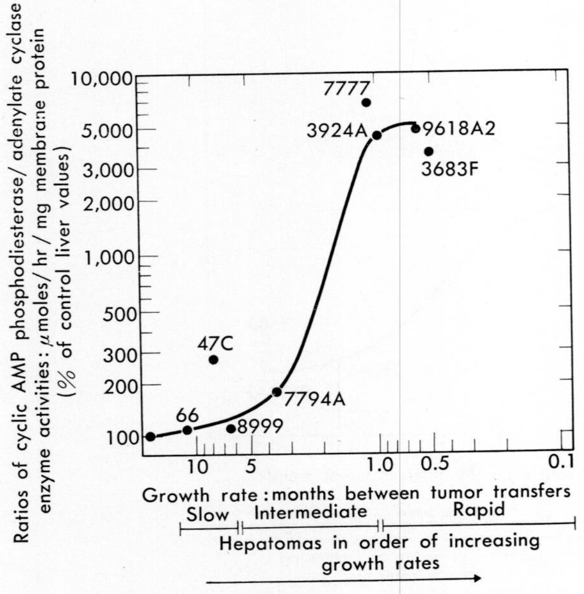

FIG. 3. Correlation of membrane cyclic AMP PDE/adenylate cyclase ratios with hepatoma growth rate.

centration in the plasma membrane of the tumors. This imbalance should confer selective advantages to the cancer cells.

So far it has not been possible to measure the cyclic AMP concentration in the plasma membrane; however, experimental evidence demonstrating that addition of cyclic AMP and theophylline to the medium of hepatoma cell cultures inhibits proliferation is in agreement with these postulates (*26*). The present data from a series of hepatomas of widely different growth rates are also in line with the report of Tomasi *et al.*, who observed increased activity of cyclic AMP PDE and decreased activity of adenylate cyclase in plasma membrane prepared from the rapidly growing Yoshida hepatoma (*15*).

4. Purine metabolism: phenotypic evidence for reprogramming of gene expression

The first enzyme that commits ribose-5-phosphate (R-5-P) to purine biosynthesis is phosphoribosylpyrophosphate (PRPP) synthetase that produces PRPP. This enzyme utilizes the product of the oxidative and non-oxidative pathways of pentose phosphate synthesis, R-5-P. Because we assume that this enzyme may have a key role in priming *de novo* purine biosynthesis, the kinetic parameters of the liver and hepatoma enzymes were worked out carefully and the enzyme was assayed in liver tumors of different growth rates (*6*). Preliminary observations demonstrated that PRPP synthetase activity was increased in the rapidly growing hepatomas (Table 4) (*6*).

TABLE 4. PRPP Synthetase Activity in Neoplastic, Regenerating, and Differentiating Rat Liver

Tissues	Growth rate (weeks)	PRPP synthetase activity (μmoles/hr)	
		per gram wet weight	per cell $\times 10^{-8}$
Normal liver (Buffalo)			
Control for 7777		9.4 ± 0.2	4.0 ± 0.1
Control for 9618A2		10.8 ± 0.5	7.2 ± 0.3
Normal liver (ACI/N)			
Control for 3924A		10.4 ± 0.2	5.7 ± 0.1
Sham operated (Wistar)		8.8 ± 0.1	4.7 ± 0.1
24-hr regenerating liver (Wistar)		10.5 ± 0.5 (120)	5.6 ± 0.2 (118)
5-Day-old liver		9.3 ± 0.3 (89)	1.5 ± 0.1 (32)[a]
Hepatomas 7777	4.1	14.0 ± 0.4 (149)[a]	5.8 ± 0.2 (147)[a]
3924A	3.8	19.5 ± 0.4 (187)[a]	11.5 ± 0.2 (202)[a]
9618A2	2.3	18.5 ± 1.1 (172)[a]	11.9 ± 0.7 (166)[a]

The data are means $\pm$ standard error of 4 experiments with percentages of corresponding control liver values in parentheses. The activities per cell are to be multiplied with the exponential given to arrive at the actual numbers. [a] Values statistically significantly different from the respective control ($P=0.05$).

This reprogramming of gene expression appears to be specific to liver neoplasia, since no similar alteration was observed in differentiating liver or in the 24-hr regenerating liver (Table 4). The increase in this enzyme activity should confer selective advantages to rapidly growing hepatoma cells.

TABLE 5. Biochemical Discriminants That Correlate with Growth Rate in Hepatomas

Carbohydrate metabolism	Nucleic acid metabolism
Glucose synthesis: decreased	DNA synthesis: increased
Pyruvate conversion to glucose	DNA content
G-6-Pase	TdR incorporation into DNA
FDPase	Formate into DNA
Phosphoenolpyruvate carboxykinase	Adenine into DNA
Pyruvate carboxylase	TdR kinase
Glucose catabolism: increased	DNA polymerase
Glycolysis	Thymidylate synthetase
HK	Deoxycytidylate deaminase
PFK	DNA nucleotidyltransferases
Pyruvate kinase	Ribonucleotide reductase
Glycolytic ATP production	DNA catabolism: decreased
Pentose phosphate pathway: increased	Thymine degradation to CO_2
C-1/C-6 oxidation of glucose	RNA synthesis: increased
Specific phosphorylating enzymes: decreased	Formate into RNA
Fructokinase and glucokinase	Dihydroorotase
Fructose metabolism: decreased	G content of rapidly sedimenting nuclear RNA
Fructose incorporation into glycogen through fructokinase	GU content of 4–7 S nuclear RNA
Thiokinase	Carbamylaspartate-transferase
Aldolase	tRNA methylase
Responsiveness to glucocorticoid stimulation: decreased	RNA catabolism: decreased
Response of gluconeogenic enzymes	Xanthine oxidase
Glycogenic response	Uricase
Isozyme shift	RNA metabolic response to glucocorticoid stimulation: decreased
High K_m isozymes: decreased	Precursor incorporation into total tumor RNA after steroid injection
Low K_m isozymes: increased	

5. Summary of malignancy-linked reprogramming of gene expression: biochemical parameters in Class 1

As discussed, the enzymatic and metabolic alterations in neoplasia that correlate with tumor growth rate are grouped in Class 1 by the molecular correlation concept. Such events in the reprogramming of gene expression appear to be linked with the different degrees of expression of malignancy in the tumor spectrum. These alterations are termed *malignancy-linked discriminants*, since they are characteristic of the degrees of malignant transformation. A summary of over 60 such malignancy-linked alterations is provided in Table 5. It may be seen that in all areas of intermediary metabolism that have received appropriate experimental and conceptual evaluation the opposing key enzymes and metabolic pathways in carbohydrate, amino acid and protein, DNA and RNA, purine and pyrimidine, ornithine, lipid, and cyclic AMP metabolism, and in respiration correlated with malignancy of the liver tumors.

Neoplastic Transformation-linked Reprogramming of Gene Expression: Class 2

The enzymatic and metabolic alterations that occur in all or nearly all tumors

Protein and amino acid metabolism	Other metabolic areas
Protein synthesis: increased	Polyamine synthesis: increased
Amino acid incorporation into protein (alanine, aspartate, glycine, serine, isoleucine, valine)	Ornithine decarboxylase
Activity of the postmicrosomal protein synthesizing system	Putrescine content
Ratio of total free amino acid to total protein content	Urea cycle: decreased
Decrease in:	Ornithine carbamyltransferase
S-adenosylmethionine synthetase	Lipid metabolism: decreased
Enzymes catabolizing amino acids: decreased	Lipid content
Tryptophan pyrrolase	α-Glycerophosphate dehydrogenase
Serotonin deaminase	Butyrate to acetoacetate
5-Hydroxytryptophan decarboxylase	Respiratory activity: decreased
Glutamate dehydrogenase	Oxygen consumption
Glutamate-oxaloacetate transaminase	Mitochondrial protein content
	Respiratory ATP production
	Membrane cyclic AMP synthesis: decreased
	Cyclic AMP PDE increased
	Adenylate cyclase decreased

are grouped by the molecular correlation concept in Class 2. Such changes indicate a link between the metabolic alterations and the neoplastic transformation of the cell that appears to be characteristic of neoplasia *per se.* These changes which occur in the very slow as well as in the very rapidly growing tumors are characteristic of the reprogramming of gene expression that might have occurred at the moment of the malignant transformation. These parameters are called *neoplastic transformation-linked discriminants* in contrast to the alterations that correlate with growth rate which are grouped in Class 1 and are called malignancy-linked discriminants.

1. Pentose phosphate biosynthesis: phenotypic evidence for reprogramming of gene expression—Class 2

In the hepatic cell R-5-P may be produced by the oxidative pathway involving the action of glucose-6-phosphate (G-6-P) and 6-phosphogluconate (6-PG) dehydrogenases, a lactonase and an isomerase. In turn, R-5-P may be utilized for recycling into glycolysis through the transketolase and transaldolase reactions and this sequence also involves an epimerase. Alternately, R-5-P may be utilized as a precursor for the biosynthesis of PRPP by the enzyme, PRPP synthetase.

Considering the major role of R-5-P in purine biosynthesis and in synthesis of

DNA and RNA, it is important to note that the *de novo* production of R-5-P may involve 2 alternate pathways. Thus, the flow of carbon may depart from glycolysis at G-6-P through the oxidative pathway and at fructose-6-phosphate through the non-oxidative pathway. In case of increased utilization of purines, RNA, and DNA, such as in cellular proliferation, the R-5-P pool would be drained by the activity of the subsequent enzyme reactions, starting with the action of PRPP synthetase, PRPP amidotransferase and the subsequent enzymes of purine biosynthesis (*29*). These postulates assume that the 2 parallel pathways of pentose phosphate production proceed in the direction of R-5-P synthesis when R-5-P is drained away for biosynthetic use. In turn, when the utilization of R-5-P declines, the pentose phosphate pool (R-5-P, ribulose-5-phosphate, and xylulose-5-phosphate) would temporarily enlarge and then the pentoses could be recycled into glycolysis through the action of transketolase and transaldolase.

In the liver when the oxidative and non-oxidative pathways operate as parallel one-way reactions their main function would be the production of NADPH and R-5-P. The NADPH would be available for the reductive biosynthetic reactions which include the action of folate reductase (*8*), and the stepping up of the production of R-5-P for purine biosynthesis (*29*). During recycling, the production of NADPH and the conservation of the carbon chain through recycling would be the predominant function. These mechanisms would be of selective advantage in the generation of energy through glycolysis and in the storage of lipids. The integration of the key enzymes of glycolysis and gluconeogenesis and the oxidative and non-oxidative pathways of pentose phosphate metabolism is shown elsewhere in this paper (Fig. 5).

Because of the possible rate-limiting nature of R-5-P production, a systematic investigation was carried out on the behavior of certain key enzymes of the oxidative and non-oxidative pathway of pentose phosphate production. In earlier work it was first reported that G-6-P dehydrogenase, the rate-limiting enzyme of the oxidative pathway, was markedly increased in the rapidly growing Novikoff hepatoma (*16*). Subsequent studies demonstrated that G-6-P dehydrogenase activity was markedly increased in practically all the hepatic tumors examined, including the Morris hepatomas and the Reuber H-35 tumor. The activity of 6-PG dehydrogenase was not related to growth rate (Class 3) (*29*).

The identification of the rate-limiting reaction is more complex for the non-oxidative pathway in which the reactions are considered completely reversible. An examination of the reported concentrations of pentose phosphate intermediates indicates that these levels are much lower than G-6-P; in particular, the concentration of erythrose-4-phosphate is less than 2% of that of G-6-P (*5*). Therefore, if we consider the non-oxidative pathway as functioning parallel with the oxidative one in the direction of R-5-P synthesis, the low level of erythrose-4-phosphate may make transaldolase activity rate-limiting in the non-oxidative production of R-5-P.

This view is supported by estimating the enzyme activities at tissue concentrations of substrate which should make the activity of transaldolase approximately as low as that of G-6-P dehydrogenase. If these kinetic conditions do operate in the hepatoma cells, then in the reprogramming of gene expression one would expect

TABLE 6. G-6-P DH, 6-PG DH, Transaldolase, and Transketolase Specific Activities in Hepatomas of Different Growth Rates (*29*)

Tissues	Growth rate (months)	Protein content	Oxidative pathway		Non-oxidative pathway	
			G-6-P DH	6-PG DH	Trans-aldolase	Trans-ketolase
Normal control liver		100	100	100	100	100
Hepatomas						
66	13.0	79[a)]	1,262[a)]	300[a)]	284[a)]	167[a)]
47C	8.0	74[a)]	727[a)]	392[a)]	228[a)]	169[a)]
8999	9.0	83	184[a)]	72[a)]	145[a)]	102
44	9.0	80	544[a)]	177[a)]	283[a)]	153
9633	4.9	77[a)]	253[a)]	43[a)]	205[a)]	110
7794A	2.8	84	3,650[a)]	284[a)]	215[a)]	220[a)]
7777	1.3	77[a)]	897[a)]	181[a)]	300[a)]	138
3924A	1.0	66[a)]	1,128[a)]	72[a)]	240[a)]	94
3683F	0.5	67[a)]	903[a)]	203[a)]	255[a)]	277[a)]
9618A2	0.4	64[a)]	4,366[a)]	608[a)]	338[a)]	266[a)]

Enzyme activities were calculated in μmoles/hr/mg protein and the protein content as mg/g wet weight. Results are expressed as percentages of corresponding control normal liver values. Four or more rats were in each group. [a)] Values statistically significantly different from the respective control ($P<0.05$).

that transaldolase activity would behave similarly to that of G-6-P dehydrogenase; *i.e.*, it would be increased in all hepatomas. Then the activity of transketolase, as of 6-PG dehydrogenase, may not relate to tumor growth rate. To test these postulates the kinetic conditions were worked out for liver and hepatoma transaldolase and transketolase, and the enzymes were measured in hepatomas of different growth rates and in various other control tissues (*7, 29*).

The behavior of the enzymes of the oxidative pathway (G-6-P and 6-PG dehydrogenases) and those of the non-oxidative pathway (transaldolase and transketolase) in the spectrum of hepatomas of different growth rates in normal liver and in regenerating liver is summarized in Table 6. The results indicate that enzymes assumed to be rate-limiting for the oxidative pathway (G-6-P dehydrogenase) and for the non-oxidative pathway (transaldolase) are increased in all tumors. In contrast, 6-PG dehydrogenase and transketolase show no relationship with growth rate; they belong to Class 3 where the molecular correlation concept groups such parameters (*17–22, 25, 28*).

Since the activities of G-6-P dehydrogenase and transaldolase are increased in all hepatomas, these enzymes are placed in Class 2 where the molecular correlation concept groups those biochemical parameters that are altered in the same direction in all tumors. Such alterations are neoplastic transformation-linked changes and they provide important discriminants in the identification of the cancer cells. The experimental evidence indicates that the first 2 enzymes of the oxidative and non-oxidative synthetic pathways of R-5-P that are involved in channeling glycolytic intermediates into pentose phosphate synthesis are increased in all hepatomas. This reprogramming of gene expression should provide increased potential activity for the pentose phosphate pathway. Isotope studies employing tissue slices from liver and hepatomas are in agreement with this suggestion (*1*, *19*).

As a result of the reprogramming of gene expression in the potential stepping up of pentose phosphate biosynthesis a selective advantage is conferred to the cancer cell.

2. Specificity to neoplasia of the increased activities of G-6-P dehydrogenase and transaldolase

Earlier work demonstrated that G-6-P dehydrogenase was unchanged in regenerating liver and low in newborn and in differentiating liver (*16*). The present results indicate that transaldolase activity was also unchanged in the regenerating liver. A systematic investigation demonstrated that the activities of hepatic G-6-P and 6-PG dehydrogenases, transaldolase, and transketolase were low in the rapidly growing differentiating liver and rose to normal adult liver levels during post-natal development (Fig. 4). It may be concluded that in the spectrum of hepatomas of different growth rates the increased activities of G-6-P dehydrogenase and transaldolase are specific to the neoplastic transformation as no similar alteration was observed in hepatic regeneration or differentiation (*29*).

Recently it was also observed in my laboratories that in primary hepatomas in man the activities of G-6-P dehydrogenase and transaldolase were markedly elevated; therefore, the reprogramming of gene expression as observed in transplantable rodent hepatomas appears to apply to primary hepatomas in man.

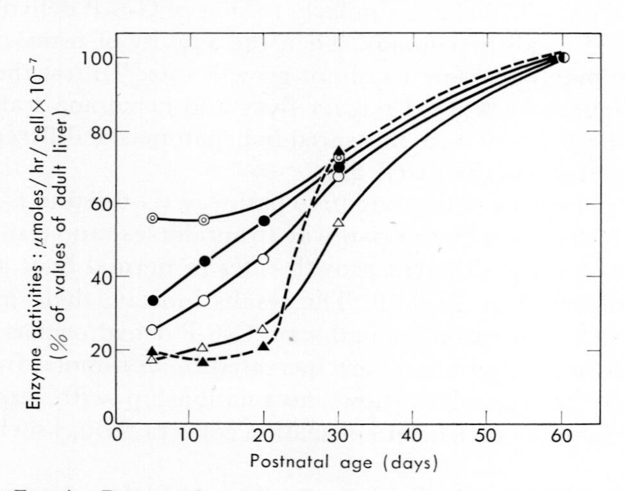

FIG. 4. Pattern of pentose phosphate enzymes in differentiation (*29*). ◎ G-6-P DH; ● transaldolase; ○ protein; △ transketolase; ▲ 6-PG DH.

3. Summary of neoplastic transformation-linked reprogramming of gene expression: biochemical parameters in Class 2

As discussed, the enzymatic and metabolic alterations that occur in all or nearly all tumors are grouped in Class 2 of the molecular correlation concept. These alterations are termed neoplastic transformation-linked discriminants because they are present even in the slowest growing, least malignant hepatomas. The ubiquitous

TABLE 7. Class 2: Parameters Altered in the Same Direction in All Hepatomas[a)]

Metabolic areas			
Carbohydrate	Amino acid, protein	Nucleic acid	Ornithine
Increased: Key enzymes of pentose phosphate synthesis G-6-P DH Transaldolase Decreased: Glycogen metab-olism Glycogen deposition Phosphoglucomutase activity Decreased: Fructose metabolism Fructose uptake Fructose to CO_2 Decreased: Carbohydrate conversion to lipid Glucose to fatty acid	Increased: Free amino acid level Decreased: Protein level Amino acid response to glucocorticoid administration	Increased: Nucleic acid anabolic processes Adenine incorporation into RNA UDP kinase Amidophosphoribosyl-transferase Decreased: Nucleic acid catabolic enzyme TdR phosphorylase Decreased: Regulatory response No increase in RNA amount after steroid injection Decreased: Orotate incorporation into RNA	Increased: Polyamine synthesis Spermidine concentration Spermine concentration Decreased: S-adenosyl methionine decarboxylase activity

[a)] Increased in all hepatomas or decreased in all hepatomas, with 1 or 2 exceptions at most.

emergence of these indicators of reprogramming of gene expression suggests that they are essential and perhaps *conditione sina qua non* for the neoplastic transformation itself. In Table 7 a few of these neoplastic conversion-linked alterations are summarized. It can be seen that in several areas of intermediary metabolism that receive appropriate conceptual and experimental exploration (carbohydrate, amino acid, protein, nucleic acid, and ornithine metabolism) there were discovered parameters that are altered in all or nearly all hepatomas.

The significance of such alterations is in the fact that they reveal alterations in gene expression that might be crucial for neoplastic transformation, thus giving us an insight into the metabolic reprogramming that provides selective biological advantages for the cancer cells. The recognition of the key enzymes and metabolic pathways that are altered in neoplastic cells should provide more specific targets in the design of anti-cancer drugs and chemotherapeutic agents.

Qualitative and Quantitative Discriminants of Neoplasia

The strategy of anti-neoplastic chemotherapy will also depend on the identification of differences in the quantitative and qualitative reprogramming of gene expression. These changes are reflected (1) in the isozyme shift that occurs for some key enzymes in neoplastic cells and (2) in the demonstration of the specificity to neoplasia of the alterations observed. This will be accomplished by a critical comparison of the biochemical pattern in rapidly growing hepatomas, and in fetal, differentiating and regenerating liver. These 2 aspects will now be discussed.

Isozyme shift

Since I have examined recently the operation and biological significance of the alterations in isozymes (*20*), I will deal with this problem very briefly. In a spectrum of hepatomas of different malignancy a metabolic imbalance is revealed not only in the activities of the opposing key enzymes and metabolic pathways but also in the isozyme pattern. The reprogramming of gene expression that results in the decrease in the production of one type of isozyme and the increase in the synthesis of another type of isozyme is of considerable interest for biochemistry and chemotherapy. The isozyme shift is manifested in the decrease in the amounts of the high K_m, hormonally and nutritionally responsive isozymes, such as the hepatic-type glucokinase and pyruvate kinase. In contrast, the low K_m isozymes such as hexokinase (HK) and muscle-type pyruvate kinase increased in parallel with tumor growth rate (*20*). A similar shift in the pattern was also demonstrated for the isozymes of cyclic AMP PDE which are localized in the cytosol (*26*). The emergence of such qualitative discriminants in the hepatomas was recognized originally in the critical examination of the isozyme pattern of key enzymes in carbohydrate metabolism (*20*).

In the case of the key glycolytic enzymes the increase in the production of isozymes that have high affinity to the substrate but lost partly or completely their responsiveness to nutritional or hormonal stimulation, feedback and allosteric control mechanisms should confer selective advantages to the cancer cells.

Specificity of Metabolic Phenotypes to Malignancy: Comparison of Hepatomas to Differentiating and Regenerating Liver

There have been claims in the literature suggesting on the basis of various analogies that the neoplastic condition is a disease of differentiation or a reestablishment of fetal conditions. If this view were to be supported by hard facts, as it seems to me that it is not, it still would not provide much conceptual progress because such views merely amount to the restatement of the problem of neoplastic transformation in terms of differentiation or embryonic mechanisms of which we know even less than of neoplasia. This would also seem to make it the more difficult to achieve therapy in the developing organisms with neoplastic conditions, yet some of the most striking treatment successes and prolongation of lives have been achieved in the clinical management of juvenile leukemias.

I already examined in some detail the specificity of the reprogramming of gene expression in neoplasia and was able to distinguish from each other by quantitative and qualitative discriminants (*25*) the biochemical pattern of normal adult liver, and rapidly growing neoplastic, newborn, fetal, and regenerating liver. Therefore, I will go through this area very concisely in light of some new material and the reader is referred to earlier analyses of this problem where the conclusion was reached that a critical examination of the biochemical phenotype provides little or no support for the contention that the neoplastic condition is a disease of differentiation, but rather points out clearly the specificity of the pattern to neoplasia. This insight in turn should be helpful in the design of anti-cancer chemotherapy.

TABLE 8. Discriminating Power of Biochemical Pattern in Glycolysis, Gluconeogenesis, and Pentose Phosphate Biosynthesis: Key Differences between Normal Liver, Rapidly Growing Neoplastic, Fetal, Newborn, and Regenerating Liver

Markers of gene expression	Livers				
	Normal (adult)	Neoplastic (rapidly growing)	Newborn (6-day-old)	Fetal (17-day-old)	Regenerating (24 hr)
G-6-Pase	100	<1	350[b]	12[b]	100[b]
FDPase	100	<1	175[b]	7.1[b]	100[b]
Phosphoenolpyruvate carboxykinase	100	<1	160[b]		100[b]
Pyruvate carboxylase	100	<1	275[b]		100[b]
HK	100	500	36	83[b]	100[b]
PFK	100	229	44	49[b]	100[b]
Pyruvate kinase	100	499	15	11[b]	100[b]
G-6-P DH	100	751	50	26[b]	100[b]
6-PG DH	100	48	18	26	100[b]
Transaldolase	100	180	33[b]		120[a]
HK/G-6-Pase	100	8,800	57[b]	690[b]	100[b]
PFK/FDPase	100	6,463	86[b]	674[b]	100[b]
G-6-P DH/6-PG DH	100	1,120	288[a]	100[b]	100[b]
Transaldolase/transketolase	100	267	179[a]		109[b]

Values are expressed as percent of normal adult liver. [a] Quantitative discriminant. [b] Qualitative discriminant.

To test the specificity to neoplasia of alterations in gene expression it is necessary to contrast the tumor metabolic pattern to a series of control tissues that exhibit similar biological properties in terms of proliferation rate. For this reason, the rapidly growing hepatomas were selected as examples because these highly malignant neoplasms display the full-blown expression of the enzymatic, isozymic, and metabolic imbalance of neoplasia. The rationale for the selection of the control tissues was discussed in detail elsewhere and the operational definitions of quantitative and qualitative discriminants were provided (*25*).

The key biochemical alterations and the ratios in the various metabolic areas are compared in Tables 8, 9, and 10. The enzyme activities were calculated in micromoles of substrate metabolized per hour per gram wet weight of tissue and were expressed as micromoles of substrate metabolized per hour per average cell. For a ready comparison, the results in these tables are provided as percentages of the per cell data, taking the values of the liver of the respective control normal adult rat as 100%.

1. *Discriminating power of the biochemical pattern in glycolysis, gluconeogenesis, and pentose phosphate biosynthesis*

As Table 8 indicates there are numerous quantitative and qualitative differences between normal liver, rapidly growing neoplastic liver and fetal, newborn, and regenerating liver. With the use of an array of quantitative and qualitative discriminants every one of the 5 tissue types can be characterized and a differential diagnosis can be made against any one of the other types of tissue.

2. Discriminating power of the biochemical pattern in pyrimidine, purine, and DNA metabolism

Table 9 indicates that there are a number of quantitative and qualitative discriminants in these metabolic areas that can readily discriminate from each other the metabolic pattern of normal, neoplastic, differentiating, or regenerating liver. By using an array of these discriminants the neoplastic liver can be clearly distinguished from any stage of the differentiating liver.

TABLE 9. Discriminating Power of Biochemical Pattern in Pyrimidine, Purine, and DNA Metabolism: Key Differences between Normal Liver and Rapidly Growing Neoplastic, Fetal, Newborn, and Regenerating Liver

Markers of gene expression	Livers				
	Normal (adult)	Neoplastic (rapidly growing)	Newborn (6-day-old)	Fetal (17-day-old)	Regenerating (24 hr)
TdR into DNA	100	3,900	1,224[a]	3,083	910[a]
TdR to CO_2	100	<0.1	68[b]	5	64[b]
TdR to DNA/TdR to CO_2	100	11,500,000	1,788[a]	54,550[a]	1,450[b]
DNA/cell	100	250	87		100
TdR phosphorylase	100	32		2.0[b]	126
Uridine phosphorylase	100	375		1.8[b]	106
Dihydrouracil DH	100	<10	50[a]		78
TdR kinase/dihydrouracil DH	100	>1,000			<10
PRPP synthetase	100	250	32[b]		122[a]

Values are expressed as percent of normal adult liver. [a] Quantitative discriminant. [b] Qualitative discriminant.

3. Discriminating power of the biochemical pattern in membrane cyclic AMP metabolism

Recent studies in my laboratories demonstrated that there are quantitative and qualitative differences in the activities of membrane adenylate cyclase and cyclic AMP PDE that can discriminate the metabolic imbalance in the hepatomas from that in differentiating or regenerating liver (Table 10). The alterations in the mem-

TABLE 10. Discriminating Power of Biochemical Pattern in Cyclic AMP Metabolism in Plasma Membrane: Key Differences between Normal Liver and Rapidly Growing Neoplastic, Newborn, and Regenerating Liver

Markers of gene expression	Livers			
	Normal (adult)	Neoplastic (rapidly growing)	Newborn (6-day-old)	Regenerating (24 hr)
Adenylate cyclase	100	19	126[b]	70[a]
Cyclic AMP PDE	100	787	76[b]	146[a]
PDE/adenylate cyclase	100	4,248	60[b]	225[b]

Activity was calculated in μmoles/hr/cell and expressed as percent of the values of normal liver of adult fed rats. [a] Quantitative discriminant. [b] Qualitative discriminant.

brane cyclic AMP system along with those in the behavior of the total adenylate cyclase and supernatant fluid cyclic AMP PDE activities, as reviewed elsewhere (*26*), provide further discriminants to distinguish biochemically the various examined tissues.

4. Key differences between rapidly growing neoplastic, differentiating, and regenerating liver

As demonstrated in Tables 8, 9, and 10 and in earlier publications a critical evaluation of the enzymic and metabolic phenotype demonstrates the vastly different and specific biochemical pattern of the cancer cells that can be clearly discriminated from that of the differentiating liver. Thus, the discriminating power of the biochemical pattern provides no support for contentions that neoplasia is a disease of differentiation or a regression to a fetal biochemical pattern. While there might be some overlapping in the metabolic pattern, it is more revealing that a critical examination pointed out numerous quantitative and qualitative differences between the biochemical phenotype of the hepatoma and that of differentiating liver. The evidence indicates that the overall molecular pattern of the cancer cell is ordered, it is linked with the degree and the nature of malignant transformation and it is specific to neoplasia (*17–22, 25, 28, 29*).

5. Evidence for the separate nature of neoplasia and differentiation

In Tables 11 and 12 further evidence is summarized showing that differentiation is not necessarily connected with the core of neoplasia. The 3 lines of evidence are biological, morphological, and metabolic. The metabolic differences were dealt with in Tables 8, 9, and 10. The biological evidence comes from observations that the specialized functions which are indicators of differentiation may be maintained, decreased, or completely lost, but also increased or even derepressed in neoplastic cells. To date this can be the most clearly demonstrated by the measurement of hormone secretion, as pointed out elsewhere (*14*). At the morphological level, it seems that the poorly differentiated or anaplastic tumors have no true equivalents in the various stages of differentiation.

As Table 11 shows, the expression of specialized endocrine functions is not necessarily lost or decreased, but may well be maintained or even increased in various tumors. The unmasking and derepressing of gene areas that are revealed in hormone secretion in non-endocrine tumors is another strong argument that the alterations in differentiation are coincidental to the core of neoplastic transforma-

TABLE 11. Evidence Indicating That Differentiation Is Not Connected with the Core of Neoplasia

Evidence	Aspects of gene expression in cancer cells
1. Biological	Specialized functions: a) maintained, b) decreased, c) lost, d) increased, e) derepressed
2. Morphological	Degrees of differentiation: (including poorly differentiated or anaplastic cells) have no organizational equivalents in fetal or post-natal development
3. Metabolic	Biochemical pattern: quantitatively and qualitatively different from that of fetal and post-natal stages of differentiation

From Weber (*30*).

TABLE 12. Relation of Expression of Differentiated Functions to Neoplasia

Behavior of differentiated functions		
Increased	Decreased	Derepressed
Insulinoma	Various solid tumors (carcinoma of the stomach, colon, *etc.*)	Thymus, pancreatic, lung tumors : ACTH
		Lung tumors : ADH
Thyroid carcinoma	Leukemias (loss of phagocytosis)	Hemangiomas (cerebellum, liver) : erythropoietin
Adrenal tumors	Lymphomas (immunol. function)	Liver, lung (large cell tumors) : gonadotrophins
Pituitary tumors	Hodgkin's (T cell function)	Trophoblastic, lung tumors : thyrotrophin
Melanotic melanoma		Renal, ovarian, lung tumors : PTH
Choriocarcinoma, *etc.*		

The expression of differentiated functions is coincidental and is not linked with the neoplastic transformation. From Weber (*30*). ACTH, adrenocorticotropic hormone; ADH, antidiuretic hormone; PTH, parathyroid hormone.

tion (Table 12). Since differentiated functions may be increased or decreased in neoplastic cells, they can be used in the diagnosis and evaluation of treatment of neoplasms, but one must carefully separate this information from the claim that neoplasia is a disease of differentiation.

Selective Biological Advantages of the Imbalance in Gene Expression for the Cancer Cells

The metabolic imbalance in carbohydrate and purine metabolism is summarized in Fig. 5 (*29*). This figure shows the increase in activities of the key glycolytic enzymes and the decrease in those of the key gluconeogenic ones. These reciprocal changes in the activities of the opposing key enzymes occur in parallel with increasing tumor growth rate. On the other hand, the activities of the key enzymes, G-6-P dehydrogenase and transaldolase, that are involved in channeling glycolytic intermediates into pentose phosphate biosynthesis are increased in all hepatomas. Consequently, in hepatic cancer cells there are biochemical discriminants that characterize the degrees of expression of neoplastic transformation and there are enzymes that are discriminants for the neoplastic transformation *per se*, because they occur in all hepatomas.

This figure also shows the integrated pattern of enzymatic alterations linking carbohydrate metabolism with purine metabolism. Thus, the utilization of R-5-P is illustrated in the *de novo* synthetic pathways of IMP. This anabolic pathway is opposed by the IMP catabolic pathway that can degrade inosinic monophosphate to uric acid. In addition to the *de novo* pathway of purine synthesis there is a so-called salvage pathway which provides for the recycling through a transferase reaction, leading from hypoxanthine to IMP. The analysis of the behavior of enzymes that may play key roles in the *de novo* synthesis (PRPP synthetase and PRPP amidotransferase) and of the salvage pathway (transferase) and the opposing catabolic enzymes (phosphatase, phosphorylase, and xanthine oxidase) is in progress. The results indicate that PRPP synthetase increases with the growth rate of liver tumors (*6*). From data in the literature (*2, 28*) it appears that xanthine oxidase activity would decrease in parallel with the increase in tumor growth rate. As a result of such a reprogram-

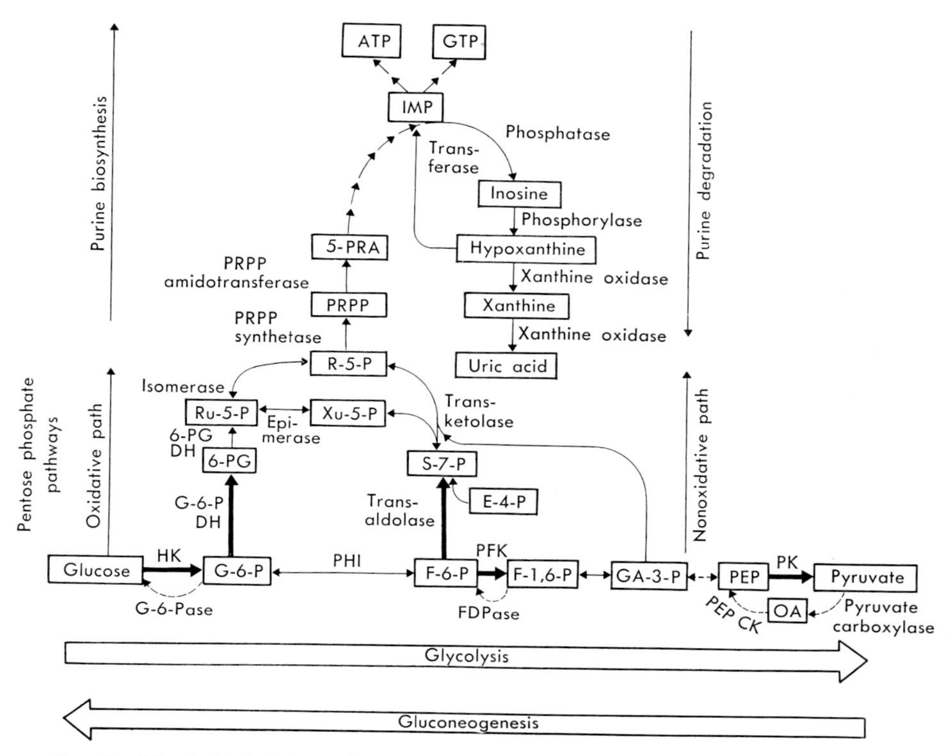

FIG. 5. Metabolic imbalance in the activities of key enzymes and metabolic pathways in glycolysis, gluconeogenesis, pentose phosphate synthesis, and purine metabolism. From Weber *et al.* (*29*). CK, carboxylase; E-4-P, erythrose-4-phosphate; F-1-P, fructose-1-phosphate; GA-3-P, glyceraldehyde-3-phosphate; OA, oxaloacetate; PEP, phosphoenolpyruvate; PK, pyruvate kinase; PRA, 5-phosphoribosylamine; Ru, ribulose; S-7-P, sedoheptulose-7-phosphate; Xu, xylulose.

ming of gene expression there is an increase in the potential to glycolize, to channel intermediates into pentose phosphate biosynthesis and to utilize them for purine biosynthesis. In turn, there is a concurrent decline in the activities of the gluconeogenic enzymes that could provide recycling and in the activities of enzymes involved in IMP degradation that could counteract IMP biosynthesis. The reprogramming of gene expression confers biological advantages to the neoplastic cells. As pointed out elsewhere, selective biological advantages are also conferred to cancer cells by the progressive predominance of the synthetic pathway over the catabolic one in pyrimidine and DNA metabolism. Through a gradually emerging imbalance in ornithine metabolism the competition of the urea cycle for carbamyl phosphate and aspartate is gradually turned off, allowing an increase in channeling of these precursors to the biosynthesis of nucleic acids and polyamines (*23, 31*).

Biochemical Pattern of Neoplastic Cells: Ordered or Random?

There have been claims by a few workers about the "diversity," and the apparently random nature of some of the alterations in hepatomas. I have pointed out the fact that while in every disease there are coincidental and random components (Class 3) the job is to identify the ordered pattern of metabolic imbalance that is linked with the nature of the disease (*17–20*). The evidence is now so extensive, as can also be seen from Tables 5 and 7, that most scientists realize the conceptual advantages in directing attention to understanding the ordered pattern of the biochemical imbalance that is linked with neoplastic transformation and the degrees of malignancy, in contrast to working on coincidental and diverse alterations that may occur.

Careful evaluation of the extensive evidence leads to the conclusion that what is important about cancer is ordered and what is not, is the random element and the diversity.

In the design of chemotherapeutic weapons the ordered pattern is the biochemical target against which selective toxicity should be and has been directed. In fact, the mechanism of action of most of the currently clinically useful anticancer drugs entails inhibition of certain key enzymes in purine and pyrimidine biosynthesis that in the hepatoma system show a definite pattern of correlation with tumor growth rate. Now interest should also be directed to those enzymes that are increased in all tumors, especially if they appear to be dispensible for the overall economy of the cell, as transaldolase might well be.

Biochemical Strategy of the Cancer Cell

1. Molecular correlates of malignancy

The molecular correlation concept provides a method for investigating, evaluating, and understanding biochemical alterations in neoplasms. In the summary Tables 5 and 7 are tabulated most of the biochemical alterations that correlate with hepatoma growth rate (Class 1) and those that occur in all hepatomas (Class 2). It was pointed out that chemotherapeutic approaches should be designed against some of these molecular targets and the success of drug therapy should be evaluated by examining the metabolic behavior of the treated tumors at the level of these enzymes. Therefore, the identification and elucidation of the pattern of metabolic imbalance and the control of gene expression that is revealed in these molecular correlates of malignancy should bear on the clinical design, use and evaluation of chemotherapy.

2. Generalization of the phenotypic evidence: reprogramming of gene expression in cancer cells

By application of the approaches of the molecular correlation concept of neoplasia, 5 generalizations may be drawn; these are presented in Table 13 which also provides some characteristic examples. In the spectrum of hepatomas of different malignancy the operation of a metabolic imbalance was revealed in measuring the activities of opposing and competing key enzymes and metabolic pathways. A shift in isozyme pattern led to a decreased responsiveness to regulation that is recognized in the decline in the susceptibility to hormonal stimulation, feedback, and allosteric control mechanisms. The relation of the biochemical alteration to malignancy indicated a link with the progressive expression of neoplastic properties. There are also biochemical alterations that are present in all transformed cells, irrespec-

TABLE 13. Generalization of the Phenotypic Evidence: Reprogramming of Gene Expression in Cancer Cells

Spheres of gene action	Indicators of reprogramming of gene expression	Examples	
		Increased	Decreased
1. Metabolic imbalance	Activities of opposing key enzymes and metabolic pathways	PFK TdR kinase PDE	FDPase Dihydrouracil DH Adenylate cyclase
2. Shift in isozyme pattern	Decrease in high K_m and increase in low K_m isozymes	HK Muscle type pyruvate kinase	Glucokinase Liver type pyruvate kinase
3. Decreased responsiveness to regulation	Decreased sensitivity to hormonal stimulation, feedback, and allosteric control		Steroid on glycogen synthesis Glucagon on adenylate cyclase
4. Relation to the degrees of expression of malignancy (biological behavior)	Biochemical alterations that are progressive and correlate with growth rate	Glycolysis TdR to DNA	Gluconeogenesis TdR to CO_2
5. Relation to neoplastic transformation	Biochemical alterations that are present in all transformed cells irrespective of the degree of malignancy	G-6-P DH Transaldolase UDP kinase	Glycogen deposition Glucose to fatty acid TdR phosphorylase

tive of the degree of malignancy; *i.e.*, parameters that emerge with the malignant transformation *per se*.

This analysis through the conceptual and experimental approaches achieved by application of the molecular correlation concept of neoplasia resulted in the recognition of an ordered and specific pattern of gene expression in the cancer cells of the hepatoma spectrum.

3. Applicability of the pattern observed in hepatomas to other neoplasms

It has already been pointed out that there is now a great deal of evidence in the literature that the correlation of biochemical alterations and tumor growth rate also applies to kidney tumors, mammary cancer, and other neoplasms of different growth rates. Therefore, the molecular correlation concept is valid not only for hepatomas, but it appears to be generally applicable to other types of neoplasms that have been subjected to appropriate investigation (*22*, *28*, *30*). Current work in my laboratories on primary hepatomas in man indicates that the metabolic imbalance discovered in the rodent transplantable hepatomas and primary liver tumors is applicable to primary liver tumors in man.

Operation of Pleiotropic Control in Neoplasia

In this context pleiotropy is defined as the operation of a coordinated, integrated array of metabolically and functionally interconnected events. In discussing pleiotropy Mayr stated that it is "The capacity of a gene to affect several different aspects of the phenotype" and he also emphasized that every gene that has been studied intensively has been found to be pleiotropic to a greater or lesser extent (*11*). The purpose of this part of the paper is to examine the phenotypic evidence supporting the concept that pleiotropic action is entailed in the reprogramming of gene expression that emerged in a spectrum of tumors of different growth rates and has been identified as an ordered pattern of biochemical imbalance. At the present stage of our scientific approaches it seems that the ordered and coordinated pattern that is linked at the biochemical, biological, and morphological aspects can be more readily analyzed at the phenotypic manifestations than an understanding can be achieved currently at the different levels of gene expression from which the integrated and inheritable regulation emanates in the cancer cells. It should be possible at this stage at least to outline the requirements for the operation of pleiotropic effects at the level of molecular genetics.

To make it easier to visualize the ordered nature of such a pattern of alterations, reference may be made here to the effect of diabetes and insulin adminstration. As analyzed elsewhere in detail, when rats are made diabetic and the insulin level drops markedly, a number of enzymes the biosynthesis of which is governed positively or negatively by insulin show marked alterations in the liver (*24*). These concurrent events include a decline in the activities of hepatic glucokinase, phosphofructokinase (PFK), pyruvate kinase, citrate cleavage enzyme, acetyl-CoA carboxylase, fatty acid synthetase, G-6-P dehydrogenase, 6-PG dehydrogenase, transaldolase, transketolase, and other enzymes and concurrently an increase in the activities of glucose-6-phosphatase (G-6-Pase), fructose-1, 6-diphosphatase (FDPase), phosphoenolpyruvate carboxykinase, pyruvate carboxylase, and other enzymes. As a result of these alterations, in part at least, the capacity of the liver for glycolysis, lipogenesis, NADPH production, and pentose phosphate biosynthesis decreased and that for gluconeogenesis increased. The administration of appropriate concentrations of insulin returned enzyme activities to normal levels. Thus, a coordinated reciprocal regulation of multi-enzyme and metabolic pathway activities is characteristic of the adaptive capacity of the organism that is based on the ability of the cell to reprogram gene expression.

These alterations are reversible. However, the altered biochemical pattern of neoplastic cells is inherited; thus, the gene expression is fixed.

In the following, we will examine the various levels at which gene expression might be regulated to result in the ordered pattern of alterations we recognized in the hepatoma spectrum.

1. Molecular genetic requirements for operation of pleiotropic effects

The 2 major alternatives involve genetic or epigenetic mechanisms. It is feasible to postulate the existence of pleiotropic genes that act as master genes or integrative genes to coordinate simultaneously at the biochemical phenotype the series of opposing and competing key enzymes in a number of opposing and competing metabolic pathways. However, it is also conceivable that regulation may take place at the epigenetic level involving coordinated processing of DNA-directed RNA species that may be endowed by coded recognition sites for an integrated processing in the nucleoplasm, at the nuclear pores, in the cytoplasm, or at the translational or post-ribosomal levels. At this place the objective is not to analyze in detail possible molecular mechanisms, but to provide some of the evidence that suggests to me

TABLE 14. Reciprocal Alterations Linked with Degrees in Expression of Neoplastic Transformation

Functions increased		Functions decreased	
Pathways	Enzymes	Pathways	Enzymes
	Carbohydrate metabolism:		
Glycolysis	HK PFK Pyruvate kinase	Gluconeogenesis	G-6-Pase FDPase Phosphoenolpyruvate carboxykinase Pyruvate carboxylase
	DNA metabolism:		
De novo pathway	Ribonucleotide reductase dCMP deaminase dTMP synthase		
Salvage pathway (TdR to DNA)	TdR kinase dTMP kinase DNA polymerase	TdR degradation (TdR to CO_2)	Dihydrothymine DH
UMP biosynthesis	Aspartate transcarbamylase Dihydroorotase	UMP degradation	Dihydrouracil DH
	Ornithine metabolism:		
Ornithine to polyamine synthesis	Ornithine decarboxylase	Ornithine to urea cycle	Ornithine carbamyltransferase
	Cyclic AMP metabolism in membrane:		
Depletion in cyclic AMP level	Cyclic AMP PDE	Cyclic AMP synthesis	Adenylate cyclase

From Weber (*30*).

that the rigorous examination of the biochemical phenotype in neoplasia leads to the assumption of the operation of pleiotropic controls.

We have 2 important manifestations of neoplasia in the hepatoma spectrum. (1) The different degrees of the neoplastic conversion are expressed to a different extent; we are dealing with hepatomas of different malignancy. (2) All the hepatomas irrespective of growth rate have undergone neoplastic transformation; they are all cancer cells. The biochemical aspects of genetic expression are found in the phenotypic alterations listed by the molecular correlation concept in Classes 1 and 2.

For instance, in the hepatoma spectrum which shows different degrees in the expression of malignant transformation there are opposing changes manifested by opposing key enzymes. In order to briefly document this I brought together in Table 14 some of these reciprocal alterations that are linked with the degrees in expression of neoplastic transformation. In parallel with increased tumor growth rate the key gluconeogenic enzymes decrease, whereas the key glycolytic enzymes increase. Similar reciprocal changes occur in DNA and ornithine metabolism and in membrane cyclic AMP metabolism.

We must pose the question by what molecular mechanisms and at what cytological levels the reprogramming of gene expression occurs that results in an inheritable, coordinated antagonistic setting of the rates of production and degradation of groups of enzymes that are functionally related and yield a meaningful, malignancy-linked, coordinated phenotype? Some of the possible cytological and molecular levels involved in the expression of pleiotropic effects are summarized in Table 15.

TABLE 15. Integrated Reciprocal Control of Gene Expression: Mechanisms for Expressing Pleiotropic Effects

Cytological levels of regulation	Molecular levels of controls	Phenotypic manifestations of gene expression
Nuclear controls		
Genetic level	1. Structural genes: present or absent	Quantitative
	a. Active or inactive	Quant.
	b. Number of genes committed	Quant.
	c. Different genes committed	Qualitative
	d. Multiple, reciprocal recognition sites, coding for regulation	Quant. and Qual.
	2. Regulator genes: present or absent	Quant.
	a. For individual enzymes	Quant.
	b. For groups of functionally limited enzymes pleiotropic, master or integrative genes: controlling groups of functionally linked enzymes	Quant. and Qual.
Epigenetic level	3. mRNA: selective degradation in nucleoplasm	Quant. and Qual.
	4. mRNA: degradation, withholding or passage at nuclear pores	Quant. and Qual.
Cytoplasmic controls	5. mRNA: selective processing or degradation	Quant. and Qual.
	6. tRNA : selective regulatory role	Quant. and Qual.
Ribosomal controls	7. mRNA: selective translation or non-processing	Quant. and Qual.
Postribosomal controls	8. Nascent enzyme: degraded by specific proteases	Quant. and Qual.

From Weber (*30*).

Control mechanisms may set the levels of the enzymes through the pleiotropic or integrative genes. The role of such master genes would be the control of relatively large groups of function-linked genes. Neoplastic transformation may be achieved through regulating the action of master genes involved in permitting, releasing or initiating cellular replication and in turn this would be expressed in pleiotropic action by coordinating the expression of a series of genes and concurrently repressing the action of opposing ones.

2. Operation of epigenetic controls

Integrated pleiotropic action may also be achieved by selective processing of nascent mRNA in the nucleoplasm, if mRNA for functionally related enzymes carried recognition sites which would selectively direct destruction or preservation of groups of mRNA. The observation that mRNA is produced as a giant macromolecule and that only part of it carries translatable information (*4*) supports the suggestion that the non-translatable parts may carry recognition sites conveying regulatory information. This non-translatable regulatory information might direct the processing of mRNA species that code for functionally related key enzymes in the nucleoplasm, at the nuclear pores, or at various cytoplasmic or ribosomal levels.

3. Translational controls

An integrated antagonistic regulation may well operate at the ribosomal level if translation or non-processing of mRNA by ribosomes may be directed by the functionally related codes carried in the recognition sites of the mRNA molecules.

Thus, the operation of pleiotropic control at the translational level would require that the mRNAs of the different enzymes that are functionally related would carry similar recognition sites. This assumption should be available for experimental testing. Even here genetic determination of the capacity to aggregate or fold at the polysome level may determine the translation or rejection of groups of mRNAs. Regulation through tRNA may also play a role.

4. Post-translational controls

It is also conceivable that regulation of gene expression is exerted at the nascent enzyme level or at the enzyme in the cytoplasm before or after it achieves its appropriate folding. Thus, groups of enzymes may be destroyed by group- or function-specific proteases that might initiate the destruction of groups of enzymes without affecting others (*9*). Subsequently, the non-specific proteases can complete the degradation.

At the present stage it is not feasible to pinpoint with much certainty the level where the pleiotropic reprogramming of gene expression is exerted in the neoplastic cells. The purpose here is to document that pleiotropic regulation operates, resulting in an ordered pattern of biochemical phenotype that suggests the emergence of malignancy-linked and neoplastic transformation-linked reprogramming of gene expression (*30*).

The sum of evidence is extensive enough to warrant a thorough investigation into the pleiotropic reprogramming of gene expression, since the phenotypic manifestations are extensively documented and provide a meaningful pattern. It should be possible to pinpoint the levels where the pleiotropic genes or subsequent control levels achieve the neoplastic conversion and concurrently with the determination of biological behavior set the integrated pattern of biochemical imbalance. An insight into the pleiotropic mechanism should also provide new targets for chemotherapeutic attack.

CONCLUSIONS

The molecular correlation concept led to a series of conceptual advances in analyzing the biochemical pattern of neoplasia (*3, 18–23, 25, 26,28–30*).

1) The role of key enzymes. It was recognized that the key enzymes and the overall pathways that antagonize each other at strategic places of intermediary metabolism are the ones that are linked with the neoplastic transformation and the degrees of malignancy.

2) Reciprocal imbalance. It was recognized that the opposing key enzymes and metabolic pathways behave antagonistically and the imbalance is linked with malignancy.

3) Relativity of malignancy. The relativity was recognized in the expression of malignant properties in the tumor spectrum. This insight permitted the elucidation of the correlation of the enzymatic and metabolic imbalance with the different degrees in the expression of malignancy in the tumor spectrum. These parameters are called malignancy-linked discriminants (Class 1).

4) Biochemical uniqueness of the neoplastic transformation. The recognition was made that certain key enzymes and biochemical parameters are altered in all hepatomas irrespective of the degree of malignancy, growth rate, or morphological differentiation; thus, these alterations represent reprogramming of gene expression that is linked with the malignant transformation *per se*. These parameters are termed neoplastic transformation-linked discriminants (Class 2).
5) Separation of essential from coincidental alterations. It was determined that certain biochemical alterations are not linked with the nature of neoplasia and they represent coincidental alterations in gene expression. They provide an apparent diversity in the biochemical pattern (Class 3).
6) Separation of neoplasia from differentiation. It was possible to separate through quantitative and qualitative discriminants the biochemical pattern of neoplastic liver from that of differentiating or rapidly growing regenerating liver. This points out the specificity of the biochemical pattern to neoplasia.
7) Essential metabolic pattern of neoplasia is ordered. It was recognized that what is important about cancer is ordered and what is not, is the apparent randomness and diversity.
8) Predictive power of the molecular correlation concept. In contrast to other cancer theories, the molecular correlation concept is able to provide precise predictions regarding the areas of metabolic and enzymatic biochemical imbalance in a tumor spectrum.
9) The role of biochemical imbalance in the diagnosis of malignancy. (a) The neoplastic transformation-linked biochemical alterations should be useful in the diagnosis of the presence or absence of neoplasia; (b) the malignancy-linked biochemical imbalance should be useful in the biochemical evaluation of the degrees of tumor malignancy.
10) Biochemical pattern as target of chemotherapeutic design. The ordered pattern of metabolic imbalance, that is linked with neoplastic transformation and with the degrees of malignancy, should provide sensitive targets for the design of chemotherapeutic weapons.

ACKNOWLEDGMENTS

This work was supported by grants from the United States Public Health Service, grants CA-05034 and CA-13526 and grants from the Damon Runyon-Walter Winchell Fund for Cancer Research.

REFERENCES

1. Ashmore, J., Weber, G., and Landau, B. R. Isotope studies on the pathways of glucose-6-phosphate metabolism in the Novikoff hepatoma. Cancer Res., *18*: 974–979, 1958.
2. DeLamirande, G., Allard, C., and Cantero, A. Purine-metabolizing enzymes in normal rat liver and Novikoff hepatoma. Cancer Res., *18*: 952–958, 1958.
3. Ferdinandus, J. A., Morris, H. P., and Weber, G. Behavior of opposing pathways of thymidine utilization in differentiating, regenerating and neoplastic liver. Cancer Res., *31*: 550–556, 1971.

4. Georgiev, G. P. The structure of transcriptional units in eukaryotic cells. Current Topics Develop. Biol., *7*: 1–60, 1972.
5. Greenbaum, A. L., Gumaa, K. A., and McLean, P. The distribution of hepatic metabolites and the control of the pathways of carbohydrate metabolism in animals of different dietary and hormonal status. Arch. Biochem. Biophys., *143*: 617–663, 1971.
6. Heinrich, P. C., Morris, H. P., and Weber, G. Increased PRPP synthetase activity in rapidly growing hepatomas. FEBS Letters, *42*: 145–148, 1974.
7. Heinrich, P. C., Morris, H. P., and Weber, G. Transaldolase, transketolase and cell proliferation. Unpublished.
8. Huennekens, F. M., Digirolamo, P. M., Fujii, K., Henderson, G. B., Jacobsen, D. W., Neef, V. G., and Rader, J. I. Folic acid and vitamin B_{12}: Transport and conversion to coenzyme forms. Adv. Enzyme Regulation, *12*: 131–153, 1974.
9. Katunuma, N., Katsunuma, T., Kominami, E., Suzuki, K., Hamaguchi, Y., Chichibu, K., Kobayashi, K., and Shiotani, T. Regulation of intracellular enzyme levels by group specific proteases in various organs. Adv. Enzyme Regulation, *11*: 37–51, 1973.
10. Klein, G., Bregula, U., Wiener, F., and Harris, H. The analysis of malignancy by cell fusion. J. Cell Sci., *8*: 659–691, 1971.
11. Mayr, E. *In;* Populations, Species, and Evolution, The Belknap Press of Harvard University Press, Cambridge, 1971.
12. Morris, H. P. and Wagner, B. P. Induction and transplantation of rat hepatomas with different growth rate. *In;* H. Busch (ed.), Methods in Cancer Research, vol. 4, pp. 125–152, Academic Press, New York, 1968.
13. Novikoff, A. B. Transplantable rat liver tumor induced by 4-dimethyl-aminoazozene. Cancer Res., *17*: 1010–1027, 1957.
14. Omenn, G. S. Ectopic polypeptide hormone production by tumors. Ann. Intern. Med., *72*: 136–138, 1970.
15. Tomasi, V., Rethy, A., and Trevisani, A. Soluble and membrane-bound adenylate cyclase of Yoshida hepatoma. *In;* J. Schultz and H. G. Gratzner (eds.), The Role of Cyclic Nucleotides in Carcinogenesis (Miami Winter Symposia), vol. 6, pp. 127–144, Academic Press, New York, 1973.
16. Weber G. and Cantero, A. Glucose-6-phosphate utilization in hepatoma, regenerating and newborn rat liver, and in the liver of fed and fasted normal rats. Cancer Res., *17*: 995–1005, 1957.
17. Weber, G. Behavior of liver enzymes during hepatocarcinogenesis. Adv. Cancer Res., *6*: 403–494, 1961.
18. Weber, G. The molecular correlation concept: studies on the metabolic pattern of hepatomas. GANN Monograph, *1*: 151–178, 1966.
19. Weber, G. and Lea, M. A. The molecular correlation concept; An experimental and conceptual method in cancer research. *In;* H. Busch (ed.), Methods in Cancer Res., vol. 2, pp. 523–578, Academic Press, New York, 1967.
20. Weber, G. Carbohydrate metabolism in cancer cells and the molecular correlation concept. Naturwissenschaften, *55*: 418–429, 1968.
21. Weber, G., Queener, S. F., and Ferdinandus, J. A. Control of gene expression in carbohydrate, pyrimidine and DNA metabolism. Adv. Enzyme Regulation, *9*: 63–95, 1971.
22. Weber, G. The molecular correlation concept: Ordered pattern of gene expression in neoplasia. GANN Monograph on Cancer Research, *13*: 47–77, 1972.

23. Weber, G., Queener, S. F., and Morris, H. P. Imbalance in ornithine metabolism in hepatomas of different growth rates as expressed in behavior of L-ornithine carbamyl transferase activity. Cancer Res., *32*: 1933–1940, 1972.
24. Weber, G. Integrative action of insulin at the molecular level. Israel J. Med. Sci., *8*: 325–343, 1972.
25. Weber, G. Ordered and specific pattern of gene expression in neoplasia. Adv. Enzyme Regulation, *11*: 79–102, 1973.
26. Weber, G. The molecular correlation concept and cyclic AMP. *In;* J. Schultz and H. G. Gratzner (eds.), The Role of Cyclic Nucleotides in Carcinogenesis (Miami Winter Symposia), vol. 6, pp. 57–94, Academic Press, New York, 1973.
27. Weber, G. Steroid action: Phenotypic evidence for reprogramming of gene expression. *In;* C. A. Villee, D. B. Villee, and J. Zuckerman (eds.), Respiratory Distress Syndrome, pp. 237–270, Academic Press, New York and London, 1973.
28. Weber, G. The molecular correlation concept: Recent advances and implications. *In;* H. Busch (ed.), The Molecular Biology of Cancer, pp. 488–520, Academic Press, New York, 1974.
29. Weber, G., Trevisani, A., and Heinrich, P. C. Operation of pleiotropic control in hormonal regulation and in neoplasia. Adv. Enzyme Regulation, *12*: 11–41, 1974.
30. Weber, G. Biochemical Strategy of the Cancer Cell. Unpublished.
31. Williams-Ashman, H. G., Coppoc, G., and Weber, G. Imbalance in ornithine metabolism in hepatomas of different growth rates as expressed in formation of putrescine, spermidine and spermine. Cancer Res., *32*: 1924–1932, 1972.

Discussion of Paper by Dr. Weber

DR. POTTER: When you base your position on what is essentially the analysis of alternative metabolic pathways and the balance between the catabolic and anabolic branches, as I have emphasized in the 1950's and in 1961, I have no disagreement with you in an overall sense. But since I have placed some emphasis on the disordered aspects of cancer biochemistry, using the phrase "the challenge of diversity," I would like to comment on your emphasis on the ordered aspects. I would like to say that order and disorder are 2 dimensions to the cancer problem, both of which must be considered, like the length and with of a house. But I would ask you, Dr. Weber, if you maintain your position, are you not implying that the therapy of cancer should be found in a single mode of treatment? Is not the disordered aspect of cancer the chief obstacle in cancer therapy, and is it possible that the ordered aspect is something in common with normal tissue?

DR. WEBER: In answer to your question, I will repeat one of the conclusions I have drawn in my lecture: what is important in cancer is ordered and what is not, is the apparent randomness and diversity. The aspect of apparent disorder represents coincidental alterations and, as is well known, in the treatment of any disease the main thrust of therapy should not be directed to taking care of coincidental symptoms and signs, but to the essential, ordered pattern of the disease. The treatment of neoplasia should employ as potential targets some of the over 70 biochemical changes that we have discovered as malignancy-linked alterations in neoplasia. In various types of cancers, the different isozyme patterns, enzyme turnover, permeability, and other metabolic factors should determine the nature and technique of single or combination chemotherapy. Depending on tissue and cellular biochemical pattern variations, drug treatment may have to be different for the different tumors. I should also point out that while the balance of anabolic and catabolic pathways has been common knowledge to biochemists, only in my laboratories have the key enzymes and their importance been recognized for neoplasia. Also, in my laboratories these pathways in carbohydrate, pyrimidine, purine, ornithine, polyamine, cyclic AMP, and other metabolic areas have been explored systematically in relation to the different degrees of malignancy and the results have been integrated into a meaningful and coherent picture with the theoretical and practical aid of the molecular correlation concept.

DR. RUTTER: You have stated that the changes occurring in the glycolysis/gluconeogenesis system (loss, of certain enzymes, retention of low K_m enzyme) (a) has an advantage (b) is correlated with malignancy. Could this simply be a correlate of growth and/or differentiation? Thus are those changes also seen in development; that is, are the malignancy correlated changes also apparent in the developing liver? And thus are not correlated with malignancy itself, I do not accept the data you present on 17-day, or neonatal liver, since the liver is in a stage of development. Have you performed experiments on the very early liver before erythropoiesis dominated the cell type?

DR. WEBER: It is a matter of record, which you will find in original publications from my laboratories since the 1950's and in my recent review articles, that the enzymatic and metabolic imbalance we discovered in the hepatoma system is specific to liver neoplasia (*e.g.*, Cancer Res., *15*: 679, 1955; *17*: 995, 1957; *19*: 763, 1959; *21*: 933, 1961; *23*: 987, 1963; *31*: 550, 1971; *31*: 1004, 1971; *32*: 1933, 1972; GANN Monograph on Cancer Research, *13*: 47, 1972; Adv. Enzyme Regulation, *9*: 63, 1971). Through an array of quantitative and qualitative discriminants the pattern in neoplasia can be readily distinguished from that observed in differentiation (Adv. Enzyme Regulation, *11*: 79, 1973).

DR. PRASAD: Do you find a difference in adenylate cyclase response to hormones in hepatomas of various growth rate?

DR. WEBER: Yes, indeed, the experimental results we published (Allen *et al.*, Cancer Res., *31*: 557, 1971) demonstrated that in the slowly growing hepatomas adenylate cyclase activity retained its sensitivity to glucagon stimulation. However, the hormone response was less in the medium-growth-rate tumors, and it was lost in the rapidly growing hepatomas. Thus, the glucagon responsiveness of hepatoma adenylate cyclase (as measured in the homogenate) correlated negatively with tumor growth rate.

DR. JOHNSON: Is there any correlation between the K_m form of cyclic AMP PDE and the growth rate of the tumor?

DR. WEBER: Yes, there is a correlation. As we observed, when the enzyme was assayed in the $100{,}000 \times g$ supernatant fluid, the high K_m cyclic AMP PDE activity decreased but the low K_m PDE isozyme increased in parallel with the increase in hepatoma growth rate (Clark *et al.*, Cancer Res., *33*: 356, 1973).

DR. JOHNSON: Could you comment on the report from a group at the University of Virginia which states that there is no correlation between the cyclic AMP levels and the growth rate of hepatomas?

DR. WEBER: The results you referred to in the literature are difficult to evaluate, since apparently in that study up to 30 sec elapsed between removal of the tumor and freezing it in liquid nitrogen. In earlier studies in my laboratories and in current work with Dr. A. Trevisani, in samples taken from tumors from anesthetized animals and freeze-clamped within 1 to 2 sec after removal from the animal, we find that in the hepatomas the total cyclic AMP level is in normal range. However, as I pointed out elsewhere, the important part of the cyclic AMP may well be that in the plasma membrane compartment (Weber, *In*; The Role of Cyclic Nucleotides in Carcinogenesis (Miami Winter Symposia), vol. 6, p. 57, Academic Press, N.Y., 1973; Weber *et al.*, Adv. Enzyme Regulation, *12*: 11, 1974). Furthermore, the behavior of cyclic AMP in ischemia is markedly different in the tumors from that observed in the normal liver.

DR. TOMKINS: Why would cells have evolved to contain a coordinated set of reactions turned on only in the malignant state? This should have been a lethal devellopment for the organism. Perhaps the "malignancy program" you describe is an exaggeration of the combination of reactions turned on during cell growth (*e.g.*, during regeneration).

DR. WEBER: The reprogramming of gene expression that occurs in neoplasia, as I pointed out in my lecture, confers selective biological advantages to cancer cells. The mechanism, in part, involves the release and amplification that is observed in normal liver regeneration; however, there are other alterations that do not have any counterpart in regeneration or in differentiation and they reveal the unique and specific reprogramming of gene expression that is linked with the degrees of neoplastic transformation (Weber *et al.*, Adv. Enzyme Regulation, *12*: 11–41, 1974).

PHENOTYPES OF TUMOR CELLS

Prof. VAN R. POTTER

Today's session will be devoted to a collection of papers, all of which bear on the general hypothesis to which I and many others subscribe, that cancer is a disease of differentiation.

I have expressed this view in the brief phrase "oncogeny is blocked ontogeny" and have described an overall program in a figure that is here presented modified from our earlier form published by P. R. Walker and V. R. Potter in Adv. Enzyme Regulation, *10*: 1972. Recent contributions from my group, not to be presented here, will appear in 4 papers in Biochem. J. in late 1973 and early 1974 (first authors: Bonney, Bonney, Hopkins, Walker).

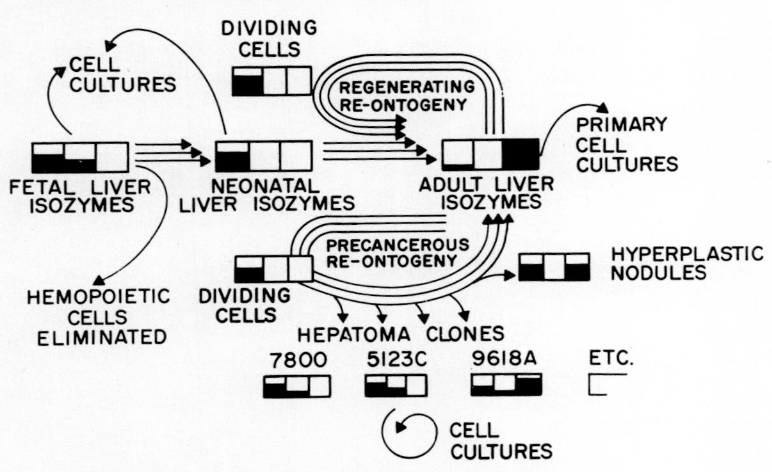

To understand what is meant by the blocked-ontogeny hypothesis it would be necessary to discuss the following subjects that are subsumed under the simplistic statement:

1. Hetero-blocked ontogeny;
2. Stringent-blocked ontogeny;
3. Leaky-blocked ontogeny;
   a. Hyperplastic nodules in liver;
   b. Some Morris hepatomas;
4. Phenotypic heterogeneity in tumor cell populations;
5. Forced inductions in leaky-blocked systems.

These topics will be presented by me on a subsequent occasion.

Enzymatic and Metabolic Alterations in Experimental Hepatomas

Sidney Weinhouse
Fels Research Institute, Temple University School of Medicine, Philadelphia, Pennsylvania, U.S.A.

Abstract: Studies of enzyme activities and isoenzyme composition in the Morris hepatomas, a series of chemically induced, transplantable rat hepatomas ranging widely in growth rate and degree of differentiation, have revealed a common pattern of isozyme alterations. In the highly differentiated, slow-growing hepatomas the sole or preponderant isozymes are the so-called liver marker types; *e.g.*, glucokinase, aldolase B, pyruvate kinase L, and the liver-type glycogen phosphorylase. With decreased differentiation and increased growth rate, these liver types decrease markedly; and in the fast-growing, poorly differentiated hepatomas, the liver-type isozymes are nearly or completely replaced by high activities of non-hepatic type isozymes. The slow-growing, well-differentiated hepatomas utilize glucose poorly and have a low rate of aerobic glycolysis; but like normal adult liver they utilize fatty acids as respiratory fuel. The fast-growing, poorly differentiated hepatomas have largely lost the ability to oxidize fatty acids, but have a high aerobic glycolysis and utilize glucose as a respiratory fuel.

Evidence from experiments in model systems points to the high levels of pyruvate kinase as a determining factor in the high aerobic glycolysis in the poorly differentiated hepatomas. It is proposed that the replacement of isozymes geared for organ function by isozymes geared primarily for efficient utilization of metabolites and that are not under host dietary or endocrine control represents a molecular basis for the uncontrolled proliferation of cancer cells.

These findings are similar to other aberrancies of protein synthesis in cancer, such as the loss of tissue-specific antigens and the appearance of tumor-associated antigens; and the "ectopic" production of polypeptide hormones by non-endocrine neoplasms. The isozyme data thus support and add functional significance to an ever-increasing body of evidence for an anomaly of gene expression in cancer, whereby certain genes that code for proteins present in normal adult tissue is switched off and genes that are normally inactive are switched on. The fact that antigens and isozymes found in tumors are also present in fetal tissues suggests that genes active in the fetal stage but inactivated during normal embryonic development are re-activated in cancer.

Although the mystery of cancer still remains unsolved, there is a growing awareness that this disease is associated with abnormalities of gene expression, manifested by misprogramming of protein synthesis. The clinical literature provides many examples of the ectopic production of polypeptide hormones in non-endocrine tumors (*10*, *16*, *33*, *37*); and the immunologic literature discloses that proteins identifiable as antigens in normal, adult differentiated tissues are lost in tumors, while neo-antigens appear that are absent from the cell of origin but may be present in fetal tissue (*1*, *4*, *5*, *14*, *25*, *43–45*). In recent years, work from our own and other laboratories (*8*, *23*, *34*, *41*, *42*, *49*) have provided additional evidence of such abnormalities and have given them a functional significance by demonstrating that neoplasia is also associated with profound alterations in the molecular structures of enzymes that appear in this condition (*8*, *23*, *34*, *41*, *42*, *49*). In the time available I would like to describe these changes and to consider with you how they may affect the metabolic behavior of neoplastic cells.

Morris Hepatomas

The experimental system comprises a series of chemically induced transplantable liver neoplasms of the rat known as the Morris hepatomas (*28–30*). Although they represent a wide spectrum of biological behavior, they may conveniently be divided into 3 classes on the basis of their degree of differentiation as judged by usual histologic criteria. The highly differentiated tumors, of which there are relatively few examples, grow very slowly. The growth rate, which can be assessed from the average time between transplantation, ranges from 3 to 4 months up to well over a year. The many well-differentiated Morris hepatomas grow somewhat more rapidly, taking from 2 to 6 months for a transplant generation; and the poorly differentiated tumors grow so rapidly as to kill their hosts in less than 1 month. The highly differentiated tumors have the normal liver chromosome number and karyotype, the well-differentiated differ slightly, and the poorly differentiated deviate markedly in chromosome karyotype and number, from those of rat liver. Respiration decreases moderately with loss of differentiation, but the most striking biochemical feature is the low or negligible glycolysis in the well-differentiated tumors; this is in decided contrast with the usual high level of glycolysis in the poorly differentiated tumors (*3*, *11*, *47*).

These hepatomas, ranging widely in their growth rate and in the degree of differentiation, yet originating from a single cell type, the parenchymal cell of the liver, obviously provide an ideal model system for exploring molecular mechanisms underlying tumor behavior. We undertook a study of their enzymology and metabolism directed to learn 2 things; first, why they differ in their glycolytic activity; and second, to ascertain to what extent enzymes playing key roles in hepatic function are lost or retained in these hepatomas.

Isozyme Alterations

Four enzymes have been studied, each of which plays a key role in hepatic carbohydrate metabolism. Each exists in multiple forms, and in each instance the liver possesses an isozyme uniquely tailored to important hepatic function (Table 1). The hepatic type of glucose-ATP phosphotransferase known as glucokinase, is under host control through its adaptive responsiveness to dietary glucose and insulin; and its kinetic properties and high specificity toward glucose make it particularly significant in liver physiology. It is accompanied by low activities of 3 other relatively nonspecific hexokinases. The hepatic aldolase, aldolase B, is essentially the sole form in adult rat liver. It is highly active toward fructose-1-phosphate, the first product of the metabolism of fructose in liver, and thus has a crucial role in the utilization of this sugar. The L-type pyruvate kinase, like glucokinase, is adaptive to diet and insulin, and also plays a key role in such hepatic functions as glucose utilization and gluconeogenesis. The fourth enzyme, glycogen phos-

TABLE 1. Enzymes Studied in Rat Liver and Hepatomas

1. Glucose-ATP phosphotransferases

   Glucose + ATP ⟶ glucose-6-phosphate + ADP

2. Aldolases

   Fructose-1, 6-diphosphate ⇌ glyceraldehyde-3-P + dihydroxyacetone-P

   Fructose-1-phosphate ⇌ glyceraldehyde + dihydroxyacetone-P

3. Pyruvate kinases

   Phosphoenolpyruvate + ADP ⟶ pyruvate + ATP

4. Glycogen phosphorylases

   n glucose-1-phosphate ⇌ glycogen + n P_i

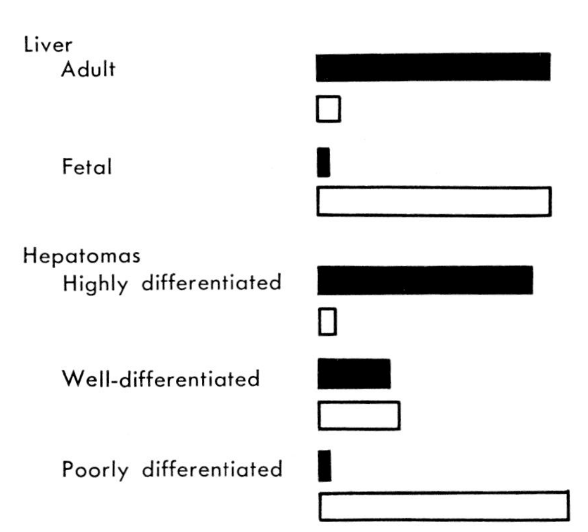

FIG. 1. Diagrammatic representation of alteration of isozyme pattern in experimental hepatomas. Data are based on experimental results obtained from glucose phosphotransferases, aldolases, pyruvate kinases, and glycogen phosphorylase and have been reported previously (*39, 40, 49*). ▬ adult liver isozyme; ▭ hepatoma isozyme.

phorylase, also exists in isozymic forms; in this instance also, the liver form has chemical and kinetic properties that are tailored to the highly regulated processes of glycogen deposition and mobilization.

All four of these enzymes displayed a common pattern of isozyme alteration, which is described diagrammatically in Fig. 1 (*38–40*, *49*). in normal adult liver, the liver form is the sole or predominant isozyme, while non-hepatic isozymes are either low or absent in activity. In all instances, the slow-growing, highly differentiated hepatomas exhibited a similar pattern, with virtually complete retention of the liver-type isozyme. As dedifferentiation proceeds, and growth rate increases, the hepatomas lose the liver-type isozyme activity and isozymes appear that are normally low or undetectable in the adult liver. With progressive dedifferentiation to the poorly differentiated state, there is not only a complete or nearly complete disappearance of the liver type isozyme; there is a replacement by a non-hepatic type isozyme, sometimes in extremely high activity compared with the original total enzyme activity in liver. It is notable that in every instance, the isozyme patterns of the poorly differentiated hepatomas resemble those of fetal liver, where, like the poorly differentiated hepatomas, the adult isozyme activity is low or absent. Data on 2 enzymes, pyruvate kinase, and phosphorylase, are detailed in the following sections.

Glycogen Phosphorylase

Glycogen phosphorylase isozymes can be distinguished readily by isoelectric focussing (*38–40*). In the top panel of Fig. 2 there is shown the focussing pattern for skeletal muscle phosphorylase. There is a sharp peak at pH 6.2; and this isozyme is largely in the b form as shown by the high activity with AMP, designated with open circles, compared with the low activity without AMP, outlined by the shaded circles. The liver isozyme shown in the second panel, focusses at pH 5.9, and is further distinguished from the muscle isozyme in requiring 0.5M sulfate for full activation, as outlined by the triangles, even in the presence of AMP shown in the open circles. In contrast, the poorly differentiated Novikoff hepatoma isozyme, which exemplifies other poorly differentiated hepatomas, has one major peak at pH 5.6, and also differs from the liver isozyme in not requiring sulfate for activation. The fourth panel shows that a well-differentiated tumor hepatoma 20, has the liver isozyme as a major peak, but this is also accompanied by a minor peak that focusses like the hepatoma peak. The fifth panel shows clearly that in 21-day fetal liver, the adult liver form is accompanied by a form which focusses like the hepatoma form. The separate identities of these 3 phosphorylase isozymes have been established by immunochemical studies (*38*) and there appears to be no doubt that the major or sole isozyme in poorly differentiated hepatomas is the same as that present in normal fetal liver.

Pyruvate Kinase—Its Role in Aerobic Glycolysis

Another striking example of the same kind is shown in Fig. 3 (*13*). Pyruvate

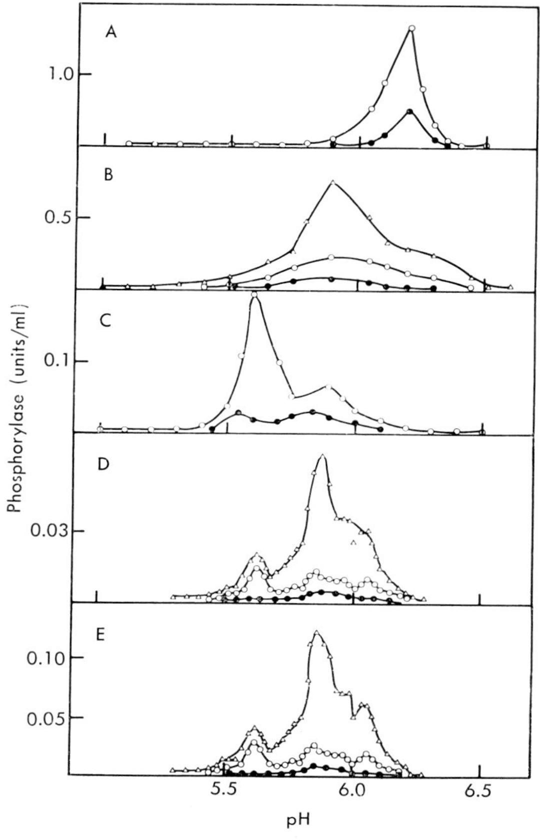

FIG. 2. Isoelectric focussing of glycogen phosphorylase isozymes of normal and neoplastic tissues (*39, 40*). A, muscle; B, liver; C, Novikoff hepatoma; D, Morris hepatoma 20; E, 21-day fetal liver. ○ with AMP; ● without AMP; △ with 0.5 M sulfate.

kinase is present in adult rat liver in a major form, which is tightly held by DEAE-cellulose and requires a concentration of approximately 0.15M Cl^- ion for elution. This isozyme has kinetic properties which endow it with unique importance in liver metabolism, and like glucokinase (*49*) is also responsive to diet and to insulin. It is accompanied by a low activity of another isozyme. As shown in the second panel, rat skeletal muscle has a single isozyme which is less tightly bound and elutes with 0.05M Cl^- ion, and appears to be identical with one of the minor liver forms. The third panel shows that the pyruvate kinase isozyme of the poorly differentiated Novikoff hepatoma is not held by the resin and is eluted in the void volume; it appears to be identical with another isozyme present also in liver at low activity. In the fourth panel we note that the major tumor isozyme is identical with a fetal liver form as well —thus we have another example of a re-activation of a fetal gene in cancer. Before returning to the subject of fetal manifestations in cancer, it will be desirable to digress to discuss the possible significance of the observed alteration of pyruvate kinase isozyme pattern in cancer metabolism.

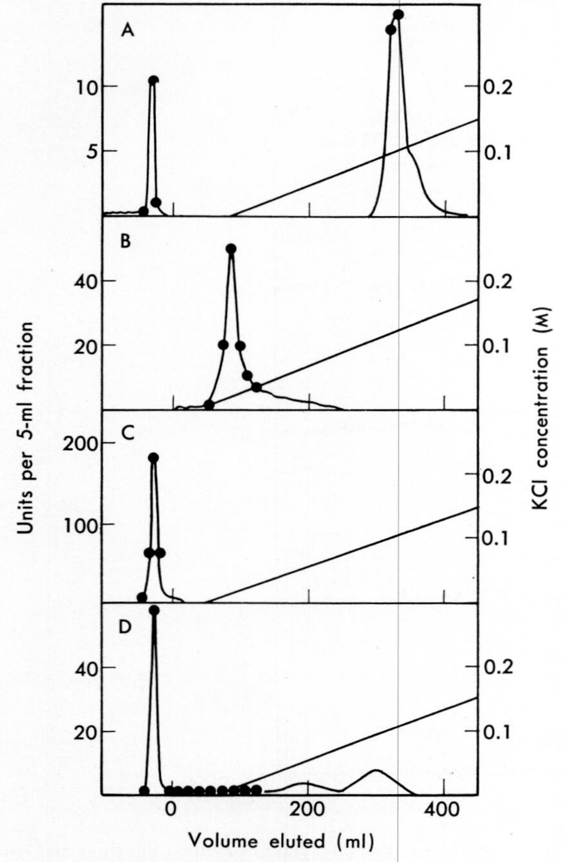

FIG. 3. Gradient elution of pyruvate kinase isozymes from DEAE-cellulose (*13*). A, adult liver or 9618A hepatoma; B, muscle; C, Novikoff hepatoma; D, 19-day fetal liver.

Figure 4 depicts how the isozymes of pyruvate kinase progress. There is essentially the normal liver pattern of high liver-type isozyme in the highly differentiated tumors. As dedifferentiation and growth rate increase, the liver-type isozyme falls to low activities in the well-differentiated; then virtually disappears in the poorly differentiated hepatomas, to be replaced by extremely high activities of the hepatoma isozyme.

The extremely high pyruvate kinase activity in the poorly differentiated tumors prompted us to examine whether this enzyme pattern might provide a clue to the high glycolytic activity of these hepatomas, as well as that of other poorly differentiated tumors. Before describing the results of our experiments, let us review briefly the background of the aerobic glycolysis of tumors. About a half-century ago Otto Warburg, an illustrious and colorful, indeed a legendary figure in cancer research, pioneered in what may be called modern cell biology by applying what were then modern techniques and concepts of physical chemistry to studies of the cancer cell (*46*). He was the first to note that slices of the most diverse tumors had one common

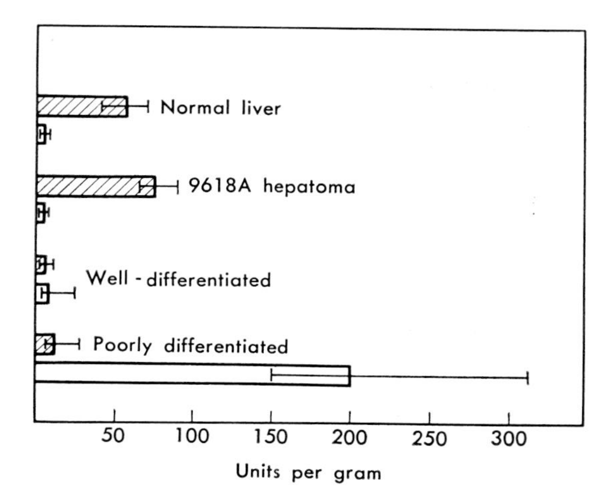

FIG. 4. Pyruvate kinase isozyme activities in rat liver and hepatomas. Type II is the normal liver isozyme and type I is the form present only in fetal liver, but also in kidney and brain. The muscle type was not present in either liver or hepatomas. [hatched box] type I; [open box] type II.

property; they produced large amounts of lactic acid from glucose. The essence of his experimental observations, somewhat oversimplified, is shown in Fig. 5. When normal tissue slices are incubated in a nutrient medium containing glucose, but in the absence of oxygen, there is a rapid and continuous utilization of glucose and production of lactic acid, a process which Warburg termed glycolysis. However, when the tissues were incubated in an oxygen atmosphere, glycolysis virtually ceased. This decrease of glycolysis brought about by oxygen was named by Warburg the Pasteur Effect, since it resembled an observation made some 100 years ago by Pasteur; namely, when yeast is grown in air, it loses the capability to ferment glucose to ethanol.

The behavior of tumor slices Warburg found, however, was quite different. Anaerobic glycolysis was very high; and although a Pasteur Effect was observed, the presence of oxygen did not lead to elimination of glycolysis, and aerobic gly-

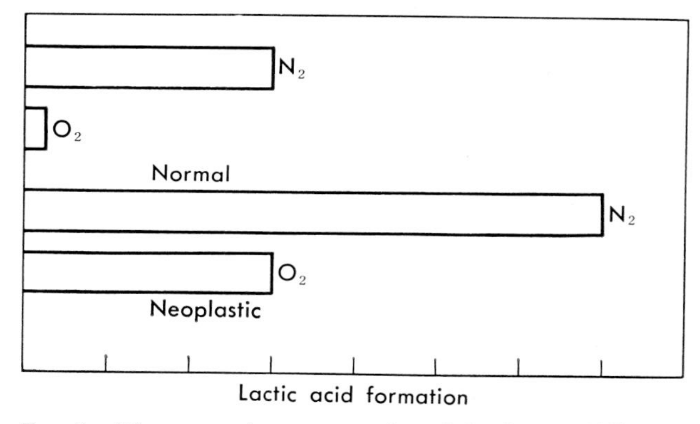

FIG. 5. Diagrammatic representation of the Pasteur Effect in normal and neoplastic tissue slices.

colysis remained high. Warburg regarded the persistence of aerobic glycolysis as a key to the neoplastic transformation. He proposed that the high aerobic glycolysis is the result of a defect in respiration; that cancer results when the cell responds to an irreversible injury to its respiratory mechanism by adopting a fermentative metabolism. According to Warburg, such cells cannot maintain the differentiated state and as undifferentiated cells they grow in uncontrolled fashion. This dictum, vigorously propounded by an illustrious figure in biology, captured the imagination of cancer researchers, and it is difficult to exaggerate the influence of this theory on the direction of cancer research for decades. Moreover, the Pasteur Effect, although still not well understood, has been a focal point for countless studies on metabolism and its regulation.

It was not until about 1950, however, when isotope tracer studies showed that tumors could oxidize glucose to CO_2 at rates comparable with those of normal tissues, that the Warburg hypothesis was seriously questioned (*48*). Although Warburg defended his theory vigorously throughout the years, many efforts to discover defects, or even substantive impairments of respiration in cancer, were unsuccessful. A variety of experimental techniques demonstrated that tumors utilize oxygen at low to moderate rates, they contain a full complement of respiratory enzymes and coenzymes, they have mitochondria which appear normal morphologically and functionally (although they may be low in number in some tumors), the citric acid cycle is operative, and they couple oxidation with ATP production. In the light of our more advanced knowledge of respiration, mitochondrial function, and the role of the mitochondria in carbohydrate metabolism, the Warburg hypothesis may now be regarded at best as a gross oversimplification of exceedingly complex regulatory mechanisms (*2*, *3*, *5*).

In considering enzymatic sites for the Pasteur Effect, an inhibitory effect of respiration on glycolysis, it is logical to consider those enzymes that carry out transfers of phosphate to and from phosphorylated intermediates, since both glycolysis and respiration are intimately coupled to phosphorylation reactions. Much attention has been focussed in recent years on phosphofructokinase because of its remarkable allosteric properties. It is strongly inhibited by ATP and citrate and is deinhibited by AMP, ADP, and P_i; numerous kinetic experiments point to feedback control on this enzyme by ATP and citrate as the major determinant of glycolytic activity in muscle and possibly also in brain (*31*, *35*).

Subsequent transphosphorylation reactions, namely at triose phosphate dehydrogenase, where inorganic phosphate is taken up into 1, 3-diphosphoglycerate; phosphoglycerate kinase, where phosphate is transferred from 1, 3-diphosphoglycerate to ADP to yield 3-phosphoglycerate and ATP; and pyruvate kinase, where phosphate is transferred from phosphoenolpyruvate to ADP to yield pyruvate and ATP are also considered to be possible sites for the Pasteur Effect. These 3 transphosphorylation reactions compete directly for ADP and P_i with the mitochondrial respiratory system, where ADP and P_i are converted to ATP by oxidative phosphorylation coupled with electon transport. They are, therefore, obvious sites for the Pasteur Effect, and indeed this was suggested 30 years ago, independently by Johnson (*20*) and Lynen (*27*).

Competition for ADP as a Determinant of Aerobic Glycolysis

To explore the possible role of pyruvate kinase as a determinant of glycolytic rate we utilized several model sytems consisting of whole, fortified homogenates of liver and hepatomas, as well as centrifugally separated cellular glycolytic and respiratory systems (*26*) and also isolated mitochondria interacting with purified pyruvate kinase in experiments currently being conducted by Gosalvez (*17*).

Pyruvate kinase and mitochondria both use ADP as a substrate, as shown in Fig. 6. During active respiration, ADP otherwise used for pyruvate kinase might be used by mitochondria; thereby lowering pyruvate kinase activity, and limiting glycolysis. This provides a plausible mechanism for the Pasteur Effect. It is also evident that when pyruvate kinase is very high, as it is in poorly differentiated tumors, it will preferentially capture ADP and thus lead to increased glycolysis.

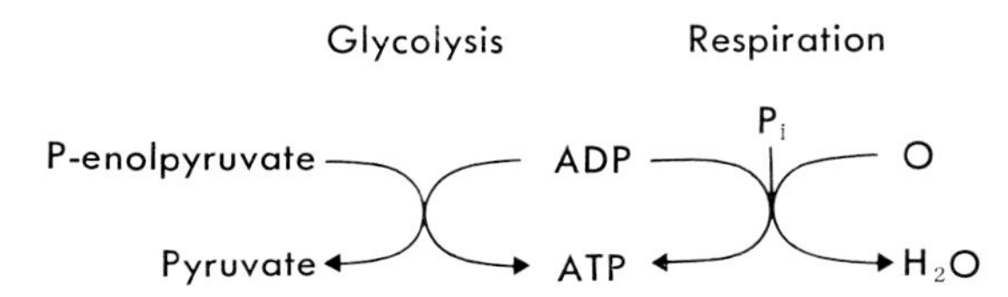

Fig. 6. Diagram illustrating competition of pyruvate kinase and mitochondria for the available ADP.

One other point is worthy of note. It has long been known that highly glycolyzing tumors, such as for example ascites cells, display respiratory inhibition in the presence of glucose, a phenomenon known as the Crabtree Effect. On the basis of this diagram, it is evident that when glycolysis is high, owing to very high pyruvate kinase levels there may be insufficient ADP to maintain optimal respiratory activity in mitochondria in which respiration is tightly coupled to oxidative phosphorylation. We found experimentally, using model systems consisting of a restricted segment of the glycolytic system together with respiratory cell fractions, that high pyruvate kinase activity will indeed inhibit mitochondrial respiration and thus simulate the Crabtree Effect (*17*); it will enhance aerobic glycolysis, and in reverse fashion, will be inhibited by a high respiratory activity, and thus simulate the Pasteur Effect (*26*). On the basis of these results, we feel that the hitherto puzzling Pasteur and Crabtree Effects, as well as the high aerobic glycolysis of tumors, may all three have their origin at this single enzymatic site.

On the basis of the foregoing results, we regard the high aerobic glycolysis of tumors, not as a *sine qua non* of cancer, but rather as a consequence of a late stage of dedifferentiation, attributable to an enhanced activity of pyruvate kinase coupled with an alteration of its molecular form.

Fatty Acid Oxidation in Hepatomas

Another metabolic process that is profoundly affected by enzymatic alterations in hepatomas is fatty acid oxidation. The well-differentiated tumors, which utilize glucose poorly, like normal liver, apparently use fatty acids for metabolic fuel. Studies from our laboratory by Bloch-Frankenthal and Langan (*7*), show that oxidation of fatty acids, either to CO_2 or to acetoacetate is high in liver and in well-differentiated hepatomas, but is negligible in poorly differentiated tumors (Table 2). β-Hydroxybutyrate dehydrogenase activity, which plays a functional role in liver in maintaining the acetoacetate-β-hydroxybutyrate equilibrium, and which is an important factor in the mitochondrial redox state, is moderately high in well-differentiated hepatomas, but negligible in poorly differentiated (*32*); and the same is true for the first step in metabolic activation of fatty acids, their conversion to acylCoA (*24*). Thus we again see a diversity of metabolic behavior. At the one extreme are those slow-growing, well-differentiated hepatomas which utilize little glucose and oxidize fatty acids; and at the other extreme are those rapidly growing, poorly differentiated hepatomas, that have largely lost the capability for fatty acid oxidation, but have acquired the ability to utilize glucose.

TABLE 2. Fatty Acid Oxidation in Rat Hepatomas

Enzyme reaction	Liver	Hepatomas	
		Well-differentiated	Poorly differentiated
Oxidation of fatty acids to CO_2	High	High	Negligible
Oxidation of fatty acids to acetoacetate	High	High	Negligible
β-Hydroxybutyrate dehydrogenase activity	High	Moderate	Negligible
Fatty acylCoA synthetase activity	High	Moderate	Low

DISCUSSION

The Morris hepatomas have taught us a number of important principles. (1) A single cell type of origin can give rise to tumors of great diversity, paralleled by diversity in the molecular structures of their enzymes. It is now clear that mere assays of total enzyme activity may obscure profound alterations in isozyme activity which undoubtedly affect the metabolism and growth of these hepatomas. (2) Those isozymes that are lost in dedifferentiation are those that participate in metabolic processes that are unique to liver. It may be considered as a general proposition that the loss of those isozymes that are under host regulation and that are geared to organ function, and their replacement by isozymes geared to the efficient utilization of metabolic fuels represents the molecular basis for the unbridled proliferation of tumors. (3) The high aerobic glycolysis commonly occurring in tumors is not an invariable result of neoplasia, but is probably a manifestation of isozyme changes occurring as a late stage of tumor differentiation. (4) Alterations of isozymes are in all likelihood a general phenomenon of neoplasia. Similar isozyme alterations in hepatomas, with expression of fetal isozymes, have been reported to occur with various aminotransferases (*19*), glutaminase (*22*), thymidine kinase (*21*); and these probably only scratch the surface.

1. Alterations in antigens

In discussing the significance of these isozyme alterations, it is important to note the striking similarity to alterations in immunologic properties. In cancer, antigens specific to the adult differentiated cells disappear, while new tumor-associated antigens appear; in many instances these neoantigens are present also in fetal tissue. The elaboration of a fetal protein, α-fetoglobulin by many human hepatomas (*10*), and the presence of a fetal antigen, the so-called carcinoembryonic antigen in tumors of the gastrointestinal tract (*14, 15*) offer the exciting promise of early clinical diagnosis. The so-called gs (group specific) antigen found in many mouse tumors induced by oncogenic C-type viruses, is also present in fetal tissues (*18*), and affords yet another example of the activation in tumors of a fetal protein. Duff and Rapp (*9*) have shown that SV-40 viral transformed hamster cells have a surface antigen that is derived from the host cell genome, and which is also found in hamster embryos; and Baldwin (*6*) has also observed fetal antigens in chemically induced rat hepatomas (Table 3).

TABLE 3. Immunologic Evidence for Abnormal Gene Expression in Cancer

1. Tumor-associated antigens
2. Loss of organ-specific antigens
3. Embryonal antigens
   a. α-Fetoglobulin
   b. Carcinoembryonic antigen
   c. Embryonic hamster antigen in SV-40 transformed cells
   d. Embryonic antigen in chemically induced hepatoma

In considering the expression of embryonic proteins in tumors, the question arises as to whether this is programmed or fortuitous. These embryonic manifestations have been described in such terms as derepressive dedifferentiation (*14*), retrogenic expression (*14, 15, 44*), retrodifferentiation (*5*), blocked ontogeny (*36*), *etc.*, implying a more or less systematic reversion to an embryonic state; that is, a reversal of normal ontogeny. Our isozyme data do indicate some order in the observed alterations. For example, many well-differentiated hepatomas have largely lost isozymes of the differentiated liver cell but do not exhibit a resurgence of the fetal isozyme; but when resurgence of the fetal form does occur, there is always a loss of the normal hepatic isozyme (*49*). These observations suggest that certain genes may have to be switched off before others are switched on. However, in general the alterations seem to be sporadic and unpredictable, and therefore suggest a disordered rather than a programmed mechanism of gene activation.

2. Ectopic protein synthesis

This conception of a disordered rather than a programmed reversion is strongly bolstered by a large and ever-growing body of clinical literature pointing to bizarre aberrations of protein synthesis in certain tumors, resulting in the ectopic production of polypeptide hormones (*10, 16, 33, 37*) (Table 4).

Such profound anomalies of gene expression are extraordinary in view of the

TABLE 4. Para-endocrine Manifestations in Cancer (3)

Ectopic hormone	Tumor site
Adrenocorticotrophic	Lung, thymus, pancreas, others
Antidiuretic	Lung, prostate, pancreas
Parathyroid	Lung, kidney, cervix, ovary, *etc.*
Erythopoietic	Kidney, cerebellum, liver
Insulin	Stomach
Gonadotrophic	Lung, esophagus

normally rigid and highly selective control of gene transcription in normal differentiated tissue, and they raise many questions concerning their significance to neoplasia. It is attractive to consider that neoplasia is initiated by some impairment of this regulatory mechanism. Once the rigid control is lost, many of the familiar patterns of neoplasia would inevitably follow; chromosome aberrations, loss of antigens, alterations in surface properties, and all of the other characteristics of tumor progression could be envisioned to result from this initial injury.

Considerable discussion surrounds the question whether cancer is due to a somatic mutation. The present work, especially in light of the alterations in antigens, points to a disorder of gene expression rather than of gene structure. Arguing also against somatic mutation is the fact that the abnormal proteins which arise in tumors are not "new" protein species, but rather are "misplaced" proteins; misplaced either in site, as exemplified by the ectopic production of polypeptide hormones; or in time, as exemplified by the resurgence of fetal proteins in the form of isozymes and antigens. If cancer results from a somatic mutation, it would appear to be a mutation of a regulatory rather than a structural gene.

It is important to ask whether these aberrations of protein synthesis are the cause or the effect of the neoplastic transformation. Neither isozyme nor antigen alterations are all or none phenomena; but rather are quantitative; many of the isozymes normally vary greatly depending on dietary or hormonal conditions. Some antigens as well as isozymes appear in regenerating as well as in fetal liver; and many of the isozyme alterations seen in poorly differentiated hepatomas also appear in ostensibly normal liver cells when these are grown *in vitro* (*12*). Many more questions can be asked and one can speculate at great length in the light of these anomalies of gene expression; but we need more information on the mechanisms of normal differentiation before we can understand these disorders of differentiation, which may lie at the heart of the neoplastic transformation.

ACKNOWLEDGMENTS

The author gratefully acknowledges the support of the National Cancer Institute for grants CA-10916 and CA-12227; and the American Cancer Society for grant BC-74. The aid of Jennie B. Shatton, Kiyomi Sato, Albert Williams, David Meranze, and Harold P. Morris is also acknowledged with appreciation.

Figure 2 is reproduced from Ref. *25* through courtesy of Science, and Figs. 4 and 5 are reproduced through the courtesy of Cancer Research.

REFERENCES

1. Abelev, G. I. α-Fetoprotein in oncogenesis and its association with malignant tumors. Adv. Cancer Res., *14*: 195–358, 1971.
2. Aisenberg, A. C. The Glycolysis and Respiration of Tumors. Academic Press, New York, 1961.
3. Aisenberg, A. C. and Morris, H. P. Energy pathways of hepatomas 5123. Nature, *191*: 1314–1316, 1961.
4. Alexander, P. Foetal "antigens" in cancer. Nature, *235*: 137–140, 1972.
5. Anderson, N. G. and Coggin, J. H., Jr. Retrogenesis: problems and prospects. *In;* N. G. Anderson, J. H. Coggin, Jr., E. Cole, and J. W. Holleman (eds.), Proceedings of the Second Conference on Embryonic and Fetal Antigens in Cancer, pp. 361–368, Oak Ridge National Laboratory, Oak Ridge, 1972.
6. Baldwin, R. W. Immunological aspects of chemical carcinogenesis. Adv. Cancer Res., *18*: 1–75, 1973.
7. Bloch-Frankenthal, L., Langan, J., Morris, H. P., and Weinhouse, S. Fatty acid oxidation and ketogenesis in transplantable liver tumors. Cancer Res., *25*: 732–736, 1965.
8. Criss, W. E. A review of isozymes in cancer. Cancer Res., *31*: 1523–1542, 1971.
9. Duff, R. and Rapp, F. Reaction of serum from pregnant hamsters with surface of cells transformed by S.V. 40. J. Immunol., *105*: 521–523, 1970.
10. Eliel, L. P. Non endocrine secreting neoplasms: clinical manifestations. Cancer Bull., *20*: 30–37, 1968.
11. Elwood, J. C., Lin, Y. C., Cristofalo, V. J., Weinhouse, S., and Morris, H. P. Glucose utilization in homogenates of the Morris hepatoma 5123 and related tumors. Cancer Res., *23*: 906–913, 1963.
12. Farber, E., Shatton, J. B., and Weinhouse, S. Unpublished.
13. Farina, F., Shatton, J. B., Morris, H. P., and Weinhouse, S. Cancer Res., *34*: 1439–1446, 1974.
14. Gold, P. Embryonic origin of human tumor-specific antigens. Prog. Exp. Tumor Res., *14*: 43–58, 1971.
15. Gold, P. and Freedman, S. O. Specific carcinoembryonic antigens of the human digestive tract. J. Exp. Med., *122*: 467–481, 1965.
16. Goodall, C. M. A review: on para-endocrine cancer syndromes. Int. J. Cancer, *4*: 1–10, 1969.
17. Gosalvez, M., Perez-Garcia, J., and Weinhouse, S. Eur. J. Biochem., *46*: 133–140, 1974.
18. Huebner, R. J., Kelloff, G. J., Sarma, P. S., Lane, W. T., Turner, H. C., Gilden, R. V., Oroszlan, S., Meyer, H., Myers, D. D., and Peters, R. L. Group-specific antigen expression during embryogenesis of the genome of the C-type RNA tumor virus: implications for ontogenesis and oncogenesis. Proc. Natl. Acad. Sci. U.S., *67*: 336–376, 1970.
19. Ichihara, A. and Ogawa, K. Isozymes of branched chain amino acid transaminase in normal rat tissues and hepatomas. GANN Monograph on Cancer Research, *13*: 181–190, 1972.
20. Johnson, M. J. The role of aerobic phosphorylation in the Pasteur Effect. Science, *94*: 100–202, 1941.

21. Jones, O. W., Taylor, A., and Stafford, M. A. Fetal gene expression in human tumor cells. J. Clin. Invest., *51*: 47a, 1972.
22. Katunuma, N., Katsunuma, T., Tomino, I., and Natsuda, Y. Regulation of glutaminase activity and differentiation of the isozyme during development. Adv. Enzyme Regulation, *6*: 227–242, 1968.
23. Knox, W. E. Enzyme Patterns in Fetal, Adult and Neoplastic Rat Tissues. S. Karger AG., Basel, 1972.
24. Langan, J., Morris, H. P., and Weinhouse, S. Unpublished.
25. Lengerova, L. Expression of normal histocompatibility antigens in tumor cells. Adv. Cancer Res., *16*: 235–272, 1972.
26. Lo, C. H., Cristofalo, V. J., Morris, H. P., and Weinhouse, S. Studies on respiration and glycolysis in transplanted hepatomas of the rat. Cancer Res., *28*: 1–10, 1968.
27. Lynen, F. The aerobic phosphate requirement of yeast. The Pasteur Effect. Ann. Chem., *546*: 120–141, 1940.
28. Morris, H. P. Some growth morphological and biochemial characteristics of hepatoma 5123 and other transplantable hepatomas. Prog. Exp. Tumor Res., *3*: 370–411, 1963.
29. Morris, H. P. Studies on the development, biochemistry and biology of experimental hepatomas. Adv. Cancer Res., *9*: 227–302, 1965.
30. Morris, H. P. and Wagner, B. P. Induction and transplantation of rat hepatomas with different growth rate (including minimal deviation hepatomas). *In;* H. Busch (ed.), Methods in Cancer Research, vol. 4, pp. 125–152, Academic Press, New York, 1968.
31. Newsholme, A. E. and Gevers, W. Control of glycolysis and gluconeogenesis in liver and kidney cortex. Vitamins and Hormones, *25*: 1–79, 1967.
32. Ohe, K., Morris, H. P., and Weinhouse, S. β-Hydroxybutyrate dehydrogenase activity in liver and liver tumors. Cancer Res., *27*: 1360–1371, 1967.
33. Omenn, G. S. Pathobiology of ectopic hormone production by neoplasms in man. *In;* H. L. Ioachim (ed.), Pathobiology Annual, pp. 177–216, Appleton, New York, 1973.
34. Ono, T. and Weinhouse, S. (eds.). GANN Monograph on Cancer Research, *13*: 1972.
35. Passoneau, J. V. and Lowry, O. H. The role of phosphofructokinase in metabolic regulation. Adv. Enzyme Regulation, *2*: 265–276, 1964.
36. Potter, V. R. Environmentally-induced metabolic oscillations as a challenge to tumor autonomy. *In*; W. J. Whelan and J. Schultz (eds.), Homologies in Enzymes and Metabolic Pathways (Miami Winter Symposia), vol. 2, pp. 291–313, North Holland, Publishing Co., Amsterdam, 1970.
37. Roof, B. S. Carpenter, B., Fink, D. J., and Gordon, G. S. Some thoughts on the nature of ectopic parathyroid hormones. Am. J. Med., *50*: 686–691, 1971.
38. Sato, K. and Weinhouse, S. Purification and characterization of the Novikoff hepatoma glycogen phosphorylase and its relations to a fetal form. Arch. Biochem. Biophys., *159*: 151–159, 1973.
39. Sato, K., Morris, H. P., and Weinhouse, S. Phosphorylase: a new isozyme in rat hepatic tumors and fetal liver. Science, *178*: 879–881, 1972.
40. Sato, K., Morris, H. P., and Weinhouse, S. Characterization of glycogen synthetases and phosphorylases in transplantable rat hepatomas. Cancer Res., *33*: 724–733, 1973.
41. Schapira, F. Isozymes in cancer. Adv. Cancer Res., *18*: 77–153, 1973.

42. Schapira, F., Reuber, M. D., and Hatzfeld, A. Resurgence of two fetal-type aldolases (A and C) in some fast-growing hepatomas. Biochem. Biophys. Res. Commun., *40*: 321–325, 1970.
43. Stanislawski-Birencwajg, M., Uriel, J., and Grabar, P. Association of embryonic antigens with experimentally induced hepatic lesions in the rat. Cancer Res., *27*: 1990–1997, 1967.
44. Stonehill, E. H. and Bendich, A. The reappearance of embryonal antigens in cancer cells. Nature, *228*: 370–371, 1970.
45. Uriel, J. Transitory liver antigens and primary hepatoma in man and rat. Pathol. Biol., *17*: 877–884, 1969.
46. Warburg, O. The Metabolism of Tumors. Arnold Constable, London, 1930.
47. Weber, G., Banerjee, G., and Morris, H. P. Comparative biochemistry of hepatomas. I. Carbohydrate enzymes in Morris hepatoma 5123. Cancer Res., *21*: 933–937, 1961.
48. Weinhouse, S. Studies on the fate of isotopically labeled metabolites in the oxidative metabolism of tumors. Cancer Res., *11*: 585–591, 1951.
49. Weinhouse, S. Glycolysis, respiration and anomalous gene expression in experimental hepatomas. Cancer Res., *32*: 2007–2016, 1972.

Discussion of Paper by Dr. Weinhouse

DR. JOHNSON: Is there evidence that the ATPase activity is altered in the hepatomas and that this could be a possible mechanism for breaking the link between glycolysis and respiration?

DR. WEINHOUSE: This seems hardly likely. ATPase levels are not unduly high in tumors, and ATP levels under both aerobic and anerobic conditions are quite high.

DR. RUTTER: Does cyclic AMP induce changes in the isoenzyme patterns, that is could these patterns simply be due to the modulation of a metabolic pathway by a general effector?

DR. WEINHOUSE: This seems a possibility and we are currently attempting to induce liver-type pyruvate kinase and hexokinase isozyme activity in cultured liver and hepatoma cells by additions of cyclic AMP and hormones to the medium; thus far without success.

We have also been unsuccessful in influencing liver-type isozymes in solid tumors *in vivo* by dietary or hormonal treatments that markedly affect these same enzymes in the liver. It would appear that either the appropriate receptor is lost or the signal does not get transmitted in the tumor cells.

DR. F. SCHAPIRA: In your opinion, at what level is the misprogramming in cancer, at the transcriptional or at the translational level?

DR. WEINHOUSE: It is probably, in my judgment due to a mutation of a regulatory gene and not of a structural gene, since the misprogramming does not involve "new" proteins. I have no evidence one way or the other, but I believe the misprogramming would be in transcription.

DR. NAKAHARA: This is a little deviation from the main stream of the discussion, but I would like to suggest that my pet enzyme catalase may be an exceptionally suitable subject in the study of the enzymatic alteration in hepatic cancer. Liver is very rich in its catalase content while liver cell cancer contains little catalase, if any at all. The alteration in the catalase activity of liver tissue during the course of hepatocarcinogenesis should yield interesting results. We, of course, do not know the real function of catalase *in vivo*, but this enzyme must be closely associated with cell respiration.

Anomalies of Differentiation in Cancer: Studies on Some Isozymes in Hepatoma and in Fetal and Regenerating Liver

Fanny SCHAPIRA, Antoinette HATZFELD, and Anne WEBER
Institut de Pathologie Moléculaire, Université de Paris, Paris, France

Abstract: Anomalies of differentiation in cancer were illustrated by the study of isozymes in cancerous tissues, especially in hepatoma. It was frequently found that the isozymic modifications are a reversion towards an embryonic pattern: the adult molecular form tends to disappear, while the fetal one appears. In some other cases, where the cancerous form of a given tissue is different from the embryonic one, this form was always found to be present in another normal tissue of the same animal: a truly new isozymic form has never been found in cancer. We have studied several enzymes with multiple molecular forms, and have compared them in hepatoma, and in fetal and regenerating liver: the extent of resurgence of fetal forms with disappearance of the adult form is roughly parallel to the rate of tumor growth.

There is a resurgence of aldolase A (muscle type) in slow-growing hepatomas, without resurgence of aldolase C (brain type) and with maintenance of aldolase B (liver type). On the contrary, in fast-growing hepatomas, fetal aldolases A and C are present, while normal adult aldolase B almost completely disappears.

Several authors have shown that in fast-growing hepatomas, liver-type pyruvate kinase levels were considerably lowered to the benefit of another form, similar but not identical to the muscle form. We have shown that this form can be found in some other normal tissues, particularly placenta and fetal liver. In slow-growing hepatomas, both forms are found.

Two isozymes of hexosaminidase (N-acetyl-β-glucosaminidase) are present in rat tissues, and may be distinguished according to their thermostability and electrophoretic migration. In fast-growing hepatomas, the hexosaminidase isozymic pattern is different from the normal liver pattern and similar to the fetal one.

In regenerating liver, after large hepatectomy, slight modifications of aldolase, pyruvate kinase, and hexosaminidase occur; they are comparable to the modifications seen in slow-growing hepatomas, but very different from the isozymic changes in fast-growing hepatomas. It seems that cell multiplication alone is not sufficient to account for the resurgence of fetal forms of enzymes in cancer. Several questions arise: is there a modification of the genome itself? Do these changes occur at the cellular level by multiplication of stem cells keeping their embryonic enzymic complement? Or at the moleculair level, by an impairment of control mechanisms occurring at the transcriptional or the post-transcriptional level?

Study of isozymic modifications in cancer may throw new light on the control mechanism in eukaryote cells.

It is not easy to define "differentiation": it may be admitted that it is the process by which some cells develop specialized functions at the expense of some others without alteration of the genome.

The concept of Greenstein (*11*) on the dedifferentiation of the cancerous tissues has been criticized, but the discovery of isoenzymes has rehabilitated it to a certain extent.

Isozymes of Aldolase

Our group was the first to find a modification of multiple forms of an enzyme—aldolase—in a cancerous tissue (human and experimental hepatomas). Moreover, we compared the new molecular forms to the fetal ones. We shall recall briefly the

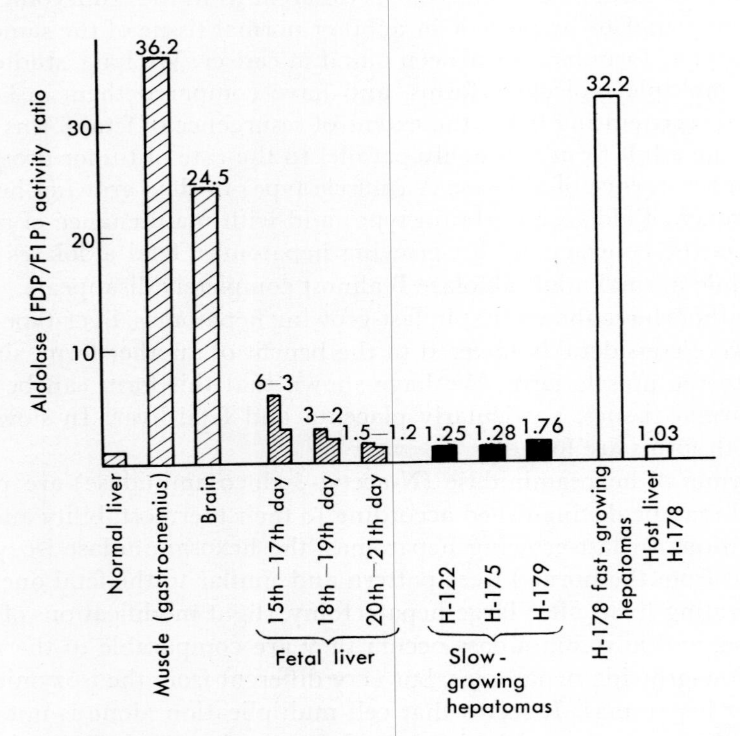

FIG. 1. Aldolase activity ratio in normal rat tissues, in fetal liver, and in transplantable hepatomas.

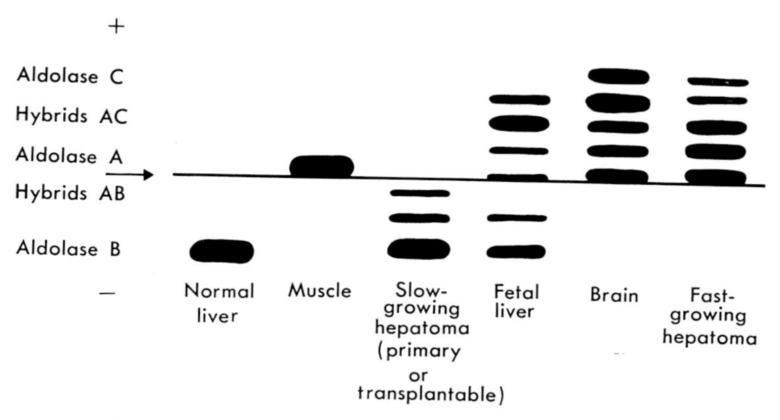

FIG. 2. Aldolase isozymes in rat tissues and hepatomas.

work which one of us began in 1962 with Schapira and Dreyfus (*41*, *42*) and continued, especially with Nordmann (*25*). We recall also the findings of Rutter *et al.* (*31*), Matsushima *et al.* (*24*), Sugimura *et al.* (*49*), Adelman *et al.* (*3*), and Ikehara *et al.* (*14*).

TABLE 1. Inhibition of FDP Aldolase Activity by Antisera

Rat tissues	Anti A (%)	Anti B (%)	Anti C (%)
Normal liver	5–10	80–90	<2
Muscle	90–95	<2	<5
Brain	85		95
Fetal liver (15th day)	30–35	60–70	30–70[a]
Fast-growing hepatomas (H178)	85–90	5	70–80

[a] On supernatant after action of anti A.

Figure 1 schematizes the kinetic results with rat hepatomas and stresses the similarities between the properties of fetal liver aldolase, and those of the aldolase of hepatomas, principally fast-growing ones. While the substrate specificity of normal adult liver aldolase (aldolase B) is characterized by an FDP/F1P* activity ratio very constant at unity, in fetal liver, this ratio is greater: in rat it is between 8 at the 15th day and 1.2 at birth. In slow-growing hepatomas, this ratio is moderately increased to between 1.1 and 2, but principally it is very high in fast-growing hepatomas (ascitic or solid): about 30. Figure 1 shows also that this ratio is comparable to that found in muscle or in brain. Figure 2 schematizes our electrophoretic results (on starch gel, at pH 7.0). Adult rat liver aldolase is characterized by a single cathodic band; muscle aldolase by a slow anodic band, and brain aldolase by 5 anodic isozymes, resulting from the hybridization of the tetrameric aldolases A and C. In fetal liver, the 3 types are found: aldolase B (cathodic band) and aldolases A and C (3 to 5 anodic bands). In slow-growing hepatomas the aldolase A is hybridized

* FDP, fructose-1,6-diphosphate; F1P, fructose-1-phosphate.

with normal aldolase B. Principally it is seen that in fast-growing hepatomas, the normal aldolase B has disappeared; it is replaced by the other 2 types, A and C.

Immunological results (Table 1) confirm the presence of the 2 types, A and C, both in fetal liver and in hepatoma. Antisera against each type were prepared by repeated injections of crystalline aldolases A, B, or C in chicken and they displayed a good type specificity (but no significant species specificity). Aldolases were isolated according to Penhoet *et al.* (*27*) with modifications. The action of antisera was tested by measuring aldolase activity in the supernatant after incubation of tissue extract with each antiserum, or with normal serum, followed by centrifugation.

It must be stressed that fetal liver does contain the 3 types as early as the 15th day of fetal life. Hematopoietic tissue does not seem to contribute significantly to the results: there is no parallelism between its evolution and the isozymic evolution. Moreover, we have found that the aldolase activity ratio does not change when the hematopoietic cells regress under the influence of a stress (*36*).

The extent of the resurgence of fetal aldolases (A alone or both A and C) and of the disappearance of normal adult aldolase B corresponds roughly to the growth rate of the hepatomas studied (*44*).

We may add that we have made analogous findings with human hepatomas, but normal adult human liver contains a small amount of aldolase C, unlike adult rat liver (*40*). It must be stressed that the increase of aldolase A is not a characteristic of cancerous tissue. In organs where aldolase A is the preponderant, characteristic isozyme of adult tissue (as, for example, muscle or spleen), we have found that aldolase A decreased after cancerization, while another type increased (*35*, *45*).

The modifications of aldolase isozymes represents at the present time the most striking example of resurgence of fetal molecular forms in cancer, but similar findings have been observed more recently for an increasing number of isozymes.

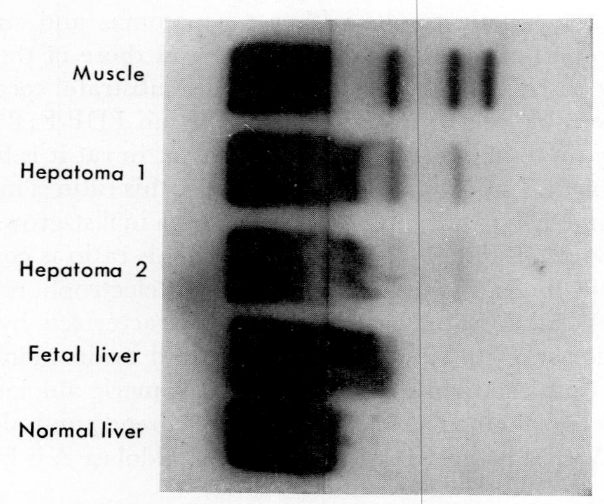

FIG. 3. LDH isozymes in rat tissues and hepatomas.

Isozymes of Lactate Dehydrogenase (LDH)

The modification of LDH in cancerous tissues have been known since the work of Pfleiderer and Wachsmuth (*28*) and of Starkweather and Schoch (*47*). Goldman *et al.* (*10*) by comparing several malignant tumors with benign tumors, demonstrated an increase of muscle-type LDH isozymes during cancerization.

The influence of muscle (M) LDH on the well-known increase of glycolytic metabolism remains controversial (*55, 58*). We pointed out that digestive and genital organs contain more M subunits at the fetal stage than at the adult one. On the other hand, fetal liver and fetal muscle contain relatively more heart (H) subunits than adult organs. Figure 3 shows the LDH pattern of a primary hepatoma induced by 3′-methyldimethylaminoazobenzene (3′-MeDAB). There is in fact an increase of H subunits and the pattern is comparable to the pattern of fetal liver (*38, 39*). We have also found an increase of H subunits in one case of rhabdomyosarcoma (*45*).

The increase of M subunits in cancerous tissues is consequently very frequently, but not always, encountered. We believe that the hypothesis of a resurgence of fetal forms fits better with some LDH modifications than the theory of a shift toward glycolysis.

Isozymes of Pyruvate Kinase

We have also applied this hypothesis to the isozymic modifications of pyruvate kinase. The multiple molecular forms of this enzyme were studied by Tanaka *et al.* (*50*), who showed that in ascitic hepatoma AH 130, pyruvate kinase was similar to the muscle enzyme. Total activity was very high, as demonstrated by Lo *et al.* (*23*). Later, Taylor *et al.* (*52*) demonstrated that pyruvate kinase of a fast-growing hepatoma was different not only from the liver enzyme, but also from the muscle enzyme. Criss (*6*), separating several forms by isoelectrofocusing, found a new form, predominant in the Novikoff hepatoma.

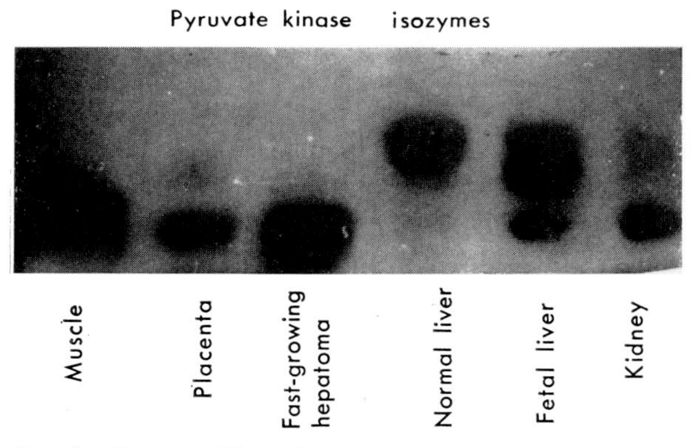

FIG. 4. Pyruvate kinase isozymes in rat tissues and hepatomas.

One of us, with Gregori (*37*), showed in 1971 that pyruvate kinase of a poorly differentiated hepatoma (Reuber H 178) resembles placental pyruvate kinase as regards its electrophoretic migration at pH 7.0 on starch gel (as shown in Fig. 4). Moreover, one of the 3 bands found in fetal liver (16th day) had the same migration as the tumor isozyme. The cancerous enzyme, like the placental enzyme, was activated by fructose diphosphate in the presence of a nonsaturating concentration of substrate phosphoenol pyruvate, in contrast to M_1 type.

On a slow-growing hepatoma we found 2 bands: the migration of one was identical to that of normal liver type and that of the other similar to the placental enzyme.

Tanaka *et al.* (*51*) and Imamura *et al.* (*16*) demonstrated that there exist 3 types of pyruvate kinase in rat tissues: type L, predominant in liver; type M_1 in muscle, heart, and brain; and type M_2 in many other tissues and in fast-growing hepatomas. We believe that the placental isozyme that we have described is identical to the M_2 type; (type K for Ibsen and Krueger (*13*). The 3 isozymes that we found in fetal liver correspond to the L and M_2 types and to a fourth one, found also in erythrocytes (*15*). In slow-growing hepatomas, the 2 bands correspond to isozyme L and isozyme M_2.

Isozymes of Hexosaminidase

We have also performed work on hexosaminidase (N-acetyl-β-glucosaminidase) the molecular forms of which have been more studied in human than in rat tissues. Our group has shown that in most rat tissues there are 2 bands, separable by electrophoresis on cellulose acetate by using as substrate a methyl umbelliferyl derivative.

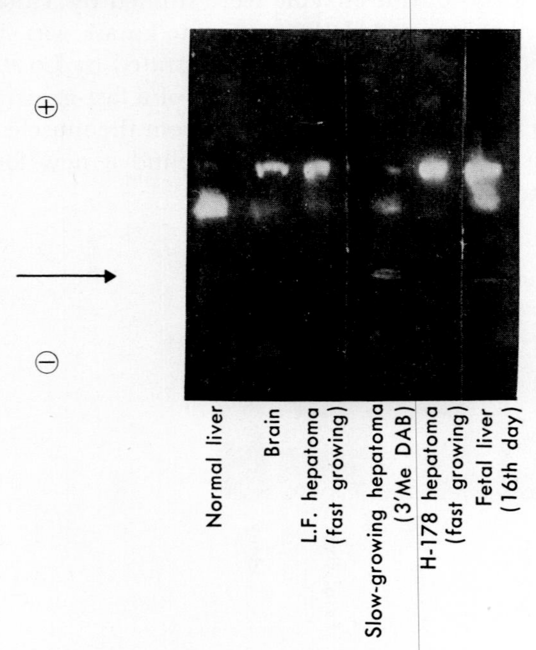

FIG. 5. Hexosaminidase isozymes in adult and fetal liver and in hepatoma.

The predominant band has the slowest migration during electrophoresis in cellulose acetate, and this cathodic band is thermostable and bound to lysosomes. The fastest band is weaker in normal liver; it is stronger in the brain, and it is strongest in fetal liver; this band is more thermolabile and more easily extractible. We found that in 2 types of transplantable fast-growing hepatomas ("LF" of Frayssinet and "H 178" of Reuber) there is always a diminution and sometimes even a disappearance of the slowest band, with a considerable increase of the fastest. Principally, the cancerous pattern is similar to the fetal one (16th day) (Fig. 5) (*56*).

Resurgence of Fetal Enzymes

From our experiments and others reported by many authors on an increasing number of enzymes, it appears that in the fastest-growing cancers, there is a diminution, and almost a disappearance of the adult-specific form of several enzymes, with

TABLE 2. Resurgence of Isozymes in Rat Hepatoma

Resurgence type	Enzyme	Form
1) Resurgence of forms found in fetal liver:	Aldolase (*41, 43, 49*)	A (muscle) and C (brain)
	LDH (*39*)	H subunits (heart) (brain)
	Pyruvate kinase (*37, 50, 51*)	M_2 (several tissues)
	Hexokinase (*46*)	II (muscle)
	Glutaminase (*19, 20*)	Phosphate-independent (kidney)
	Hexosaminidase (*56*)	A (brain)
	Phosphorylase (*33*)	
	Alcohol dehydrogenase (*4*)	
	Uridine kinase (*22*)	
	Thymidine kinase (*17*)	
2) Resurgence of forms found in another tissue of the same species:	Branched-chain amino acid transaminase (*26*)	Brain
	Fructose diphosphatase (*32*)	Muscle
	Glycogen synthetase (*34*)	Muscle
	Alkaline phosphatase (*8*)	Placenta
	Phosphofructokinase (*51*)	Muscle, brain, and several other tissues

a concomitant increase of the less specific forms; moreover there is sometimes a resurgence of some fetal forms which disappeared in the adult stage. In some cases, the "abnormal" form is found, not in the normal adult tissue, or in this tissue at the fetal stage, but in another tissue of the same species. Table 2 summarizes the main findings of several authors on the resurgence of isozymes in hepatoma. An important fact arises: all the abnormal new molecular forms found in hepatoma are in fact neither abnormal nor new, but may be found, either in fetal liver, or in another tissue of the same species, or both. Very often, the fetal type is also found in placenta (which, moreover, is an embryonic tissue) or in muscle, or brain. It is possible that, in cases where the abnormal form was not found in the fetal liver, this is because it disappears too early to be detected.

TABLE 3. Aldolase Activities in Regenerating Liver

Time after surgery	Aldolase activities[a] (in IU/g)		Activity ratio (FDP/FIP)
	FDP	FIP	
18 hr	17.4	15.9	1.13
24	16.4	13.4	1.20
48	16.8	14.6	1.10
3–4th days	20.8	18.1	1.14
5–6th	18.5	16.9	1.10
7th	16.7	14.0	1.20
Normal liver [a),b)] (after a sham operation)	23.1±3.3	21.5±3.2	1.065±0.077

a) Mean of 3 experiments. b) Mean of 15 experiments ± standard deviation.

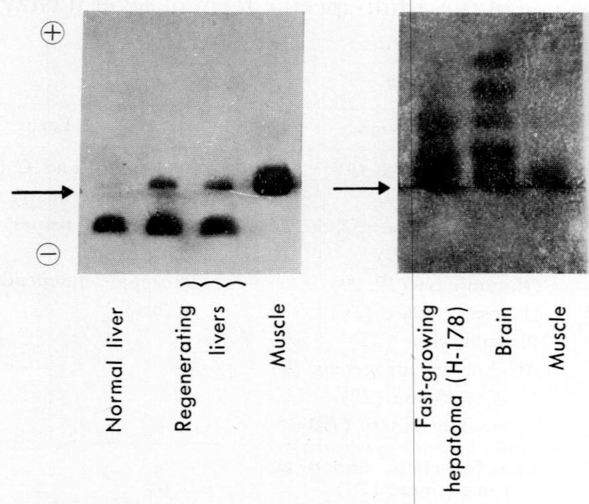

FIG. 6. Aldolase isozymes in regenerating and cancerous liver.

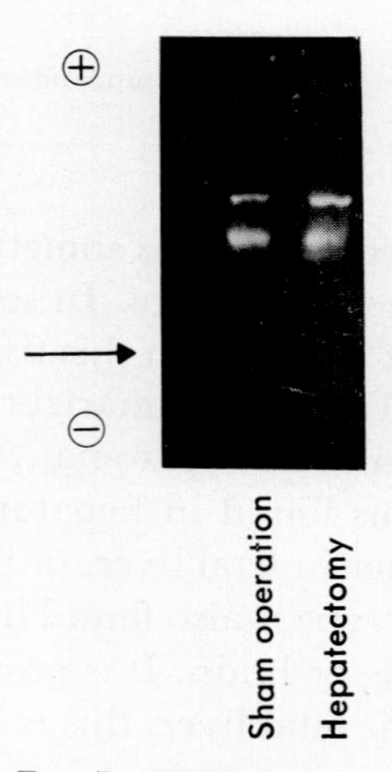

FIG. 7. Hexosaminidase isozymes in regenerating liver after hepatectomy.

In order to try to explain the mechanism(s) of this resurgence of enzymes in cancer—which we may compare to the resurgence of fetal antigens, principally α-fetoprotein (AFP) (*1, 2*) and carcinoembryonic antigen (*9*), we have searched for similar isozymic modifications in regenerating liver.

Changes in Regenerating Liver

Rat liver regeneration was studied from 6 hr to 7 days after extensive hepatectomies according to the technique of Higgins and Anderson (*12*), and also after injection of carbon tetrachloride. Comparison of livers after hepatectomies were made with livers after a sham operation.

Table 3 shows the aldolase activity ratio in regenerating liver at different times after hepatectomy. It is noticeable that the increase of this ratio is very slight (although significant); its moderation is confirmed by the electrophoretic results, of which an example is given in Fig. 6. There is a moderate increase of aldolase A, confirmed by the immunological results. Principally there is no resurgence of aldolase C, unlike the pattern seen in fast-growing hepatomas and in fetal liver.

For hexosaminidase, the results are comparable, regardless of whether regeneration was induced by hepatectomy (Fig. 7) or by CCl_4 (Fig. 8). The slight increase of the fastest band which we occasionally found was also sometimes present after a sham operation.

Results for pyruvate kinase were similar. We studied the electrophoretic pattern after partial hepatectomy (from 18 hr to 7 days) and after CCl_4 injection. Figure 9 gives an example of the pattern seen in regenerating liver: we found only a strengthening of the M_2 band very similar to the pattern of pyruvate kinase in slow-growing hepatomas, and very different from that in fast-growing ones. Im-

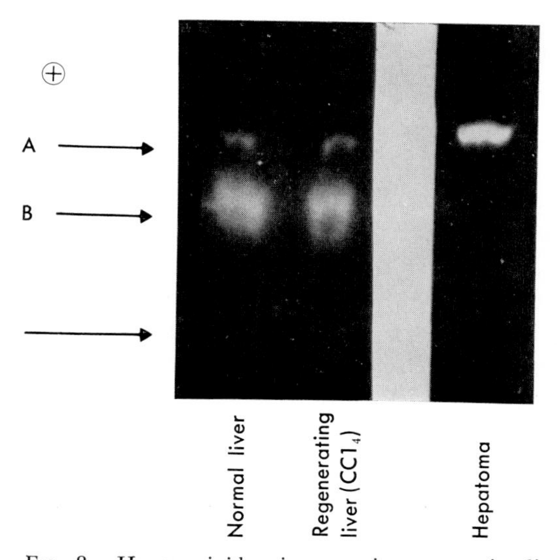

FIG. 8. Hexosaminidase isozymes in regenerating liver after CCl_4 injection.

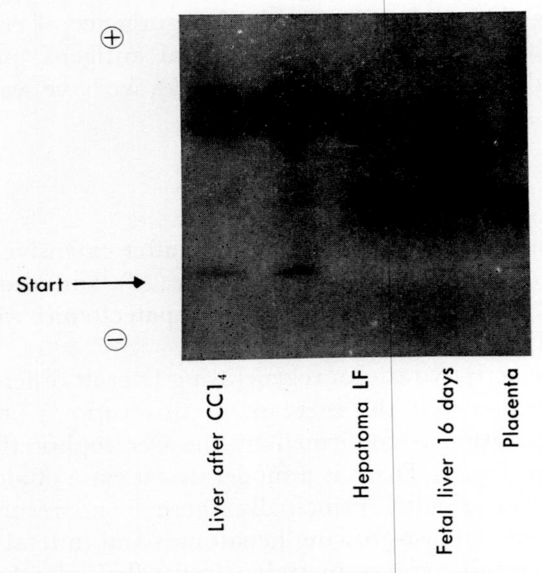

FIG. 9. Pyruvate kinase isozymes in regenerating liver.

munological studies, using anti M-pyruvate kinase antiserum, confirmed that 48 hr after partial hepatectomy, M_2-type pyruvate kinase was moderately increased.

These findings do not agree completely with those of Suda *et al.* (*48*) who found a considerable increase of the M_2 band in regenerating liver. However, Bonney *et al.* (*5*) have shown recently that this increase remains relatively moderate.

We would like to point out that for several other enzymes also, only slight modifications are generally found in regenerating liver. Generally speaking, regenerating liver differs not only from fast-growing hepatomas, but also from fetal liver, and resembles slow-growing hepatomas.

DISCUSSION

It seems that cell multiplication alone is not sufficient to account for the resurgence of fetal forms of enzymes in cancer. Comparison has to be made with the results of Abelev (*2*) on the contrast between the very moderate resurgence of AFP in regenerating liver, and the considerable synthesis of this antigen in some hepatomas.

Recently, Uriel *et al.* (*54*) have shown that synthesis of AFP seems to occur only in a small number of "transitional" liver cells, distinct from either normal or neoplastic ones. These cells synthesized AFP in both immature and tumoral livers. Perhaps there are also some specialized cells which synthesize fetal isozymes in hepatoma; these "transitional" cells would proliferate during carcinogenesis, as perhaps during regeneration. However, the fact remains that the resurgence of fetal forms is moderate and transitory and the synthesis of adult forms seems normal in regenerating liver.

On the other hand the synthesis of adult forms decreases and even seems to cease in the fastest-growing hepatomas, while that of fetal forms takes place.

We plan to investigate whether the same cells synthesize both adult and fetal aldolases in hepatoma, as indicated by A-B hybrids. If not, the anomaly would be located at the cellular level, and it must be supposed that, during carcinogenesis, stem cells or transitional cells would multiply without complete maturation.

If, on the contrary, the same cells synthesize adult and fetal isozymes, at what level is the anomaly? Is there a modification of the genome itself? Chromosomal anomalies in cancer are well-known, but their significance remains controversial. As concluded by Pierce (*29*), it seems probable that the karyotypic changes are the result of carcinogenesis and not its cause. It must be recalled that a truly new isozymic form has never been found in cancer. The translational phase might be involved, as suggested by the experiments of Uenoyama and Ono (*53*).

If the impairment is located at the transcriptional level, we have to consider the role of nonhistone proteins since they have been involved in the control of the synthesis of mRNA (*18*, *21*). We proposed in 1963 that genes which code for the adult type would be repressed, while genes coding for fetal forms would be derepressed (*42*). Weinhouse (*57*) noted that repression generally preceeds derepression. Potter (*30*) supposed that in oncogeny, the switch on and off of certain genes would be reversed. We recall that the concepts of "repression" and "derepression" do not apply in the same way to eukaryotes as they do to prokaryotes (*7*).

In conclusion, the appearance in cancerous tissues of many fetal forms of enzymes, at the expense of the adult forms, is a phenomenon which seems general and is probably not due to cell multiplication alone. However, its significance remains obscure. We plan to investigate whether it occurs at the cellular or at the molecular level.

ACKNOWLEDGMENT

The authors thank Claudine Gregori for his technical collaboration of this work.

REFERENCES

1. Abelev, G. J. Study of the antigenic structure of tumors. Unio. Int. Contra Cancrum., *19*: 80–82, 1963.
2. Abelev, G. J. Alpha-fetoprotein in ontogenesis and its association with malignant tumors. Adv. Cancer. Res., *14*: 295–358, 1971.
3. Adelman, R. C., Morris, H. P., and Weinhouse, S. Fructokinase, triokinase and aldolases in liver tumors of the rat. Cancer Res., *27*: 2408–2413, 1967.
4. Bertolotti, R. and Weiss, M. C. Expression of differentiated functions in hepatoma cell hybrids. VI. Extinction and re-expression of liver alcohol dehydrogenase. Biochimie, *54*: 195–201, 1972.
5. Bonney, R. J., Hopkins, H. A., Walker, P. R., and Potter V. R. Glycolytic isoenzymes and glycogen metabolism in regenerating liver from rats on controlled feeding schedules. Biochem. J., *136*: 115–124, 1973.
6. Criss, W. E. A new pyruvate kinase isozyme in hepatomas. Biochem. Biophys. Res. Commun., *35*: 901–905, 1969.

7. Dreyfus, J. C. The application of bacterial genetics to the study of human genetic abnormalities. Prog. Med. Genetics, *6*: 189–200, 1969.
8. Fishman, W. H., Inglish, N. R., Green, S., Anstiss, C. L., Gosh, N. K., Reif, A. E., Rustigian, R., Krant, M. J., and Stolbach, L. L. Immunology and biochemistry of Regan isoenzyme of alkaline phosphatase in human cancer. Nature, *219*: 697–699, 1968.
9. Gold, P. and Freedman, S. O. Demonstration of tumor-specific antigens in human colonic carcinomata by immunological tolerance and absorption techniques. J. Exp. Med., *121*: 439–462, 1965.
10. Goldman, R. O., Kaplan, N. O., and Hall, C. Lactic dehydrogenase in human neoplastic tissue. Cancer Res., *24*: 389–399, 1964.
11. Greenstein, J. P. Biochemistry of Cancer, 2nd ed., Academic Press, New York, 1954.
12. Higgins, G. M. and Anderson, R. M. Experimental pathology of the liver. I. Restoration of the liver of the white rat following partial surgical removal. Arch. Pathol., *12*: 186–202, 1931.
13. Ibsen, K. H. and Krueger, E. Distribution of pyruvate kinase isozymes among rat organs. Arch. Biochem. Biophys., *157*: 500–513, 1973.
14. Ikehara, Y., Endo, H., and Akada, Y. The identity of the aldolases isolated from rat muscle and primary hepatoma. Arch. Biochem. Biophys., *136*: 491–497, 1970.
15. Imamura, K. and Tanaka, T. Multimolecular forms of pyruvate kinase from rat and other mammalian tissues. I. Electrophoretic studies. J. Biochem., *71*: 1043–1051, 1972.
16. Imamura, K., Taniuchi, K., and Tanaka, T. Molecular forms of pyruvate kinase. II. Purification of M^2-type pyruvate kinase from Yoshida ascites hepatoma 130 cells and comparative studies on the enzymological and immunological properties of the three types of pyruvate kinase, L, M_1, and M_2. J. Biochem., *72*: 1001–1015, 1972.
17. Jones, O. W., Taylor, A., and Stafford, M. A. Fetal gene expression in human tumor cells. J. Clin. Invest., *51*: 47a, 1972.
18. Kamiyama, M., Dastugue, B., Defer, N., and Kruh, J. Liver chromatic nonhistone proteins. Partial fractionation and mechanism of action on RNA synthesis. Biochim. Biophys. Acta, *277*: 576–583, 1972.
19. Katunuma, N., Katsunuma, T., Tomino, J., and Matsuda, Y. Regulation of glutaminase activity and differentiation of the isozyme during development. Adv. Enzyme Regulation, *6*: 227–242, 1968.
20. Knox, W. E., Tremblay, G. C., Spanier, B. B., and Friedall, G. H. Glutaminase activities in normal and neoplastic tissues of the rat. Cancer Res., *27*: 1456–1458, 1967.
21. Kruh, J. RNA and the control of gene expression in animal cells. Rev. Eur. Et. Clin. Biol., *17*: 739–744, 1972.
22. Krystal, G. and Webb, T. E. Multiple forms of uridine kinase in normal and neoplastic rat liver. Biochem. J., *124*: 943–947, 1971.
23. Lo, C. H., Farina, F., Morris, H. P., and Weinhouse, S. Glycolytic regulation in rat liver and hepatomas. Adv. Enzyme Regulation, *6*: 453–464, 1968.
24. Matsushima, T., Kawabe, S., Shibuya, M., and Sugimura, T. Aldolase isozymes in rat tumor cells. Biochem. Biophys. Res. Commun., *30*: 565–570, 1968.
25. Nordmann, Y. and Schapira, F. Muscle-type isoenzymes of liver aldolase in hepatomas. Eur. J. Cancer, *3*: 247–250, 1967.

26. Ogawa, K., Yokojima, A., and Ichihara, A. Transaminases of branched-chain amino acids. VII. Comparative studies on isozymes of ascites hepatoma and various normal tissues of rat. J. Biochem., *68*: 901–911, 1970.
27. Penhoet, E. E., Kochman, U., and Rutter, W. J. Isolation of fructose diphosphate aldolases A, B and C. Biochemistry, *8*: 4391–4395, 1969.
28. Pfleiderer, G. and Wachsmuth, E. O. Alters und Funktionsabhängige. Differenzierung der Lactatedehydrogenase menschlichen Organe. Biochem. Zeitschr., *334*: 185–198, 1961.
29. Pierce, G. B. Differentiation of normal and malignant cells. Fed. Proc., *29*: 1248–1254, 1970.
30. Potter, V. R. Recent trends in cancer biochemistry: the importance of studies on fetal tissue. Canad. Cancer Conf., *8*: 9–30, 1969.
31. Rutter, W. J., Richards, O. C., Woodfin, B. M., and Weber, C. S. Enzyme variants and metabolic diversification. Adv. Enzyme Regulation, *1*: 39–56, 1963.
32. Sato, K. and Tsuiki, S. Fructose 1,6-diphosphatase of mouse Ehrlich ascites tumor and its comparison with the enzymes of liver and skeletal muscle of the mouse. Biochim. Biophys. Acta, *159*: 130–140, 1968.
33. Sato, K., Morris, H. P., and Weinhouse, S. Phosphorylase: A new isozyme in rat hepatic tumors and fetal liver. Science, *178*: 879–881, 1972.
34. Sato, K., Morris, H. P., and Weinhouse, S. Characterization of glycogen synthetases and phosphorylases in transplantable rat hepatomas. Cancer Res., *33*: 724–733, 1973.
35. Schapira, F. Aldolase isozymes in cancer. Eur. J. Cancer, *2*: 131–134, 1966.
36. Schapira, F. Unpublished results.
37. Schapira, F. and Gregori, C. Pyruvate kinase de l'hépatome, du placenta et du foie foetal de rat. C. R. Acad. Sci. Paris, Série D, *272*: 1169–1172, 1971.
38. Schapira, F. and Josipowicz, A. Anomalies de type foetal des isozymes de l'aldolase et de la lactico-deshydrogénase dans des hépatomes ascitiques. Compt. Rend. Soc. Biol., *164*: 31–36, 1970.
39. Schapira, F. and De Néchaud, B. Présence anormale d'isozymes de la lactico-déshydrogénase type coeur dans des hépatomes expérimentaux. Compt. Rend. Soc. Biol., *162*: 86–89, 1968.
40. Schapira, F. and Nordmann, Y. Présence de trois types d'aldolase dans le foie humain. Clin. Chim. Acta, *26*: 189–195, 1969.
41. Schapira, F., Schapira, G., and Dreyfus, J. C. Type foetal de l'aldolase hépatique dans un cas d'hépatome humain. Compt. Rend. Acad. Sci., *254*: 3143–3145, 1962.
42. Schapira, F., Dreyfus, J. C., and Schapira, G. Anomaly of aldolase in primary liver cancer. Nature, *200*: 995–997, 1963.
43. Schapira, F., Reuber, M. D., and Hatzfeld, A. Resurgence of two fetal-type aldolases (A and C) in some fast-growing hepatomas. Biochem. Biophys. Res. Commun., *40*: 321–327, 1970.
44. Schapira, F., Hatzfeld, A., and Reuber, M. D. Fetal pattern of aldolase in transplantable hepatomas. Cancer Res., *31*: 1224–1230, 1971.
45. Schapira, F., Micheau, C., and Junien C. Foetal enzyme pattern of two alveolar rhabdomyosarcomas. Rev. Eur. Et. Clin. Biol., *17*: 896–899, 1972.
46. Shatton, J. B., Morris, H. P., and Weinhouse, S. Kinetic electrophoretic and chromatographic studies on glucose ATP phosphotransferases in rat hepatomas. Cancer Res., *29*: 1161–1172, 1969.

47. Starkweather, W. H. and Schoch, H. K. Some observations on the lactate dehydrogenase of human neoplastic tissue. Biochim. Biophys. Acta, *62*: 440–442, 1962.
48. Suda, M., Tanaka, T., Yanagi, S. Hayashi, S., Imamura, K., and Taniuchi, K. Differentiation of enzymes in the liver of tumor-bearing animals. GANN Monograph on Cancer Research, *13*: 79–93, 1972.
49. Sugimura, T., Sato, S., and Kawabe, S. The presence of aldolase C in rat hepatomas. Biochem. Biophys. Res. Commun., *39*: 626–630, 1970.
50. Tanaka, T., Harano, Y., Morimura, H., and Mori, R. Evidence for the presence of two types of pyruvate kinase in rat liver. Biochem. Biophys. Res. Commun., *21*: 55–60, 1965.
51. Tanaka, T., Imamura, K., Aun, T., and Taniuchi, K. Multimolecular forms of pyruvate kinase and phosphofructokinase in normal and cancer tissues. GANN Monograph on Cancer Research, *13*: 219–234, 1972.
52. Taylor, C. B., Morris, H. P., and Weber, G. A comparison of the properties of pyruvate kinase from hepatoma 3124-A, normal liver and muscle. Life Sci., *8*: 635–644, 1969.
53. Uenoyama, K. and Ono, T. Post-transcriptional regulation of catalase synthesis in rat liver and hepatoma: Factors activating and inhibiting catalase synthesis. J. Mol. Biol., *74*: 439–452, 1973.
54. Uriel, J., Anssel, C., Bouillon, O., De Néchaud, B., and Loisillier, F. Localisation of rat liver α-fetoprotein by cell affinity. Labelling with tritiated oestrogens. Nature New Biol., *244*: 190–192, 1973.
55. Vesell, E. S. and Pool, P. E. Lactate and pyruvate concentrations in exercised ischemic canine muscle: relationship of tissue substrate level to lactate dehydrogenase isozyme pattern. Proc. Natl. Acad. Sci. U.S., *55*: 756–762, 1966.
56. Weber, A., Poenaru, L., Lafarge, C., and Schapira, F. Modification of hexosaminidase isozymes in rat hepatoma. Cancer Res., *33*: 1925–1930, 1973.
57. Weinhouse, S. Glycolysis, respiration and anomalous gene expression in experimental hepatomas. Cancer Res., *32*: 2007–2016, 1972.
58. Wuntch, T., Chen, R. F., and Vesele, E. S. Lactate dehydrogenase isozymes: kinetic properties at high enzyme concentrations. Science, *167*: 63–65, 1970.

Discussion of Paper by Drs. F. Schapira et al.

DR. SUGIMURA: How many strains of fast-growing hepatoma did you investigate to prove the presence of aldolase C? Did cells of the fast-growing hepatomas contained aldolase C? In our experience, only 3 out of about 20 strains of fast-growing hepatoma which we keep here possessed aldolase C.

DR. F. SCHAPIRA: We have tested the following fast-growing hepatomas: Reuber H 178 (solid, transplantable), Frayssinet LF (solid, transplantable), and Zajdela (ascitic, transplantable). All these hepatomas contained aldolase C (as proven by electrophoresis on starch gel, and inhibition by anti-aldolase C antiserum) but in very variable amounts. Slow-growing hepatomas, primary (3′-MeDAB) or transplantable (Reuber H 122, H 175, and H 189) did not contain aldolase C.

DR. PAUL: By what criteria are fast-growing hepatomas (such as H 178) identified as originating from adult parenchymal hepatocytes?

DR. F. SCHAPIRA: This is the general problem of the origin of cancer cells (from specialized stem cells or not). The case of Reuber hepatoma H 178 does not differ (from the point of view of its cell origin) from that of other hepatomas studied by different authors. It was induced by N_2-fluorenyl diacetamide, and histologically controlled.

DR. WEBER: You showed in many slides that the molecular forms of aldolase, lactic dehydrogenase, pyruvate kinase, hexosaminidase in rapidly growing hepatomas resemble the molecular forms observed in brain and are not similar to that of fetal tissues. Yet in your concluding slide and remarks you stated that in the rapidly growing hepatomas there is a resurgence of the fetal forms. How do you reconcile this contradiction between your data and conclusion?

DR. F. SCHAPIRA: For aldolase, it may be admitted that fetal and brain types are identical if we suppose that aldolases A (muscle type) and C (brain type), which are more abundant in liver at the 15th day of fetal life than at the 20th day, are the only forms present at an earlier stage (aldolase B, adult liver type, is continually increasing during fetal life). For lactic dehydrogenase, the hepatoma pattern resembles the fetal pattern (at the 16th day) more than the brain pattern. Principally, for pyruvate kinase the specific fetal form is the M_2 type. It is predominant in the fetal brain, but in the adult brain, M_1 type (muscle type) becomes predominant. In fast-growing hepatomas, M_2 type (and not M_1 type) is found.

DR. OSTERTAG: Could one not reverse the question of the resurgence of fetal characteristics in cancerous tissue and ask the question: is it perhaps that some few retained cells, otherwise normal, but still with fetal or poorly fetal differentiation in the adult are more responsive to cancerous changes?

DR. F. SCHAPIRA: I may agree with your hypothesis. We have begun experiments in order to investigate whether or not the same cells synthesize adult and fetal isozymes (which could partly answer this question).

Deviations in Patterns of Multimolecular Forms of Pyruvate Kinase in Tumor Cells and in the Liver of Tumor-bearing Animals

Takehiko TANAKA,*[1] Susumu YANAGI,*[1] Kiichi IMAMURA,*[1] Atsumori KASHIWAGI,*[1] and Nobuyuki ITO*[2]

Department of Nutrition and Physiological Chemistry, Osaka University Medical School, Osaka, Japan[1] *and Cancer Research Institute, Nara Medical University, Nara, Japan*[2]

Abstract: At least 3 types of pyruvate kinase (L, M_1, and M_2) are present in mammalian tissues. The M_2 type is widely distributed in glycolytic tissues and in tumor cells such as ascites hepatoma AH-130, Walker carcinosarcoma-256, and in primary hepatomas induced by various carcinogens. The M_1 and L types are present only in a few tissues with specialized physiological functions of glycolytic system, such as the skeletal muscle and the liver. The M_2 type is predominant fetal muscle, brain, and liver, but during differentiation from the fetus to the adult, (1) the M_2 type disappears or decreases, (2) the L type becomes predominant in the liver, and (3) the M_1 type becomes predominant in muscle and brain. In contrast, the L type decreases or disappears, and the M_2 type increases and becomes predominant in hepatoma cells. These results, therefore, strongly suggest that M_2 type pyruvate kinase is the prototype of the pyruvate kinase isoenzymes, while the L and M_1 types are differentiated types.

It can be concluded from isoenzyme patterns of glycolytic key enzymes that the differentiated type homotopically decreases or disappears and sometimes ectopically appears, though it occurs very rarely, while the prototype necessarily appears or increases and becomes predominant in cancer cells.

The deviation in isoenzyme patterns of glycolytic key enzymes in the liver of tumor-bearing animals is similar to those in embryonic and regenerating liver. Results of parabiotic experiments suggested that some humoral factor elevated the level of pyruvate kinase M_2 in the liver of tumor-bearing animals. Increase in M_2 type activity is also induced by injection of cell-free extract of tumor cells. The effective factor is bound to the nuclear fraction of a tumor cell, and seems to be the non-histone nuclear protein.

It is generally accepted that there are at least 3 irreversible and regulatory steps in glycolytic pathway between glucose and pyruvate: hexokinase, phosphofructokinase, and pyruvate kinase. Multimolecular forms of hexokinase were reported from several laboratories (*7, 23, 25*). Four types of phosphofructokinase were identified in rat tissues by Tanaka *et al.* (*19*) and by Kurata *et al.* (*8*).

In 1965, it was reported from our laboratory that there were at least 2 types of pyruvate kinase in mammalian tissues which were named L type and M type (*22*). At that time, it was still uncertain whether liver M type was the same as the muscle enzyme or not, though both enzymes were immunochemically indistinguishable (*21*).

In 1968, Susor and Rutter (*16*) reported that there were at least 3 types of pyruvate kinase in mammalian tissues. Shortly after that, we observed similar results (*20*). We have purified very extensively each of these 3 enzymes and have obtained a preparation which showed only a single component of protein by disc-gel electrophoresis and by ultracentrifugal analysis. The properties of these 3 types

TABLE 1. Summary of 3 Types of Pyruvate Kinase (*5*)

Type	M_1	M_2	L
Conditions when increased activity in liver observed		Regenerating liver, liver of newborn child, liver of tumor-bearing animals, liver tumors	Insulin, high carbohydrate diet
Distribution	Muscle, heart, brain, intestine	Kidney, spleen, lung, testis, ovary, stomach, adipose tissue, liver, intestine, heart, brain tumors	Liver, kidney, intestine, erythrocyte?
Molecular weight	250,000	216,000	208,000
Anti-M_1 Anti-M_2 Anti-L	Neutralized Neutralized Not neutralized	Neutralized Neutralized Not neutralized	Not neutralized Not neutralized Neutralized
S value	9.5	9.7	10.1
K_m for PEP ($\times 10^{-4}$ M)	0.75	4.0	8.3
Hill coefficient for PEP	1.0 (hyperbolic)	1.4–1.5 (sigmoidal)	2.0 (sigmoidal)
Activation by FDP	– (hyperbolic)	+ (hyperbolic)	+ (hyperbolic)
ATP inhibition K_i (APP) for ATP ($\times 10^{-3}$ M)	+ 3.0	+ 2.5	++ 0.15 (cooperative)
Reversal of ATP inhibition by Mg^{2+}	Complete	Partial	None
Inhibition by PCMB	Weak	Medium	Strong
Inhibition by L-Ala, L-Phe, L-Try K_i (APP) for L-Ala K_i (APP) for L-Phe ($\times 10^{-3}$ M)	 – – –	 + 0.6 0.5	 + 1.0 (cooperative) 5

PEP, phosphoenol pyruvate; FDP, fructose-1, 6-diphosphate; PCMB, *p*-chloromercuribenzoate. K_i (APP), apparent K_i.

of pyruvate kinase are summarized in Table 1 (*5*). Three types of pyruvate kinase differ from each other in their tissue distribution pattern, immunochemical reactivities, and kinetic properties. Generally speaking, L type shows the most allosteric regulatory properties and M_1 type the least. M_1 type and M_2 type show very similar immunochemical reactivities.

Pyruvate Kinase Patterns in Tumor Cells

Figure 1 shows one of typical electrophoretic zymograms of pyruvate kinase (*6*). The enzyme activity on the polyacrylamide gel plate was detected by Rutter's method. It is clearly seen that there are at least 3 major peaks, L, M_1, and M_2. To see whether the fractions separated were artifacts, the 3 kinds of mixture of various

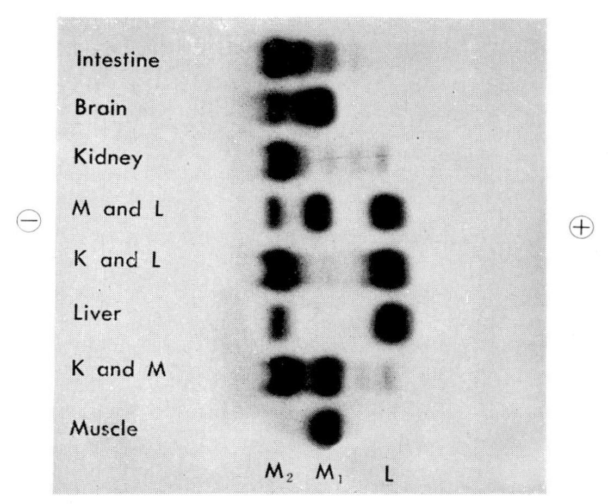

FIG. 1. Zymogram of pyruvate kinases from adult rat tissues. Supporting medium: 3.54% acrylamide gel. Buffer: 10 mM Tris-HCl, 5 mM $MgSO_4$, 0.5 mM dithiothreitol, 0.5 mM fructose-1, 6-diphosphate, pH 8.2. Voltage: 40 V/cm. Time: 3 hr.

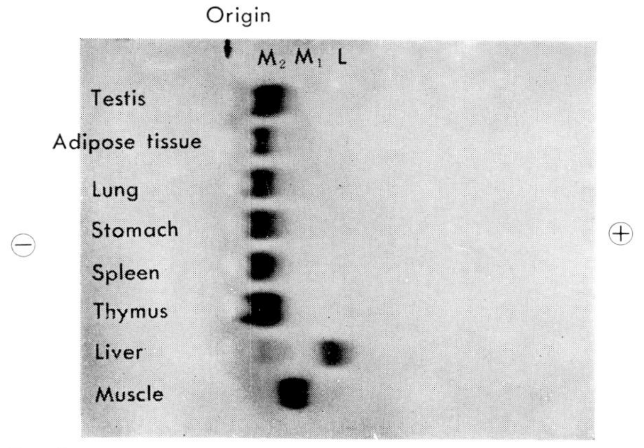

FIG. 2. Electrophoretic zymogram of pyruvate kinase of rat tissues. Supporting medium: 3.75% acrylamide gel. Buffer: 10 mM Tris-HCl, 5 mM $MgSO_4$, 0.5 mM dithiothreitol, 0.5 mM fructose-1, 6-diphosphate, pH 8.1. Voltage: 22 V/cm. Time: 3 hr.

tissues were submitted to electrophoresis and each spot moved at its own electrophoretic rate.

Figure 2 also shows the electrophoretic zymogram of various tissues. These tissues contain only M_2 type. It is clear that L and M_1 are distributed only in a limited number of specific tissues.

Figure 3 shows the electrophoretic patterns of the pyruvate kinase zymograms of fetal tissues obtained on the 17th day of pregnancy together with those of mother rat. A characteristic difference between the fetal and parental tissues is that M_2 type pyruvate kinase is predominant in the muscle, brain, and liver of the fetus, while in the mother, M_1 type is predominant in muscle and brain, and L type in liver. On the other hand, as shown in Fig. 4, in the rat 3 hr after birth, M_1 type has already appeared and M_2 type decreased in the skeletal muscle, M_2 type has decreased and M_1 type increased in the brain, and M_2 type decreased and L type increased in the liver. Figure 5 shows the time course of the level of pyruvate kinase in a regenerating liver (*15*). Activity of M_2 type pyruvate kinase reached a maximum level 2 days after resection of two-thirds of the liver tissues and finally reached

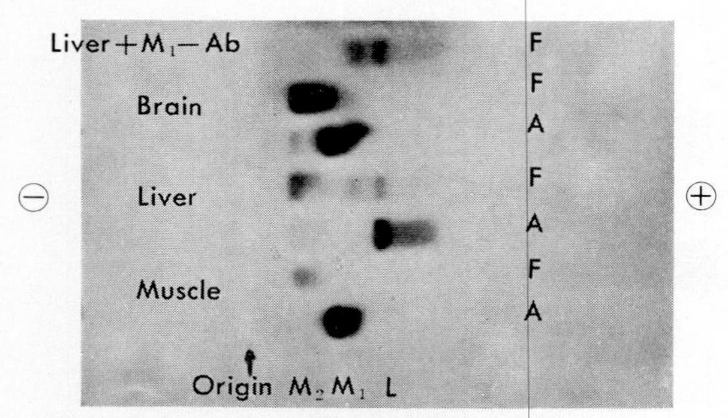

Fig. 3. Electrophoretic zymogram of pyruvate kinase of adult (A) and fetal (F) rats tissues (*6*). Supporting medium: 3.75% acrylamide gel. Buffer: 10 mM Tris-HCl, 5 mM $MgSO_4$, 0.5 mM dithiothreitol, 0.5 mM fructose-1, 6-diphosphate, pH 8.1. Voltage: 40 V/cm. Time: 2 hr.

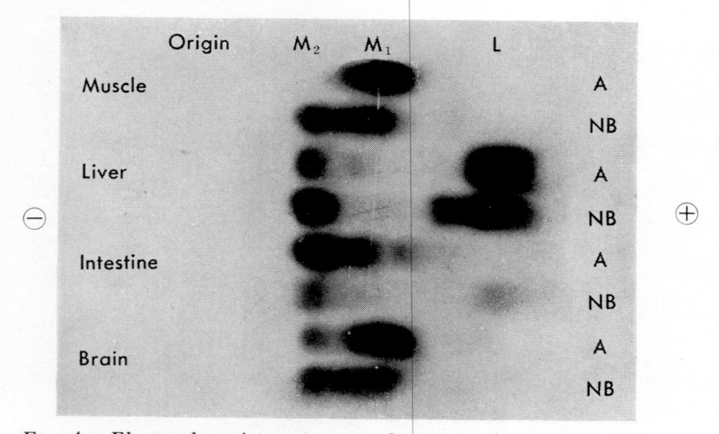

Fig. 4. Electrophoretic zymogram of pyruvate kinase of adult (A) and new born (NB) rats tissues (*6*). Supporting medium: 3.75% acrylamide gel. Buffer: 10 mM Tris-HCl, 5 mM $MgSO_4$, 0.5 mM dithiothreitol, 0.5 mM fructose-1, 6-diphosphate, pH 8.1. Voltage: 40 V/cm. Time: 2 hr.

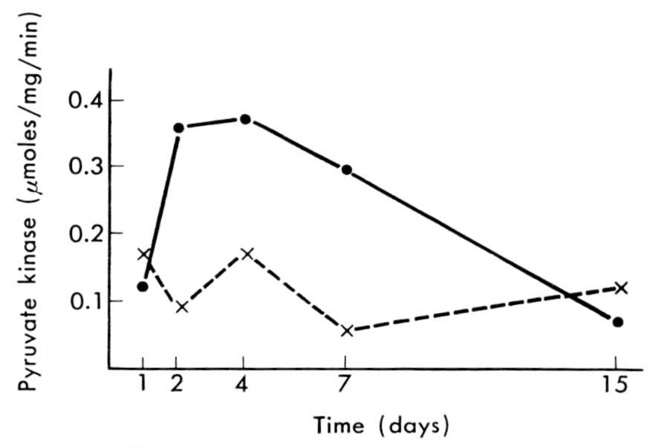

FIG. 5. Changes in pyruvate kinase L and M_2 type activities after partial hepatectomy (*15*). —— M_2 type; --- L type.

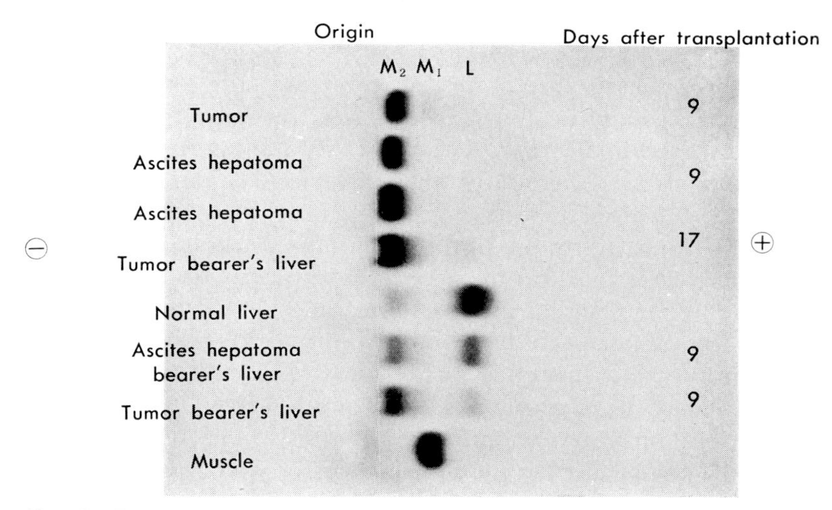

FIG. 6. Electrophoretic zymogram of pyruvate kinase of normal and cancer bearer's tissues, and cancer cells of rat. Supporting medium: 3.75% acrylamide gel. Buffer: 10 mM Tris-HCl, 5 mM $MgSO_4$, 0.5 mM dithiothreitol, 0.5 mM fructose-1, 6-diphosphate, pH 8.1 Voltage: 22 V/cm. Time: 3 hr.

a normal range about 2 weeks after hepatectomy. Activity of L type did not show any change during this regenerating process. Figure 6 shows the electrophoretic patterns of pyruvate kinase in crude extracts of cancer cells, and normal non-cancerous liver tissues of tumor-bearing animals. Tumor strain used in this experiment was Walker carcinosarcoma and hepatoma was Yoshida ascites hepatoma AH-130. It can clearly be seen that Walker tumor and Yoshida ascite hepatoma cells contain only a single type pyruvate kinase (M_2 type). In normal tissues of the liver of tumor-bearing animal, the L type is gradually replaced by M_2 type with lapse of time after transplantation of the cancer or with growth of cancer cells (*13*). This metabolic dedifferentiation in the liver of tumer-bearing animals will be discussed later more in detail.

TABLE 2. Pyruvate Kinase Activity in Normal Rat Liver, Nodular Hyperplasia, and Hepatomas (*26*)

Tissue	No.	Pyruvate kinase activicy (U/g tissue)			
		Total	L type	M_2 type	M_2 type (%)
Normal liver	16	42.3±7.1	34.3±9.3	7.9±3.2	20.3±10.8
Nodular hyperplasia	14	40.5±4.6	23.7±7.0	16.8±4.4	42.2±12.7
		12.1	3.4	8.6	71.1
		17.1	4.4	12.7	71.0
Well-differentiated hepatoma	5	21.3	5.4	15.9	74.5
		28.2	11.3	16.9	60.0
		23.3	6.1	17.1	73.5
		48.5	30.9	17.6	36.2
Highly differentiated hepatoma	3	38.1	10.4	27.7	72.8
		44.0	15.1	28.8	60.0
Poorly differentiated hepatoma	2	125.4	10.9	114.5	91.3
		129.1	11.2	117.9	91.3

Table 2 shows the isozyme pattern of pyruvate kinase of primary hepatomas and nodular hyperplasia of rat liver (*26*). N, N-nitrosomorpholine was used for the production of both nodular hyperplasia and hepatoma. A marked difference in pyruvate kinase activities was found among various hepatomas. In Table 2, the hepatomas are arranged in the order of total activity of pyruvate kinase. The hepatomas are divided into 3 groups, In the first group, total activity is reduced to less than one-half of the normal. L type activity is markedly decreased and M_2 type is increased about twice. Consequently, activity of M_2 type to total is about 70%. In the second group, total activity is in the normal range, but that of L type is decreased and M_2 type is increased significantly. Hyperplastic nodules showed essentially the same pattern of activity as this group but to a lesser extent. In the third group, total activity increased remarkably and this increase was due to the increase in M_2 activity. Histological examination showed the first group to be well-differentiated hepatoma, the second group as highly differentiated hepatoma, and the third group as poorly differentiated hepatoma. Weinhouse *et al.* (*2*) pointed out that there was a good correlation between the level of M_2 pyruvate kinase and growth rate. The present results on primary hepatoma show that the ratio of M_2 type activity to the total is a better index of the degree of differentiation.

Pyruvate Kinase Patterns in Liver of Tumor-bearing Animals

It is still unknown what mechanism caused such a deviation in the isozyme pattern of cancer cells. One clue to this question can be obtained by the different way of approach. Greenstein (*3*) pointed out that metabolism in tumor-bearing host shifted to a pattern similar to that in tumor cells. On the line of this idea, Nakahara (*9*) discovered toxohormone which was extracted from tumor cells and lowered the level of catalase in the liver. We have been studying the deviation of isozyme pattern in normal liver of tumor-bearing animals (*15*). Figure 7 shows the pyruvate kinase isozyme pattern in the liver of tumor-bearing rats. M_2 type pyruvate kinase in the

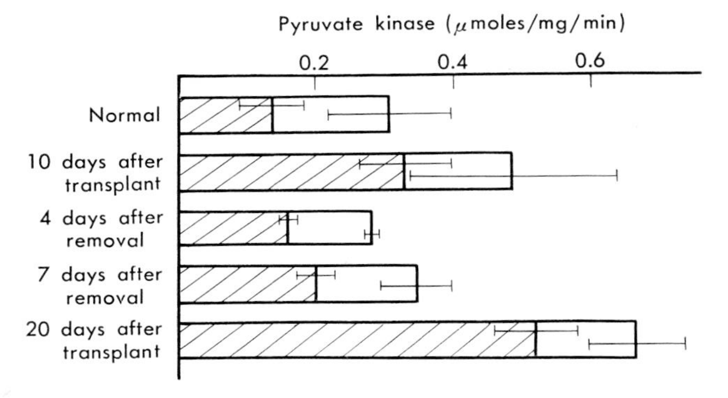

FIG. 7. Changes in pattern of pyruvate kinase L and M types after transplantation and removal of Walker's tumor. ▨ M type; □ L type.

liver of tumor-bearing animals increases with lapse of time after transplantation of Walker carcinosarcoma (*15*). When a solid tumor in the femoral muscle was removed 10 days after transplantation, the level of M_2 type enzyme returned rapidly to the normal range. The deviation in isozyme pattern was observed not only in pyruvate kinase but also in hexokinase under the same condition. Ikehara *et al.* (*4*) also reported a similar deviation in aldolase isozyme pattern in the liver of cancer patients. Several series of experiments suggest the presence of a humoral factor which mediates the same message from tumor cells to the liver and causes deviation in the isozyme pattern of pyruvate kinase. These experiments include perfusion of normal liver with blood taken from tumor-bearing animals (*14*), parabiotic experiments (*15*), and direct injection of cell-free tumor extract (*17*). Figure 8 shows a parabiotic couple of normal and tumor-bearing animals (*15*). Walker carcinosarcoma was transplanted into the left femoral muscle of one of partners of parabiotic couples. The level of pyruvate kinase M_2 type increased not only in the liver of transplanted partner but also in that of the intact partner. Similar deviation in isozyme pattern was also observed in the liver of various strains of mice transplanted with Ehrlich ascites tumor cells (Table 3). Table 4 shows the effect of injection with particulate fraction of Ehrlich ascites tumor cells on the pattern of pyruvate kinase isozyme. As clearly seen here, M_2 type increased and L type decreased in the liver of a mouse injected the particulate fractions of a tumor. This factor can be solubilized by super-

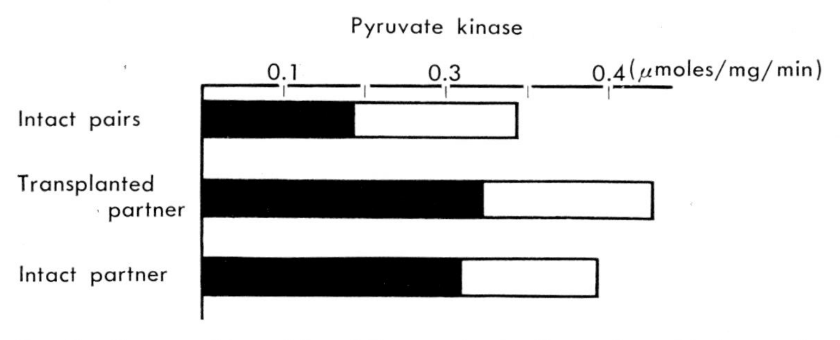

FIG. 8. Pyruvate kinase M and L types in the livers of parabiotic couples of normal and Walker's carcinosarcoma-bearing rats. ■ M type; □ L type.

TABLE 3. Pyruvate Kinase Activity in Liver of Tumor-bearing Mice (*17*)

Strain	Control			
	No.	Total	M type	M type (%)
ddO	6	0.472±0.070	0.284±0.062	59.7± 6.2
DDD	6	0.545±0.032	0.237±0.020	43.3± 4.0
ddN	6	0.528±0.028	0.198±0.030	37.4± 5.2
dd/s	6	0.510±0.048	0.188±0.073	36.3±11.7
KK	6	0.359±0.027	0.161±0.031	44.8± 6.9
A-Jax	6	0.451±0.051	0.189±0.063	41.6±11.7
NA_2	6	0.479±0.028	0.131±0.024	27.4± 5.1
C3H	6	0.397±0.033	0.068±0.020	20.8± 2.4
129	6	0.628±0.024	0.125±0.026	19.8± 4.0
ICR	6	0.615±0.061	0.067±0.012	11.2± 2.6
CF#1	5	0.513±0.038	0.045±0.043	9.5± 9.4

Data are expressed as the mean value ± standard deviation. Enzyme activities were assayed 4 days after inoculation of tumor cells. Pyruvate kinase activity: U/mg protein.

TABLE 4. Effect of Cell-free Preparation of Ehrlich Ascites Tumor Cell on Mouse Liver Pyruvate Kinase Activity (*27*)

Fraction	No.	Pyruvate kinase activity (U/g liver)		
		Total	L type	M type
I. Control saline	3	35.5±2.2	29.7±2.9	5.7±1.1
II. Supernatant	4	36.7±6.2	23.7±9.2	9.3±3.4
III. Precipitate	5	34.5±5.6	24.1±5.4	10.3±0.5
Statistical significance (*t*-test)				
I/II		N.S.	N.S.	$P<0.3$
I/III		N.S.	$P<0.3$	$P<0.005$

N.S.: not significant.

sonication. Since the active fraction was precipitated by centrifugation at $100{,}000\times g$ for 60 min in a medium of 2.0 M sucrose solution, this factor must be associated with the cell nucleus.

The nuclear preparation was further fractionated into several fractions by differential extraction of nuclear proteins described by Busch (*1*) as shown in Fig. 9. The factor was not extracted from nuclei with 2.0 M NaCl solution (Table 5). The factor is solubilized by a dilute alkali solution as shown in Table 6. These experiments suggest that the factor might be a non-histone residual protein of the nucleus (*27*). Homogenates of several normal tissues so far tested did not show such a significant activity to increase pyruvate kinase M_2 of the liver.

DISCUSSION

There are many papers refering to deviation of isozyme patterns in tumor cells. Even when limited to glycolytic enzymes, isozyme patterns of aldolase (*2*, *12*),

	Tumor-bearer		
No.	Total	M type	M type (%)
6	0.584±0.122	0.405±0.097	69.0±3.1
6	0.654±0.078	0.321±0.034[a]	50.7±6.7
6	0.560±0.048	0.263±0.052	46.7±7.5
6	0.713±0.026[a]	0.375±0.054[a]	52.8±2.8[a]
6	0.403±0.009[b]	0.244±0.031[a]	60.6±3.5
6	0.653±0.036[a]	0.409±0.014[a]	63.3±5.1[b]
6	0.507±0.072	0.285±0.035[a]	57.2±6.6[a]
6	0.455±0.047	0.190±0.036[a]	41.5±5.9[a]
6	0.756±0.080	0.499±0.067[a]	55.5±2.0[a]
6	0.688±0.045	0.234±0.049[a]	34.3±8.2[a]
4	0.685±0.083	0.216±0.047[b]	31.4±6.0[b]

[a] $P<0.001$. [b] $P<0.01$.

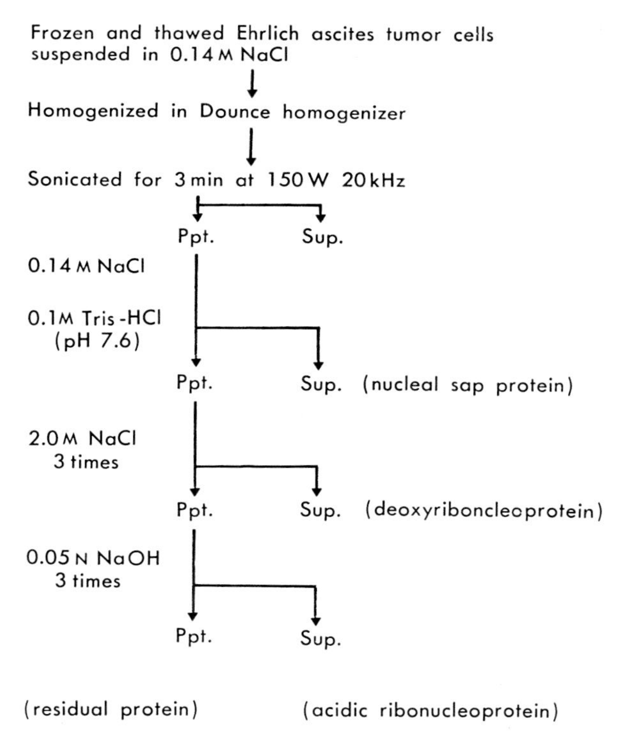

FIG. 9. Fractionation of M_2 factor.

hexokinase (*2, 11*), phosphofructokinase (*8, 18*), and pyruvate kinase (*2, 14, 24*) are reported to deviate in tumor cells. Isozymes of glycolytic key enzymes can be classified into 2 groups, the prototype and differentiated type. Prototype isozyme may also be called a carcinoembryonic or fetal type, but, since this type is also distributed

TABIE 5. Preparation of the M_2 Factor (*27*)
Fractionation with sodium chloride

Fraction	No.	Pyruvate kinase activity (U/g liver)		
		Total	L type	M type
I. Control saline	5	45.6±4.7	36.2±2.6	9.5±2.3
II. 0.14 M sup. 202.4 mg.	5	43.2±4.6	31.5±5.6	11.5±3.2
III. 2.0 M sup. 57.2 mg.	5	39.3±2.0	25.4±2.4	13.6±3.9
IV. 2.0 M ppt. 78.6 mg.	5	44.7±4.2	28.9±2.3	15.5±2.6
Statistical significance (*t*-test)				
I/II		N.S.	N.S.	N.S.
I/III		$P<0.05$	$P<0.001$	$P<0.2$
I/IV		N.S.	$P<0.005$	$P<0.01$

N.S. : not significant.

TABLE 6. Preparation of the M_2 Factor (*27*)
Extraction with dilute alkali solution

Fraction	No.	Pyruvate kinase activity (U/g liver)		
		Total	L type	M type
I. Control saline	5	56.8±8.3	46.9±8.2	10.0±1.9
II. Alkaline ppt. 24.8 mg	5	50.4±6.2	35.7±7.1	14.6±3.4
III. Alkaline sup. 19.4 mg	5	52.2±2.2	35.8±2.7	16.3±2.1
Statistical significance (*t*-test)				
I/II		$P<0.3$	$P<0.1$	$P<0.05$
I/III		N.S.	$P<0.05$	$P<0.005$

N.S. : not significant.

often in adult tissues, it would be better to call this type a prototype. Pyruvate kinase M_2 is a prototype, and L and M_1 types are differentiated type. According to the experiments reported by Weinhouse *et al.* (*2*) and by Sugimura *et al.* (*11*) hexokinase II is a prototype and glucokinase (hexokinase IV) is a differentiated type. It was reported from this laboratory that IV type is a prototype in the case of phosphofructokinase. However, arguments against this idea has been made by Kurata *et al.* (*8*), and according to their results, II type is a prototype. The discrepancy between these results has not yet been solved.

Figure 10 shows the tentative working hypothesis on our experimental observations on deviations of isozyme patterns in tumor cells and in the liver of tumor-bearing animals. In tumor cells, isozyme patterns deviate to suppress homotopic differentiation pattern and to stimulate prototype pattern. Deviation toward this way can be called dedifferentiation. The other type of deviation in isozyme pattern is also reported, though it cannot necessarily be observed in all tumor cells. This is the way of ectopic appearence of differentiated type in tumor cells. The typical example of the latter type of deviation was reported by Miyaji *et al.* (*10*). They observed appearance of glucokinase in Rhodamine sarcoma which is thought to

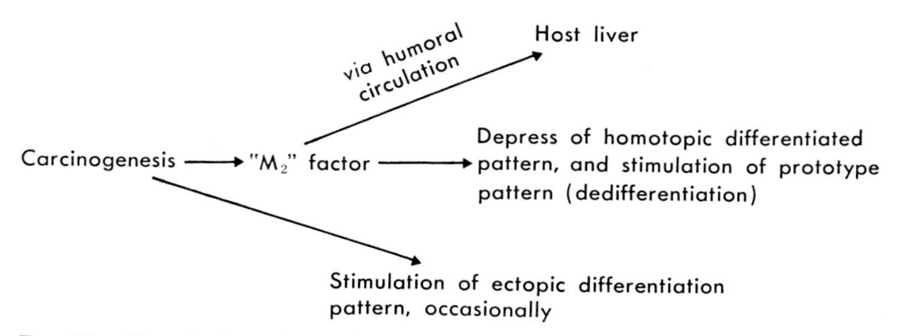

FIG. 10. Hypothesis on the regulation of isozyme patterns.

originate from fibroblast and not from hepatocyte. Dedifferentiated deviation in isozyme pattern similar to that in tumor cells was also observed in the liver of tumor-bearing animals (*4*, *13*, *15*), and a humoral factor in tumor cells showed the activity of causing such a deviation in normal animals (*17*). It could be assumed that this factor may be acting not only on the liver of the tumor-bearing host but also inside the tumor cells in the same way to cause a deviation in isozyme patterns.

REFERENCES

1. Busch, M. Isolation and purification of nuclear proteins. Methods Enzymol. *B12*: 65–84, 1968.
2. Farima, F. A., Adelman, R. C., Morris, H. P., Lo, C. H., and Weinhouse, S. Metabolic regulation and enzyme alterations in the Morris hepatomas. Cancer Res., *28*: 1897–1900, 1968.
3. Greenstein, G. P. Biochemistry of Cancer, Academic Press, New York, 1947.
4. Ikehare, Y., Inokuchi, K., and Endo, H. Aldolase isozyme patterns in the liver of patients with malignant tumor in the stomach or esophagus. Gann, *60*: 449–459, 1969.
5. Imamura, K., Taniuchi, K., and Tanaka, T. Multimolecular forms of pyruvate kinase. II. Purification of M_2-type pyruvate kinase from Yoshida ascites hepatoma 130 cells and comparative studies on the enzymological and immunological properties of the three types of pyruvate kinase, L, M_1, and M_2. J. Biochem., *72*: 1001–1015, 1972.
6. Imamura, K. and Tanaka, T. Multimolecular forms of pyruvate kinase from rat and other mammalian tissues. I. Electrophoretic studies. J. Biochem., *71*: 1043–1051, 1972.
7. Katzen, M. H. and Schimke, K. T. Multitype forms of hexokinase in the tissue distribution, age-dependency and properties. Proc. Natl. Acad. Sci. U.S., *54*: 1218–1225, 1965.
8. Kurata, N., Matsushima, T., and Sugimura, T. Multiple forms of phosphofructokinase in rat tissues and rat tumors. Biochem. Biophys. Res. Commun., *48*: 473–479, 1972.
9. Nakahara, W. and Fukuoka, F. A toxic cancer tissue constituent as evidence by its effect on liver catalase activity. Jap. Med. J., *1*: 271, 1948 (in Japanese).
10. Miyaji, K. Hexokinase isozymes in diabetes and in cancer. Naika (Internal Medicine), *32*: 815–820, 1972 (in Japanese).

11. Sato, S., Matsushima, T., and Sugimura, T. Hexokinase isozyme patterns of experimental hepatomas of rats. Cancer Res., *29*: 1437–1443, 1969.
12. Schapira, F., Dreyfus, G. C., and Schapira, G. Anomaly of aldolase in liver cancer. Nature, *200*: 995–997, 1963.
13. Suda, M., Tanaka, T, Yanagi, S., Hayashi, S., Imamura, K., and Taniuchi, K. Dedifferentiation of enzymes in the liver of tumor-bearing animals. GANN Monograph on Cancer Research, *13*: 79–93, 1972.
14. Suda, M., Tanaka, T., Sue, F., Kuroda, Y., and Morimura, H. Rapid increase of pyruvate kinase (M-type) and hexokinase in normal rat liver by perfusion of the blood of tumor-bearing rat. GANN Monograph, *4*: 103–112, 1968.
15. Suda, M., Tanaka, T., Sue, F., Harano, Y., and Morimura, H. Dedifferentiation of sugar metabolism in the liver of tumor-bearing rat. GANN Monograph, *1*: 127–141, 1966.
16. Susor, W. A. and Rutter, W. J. Some distinctive properties of pyruvate kinase purified from rat liver. Biochem. Biophys. Res. Commun., *30*: 14–20, 1968.
17. Tanaka, T., Yanagi, S., Miyahara, M., Kaku, R., Imamura, K., Taniuchi, K., and Suda, M. A factor responsible for the metabolic deviations in liver of tumor-bearing animals. Gann, *63*: 552–562, 1972.
18. Tanaka, T., Imamura, K., Ann, T., and Taniuchi, K. Multimolecular forms of pyruvate kinase and phosphofructokinase in normal and cancer tissues. GANN Monograph on Cancer Research, *13*: 219–234, 1972.
19. Tanaka, T., Ann, T., and Sakaue, Y. Studies on multimolecular forms of phosphofructokinase in rat tissues. J. Biochem., *69*: 609–612, 1972.
20. Tanaka, T., Imamura, K., Ann, T., and Taniuchi, K. Abstr. Papers 8th Int. Congr. Biochem., 227–229, 1970.
21. Tanaka, T., Harano, Y., Sue, F., and Morimura, H. Crystallization, characterization and metabolic regulation of two types of pyruvate kinase isolated from rat tissues. J. Biochem., *62*: 71–91, 1967.
22. Tanaka, T., Harano, Y., Morimura, H., and Mori, R. Evidence for the presence of two types of pyruvate kinase in rat liver. Biochem. Biophys. Res. Commun., *21*: 55–60, 1965.
23. Vinuela, E., Salas, M., and Sols, A. Glucokinase and hexokinase in liver in regulation to glycogen synthesis. J. Biol. Chem., *238*: PC1175–1177, 1963.
24. Walker, P. R. and Potter, V. R. Isozyme studies on adult, regenerating, precancerous and developing liver in relation to findings in hepatoma. Adv. Enzyme Regulation, *10*: 339–364, 1972.
25. Walker, D. G. On the presence of two soluble glucose-phosphorylating enzymes in adult liver and the development of one of these after birth. Biochim. Biophys. Acta, *77*: 209–226, 1963.
26. Yanagi, S., Makiura, S., Kawamoto, Y., Matsumura, K., Hirao, K., Ito, N., and Tanaka, T. Isozyme patterns of pyruvate kinase in various primary tumors induced during the process of hepatocarcinogenesis. Cancer Res., in press.
27. Yanagi, S., Kashiwagi, A., Tanaka, T., and Suda, M. Existence of a factor (M_2-factor) inducing deviation of the isozyme pattern of pyruvate kinase in the liver of tumor-bearing mice. Unpublished.

Discussion of Paper by Drs. Tanaka et al.

DR. WEBER: The data you presented confirm the earlier reports you did not refer to, of Taylor, Morris, and Weber in Life Science, Advance in Enzyme Regulation *etc.* They also confirm my thesis published years ago stating that in hepatoma, there emerges an isozyme shift which results in the gradual decrease in liver type, regulatory isozyme and the emergence of non-liver type isozymes (Weber, reference to be cited).

My question refers to your claims in one of your tables stating that the liver type pyruvate kinase is inhibited by L-Ala and L-Phe in a "cooperative fashion." I reported first the inhibition of liver pyruvate kinase by L-Ala and L-Phe and presented these results also in 1965 in Osaka where you were present. What is your evidence now that this inhibition is not competitive, but cooperative?

DR. TANAKA: You have emphasized the existence of tumor-specific pyruvate kinase in your article in Life Science, 1969. However we could not find such tumor-specific enzyme. L type pyruvate kinase is cooperatively inhibited by alanine as shown in Fig. 11. This figure shows the effect of alanine concentration on L type

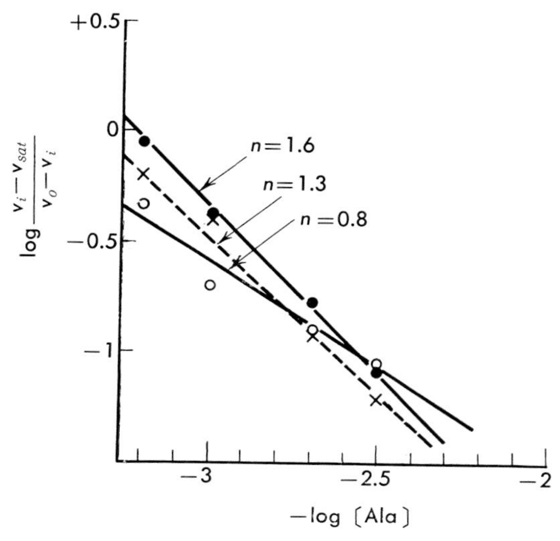

FIG. 11. Effect of alanine concentration on L type pyruvate kinase activity at different levels of PEP. ○ 0.1 mM; × 0.5 mM; ● 1 mM PEP.

pyruvate kinase activity at different levels of PEP. Hill constant for alanine at 0.1 mM PEP is 1.6.

DR. NAKAHARA: Do you think that the nuclear non-histone fraction can be released from the nuclear structure and excreted from the tumor cell into general circulation? This is important in determining the humoral role of the non-histone component in producing any systemic effect on the host.

DR. TANAKA: The question made by Dr. W. Nakahara is very hard to answer. I think that nuclear fragments may be released into blood after cellular autolysis and then reach target organs such as the liver.

DR. WEINHOUSE: The observation that livers of tumor-bearing hosts have increased M_2 pyruvate kinase adds a new dimension to the possible effects of tumors on the host and is very important. Are these changes accompanied by histologic changes that might indicate alteration of the number of different cell types?

DR. TANAKA: It has been well-known that hematopoietic cells increase in the liver of tumor-bearing animals. However, we have observed by the fluorescence antibody method that M_2 type increased in liver parenchymal cells.

Alteration and Significance of Glucosamine 6-Phosphate Synthetase in Cancer

Shigeru TSUIKI, Taeko MIYAGI, Kunimi KIKUCHI, and Hisako KIKUCHI
Biochemistry Division, Research Institute for Tuberculosis, Leprosy and Cancer, Tohoku University, Sendai, Japan

Abstract: Glucosamine 6-phosphate synthetase (L-glutamine: D-fructose 6-phosphate amidotransferase), the rate-limiting enzyme for UDP-N-acetylglucosamine synthesis, increases strikingly in hepatocarcinogenesis. The enzymes purified from rat liver and a hepatoma (Yoshida sarcoma) were different from each other kinetically, chromatographically, and immunologically.

In the rat, the enzyme is rich in tissues that undergo extensive proliferation as well as in those engaged in glycoprotein secretion. Isoelectric focusing studies have revealed that there are at least 3 isozymic forms of the enzyme. The form having an isoelectric point (pI) at 5.0 is present only in adult liver and appears to be the liver-specific form. The form with a pI of 4.1 has been found in early embryo and adult brain and is referred to as the embryonic form. It is present in tumors as a minor component.

The form having a pI at 4.5 is the major tumor form and is also present in proliferating tissues including late-fetal and regenerating liver. The form strikingly resembles the enzymes associated with DNA synthesis with respect to tissue distribution, developmental change, subcellular localization, and response to hydrocortisone. The possible association of this form with mucopolysaccharide synthesis has been suggested.

It is increasingly apparent that alterations in cell-surface glycoproteins are associated with neoplastic transformation (*8*, *30*, *32*). Evidence from tissue-culture systems has further suggested that these macromolecules may play a significant role in determining the growth behavior of the cells (*5*, *6*, *27*). It is therefore of prime importance to elucidate how the systems for cellular glycoprotein biosynthesis are affected upon neoplastic transformation. In attempts to answer this question, we have investigated the properties of glucosamine 6-phosphate synthetase (L-glutamine: D-fructose 6-phosphate amidotransferase) from normal and neoplastic tissues. The purpose of the present communication is to report the results, some parts of which have already been published elsewhere (*11*, *13*, *14*, *23*, *25*, *29*).

Glucosamine 6-phosphate synthetase catalyzes the formation of glucosamine 6-phosphate according to the following equation:

Fructose 6-phosphate + glutamine ⟶
glucosamine 6-phosphate + glutamic acid.

The enzyme is subject to feedback inhibition by UDP-N-acetylglucosamine (*18, 19, 25*) and hence plays an important regulatory role in its biosynthesis. This sugar nucleotide is the direct precursor of the N-acetylglucosamine unit of glycoproteins and mucopolysaccharides.

We have also studied the properties of UDP-N-acetylglucosamine 2′-epimerase from normal and neoplastic tissues (*15, 16*). The enzyme catalyzes the reaction,

UDP-N-acetylglucosamine + H_2O ⟶ UDP + N-acetylmannosamine,

and occupies a rate-controlling position in the biosynthesis of CMP-N-acetylneuraminic acid (*16, 19*), the precursor of the N-acetylneuraminic acid unit of glycoproteins. Some of the results obtained in these studies will be presented in this communication.

Amidotransferase and Epimerase Activities in Neoplastic and Fetal Liver*

To start with, the amidotransferase and epimerase activities of rat liver and hepatomas were compared. The tumors examined were 3 ascites hepatoma strains (Yoshida sarcoma, AH-66F and AH-130) and Morris hepatoma 5123D which was supplied by Dr. M. Watanabe of this Institute. Yoshida sarcoma has been shown recently to be of hepatoparenchymal origin (*12, 33*).

The results are summarized in Fig. 1, which shows that upon neoplastic trans-

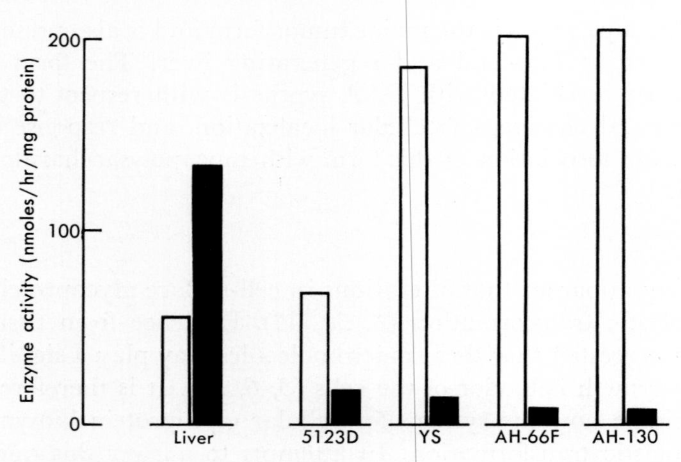

FIG. 1. Amidotransferase (□) and epimerase activities (■) of rat liver and hepatomas (*14, 15*). YS stands for Yoshida sarcoma.

* Glucosamine 6-phosphate synthetase and UDP-N-acetylglucosamine 2′-epimerase are referred to as amidotransferase and epimerase, respectively.

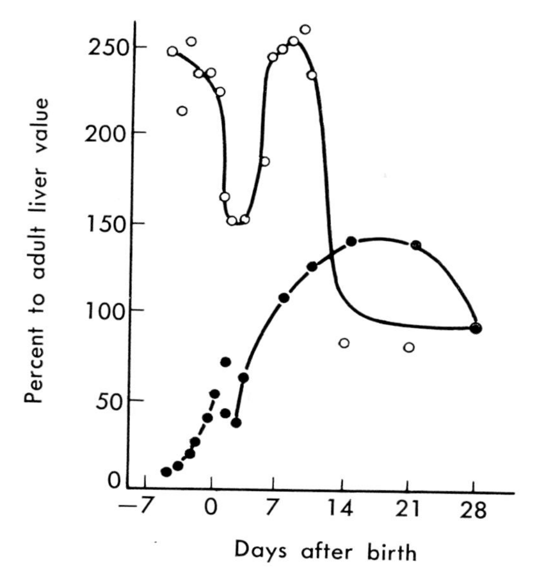

FIG. 2. Amidotransferase and epimerase activities of rat liver during late-fetal and neonatal development. ○ amidotransferase; ● epimerase.

formation, amidotransferase activity rose strikingly, whereas that of epimerase fell to very low levels. This inverse behavior was felt to be puzzling, since the two enzymes were thought to be associated with the same metabolic function, *i.e.*, synthesizing cellular glycoproteins.

Similar differences between the 2 enzymes have been observed for fetal liver. As shown in Fig. 2, five days before birth, the level of amidotransferase was about twice the adult level, but the presence of epimerase was almost undetectable. The latter enzyme began to rise around birth.

These results suggested that mechanisms might operate in determining the amidotransferase level which were more complex than those for epimerase, and prompted us to investigate the properties of amidotransferase from normal and neoplastic tissues.

Properties of Amidotransferase from Rat Liver and Yoshida Sarcoma

The idea that amidotransferase might exist in multiple molecular forms first arose when the tumor enzyme was found to be more sensitive to inhibition by UDP-N-acetylglucosamine than the liver enzyme. The data presented in Fig. 3, however, were for relatively crude preparations; attempts were therefore made to purify the enzyme. A typical purification protocol for rat liver amidotransferase is shown in Table 1. Amidotransferase was also purified from Yoshida sarcoma by essentially the same procedure. Using highly purified preparations, the above observations have been confirmed.*

* An extensive inhibition could be observed only when glucose 6-phosphate was present. For the reason, see Ref. *25* or *29*.

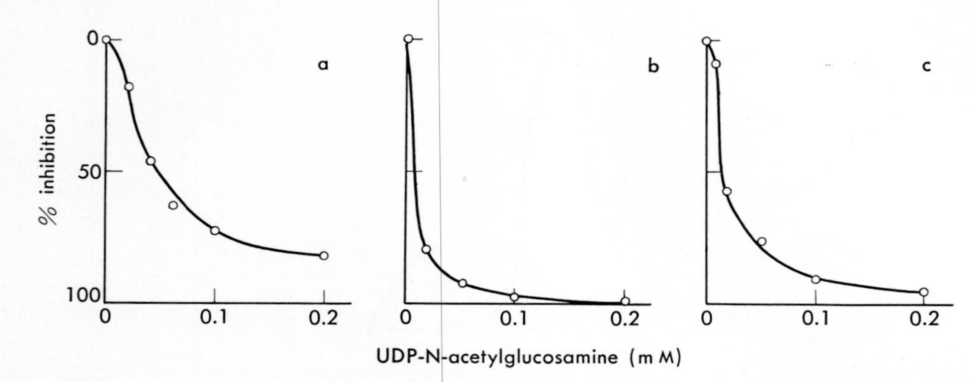

FIG. 3. Inhibition of rat liver and hepatoma amidotransferase by UDP-N-acetylglucosamine (*14*). a, liver; b, Yoshida sarcoma; c, AH-66F.

TABLE 1. Purification of Amidotransferase from Rat Liver (*25*)

Purification step	Total activity (units)[a]	Specific activity (units/mg)	Purification (×)	Recovery (%)
105,000×*g* supernatant	22,700	25.3	(1)	(100)
$(NH_4)_2SO_4$ precipitate	17,950	58.8	2	80
DEAE-Sephadex column	15,900	663.0	26	71
Hydroxylapatite column	6,750	6,062.9	240	30

[a] nmoles/hr.

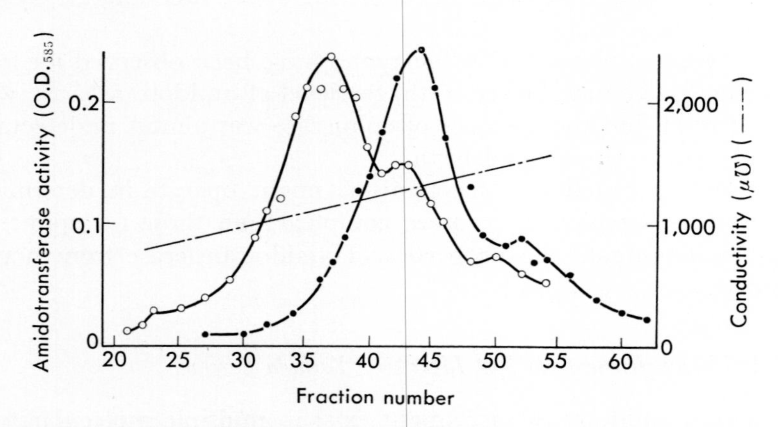

FIG. 4. DEAE-Sephadex column chromatography of rat liver and Yoshida sarcoma amidotransferase. For the experimental conditions, see Ref. *25* or *29*. ○ liver; ● Yoshida sarcoma.

Figure 4 compares the elution patterns of liver and tumor amidotransferases from a DEAE-Sephadex column. A linear gradient of KCl from 50 to 500 mM added to 50 mM sodium phosphate (pH 7.5) was applied and the tumor enzyme was generally eluted at a concentration of KCl greater than that required for the liver enzyme.

These results suggested that the tumor enzyme might be of a more acidic nature

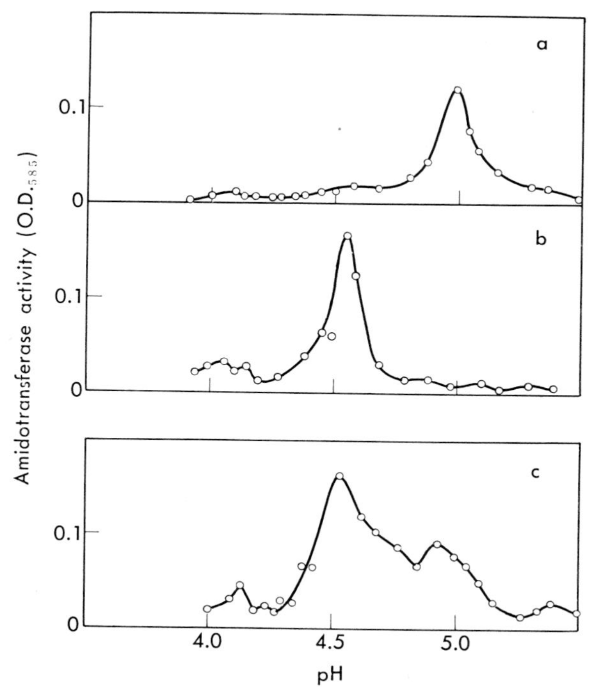

FIG. 5. Isoelectric focusing pattern of rat liver and Yoshida sarcoma amidotransferase. In these experiments, an LKB 8101 column with a volume of 110 ml and Ampholine in the pH range from 3 to 6 were used. A 0–45% (w/v) sucrose density gradient containing 1% Ampholine was prepared and a 1.5-ml sample was applied. The voltage was set initially at 300 V and increased to 700 V after 12 hr. Focusing was then conducted for 28 hr while the column was kept at 1°–3°C. Fractions of 1.5 ml were collected, and the pH (12°C) and enzyme were assayed in each fraction. The enzymes analyzed were from the $(NH_4)_2SO_4$ precipitate fraction, containing about 20 mg protein/ml, but the same results were obtained using more purified preparations. a, liver; b, Yoshida sarcoma; c, mixed.

and this was substantiated by an isoelectric focusing study. As shown in Fig. 5, the major peak of the tumor pattern had an isoelectric point (pI) at 4.5 and was clearly different from the liver enzyme, with a pI of 5.0. The tumor pattern had an additional peak at pH 4.1, which was not found in the liver pattern. When a mixture of the liver and tumor enzymes was examined, an isoelectric focusing pattern emerged which was additive for the 2 enzymes (the bottom graph).

Further evidence that the tumor enzyme differs from the liver enzyme has been obtained by an immunological study, shown in Fig. 6. An antibody to liver amidotransferase, prepared by injection of the purified enzyme into rabbits, neutralized the liver enzyme, but there was only a slight effect on the tumor enzyme. An antibody to the tumor enzyme, on the other hand, neutralized the tumor enzyme but neutralized the liver enzyme only slightly.

Table 2 summarizes the results of our studies on rat liver and Yoshida sarcoma amidotransferase. Except for a difference in sensitivity to UDP-N-acetylglucosamine inhibition, the kinetic properties of the 2 enzymes were similar. However, the re-

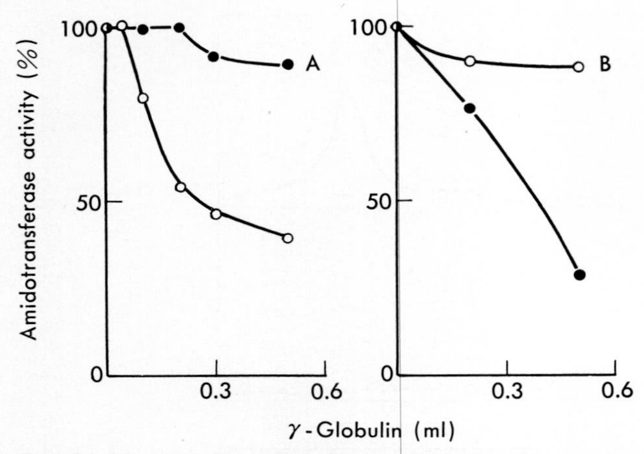

FIG. 6. Neutralization of amidotransferase by antibody. An antibody against rat liver or Yoshida sarcoma amidotransferase was prepared by injecting an adult albino rabbit with the appropriate enzyme purified as described above. Five intramuscular followed by 4 intravenous injections were given at 1-week intervals. The rabbit was then bled and the antiserum was fractionated with $(NH_4)_2SO_4$ to obtain a γ-globulin fraction. For neutralization, (○) rat liver or (●) Yoshida sarcoma amidotransferase was assayed after incubation with A, anti-liver or B, anti-Yoshida sarcoma amidotransferase at 37°C for 10 min.

TABLE 2. Summary of the Properties of Amidotransferase from Rat Liver and Yoshida Sarcoma

Properties	Amidotransferase	
	Liver	Tumor
Molecular weight	380,000	410,000
K_m (mM) for fructose 6-phosphate	0.31	0.42
K_m (mM) for glutamine	0.69	0.71
K_i (mM) for UDP-N-acetylglucosamine	0.45	0.05
Optimal pH	7.4	7.4
Elution from DEAE-Sephadex column	900–1,000[a]	1,050–1,150[a]
pI	5.0	4.5[b]
Neutralization by		
Anti-L	+	Slight
Anti-T	Slight	+

[a] The concentration of KCl required for elution from a DEAE-Sephadex column expressed in terms of conductivity ($\mu\mho$). [b] The pI of the major peak.

sults of isoelectric focusing and the immunological data leave little doubt that the tumor enzyme is a second form, probably arising at a gene locus different from that for the liver enzyme.

Developmental Pattern of Amidotransferase Isozymes

To further elucidate the nature and possible physiological function of the amidotransferase isozymes appearing in tumor tissues, the embryonic development of amidotransferase isozymes was studied by means of isoelectric focusing. The re-

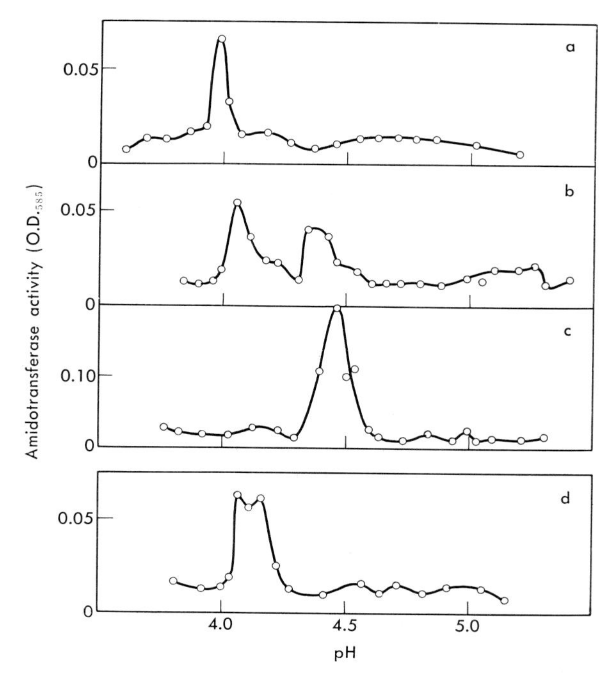

FIG. 7. Isoelectric focusing pattern of amidotransferase of embryonic tissues and brain. For the detailed procedure, see the legend to Fig. 5. a, whole embryo (12-day); b, whole embryo (14-day); c, fetal liver (19-day); d, adult brain.

sults are shown in Fig. 7. It can be seen that in the embryo at the 12th day of gestation, the amidotransferase form with a pI at 4 was present almost exclusively. This embryonic form was still present in 14-day embryo, but was absent from late-fetal as well as adult liver (Fig. 5), showing that it is eliminated from liver with the progress of development. In contrast, in the brain, the embryonic isozyme appears to remain present throughout development.* The minor component of tumor amidotransferase having a pI at 4.1 (Fig. 5) is probably identical with this isozyme.

The finding that the embryonic isozyme becomes evident in liver upon neoplastic transformation may support the concept that dedifferentiation is one of the most obvious characteristics of neoplasia (*31*). However, even in highly deviated ascites hepatomas such as Yoshida sarcoma, this isozyme is not a major component of the amidotransferase.

Another interesting finding to emerge from these isozyme studies is the marked difference between late-fetal (Fig. 7) and adult liver (Fig. 5). The predominant form of amidotransferase of 19-day fetal liver exhibited a pI value which was identical with that of the major tumor isozyme. Since its presence in adult liver is doubtful, this form must be eliminated with increasing age. Only a minute amount of the adult-liver isozyme, on the other hand, was found in the fetal liver. This isozyme

* The predominance of this isozyme in brain is reflected also by the data shown in Table 7 ; brain amidotransferase required the highest concentration of KCl for elution from a DEAE-Sephadex column.

evidently increases with increasing age, most probably in parallel with epimerase (Fig. 2).

It was pointed out earlier that in late-fetal and neoplastic liver, there was a puzzling discrepancy of activity between amidotransferase and epimerase (Figs. 1 and 2). To this problem, a solution can now be provided by suggesting that epimerase lacks an isozymic form that corresponds to the major tumor (or late-fetal) form of amidotransferase. Further investigations were therefore undertaken to determine if this isozyme unique to amidotransferase could be related to some special cellular function.

Tissue Distribution Pattern of Amidotransferase

To obtain the activity distribution data shown in Table 3, rat tissues were homogenized and 105,000 $\times g$ supernatants were prepared. For tumors and spleen, however, the measured activities were much smaller than the actual ones because of inhibitory factors in the supernatants (*13*, *14*). For this reason, assays were also made in the fraction precipitated from the supernatant at 40–60% saturation of $(NH_4)_2SO_4$. On the other hand, the activities from submaxillary gland and intestine were markedly lower after the $(NH_4)_2SO_4$ precipitation. This was due to an amidotransferase inactivator present in these tissues which could not attack the enzyme unless it was freed from the supernatant (*24*).

It can be seen that amidotransferase was present at high levels in liver, submaxillary gland, intestine and fetal cartilage. This is quite reasonable, because these

TABLE 3. Distribution of Amidotransferase in Rat Tissues

Tissues	Specific activity (nmoles/hr/mg protein)	
	105,000 $\times g$ supernatant	$(NH_4)_2SO_4$ precipitate
Liver	23.2	82.0
Liver, regenerating (48 hr)	75.7	324.3
Liver, fetal (19-day)	78.8	183.9
Liver, neoplastic (YS[a])	28.0	178.8
Submaxillary gland	125.5	52.0
Intestine	38.1	14.3
Fetal cartilage (19-day)	82.5	243.0
Brain	34.3	74.2
Lung	6.4	16.8
Kidney	1.0	
Heart	0	
Skeletal muscle	0	
Testis	31.8	192.3
Spleen	6.8	57.4
Spleen, newborn (6-day-old)	42.0	150.6
Thymus	14.5	99.7
Bone marrow	7.3	14.8
Whole embryo (15-day)	92.7	257.0

[a] Yoshida sarcoma.

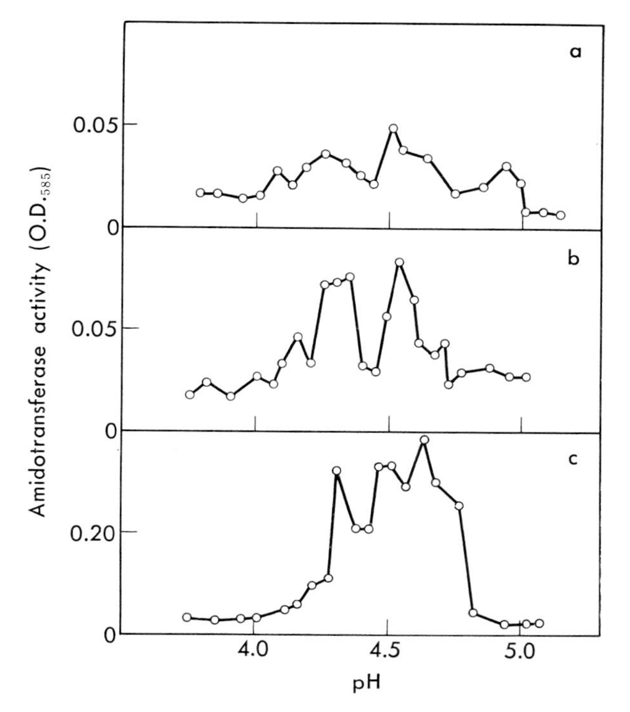

FIG. 8. Isoelectric focusing pattern of amidotransferase from several proliferating tissues. The procedure was that described in the legend to Fig. 5. a, spleen (newborn 6-day-old); b, testis; c, regenerating liver (48 hr).

tissues synthesize and export large amounts of glycoproteins or mucopolysaccharides. Certain other tissues such as kidney, heart, and skeletal muscle contained no or little amidotransferase. These tissues do not export glycoproteins.

There is, however, a third group of tissues, which is not engaged in extensive glycoprotein production but which nevertheless contains high amidotransferase activity. They include testis, newborn spleen and thymus. Regenerating liver (see also Ref. *1*) and 15-day embryo were also found to be rich in amidotransferase and the value in the former was even greater than that in fetal and neoplastic liver. These tissues are all proliferating, and it is therefore evident that at least a fraction of the amidotransferase activity distributed in rat tissues has a positive correlation with cell proliferation. A similar conclusion has been reached recently by other investigators (*26*).

The isozymic patterns of newborn spleen, testis, and regenerating liver were subsequently studied by means of isoelectric focusing. The results are shown in Fig. 8. Although they are only preliminary and therefore more or less inconsistent, it is certain that the major tumor form, which is also the preponderant form in late-fetal liver, is a major isozyme of these tissues.

Correlation of the Major Tumor Isozyme of Amidotransferase with the DNA-synthesizing System

The developmental patterns of amidotransferase activity in several tissues are shown in Fig. 9. In the thymus, and to a lesser degree in the spleen, there was a characteristic pattern present, reaching a peak in activity about 5 to 7 days after birth, then declining rather abruptly. The pattern is closely similar to that described for the enzymes associated with DNA synthesis, such as aspartate transcarbamylase (*10*), ribonucleotide reductase (*7*), and thymidine kinase (*21*). As was demonstrated earlier, a transition of the major tumor-type to adult-type isozyme occurs in liver during late-fetal and neonatal development. The rather complicated developmental pattern found in this tissue (Fig. 2) probably reflects this transition.

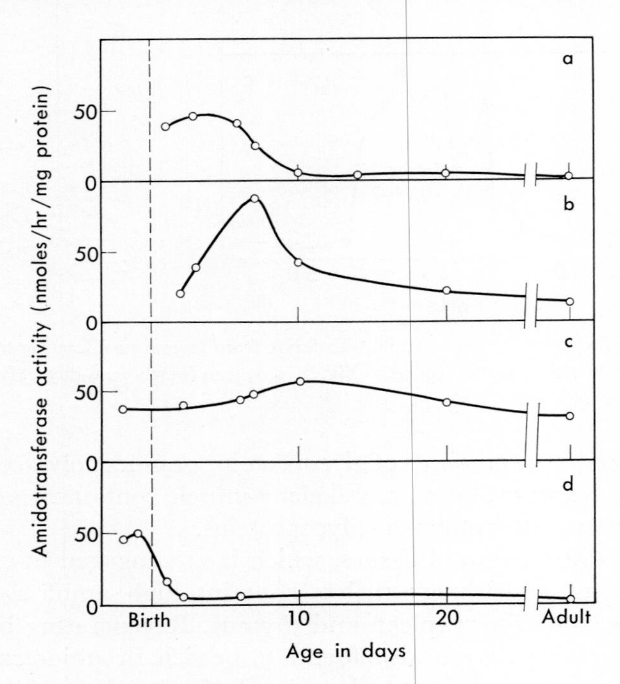

FIG. 9. Amidotransferase in serveral tissues during neonatal development. The enzyme was assayed in 105,000 $\times g$ supernatant. a, spleen; b, thymus; c, brain; d, kidney.

TABLE 4. Effect of Hydrocortisone on Amidotransferase in Different Tissues[a)]

Tissues	Amidotransferase (nmoles/hr/mg protein)	
	Control	Hydrocortisone
Liver	25.8	20.4
Submaxillary gland	120.5	100.0
Brain	23.7	21.6
Yoshida sarcoma	26.0	13.0

[a)] A rat bearing Yoshida sarcoma was injected with hydrocortisone (5 mg/100 g body weight) intraperitoneally once a day. Assays were made after 3 days. The control rat received injections of saline.

TABLE 5. Subcellular Localization of Amidotransferase Activity in Rat Liver and a Hepatoma[a]

Fraction	Activity (nmoles/hr)	
	Liver	AH-66F
600×g precipitate	0	0
10,000×g precipitate	0	
105,000×g precipitate	0	
105,000×g supernatant (postmicrosomal supernatant)	6,466	1,469
Centrifugation of postmicrosomal supernatant		
Supernatant	1,062	19
Pellet	682	2,252

[a] For the detailed procedure, see text.

It has been reported that the level of thymidine kinase in thymus, fetal liver, and neonatal spleen is decreased by hydrocortisone administered *in vivo* (*9*). Similar observations have been made for the amidotransferase activity of rat liver (*2*). Table 4 shows that *in vivo* administration of hydrocortisone decreased the amidotransferase activity of tumor tissue much more extensively than that of liver, submaxillary gland or brain. It therefore appears that the hormone may preferentially affect the tumor isozyme.

The subcellular localization of amidotransferase activity in liver and a hepatoma (AH-66F) was studied by a differential centrifugation procedure developed by Baril *et al.* (*3, 4*). As shown in Table 5, the bulk of the activity present in the original hepatoma extract was recovered in the pellet fraction obtained by centrifugation of the postmicrosomal supernatant for 15 hr at 78,000×g. Under the same conditions, typical cytosol enzymes such as lactate dehydrogenase and glucose 6-phosphate dehydrogenase were not precipitated. The pellet has been shown by Baril *et al.* (*3, 4*) to contain membrane elements, with which DNA-synthesizing enzymes such as DNA polymerase (*3, 4*), thymidine kinase, and ribonucleotide reductase (*7*) are associated. It is of particular interest that only a small fraction of the liver activity was precipitated with this pellet.

These data, when considered together with those presented in Fig. 9 and Table 4, demonstrate that there is a positive correlation between the major tumor isozyme of amidotransferase and the DNA-synthesizing system. The adult-liver isozyme and the brain (embryonic) isozyme appear to possess no such correlation.

DISCUSSION

The studies described above have led to the conclusion that there are multiple isozymic forms of amidotransferase (glucosamine 6-phosphate synthetase) in rat tissues. The 3 isozymes that have been characterized are listed in Table 6. There are some indications of the existence of still other forms (for instance, Fig. 8), but no systematic studies have yet been carried out on them.

The almost exclusive adult form of liver amidotransferase has a pI at 5.0. As no other tissues tested were found to contain this isozyme (except for a minute

TABLE 6. Isozymic Forms of Amidotransferase

pI 5.0 isozyme
Present in adult liver (the liver-specific form)
pI 4.5 isozyme
Present in tissues that proliferate rapidly (the form associated with cell replication)
pI 4.1 isozyme
Present in early embryo and brain (the embryonic form)

quantity found in late-fetal liver), it must be associated with some differentiated function characteristic of liver, possibly synthesizing blood glycoproteins. A similarity may be expected between this isozyme and epimerase as regards the time of developmental appearance and the mode of regulation.

The form with a pI of 4.1 is preponderant in early embryo and adult brain and has been referred to as the embryonic form. This isozyme also becomes evident in hepatomas, but appears to remain a minor component.

In addition to this embryonic form, neoplastic tissues contain a third form of amidotransferase activity, with a pI of 4.5. It is also the major form in proliferating tissues including late-fetal and regenerating liver. A striking similarity has been observed between this isozyme of amidotransferase and the enzymes associated with DNA synthesis with respect to (1) tissue distribution, (2) developmental change, (3) subcellular localization, and (4) response to hydrocortisone. Therefore, the isozyme must have a close correlation with DNA synthesis. Its probable association with membraneous structure is also of particular interest.

Although the functional role of this isozyme is not yet clear, attention may be directed to the possibility of its association with mucopolysaccharide synthesis. Figure 10 shows that a major amidotransferase component of 19-day fetal cartilage, a mucopolysaccharide-producing tissue, has a pI at 4.5, the value characteristic of the major tumor form. The enzyme from fetal cartilage was again chromatographically indistinguishable from the tumor enzyme, as shown in Table 7. Furthermore, mucopolysaccharides, unlike glycoproteins, do not require UDP-N-acetylglucosamine 2′-epimerase activity for synthesis and this is in harmony with our finding that epimerase but not amidotransferase is deleted in tumors (Fig. 1).

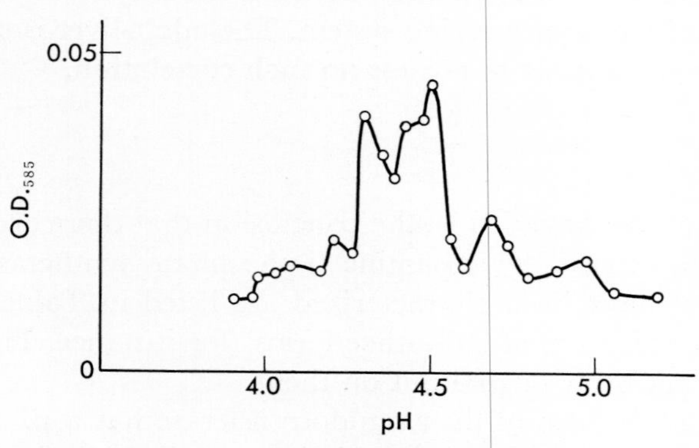

FIG 10. Isoelectric focusing pattern of fetal (19-day) cartilage amidotransferase.

TABLE 7. Chromatographic Properties of Amidotransferase from Different Tissue Sources

Tissues	Conductivity ($\mu\mho$)[a]
Liver	900–1,000
Submaxillary gland	900–1,000
Fetal cartilage (19-day)	1,100–1,200
Fetal liver (19-day)	1,070–1,200
Yoshida sarcoma	1,050–1,150
AH-66F	1,050–1,100
Brain	1,300

[a] The concentration of KCl required for elution from a DEAE-Sephadex column expressed in terms of conductivity. For the detailed procedure, see Ref. *25* or *29*.

If this possibility is correct, one could offer rational support to the currently popular view that mucopolysaccharides, as constituents of the cell-surface membrane or coat, play an important role in controlling cell growth and differentiation (*17, 20, 22, 28*).

ACKNOWLEDGMENT

This work was supported by a scientific research grant from the Ministry of Education of Japan.

REFERENCES

1. Akamatsu, N. and Maeda, H. Formation of glucosamine 6-phosphate in regenerating rat liver. Biochim. Biophys. Acta, *244*: 311–321, 1971.
2. Anastasiades, T. The effect of cortisone on hexosamine metabolism in the rat. Biochim. Biophys. Acta, *244*: 167–177, 1971.
3. Baril, E. F., Brown, O. E., Jenkins, M. D., and Laszlo, J. Deoxyribonucleic acid polymerase associated with rat-liver ribosomes and smooth membranes. Purification and properties of the enzyme. Biochemistry, *10*: 1981–1992, 1971.
4. Baril, E. F., Jenkins, M. D., Brown, O. E., and Laszlo, J. DNA polymerase activities associated with smooth membranes and ribosomes from rat liver and hepatoma. Science, *169*: 87–89, 1970.
5. Burger, M. M. Proteolytic enzymes initiating cell division and escape from contact inhibition. Nature, *227*: 170–171, 1970.
6. Burger, M. M. and Noonan, K. D. Restoration of normal growth by covering of agglutinin sites on tumor cell surface. Nature, *228*: 512–515, 1970.
7. Elford, H. L. Functional regulation of mammalian ribonucleotide reductase. Adv. Enzyme Regulation, *10*: 19–38, 1972.
8. Funakoshi, I. and Yamashina, I. Isolation and analysis of glycopeptides from microsomes of an ascites hepatoma, AH66. J. Biochem., *72*: 459–468, 1972.
9. Greengard, O. and Machovich, R. Hydrocortisone regulation of thymidine kinase in thymus involution and hepatopoietic tissues. Biochim. Biophys. Acta, *286*: 382–396, 1972.
10. Herzfeld, A. and Knox, W. E. Aspartate transcarbamylase concentrations in relation to growth rates of fetal, adult, and neoplastic rat tissues. Cancer Res., *32*: 1842–1847, 1972.

11. Ikeda, Y. and Tsuiki, S. Occurrence and properties of a natural activator of L-glutamine: D-fructose 6-phosphate amidotransferase. Sci. Rep. Res. Inst. Tohoku Univ.-C, *19*: 53–63, 1972.
12. Isaka, H., Umehara, S., Hirai, H., Tsukada, Y., and Watabe, H. Isolation of α-fetoprotein-producing cells from Yoshida ascites sarcoma and its clone. Gann, *64*: 133–138, 1973.
13. Kikuchi, H., Ikeda, Y., and Tsuiki, S. Inhibition of L-glutamine: D-fructose 6-phosphate aminotransferase by methylglyoxal. Biochim. Biophys. Acta, *289*: 303–310, 1972.
14. Kikuchi, H., Kobayashi, Y., and Tsuiki, S. L-Glutamine: D-fructose 6-phosphate amidotransferase in tumors and the liver of tumor-bearing animals. Biochim. Biophys. Acta, *237*: 412–421, 1971.
15. Kikuchi, K., Kikuchi, H., and Tsuiki, S. Activities of sialic acid-synthesizing enzymes in rat liver and rat and mouse tumors. Biochim. Biophys. Acta, *252*: 357–368, 1971.
16. Kikuchi, K. and Tsuiki, S. Purification and properties of UDP-N-acetylglucosamine 2′-epimerase from rat liver. Biochim. Biophys. Acta, *327*: 193–206, 1973.
17. Kojima, K. and Maekawa, A. A difference in the surface membrane architecture between two cell types of rat ascites hepatomas. Cancer Res., *32*: 847–852, 1972.
18. Kornfeld, R. Studies on L-glutamine: D-fructose 6-phosphate amidotransferase. I. Feedback inhibition by uridine diphosphate-N-acetylglucosamine. J. Biol. Chem., *242*: 3135–3141, 1967.
19. Kornfeld, S., Kornfeld, R., Neufeld, E. F., and O'Brien, P. J. The feedback control of sugar nucleotide biosynthesis in liver. Proc. Natl. Acad. Sci. U.S., *52*: 371–379, 1964.
20. Lie, S. O., McKusik, V. A., and Neufeld, E. F. Stimulation of genetic mucopolysaccharides in normal human fibroblasts by alteration of the pH of the medium. Proc. Natl. Acad. Sci. U.S., *69*: 2361–2363, 1972.
21. Machovich, R. and Greengard, O. Thymidine kinase in rat tissues during growth and differentiation. Biochim. Biophys. Acta, *286*: 375–381, 1972.
22. Makita, A. and Shimojo, H. Polysaccharides of SV 40-transformed green monkey kidney cells. Biochim. Biophys. Acta, *304*: 571–574, 1973.
23. Miyagi, T. A comparative study of L-glutamine: D-fructose 6-phosphate amidotransferase. Kokensi, *25*: 105–121, 1972 (in Japanese).
24. Miyagi, T., Kikuchi, H., and Tsuiki, S. Studies on the substance that specifically inactivates glutamine: fructose 6-phosphate amidotransferase. Seikagaku, *44*: 506, 1972 (in Japanese).
25. Miyagi, T. and Tsuiki, S. Effect of phosphoglucose isomerase and glucose 6-phosphate on UDP-N-acetylglucosamine inhibition of L-glutamine: D-fructose 6-phosphate aminotransferase. Biochim. Biophys. Acta, *250*: 51–62, 1971.
26. Richards, T. C. and Greengard, O. Distribution of glutamine hexose phosphate aminotransferase in rat tissues. Biochim. Biophys. Acta, *304*: 842–850, 1973.
27. Roseman, S. The synthesis of complex carbohydrates by multiglycosyl-transferase systems and their potential function in intercellular adhesion. Chem. Phys. Lipids, *5*: 270–297, 1970.
28. Satoh, C., Duff, R., Rapp, F., and Davidson, E. A. Production of mucopolysaccharides in normal and transformed cells. Proc. Natl. Acad. Sci. U.S., *70*: 54–56, 1973.

29. Tsuiki, S., Sato, K., Miyagi, T., and Kikuchi, H. Isozymes of fructose 1,6-diphosphatase, glycogen synthetase, and glutamine: fructose 6-phosphate amidotransferase. GANN Monograph on Cancer Research, *13*: 153–165, 1972.
30. Warren, L., Fuhrer, J. P., and Buck, C. A. Surface glycoproteins of normal and transformed cells: A difference determined by sialic acid and a growth-dependent sialyltransferase. Proc. Natl. Acad. Sci. U.S., *69*: 1838–1842, 1972.
31. Weinhouse, S., Shatton, J. B., Criss, W. E., Farina, F. A., and Morris, H. P. Isozymes in relation to differentiation in transplantable rat hepatomas. GANN Monograph on Cancer Research, *13*: 1–18, 1972.
32. Wu, H. C., Meezan, Z., Black, P. H., and Robbins, P. W. Comparative studies on the carbohydrate-containing membrane components of normal and virus-transformed mouse fibroblast. I. Glucosamine-labeling patterns in 3T3, spontaneously transformed 3T3, and SV 40-transformed 3T3 cells. Biochemistry, *8*: 2509–2517, 1969.
33. Yoshida, T. Comparative studies of ascites hepatomas. Methods Cancer Res., *6*: 97–157, 1971.

Discussion of Paper by Drs. Tsuiki et al.

DR. HOLTZER: Do you think in chondrosarcoma there would be a shift in the intracellular localization of either the UDP-N-acetylglucosamine or the amidotransferase?

DR. TSUIKI: Chondrosarcoma might not be suitable for this kind of study since mucopolysaccharide synthesis in this tumor might be associated not only with the proliferating ability of the cells but also with their differentiated function. I simply suppose that the amidotransferase might be distributed both in the membrane (for proliferation) and in the cytoplasm (for differentiated function).

DR. DUBE: It is very interesting that the fetal-liver and tumor enzymes are very similar and different from the adult liver enzyme. I wonder if the fetal enzyme could be converted to the adult form by a specific proteolytic cleavage.

DR. TSUIKI: There are several reasons to believe that this is not the case: 1) we have never observed that the 2 forms are inter convertible; 2) if we mix crude preparations of the 2 isozymes and subsequently carry out fractionation, we obtain the two isozymes again in unchanged proportions; 3) the 2 isozymes are equally sensitive to proteolysis resulting in a gradual loss of enzyme activity.

Enzymological Changes in Abnormal Differentiation: Intestinal Metaplasia in Human Gastric Mucosa: A Possible Precancerous Change

Takashi Sugimura, Takashi Kawachi, Kikuko Kogure,
Akira Tokunaga, Noritake Tanaka, Koji Sasajima, Yasuo Koyama,
Teruyuki Hirota, and Ryozo Sano

National Cancer Center Research Institute, Tokyo, Japan

Abstract: Human gastric cancer frequently arises from an area of intestinal metaplasia of the stomach mucosa. Enzymes specific to normal intestinal epithelium are observed in the intestinalized and abnormal mucosal epithelium of the stomach. The presence of disaccharidases such as sucrase, maltase, trehalase, and lactase can be visualized by spreading each disaccharide substrate on the mucosal membrane surface of the stomach and covering it with pieces of "Tes-Tape." About 130 cases in which the stomach was removed surgically from patients suffering from stomach carcinoma, stomach ulcer, and other diseases were subjected to this test. The area giving a positive reaction for maltase in the intestinal metaplasia was wider than that of sucrase activity. Trehalase was sometimes observed in the intestinal metaplasia with a concurrent presence of alkaline phosphatase. Lactase-positive reaction was also found in some stomachs. Carcinoma was often found in the area where disaccharidase reaction was positive but the carcinoma itself had no disaccharidase activity. The possible significance of such abnormal differentiation will be discussed in relation to the development of cancer.

Stomach cancer is the most common neoplasia in Japan, especially in men, in whom it accounts for about half the total deaths from cancer (*14*). Many studies have been published on the pathogenesis of stomach carcinoma, mainly from the morphological viewpoint (for a review, see Ref. *4*).

Intestinalization or intestinal metaplasia is defined as the appearance of intestinal-type epithelial cells in the stomach mucosa (*5*), and human gastric cancer often arises from areas of intestinal metaplasia in the gastric mucosa (*3, 6*). It is reported that intestinalization occurs more frequently in the stomachs of Japanese than of Western people (*2*).

In our laboratory, we are studying the problem of intestinal metaplasia on the basis of biochemical and enzymological parameters. Intestinal epithelium cells have so-called "intestine-specific enzymes" which are found in the intestinal epithelium but not in normal epithelium of the gastric mucosa. For instance, the enzymes sucrase, maltase, trehalase, lactase, and alkaline phosphatase are normally present in the intestinal epithelium but not the gastric epithelium. Studies on alkaline phosphatase were reported by Stemmermann and Hayashi (*9*).

This paper deals with the appearance of the four disaccharidases, sucrase, maltase, trehalase and lactase, in the stomach mucosa. The Tes-Tape method was used to detect the presence of disaccharidases, which yielded glucose from their respective substrates. We have already reported the principle of the Tes-Tape method (*13*). This method is very quick, easy, and reproducible.

We investigated 134 cases of resected stomach using this method and compared the various patterns of intestine-specific enzymes with histopathological findings. In this paper the relation between intestinalization revealed by tests for sucrase and maltase and the occurrence of cancer is analysed and discussed.

Methods for Detection of Disaccharidases in Gastric Mucosa with Tes-Tape

Briefly, the method used to detect disaccharidases was as follows. Specimens were obtained by gastrectomy from patients with gastric carcinomas, gastric sarcomas, gastric ulcers, and duodenal ulcers. The excised stomach was quickly brought to the laboratory and usually opened along the greater curvature. The mucosal surface was thoroughly washed with cold saline and spread out on a thick glass plate. Then a solution of 5% sucrose, maltose or trehalose in 10 mM sodium phosphate buffer (pH 6.5) was sprayed over the whole mucosal surface. The stomach was kept in an incubator at 37°C for 5 min to allow formation of glucose from disaccharides by the action of disaccharidase in the mucosa. Then the mucosal surface was quickly covered with pieces of Tes-Tape. In areas where disaccharidases were present the Tes-Tape turned green. A color photograph was then taken. The whole procedure to detect a single disaccharidase required only 15 min, and after washing off the Tes-Tape strips with saline, the specimen is again ready for another test using a different kind of disaccharide.

Detection of Intestinal Metaplasia and Stomach Cancer

Figure 1 shows the gross appearance of a resected stomach from a 62-year-old man with stomach carcinoma. The stomach carcinoma was located in the antrum, and diagnosed histologically as a tubular adenocarcinoma. As shown in Fig. 2, using sucrose almost the entire field of the resected stomach, except for part of the corpus, gave a green color with Tes-Tape. It should be noted that, though surrounded by

TABLE 1. Origins of the Cancers Examined

Origin	No. of cases
Intestinalized areas[a]	96
Nonintestinalized areas	21

[a] Intestinalized areas were defined as areas showing sucrase activity with Tes-Tape.

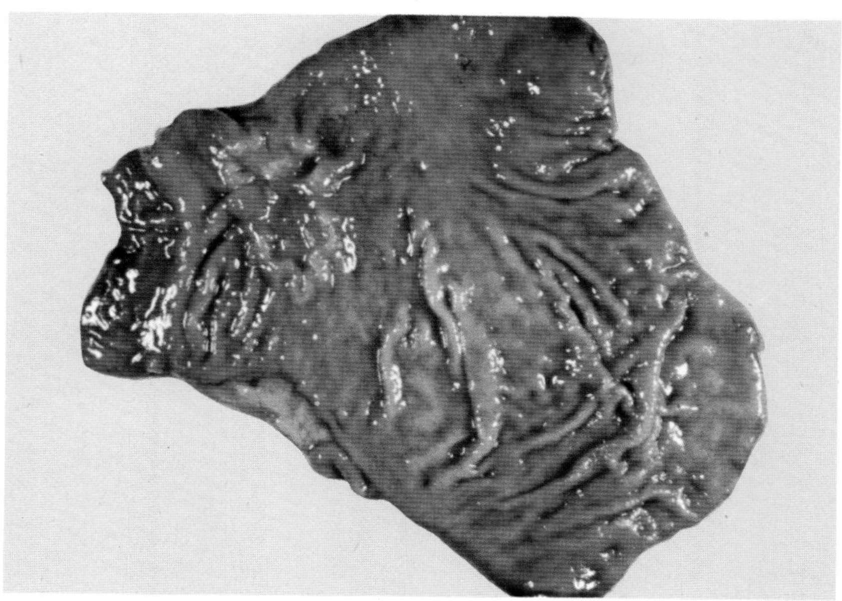

FIG. 1. Gross appearance of the stomach of a 62-year-old man with carcinoma of the antrum.

FIG. 2. The same specimen as in Fig. 1 demonstrating intestinal metaplasia using the sucrase test. The area of carcinoma showed no sucrase activity.

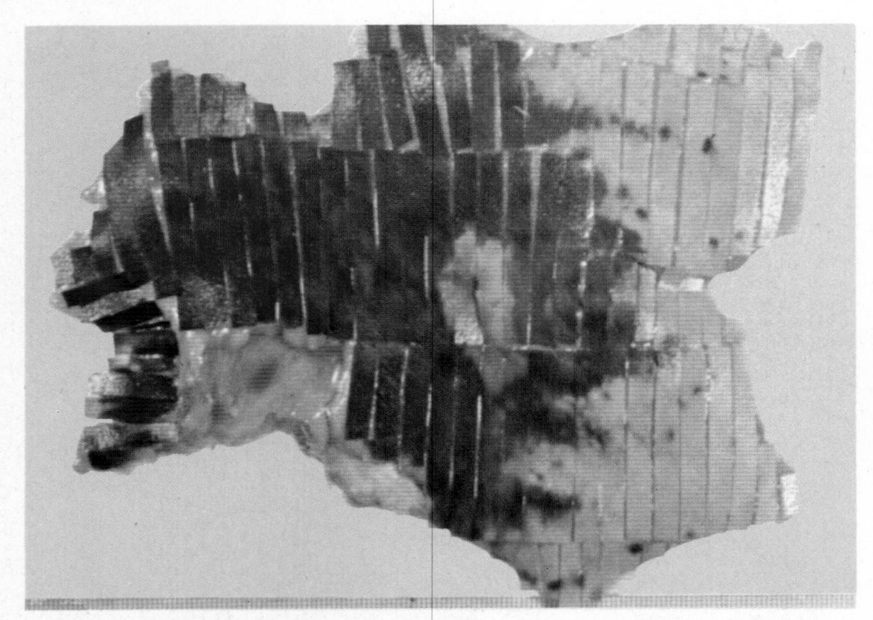

FIG. 3. The stomach of a 72-year-old woman with an area of intestinal metaplasia giving a positive sucrase reaction.

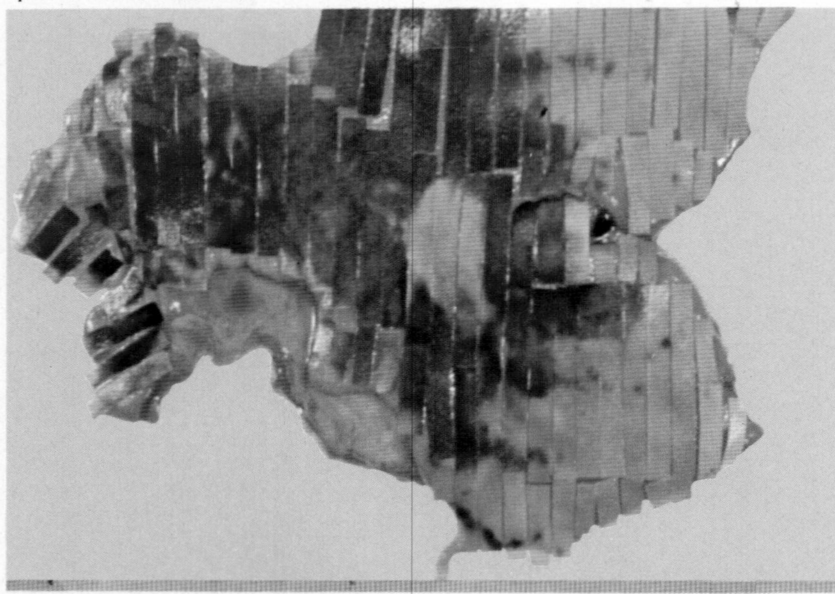

FIG. 4. The same specimen as in Fig. 3 showing a positive trehalase reaction in the area of intestinal metaplasia.

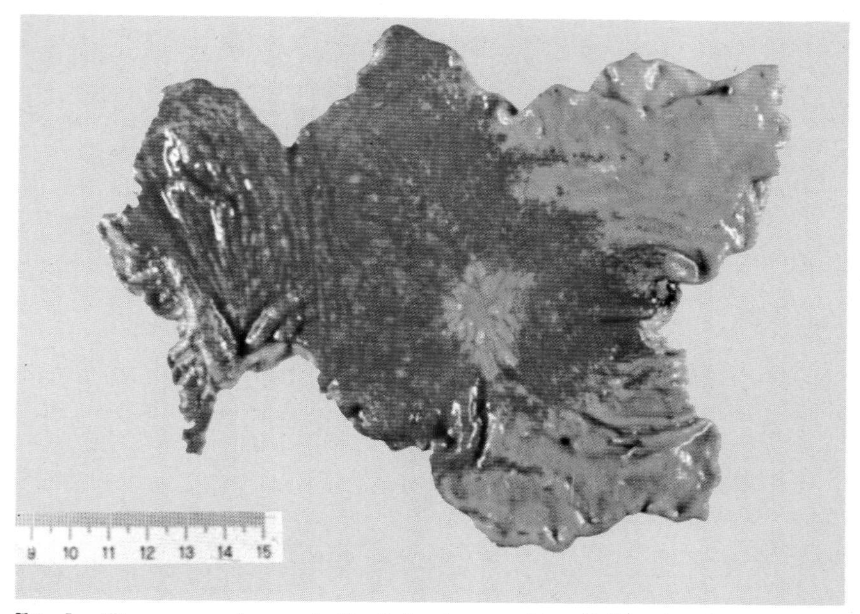

FIG. 5. The same specimen as in Fig. 3 showing a positive alkaline phosphatase reaction in the area of intestinal metaplasia.

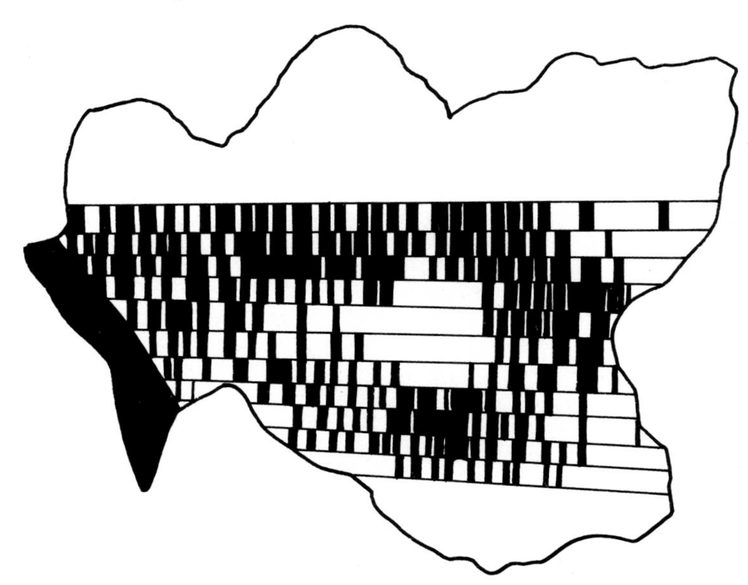

FIG. 6. The same specimen as in Fig. 3. Histopathological distribution of intestinal metaplasia with Paneth cells.

FIG. 7. The stomach of a 57-year-old man with intestinal metaplasia giving a positive sucrase reaction. A carcinoma is localized in the greater curvature of the ventral wall of the antrum.

FIG. 8. The same specimen as in Fig. 7 showing a positive trehalase reaction in only some of the sucrase-positive areas of intestinal metaplasia.

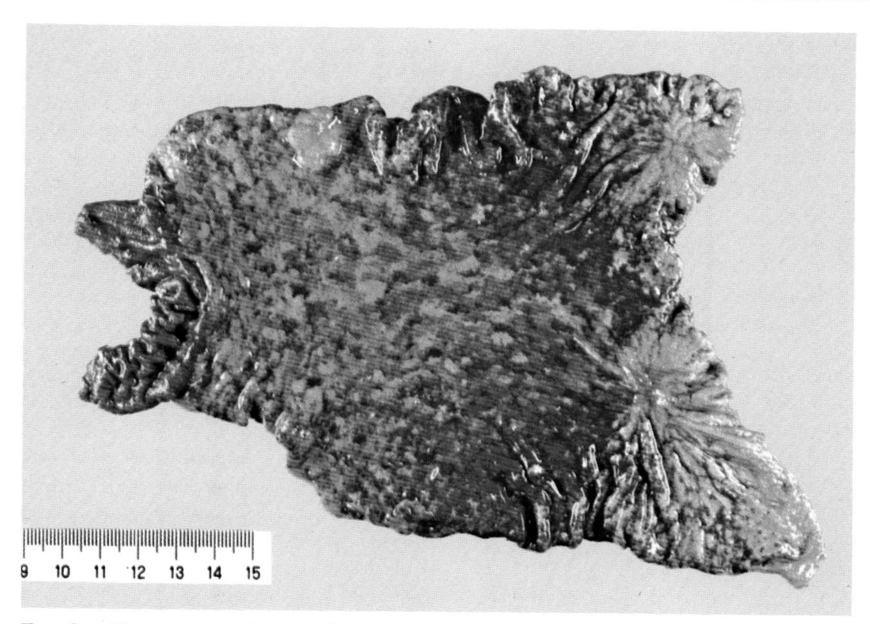

FIG. 9. The same specimen as in Fig. 7 showing a positive alkaline phosphatase reaction in some parts of the sucrase-positive areas of intestinal metaplasia.

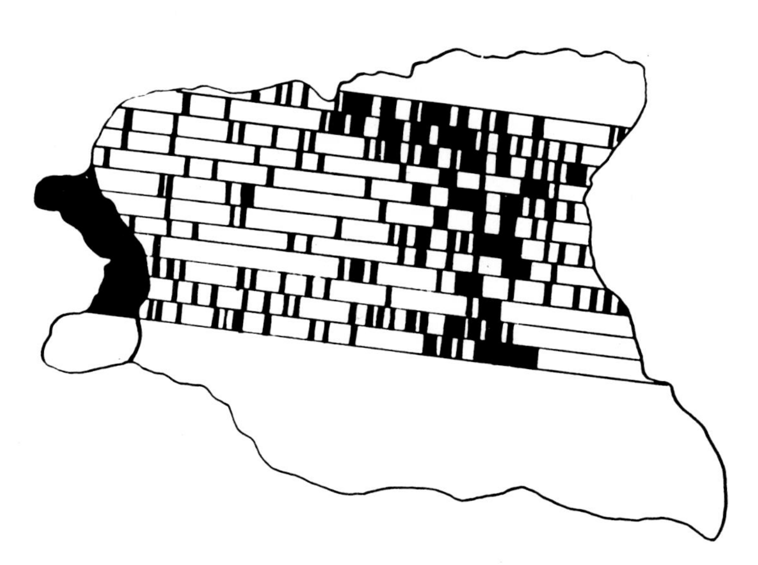

FIG. 10. The same specimen as in Fig. 7. Histopathological distribution of intestinal metaplasia with Paneth cells.

tissue giving a positive reaction for sucrase, the cancer area gave a negative reaction. As summarized in Table 1, 117 cases of stomach carcinoma all gave a negative reaction for sucrase, but in 96 cases the cancer areas were surrounded by tissue giving a positive reaction for sucrase. This indicates that about four-fifths of the cancers arose from areas giving a positive reaction for sucrase.

Presence of 2 Types of Intestinalization with Different Patterns of Intestinal Enzymes

To investigate so-called "intestinalization" in more detail, tests for 4 kinds of intestine-specific enzymes were made on each specimen. The intestine-specific enzymes concerned were sucrase, maltase, and trehalase, and they were all examined using Tes-Tape (*13*). Alkaline phosphatase was also examined using fixed specimens (*9*). Histological examinations of mucosal specimens were done by cutting the stomach into 5 mm wide slices and staining 5 μm thick sections of the slices with H-E. The sections were examined for the presence of goblet cells and Paneth cells. An example of the pattern of intestinalization deduced from microscopic examination of histological preparations is shown in Fig. 6.

TABLE 2. Characteristics of the 2 Types of Intestinal Metaplasia

Character	Type A	Type B
Sucrase	+	+
Maltase	+	+
Goblet cells	+	+
Trehalase	+	−
Alkaline phosphatase	+	−
Paneth cells	+	−

Two types of intestinalization were found. As shown in Figs. 3, 4, 5, and 6, in some cases, the areas giving positive reactions for sucrase and trehalase and for alkaline phosphatase coincided very well. Areas giving a positive reaction for maltase also coincided very well with sucrase-positive areas. Moreover, goblet cells and Paneth cells were found throughout the areas giving a positive sucrase reaction. As shown in Table 2, intestinalization with positive reactions for sucrase, maltase, trehalase, and alkaline phosphatase and with goblet cells and Paneth cells was defined as type A intestinalization.

In some cases, areas of intestinalization gave negative reactions for trehalase and alkaline phosphatase but definite positive reactions for sucrase and maltase. These areas contained goblet cells, but not Paneth cells. This type of intestinalization, with positive reactions for maltase and sucrase and with goblet cells, but with negative reactions for trehalase and alkaline phosphatase and no Paneth cells, was defined as type B intestinalization.

Sometimes both type A and type B intestinalization were found in a single stomach. Examples are shown in Figs. 7, 8, 9, and 10. As shown in Fig. 7, almost the entire area of the resected stomach of a 57-year-old man gave a positive reaction for sucrase. However, as shown in Fig. 8, the trehalase reaction was only positive on the oral side of areas showing a positive sucrase reaction. It is interesting that, as shown in Fig. 9, the alkaline phosphatase reaction was only positive in areas where the trehalase reaction was positive. Furthermore, areas giving positive reactions for trehalase and alkaline phosphatase were found to contain Paneth cells, as shown in Fig. 10. In general, when both types A and B intestinalization were found in the same stomach, type A intestinalization was in the antral region and type B intestinalization in the oral region.

Among the specimens examined, one-third showed type A intestinalization, one-third type B intestinalization and one-third showed both types. Thus the frequencies of cases of cancer originating from types A and B were not significantly different.

Future Studies on Intestinalization

Type A intestinalization seems to be a more complete form than type B intestinalization, since more specific intestinal enzymes are found in this form than in type B. However, it is still uncertain whether type B is a precursor state of type A, or a degenerated state of the latter.

As previously reported (*11*, *12*), we sometimes observed intestinalization in the stomach of rats on administration of N-methyl-N′-nitro-N-nitrosoguanidine, though it was not common. The presence of goblet cells has been noticed in the stomachs of dogs which received N-methyl-N′-nitro-N-nitrosoguanidine (*8*).

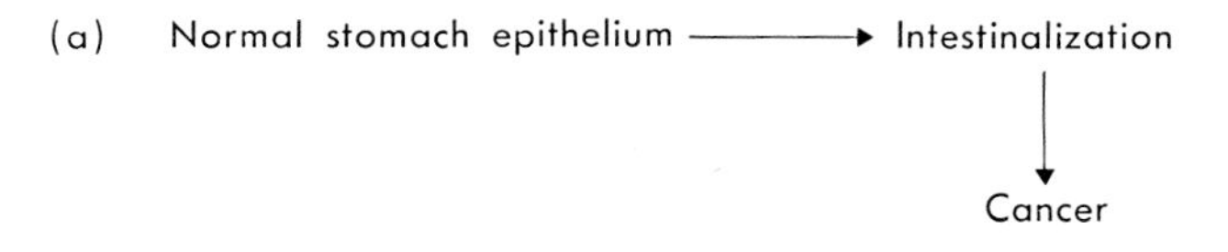

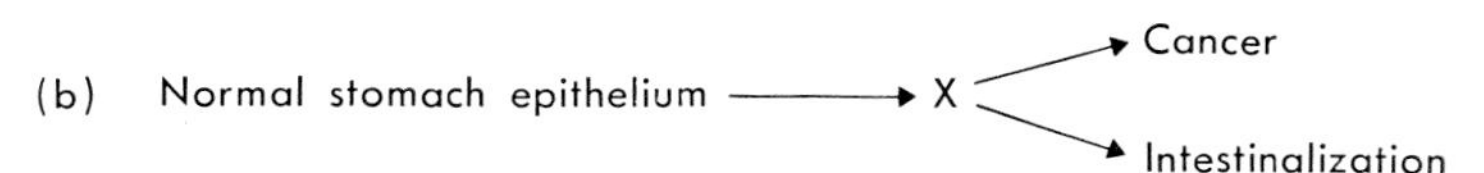

FIG. 11. Intestinalization and cancerization.
a: intestinalization *is* an obligatory process in cancerization; b: intestinalization *is not* an obligatory process in cancerization.

There is much debate about whether intestinalization is a necessary step in the cancerization of gastric mucosa cells (*1*, *7*). The idea that carcinogenesis is closely related to the process of mutagenesis may be valid, but still the idea persists that cancer cells may result from abnormal differentiation-disdifferentiation (*10*). Intestinalization might be understood in 2 different ways. Namely, shown schematically in Fig. 11a indicates the possibility that intestinalization is an obligatory process in cancerization of the stomach mucosa cells, at least in some cases, while Fig. 11b indicates the possibility that the stomach epithelium could be converted to a metastable, unidentified state, "X," from which both cancer and intestinalization may arise. However, intestinalization represents a stabilized, differentiated status, in which cells may be free from the risk of cancerization. At present, it is uncertain which possibility is more likely.

It would be very useful if *in vivo* studies could be made on intestinalization, by observation of the sucrase or maltase reaction in the gastric mucosa using a fiberscope, since such studies would provide much information on the relation between intestinalization and cancerization.

ACKNOWLEDGMENTS

This work was supported by grants from the Ministry of Health and Welfare, the Ministry of Education of Japan, the Society for Promotion of Cancer Research and the Japanese Association for the Study of Metabolism and Disease.

Data described in this article will be published in detail, together with further material, in Journal of the National Cancer Institute (T. Kawachi *et al.*, *53*: 19–30, (1974)).

REFERENCES

1. Feit, J., Svedjda, J., and Sochorova, M. Experimental intestinal metaplasia of gastric mucosa of rats and its relationship to carcinoma. Neoplasma, *14*: 285–290, 1967.
2. Imai, T., Kubo, T., and Watanabe, H. Chronic gastritis in Japanese with reference to high incidence of gastric carcinoma. J. Natl. Cancer Inst., *47*: 179–195, 1971.
3. Jarvi, O. and Lauren, P. On the role of heterotopias of the intestinal epithelium in the pathogenesis of gastric cancer. Acta Pathol. Microbiol. Scand., *29*: 26–44, 1951.
4. Kuru, M. Atlas of Early Carcinoma of the Stomach. Nakayama Shoten, Tokyo, 1967.
5. Magnus, H. A. Observations on the presence of intestinal epithelium in the gastric mucosa. J. Pathol. Bacteriol., *44*: 389–398, 1937.
6. Morson, B. C. Carcinoma arising from areas of intestinal metaplasia in the gastric mucosa. Brit. J. Cancer, *9*: 377–385, 1955.
7. Planteydt, H. T., Leemhuis, M. P., and Willighagen, R. G. J. Enzyme histochemistry of gastric tumours in animals. J. Pathol. Bacteriol., *83*: 31–38, 1962.
8. Shimosato, Y., Tanaka, N., Kogure, K., Fujimura, S., Kawachi, T., and Sugimura, T. Histopathology of tumors of canine alimentary tract produced by N-methyl-N′-nitro-N-nitrosoguanidine, with particular reference to gastric carcinomas. J. Natl. Cancer Inst., *46*: 1053–1070, 1971.
9. Stemmermann, G. N. and Hayashi, T. Intestinal metaplasia of the gastric mucosa, a gross and microscopic study of its distribution in various states. J. Natl. Cancer Inst., *41*: 627–634, 1968.
10. Sugimura, T. Decarcinogenesis, a newer concept arising from our understanding of the cancer phenotype. *In;* W. Nakahara (ed.), Chemical Tumor Problem, pp. 269–284, Japanese Society for the Promotion of Science, Tokyo, 1970.
11. Sugimura, T., Fujimura, S., and Baba, T. Tumor production in the glandular stomach and alimentary tract of the rat by N-methyl-N′-nitro-N-nitrosoguanidine. Cancer Res., *30*: 455–465, 1970.
12. Sugimura, T., Fujimura, S., Kogure, K., Baba, T., Saito, T., Nagao, M., Hosoi,

H., Shimosato, Y., and Yokoshima, T. Production of adenocarcinomas in glandular stomach of experimental animals by N-methyl-N′-nitro-N-nitrosoguanidine. GANN Monograph, *8*: 157–196, 1969.

13. Sugimura, T., Kawachi, T., Kogure, K., Tanaka, N., Kazama, S., and Koyama, Y. A novel method for detecting intestinal metaplasia of the stomach with Tes-Tape. Gann, *62*: 237, 1971.
14. Wynder, E. L., Kmet, J., Dungal, N., and Segi, M. An epidemiological investigation of gastric cancer. Cancer, *16*: 1461–1496, 1963.

Discussion of Paper by Drs. Sugimura et al.

DR. POTTER: As for the reason for the high incidence of stomach cancer in Japan, it seems that there may be a genetic factor in the Japanese population.

DR. WEINHOUSE: Although stomach cancer is now low in Western populations, this has been the result of a marked decline over the past few decades. This would argue against a genetic factor and for an environmental factor.

DR. SUGIMURA: Yes, environmental factor(s) might be more important than genetic factor(s). Japanese immigrants to Hawaii and California showed a much lower incidence of stomach carcinoma.

DR. RUTTER: I wonder whether there are incidences of intestinalization in embryonic organs, especially the stomach. In addition, could glucocorticoids induce these enzymes in the stomach (it is known that glucocorticoids induce those enzymes in the gut)?

DR. SUGIMURA: Even in the normal intestine of rats, lactase activity was the only disaccharidase found in the embryonal stage. Maltase, sucrase, and trehalase could not be detected in the intestine of embryos. Thus, it is hard to expect the presence of intestinalization as found in human abnormal stomach. No information is available on the induction of disaccharidase by glucocorticoids in the stomach. However, this is not likely, since this hormone induced pepsinogen in newborn rat stomach.

DR. WEBER: I see many important advances in the stimulating paper Dr. Sugimura presented. For instance, the use of strip tests to visualize rapidly the intestinalization and thus neoplastic transformation. Moreover, the experimental production of stomach cancer and intestinalization in rats is also an important breakthrough.

DR. SUGIMURA: Thank you very much.

DR. ONO: You have shown that stomach cancer arises in the area of intestinal metaplasia, but none of the cancer tissue has the marker enzymes of intestine. That seems to deny the possibility that the cancer arose from intestinalized cells. How do you explain this?

DR. SUGIMURA: Some stomach carcinomas showed maltase activity, although they are rather rare. Very early stomach carcinoma was claimed to have microvilli in an incomplete form. Such evidence may indicate the possible origin of cancer as intestinalized cells. Moreover, even in normal intestine, maturated cells on the top of villi had these enzymes but dividing cells on the crypt did not. Thus, the absence of these enzymes in stomach cancer does not necessarily rule out possible precursor cells in the intestinalized area for stomach cancer.

DR. JOHNSON: Is intestinalization also a problem in diabetes?

DR. SUGIMURA: No, not directly, but there is an interesting aspect: namely persons with intestinalization can absorb glucose in the stomach immediately after the ingestion of sucrose. By checking blood glucose level, we may predict the presence of intestinalization in the stomach if the glucose level rises rapidly after the administration of sucrose.

DR. PIERCE: Metaplasia as viewed in the cervix can be detected and followed by painting with iodine to follow glycogen distribution. These lesions are reversible. I have thought of intestinalization as another example of modulation. Your use of the term "disdifferentiation" might imply that intestinalization is a permanent change. Do you have specific data on this point?

DR. SUGIMURA: We are attempting to obtain this information by following patients with intestinalization by serial biopsy. As you are aware this is very difficult information to obtain.

DR. DUBE: I enjoyed your fascinating lecture. I think Dr. Sugimura should be congratulated for developing an elegantly simple diagnostic method for stomach cancers. Is it possible to extend your simple method for the *in vivo* detection of intestinalization?

DR. SUGIMURA: Thank you. We are developing a new method to detect intestinalization by the same principle *in vivo*. We hope it will be possible soon. One difficult point is the rapid absorption of glucose from the surface epithelium to villi.

CONTROL OF DIFFERENTIATION

INDUCTION OF DIFFERENTIATION

Enzyme Induction in Primary Cultures of Rat Liver Parenchymal Cells

Michael W. PARIZA, Joyce E. BECKER, James D. YAGER, JR.,*[1]
Robert J. BONNEY,*[2] and Van R. POTTER

McArdle Laboratory for Cancer Research, University of Wisconsin, Madison, Wisconsin, U.S.A.

Abstract: The present report brings together previous experiments with 1) entrained rats on controlled feeding and lighting schedules; 2) regenerating rat liver; 3) serially transplanted hepatoma cultures; and 4) the preparation of rat liver cell suspensions. Over 300×10^6 single parenchymal rat liver cells could be obtained from one rat following collagenase perfusion. Replicate cultures were set up with about 1,000 cells/mm^2. About 50% of the cells were attached and contiguous within 24 hr of culturing using Ham's F-12 medium supplemented with fetal calf serum, or using a modification of serum-free Waymouth's MB 752/1 in conjunction with collagen-coated plates. The inductions of tyrosine aminotransferase (TAT) and ornithine decarboxylase (ODC) activities in the cells were studied. In addition, the uptake of ^{3}H-TdR into DNA by cells isolated from entrained rats given partial hepatectomies at a prescribed time was investigated. The significance of our findings is discussed with reference to the work of other investigators.

The introduction of the *in situ* collagenase liver perfusion technique (*2, 6, 8*) represents a marked advance in methodology for the isolation of single cell suspensions of adult hepatocytes. The technique is superior to methods involving mechanical stress alone, chelating agents, or treatment of chopped liver with proteolytic enzymes (*14, 17, 31*). Cell suspensions prepared by collagenase digestion of liver connective tissue have been used immediately after isolation in metabolic studies (*2, 14, 16, 31*), or to prepare serial cultures of growing cells (*15*). However, there appear to be fundamental problems associated with each of these protocols. The immediate use of single cell suspensions in regulatory studies is questionable because the cells appear to have lost many *in vivo* properties. Thus, freshly isolated hepatocytes are unable to retain many small molecules, *i.e.*, amino acids (*31*) and

*[1] Present address: Department of Biological Sciences, Dartmouth College, Hanover, New Hampshire, U.S.A.
*[2] Present address: New York State Department of Health, New York, U.S.A.

inorganic ions (*2*), and the mitochondrial enzyme, lactic acid dehydrogenase, diffuses from the cells into the medium following cell isolation (*16*). It has also been found that ATP levels in freshly isolated hepatocytes are depressed in comparison with whole liver or isolated hepatocytes after 1, 2, or 4 days in monolayer culture (*3*), a finding which would be expected to have profound consequences for the regulation of glycolysis, gluconeogenesis, and intermediary metabolism in general. Indeed, Berry and Friend (*2*) reported the synthesis of glucose from lactate in freshly isolated hepatocytes to be 50% of that in perfused liver, an observation consistent with a depressed ATP level. Hormone receptors may also be damaged by the perfusion technique (*3*), and data to be presented in this communication indicate that ornithine decarboxylase (ODC) is not inducible by insulin during the first 6 hr after cell isolation but is in fact insulin inducible at 24 hr. On the other hand, the isolation and propagation of dividing "parenchymal" cells also seems to be an unsatisfactory protocol for studying metabolic regulation in adult liver cells. This is because dividing "liver" cells appear to lose differentiated functions (*17, 26, 28, 34, 35*) and therefore are unsuitable models for the study of regulatory mechanisms in adult liver cells.

For the past year, our laboratory has investigated the properties of hepatocytes isolated with the collagenase perfusion technique and cultured in monolayers for short periods of time, *i.e.*, less than 5 days. This method has several advantages. Firstly, the cells are given time to "recover" from the trauma inherent in the isolation procedure, and to adjust to new environmental conditions. Secondly, cells unable to attach to the culture surface can be discarded or studied separately. Thirdly, experiments lasting several days (instead of a few hours) can be conducted, thus broadening investigative possibilities.

The value of data obtained with primary monolayer cultures of adult hepatocytes is dependent upon 2 factors: the "viability" of the cells in culture, and the expression of adult liver parenchymal cell characteristics by the cells. The "viability" of a differentiated hepatocyte is particularly difficult to evaluate, since, by definition, such a cell does not divide. Thus, the traditional measure of cell viability, cell division, is not applicable. Therefore, in establishing "viability" for an adult liver cell one must consider other factors which fall into the category of *functional ability*, *i.e.*, membrane permeability and the active transport system, the ability to phosphorylate, the ability to synthesize macromolecules, the ability to respond to hormonal induction, the ability to be induced to de-differentiate and subsequently divide, and so forth. In addition, it is generally considered that viability and the ability to attach to a suitable cell culture surface (*i.e.*, tissue culture dish) are closely related, although the relationship is not immediately clear.

In addition to functional ability common to all normal cells of higher animals, it is reasonable to expect isolated hepatocytes to resemble, morphologically and biochemically, parenchymal cells as they appear *in vivo*. It is of the greatest importance that the isolated hepatocytes be non-dividing, that they possess adult liver isozymes, and that they perform functions of the *in vivo* adult liver, *e.g.*, gluconeogenesis, the capacity to synthesize arginine from ornithine, the synthesis of liver specific products such as albumin, *etc*. Of considerable practical importance is the number of cells that can be obtained from one liver. One is interested in culturing a representative population of adult hepatocytes, and as the number of parenchymal cells obtained from one liver increases, the probability of the cells in question being representative of all of the parenchymal cells also increases. Additionally, it is important to have a reasonable measure of the percent of contaminating non-parenchymal cells. Ideally, one would like to have no non-parenchymal cells present in the cultures, but, lacking that, the number of contaminating cells should be kept minimal.

The experiments to be reported in this communication were all performed with parenchymal liver cells obtained from rats which were maintained under controlled conditions of feeding and lighting and killed at specified times, according to procedures previously established in this laboratory (*37*). Such procedures virtually assure that animals need not be unduly stressed (by fasting, *etc.*) prior to cell isolation. In addition, use can be made of known diurnal oscillations in enzyme levels (*12*, *29*, *36*, *37*) and in DNA synthesis following partial hepatectomy (*13*), to suggest useful experimental protocols and to standardize results of one experiment with those of another.

In beginning studies on primary cultures of adult rat hepatocytes, the following working assumptions were made:

1) That the phenotypic properties of parenchymal liver cells in culture will depend on the culture medium, *i.e.*, nutrients, hormones, and metabolic signals.
2) That the cells cannot synthesize their complete repertoire of enzyme patterns at a maximum rate simultaneously.
3) That cycling of enzyme syntheses is necessary for survival.
4) That a 24-hr cycle might prove to be optimum.
5) That freshly isolated cells are injured cells, and that the first problem they face in culture is repair of their injury.
6) That repair of injury will be facilitated in monolayers in which each cell is in contact with 2 or more neighboring cells.

Experimental Parameters

1. Cells and culture techniques

One of the main advantages of the collagenase perfusion technique (*2*, *6*, *8*) is that it permits the isolation of high yields of virtually pure parenchymal cell suspensions. In our hands, $300–600 \times 10^6$ cells can be obtained from the liver of one 200–350 g adult rat, and about 50% will attach to a suitable cell culture surface and remain attached for a minimum of 24 hr under a variety of experimental conditions. This recovery of parenchymal cells is roughly 20–50% of the total hepatocyte population (*4*), and because about half of these cells attach and form monolayers in culture, our experiments are performed with about 10–25% of the total parenchymal cells present in an adult rat liver.

The average diameter of cells in the culture inocula was 27.7 μm, with a range of 19.5–33.2 μm (*6*). These values are in good agreement with literature values for hepatic cord cells (*6*). Virtually no cells having diameters less than 20 μm were observed, thus eliminating significant contamination from bile duct or Kupffer cells (*6*). Electron micrographs of the isolated cells showed that they contained intact cellular and nuclear membranes, rough and smooth endoplasmic reticula, and a large number of mitochondria, which is characteristic of adult liver (*6*).

Bonney *et al.* (*6*) studied isoenzyme patterns in the isolated parenchymal cells and in the non-parenchymal cells that are selectively removed from the cell suspensions during isolation. They observed adult rat liver isoenzyme patterns of pyruvate kinase and aldolase in the parenchymal cells as well as very high concentrations of glucokinase compared with hexokinase. In contrast, the non-parenchymal cells (those removed from the cell suspensions used to prepare primary monolayer cultures) contained more hexokinase than glucokinase, and isoenzyme patterns not characteristic of adult rat liver.

In culture, the isolated cells morphologically resemble adult parenchymal hepatocytes (Fig. 1). Cells having morphologies not characteristic of adult liver cells are almost never observed. When cells were cultured under conditions conducive to cell division, *i.e.*, in serum-containing medium, virtually no mitotic figures were observed during the first 3 days. Additionally, when parenchymal cells prelabeled *in vivo* with ^{3}H-TdR were cultured in unlabeled serum-containing medium no reduction in specific activity of DNA was observed during the first 6 days in culture (*4*).

Cell attachment has been studied using 2 different media: Ham's F-12 medium (*10*) supplemented with fetal calf serum, and serum-free Waymouth's MB 752/1 (*38*) supplemented with bovine serum albumin and oleate (*21*) and with serine

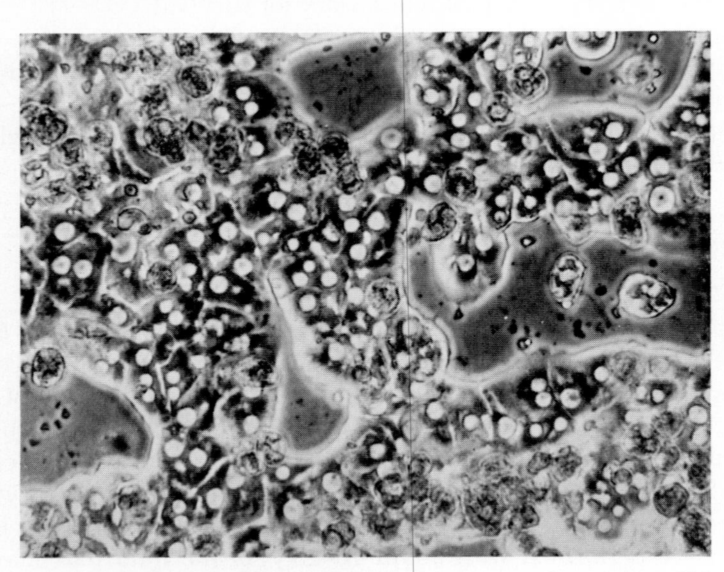

FIG. 1. Adult liver parenchymal cells in primary monolayer culture 21 hr after isolation. The incubation medium was HI-WO/BA+13 mUnits/ml insulin (0.5 μg/ml), and the attachment stratum was a collagen-coated Falcon plastic petri plate. The cells were observed under phase contrast (×280). Note the prominent round nuclei, and the presence of several bi-nucleated cells.

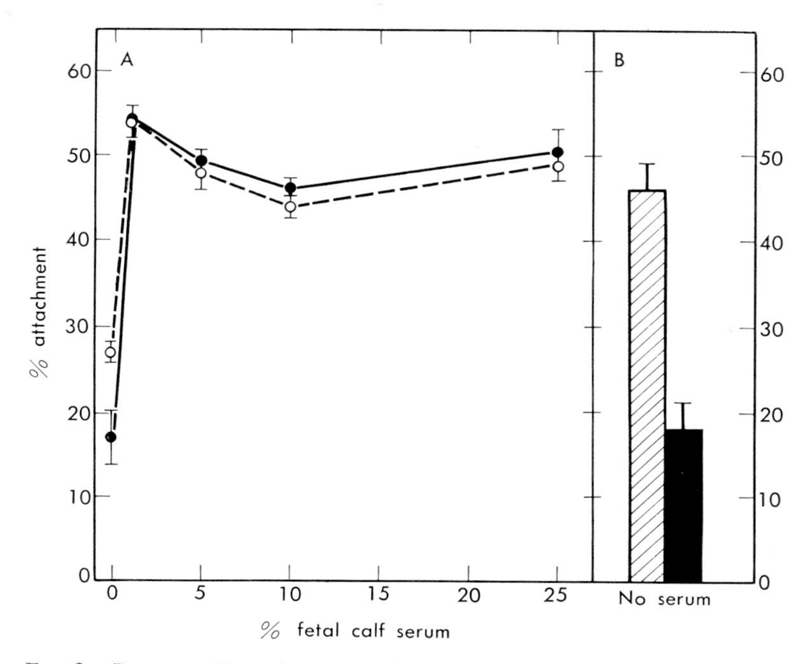

FIG. 2. Percent cell attachment at 24 hr as a function of percent fetal calf serum in Ham's F-12 medium+13 mUnits/ml insulin (0.5 μg/ml), or as a function of plastic or collagen-coated plastic plates when the medium was serum-free HI-WO/BA+13 mUnits/ml insulin (0.5 μg/ml). Points represent averages of 3 plates. A: F-12+insulin. ● plastic alone; ○ collagen-coated plastic. B: HI-WO/BA+insulin. ▨ collagen-coated plastic; ■ plastic alone.

(0.5 mM), alanine (0.4 mM), and glutamine to 6.8 mM (we shall refer to this modified medium as HI-WO/BA). Both media routinely contained 13 milliunits of insulin per ml (0.5 μg/ml), as did the perfusion medium.

Figure 2 shows the percent cell attachment in 24 hr as a function of percent fetal calf serum (in Ham's F-12 medium) and/or the presence or absence of collagen on the culture surface (Falcon plastic petri dish). The cell inoculum used was that found previously to allow maximum percent attachment in 24 hr, *i.e.*, 1,000 cells/mm^2 (*4*; unpublished observations). Using Ham's F-12 medium without serum, only 17% of the cells attached to plastic alone, while 27% attached to collagen-coated plastic. When serum to 1% was added to Ham's F-12 medium 54% of the inoculum attached, and cell attachment appeared independent of added collagen. The results were not appreciably altered when serum was added to 5, 10, or 25% of the same medium (Fig. 2). These findings are an extension of earlier work reported from this laboratory (*4*) where cell attachment in 4 hr was found to be maximal in Ham's F-12 medium +25% fetal calf serum. In the present experiment, attachment was assessed after 24 hr in culture, and under these conditions 1% fetal calf serum in Ham's F-12 medium was sufficient to give maximum cell attachment (Fig. 2). It is of interest that Bissell *et al.* (*3*) reported that 10% fetal calf serum was necessary to achieve optimal cell attachment of normal adult hepatocytes to plastic culture surfaces, thus supporting our observation that cell attachment to uncoated plastic was dependent upon fetal calf serum.

Cell attachment in serum-free HI-WO/BA differed from that observed in Ham's F-12 medium. Forty-eight percent of the inoculum attached to collagen-coated plastic plates when HI-WO/BA was used as the suspending medium, while only 18% of inoculum in HI-WO/BA attached to uncoated plastic surfaces (Fig. 2).

The retention of attached cells for more than 24 hr appears to be the same in both HI-WO/BA (collagen-coated plates) and in F-12+fetal calf serum (plastic alone). About 10% or less of the unattached cells detach and are lost per 24-hr period (*4*; unpublished observations).

Primary liver cell cultures prepared as described above offer a unique opportunity for studying essentially all facets of metabolic regulation in adult hepatocytes. We have begun studies on the regulation of tyrosine aminotransferase (TAT) activity and ODC activity in these cells. In addition, we are investigating the synchronous division of hepatocytes following partial hepatectomy by culturing cells from entrained animals operated on at prescribed times (*13*) and killed at various intervals following partial hepatectomy.

2. *Enzyme studies*

The inducibility of TAT activity by glucocorticoids in whole rats (*18*), in perfused liver (*40*), in hepatoma cells (*1, 7, 27, 30, 33*), in long-term dividing cultures of rat liver cells (*9*), and in parenchymal cell suspensions during short-term incubation (less than one day) (*14*) is well documented and is thought to be dependent upon *de novo* synthesis of TAT mRNA (*7*). Our initial studies on TAT inducibility in cultured primary adult hepatocytes indicate that glucocorticoids induce an increase in TAT activity in these cells as well.

The cells were allowed to attach for 24 hr to uncoated plastic petri plates in the presence of Ham's F-12+25% fetal calf serum+insulin, and then the medium was changed to fresh Ham's F-12 with or without 25% fetal calf serum (no insulin in either case). Twenty-four hours later, the medium was again changed to fresh Ham's F-12, with or without 25% fetal calf serum, and dexamethasone to 10^{-5}M was added to some of the plates. The cells were harvested after 6 hr of incubation and the levels of TAT activity determined. As can be seen in Fig. 3, the level of TAT activity induced by dexamethasone in the presence or absence of serum was essentially identical, *i.e.*, about 2.5 times control levels. Other studies have indicated that similar levels of TAT activity were induced by hydrocortisone or dexamethasone on day 2 or day 3 of culture in F-12+25% fetal calf serum+insulin, or in HI-WO/BA+insulin (*4*; unpublished observations).

ODC is an important enzyme in intermediary metabolism, being the rate-controlling enzyme in polyamine synthesis. It has the shortest half-life of any known enzyme (*32*), and thus control of its activity is apparently very important in metabolic regulation. In addition, ODC activity is known to exhibit diurnal oscillations in the livers of rats entrained to specified feeding and lighting regimens (*12, 36*).

In the experiments to be described, cells were allowed to attach to collagen-

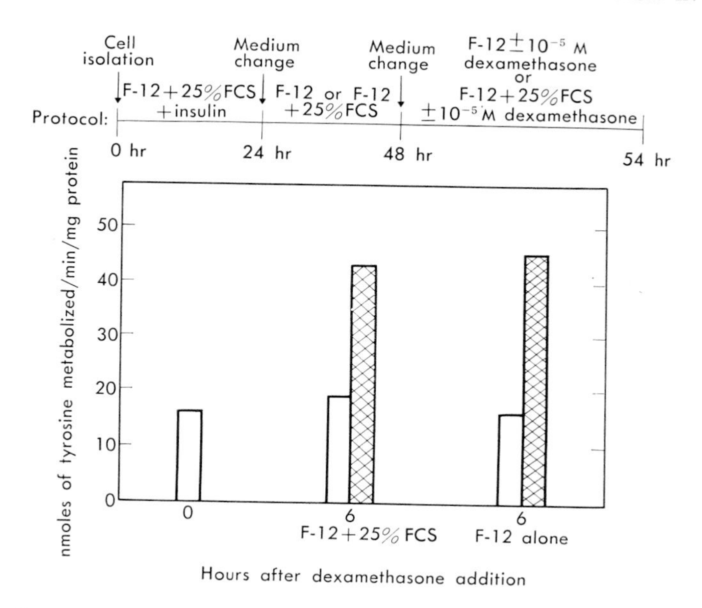

FIG. 3. Effect of dexamethasone treatment for 6 hr on TAT activity in cell cultures following 48-hr pre-treatment as indicated. Bar heights are averages of values determined from 2 plates. The range was 6% or less in each instance. FCS, fetal calf serum. ▩ 10^{-5} dexamethasone; □ no hormone.

coated plastic culture plates in HI-WO/BA+insulin. Throughout the course of the experiments, cells were exposed only to serum-free media, thereby eliminating the possibility of interference from factors present in normal fetal calf serum known to stimulate ODC activity *in vivo* (*20*).

Initial studies were designed to follow the activity of ODC in the parenchymal cell cultures under various nutritional conditions. Figure 4 shows that the initial level of ODC in the isolated hepatocytes was zero, *i.e.*, undetectable activity in our assay system. After 24 hr in culture in HI-WO/BA+insulin, the specific activity of ODC was 38 pmoles L-ornithine metabolized per hr per μg DNA. The cultures were then divided into 2 sets, and the medium of one set changed to fresh HI-WO/BA+insulin, while the other set was exposed to Swim's S-77 medium (*22*) with 4 mM glutamine (S-77) plus insulin. At 48 hr, the level of ODC activity in the cultures incubated in HI-WO/BA+insulin had increased to 79 pmoles L-ornithine metabolized per hr per μg DNA. The cells incubated in S-77+insulin, however, had no detectable ODC activity. To ascertain if this effect was due to nutritional factors alone, the 2 sets were subdivided, and the medium from one sub-set of each set was changed to HI-WO/BA without insulin. The other sub-set from each set was incubated in S-77 without insulin (Fig. 4). In the absence of insulin, cells previously incubated in HI-WO/BA+insulin rapidly lost ODC activity (Fig. 4). The cells exposed to S-77 lost ODC activity most rapidly, declining to 26 pmoles L-ornithine metabolized per hr per μg DNA after 1 hr of incubation, and to an undetectable level by 2 hr. Cells incubated in HI-WO/BA also lost ODC activity, but at 6 hr the ODC level was still 17 pmoles L-ornithine metabolized per hr per μg

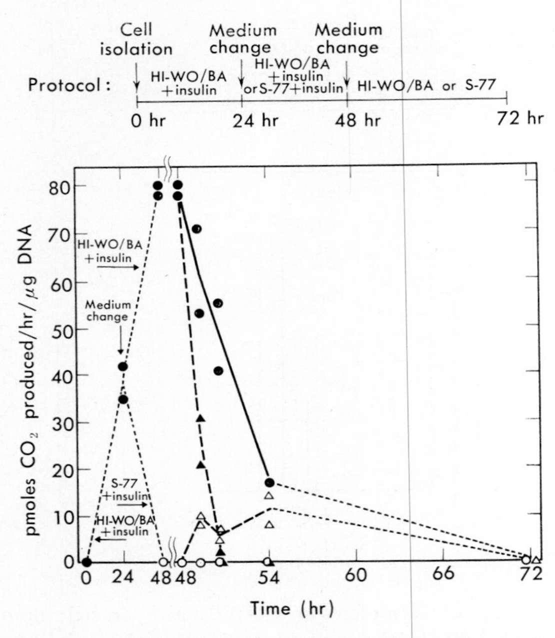

FIG. 4. ODC activity in cultures incubated in medium+13 mUnits/ml insulin (0.5 μg/ml), and in medium when insulin was removed. Points represent individual plates. ● HI-WO/BA+insulin to HI-WO/BA; ▲ HI-WO/BA+insulin to S-77; ○ S-77+insulin to S-77; △ S-77+insulin to HI-WO/BA.

DNA (Fig. 4). Cells incubated between the 24th and 48th hr in culture in S-77+insulin showed no increase from undetectable levels of ODC activity when exposed to S-77 without insulin. However, similarly pretreated cultures exposed to HI-WO/BA without insulin gained measurable ODC activity, rising from 0 to 9 pmoles L-ornithine metabolized per hr per μg DNA after 1 hr of incubation. This level was maintained through the 6th hr of incubation (Fig. 4). By the 24th hr of incubation (72 hr in culture) in insulin-free medium all of the cultures had lost detectable ODC activity.

The protocol outlined in Fig. 4 offered an opportunity to study the effects of hormones on the induction of ODC activity. Cells were allowed to attach for 24 hr in HI-WO/BA+insulin on collagen-coated plates, and then transferred to S-77+insulin for 24 hr. The cultures were divided into 6 sets, and each set was exposed to one of the following media: S-77 alone, HI-WO/BA alone, HI-WO/BA+0.05 μg (1.3 mUnits) insulin per ml, HI-WO/BA+0.50 μg (13 mUnits) insulin per ml, HI-WO/BA+10^{-6}M hydrocortisone, or HI-WO/BA+10^{-5}M hydrocortisone. Figure 5A and 5B show the results of this experiment. Incubation in S-77 alone resulted in no detectable ODC activity, while incubation in HI-WO/BA alone resulted in detectable ODC activity in 1 hr (5 pmoles L-ornithine metabolized per hr per μg DNA) which continued to increase through the 2nd hr to 15 pmoles L-ornithine per hr per μg DNA. As in the previous experiment (Fig. 4), ODC activity was maintained at a relatively constant level through the 6th hr, when the specific

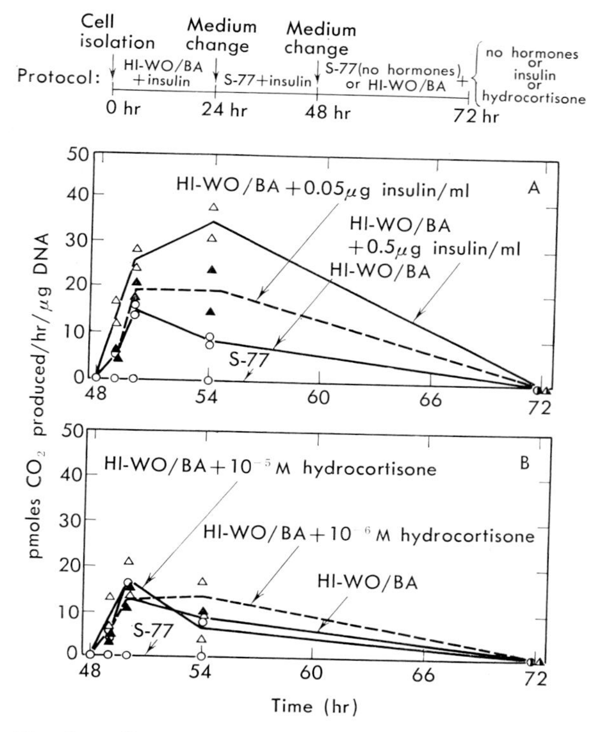

FIG. 5. Effects of hormones on ODC activity. A: induction of ODC activity by insulin; B: lack of induction of ODC activity by hydrocortisone. Points represent individual plates.

activity was 9 pmoles L-ornithine metabolized per hr per μg DNA (Fig. 5A and 5B). In the presence of insulin, a marked stimulation of ODC activity over that observed in S-77 or HI-WO/BA alone was recorded. HI-WO/BA+0.05 μg insulin/ml gave a 2-fold increase in ODC activity over that of HI-WO/BA alone, and HI-WO/BA+0.50 μg insulin/ml induced a level of ODC activity approximately 3.5 times that of HI-WO/BA alone (Fig. 5A). In contrast, addition of 10^{-6} or 10^{-5}M hydrocortisone to HI-WO/BA resulted in no appreciable increase in ODC activity over that observed with HI-WO/BA alone (Fig. 5B). In both Fig. 5A and 5B, undetectable levels of ODC activity were observed after 24 hr (72 hr in culture) regardless of treatment.

The results of the experiment shown in Fig. 5A indicated that ODC activity was induced by insulin if the cells were incubated in HI-WO/BA medium. With this in mind, an experiment was designed to follow the activity of ODC from the time of cell isolation in the constant presence of HI-WO/BA+insulin. The results are shown in Fig. 6. It was found that no ODC activity was detected during the first 6 hr of incubation even though insulin was constantly present. At 18 hr, however, the level of ODC activity was about 45 pmoles L-ornithine metabolized per hr per μg DNA. At 24 hr, the specific activity of ODC was down slightly (from 18 hr), but upon medium change the level of activity immediately began to rise. By 42.5 hr in culture, the cells had a level of ODC activity of about 270 pmoles

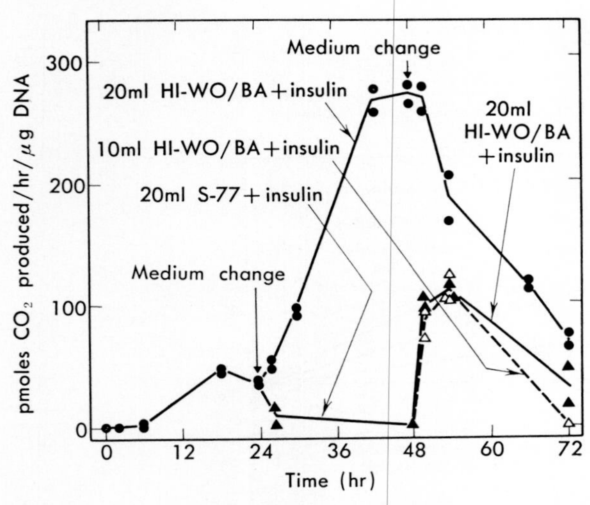

FIG. 6. ODC activity in cultures incubated continuously in HI-WO/BA+13 mUnits/ml insulin (0.5 μg/ml), and induction of ODC activity in 10 ml and 20 ml HI-WO/BA+13 mUnits/ml insulin (0.5 μg/ml) following incubation as indicated. Points represent individual plates.

L-ornithine metabolized per hr per μg DNA. This level appeared to be a maximum, and did not increase even after the medium was changed at 48 hr. Following the medium change, the level of ODC activity was maintained for about 2 hr, and then began to rapidly decline, reaching a level at 72 hr about 1.5 times higher than the level at 24 hr (Fig. 6).

Figure 6 also depicts an experiment following the basic protocol used in the experiment shown in Fig. 5. After 24 hr in HI-WO/BA+insulin, the medium on some of the cultures was changed to S-77+insulin. At 48 hr, the cells were exposed again to HI-WO/BA+insulin, and there was an immediate increase in ODC activity. This experiment shows that the cells (in Fig. 6) were capable of HI-WO/BA+insulin induction after 48 hr in culture, so the decline in ODC activity seen in the cultures constantly exposed to HI-WO/BA+insulin for 48 hr was apparently not due to a loss of cell functional ability. Experiments have not yet been conducted beyond 72 hr in culture, but we view investigation of ODC inducibility after this time as important in determining the overall utility of the primary parenchymal cell culture technique in enzyme regulation studies.

Another interesting point raised by the data in Fig. 6 is that cells incubated in S-77+insulin between the 24th and 48th hr, and then shifted to 20 ml of HI-WO/BA+insulin, maintained detectable levels of ODC for 24 hr (72 hr in culture). In contrast, cells incubated in 10 ml of HI-WO/BA+insulin reached the same levels of ODC activity at 2 and 6 hr of incubation (50 and 54 hr in culture) but lost all detectable ODC activity by the 24th hr. Previous induction experiments (Figs. 4, 5A, and 5B) had been performed with 10 ml of media, thus raising the possibility that the complete loss of ODC activity after 24 hr (72 hr in culture in Figs. 4, 5A, and 5B) was due to the experimental protocol. The data of Fig. 6 suggest that either a metabolite necessary for the maintenance of ODC activity is depleted from 10 ml of medium in 24 hr, or an inhibitor affecting ODC activity reaches a critical concentration in 24 hr in 10 ml of medium. These possibilities are currently being explored.

3. Induction of DNA synthesis

One of our long-term goals is to establish conditions whereby cultured primary hepatocytes can be given certain metabolic "signals" to divide, and then "signals" to cease division, thus mimicking regenerating liver *in vivo*. Initial investigations have centered around isolating cells from optimally entrained rats at various times following partial hepatectomy. Hopkins *et al.* (*13*) have shown that rats entrained to specified feeding and lighting schedules will follow predictable and well coordinated patterns of liver DNA synthesis after partial hepatectomy. Rats entrained to an 8+16 feeding regimen combined with a 12-hr dark/12-hr light schedule (*37*) and partially hepatectomized at the end of the dark period will show a well-defined period of DNA synthesis peaking *in vivo* 23 hr post operation (*13*). Using this pro-

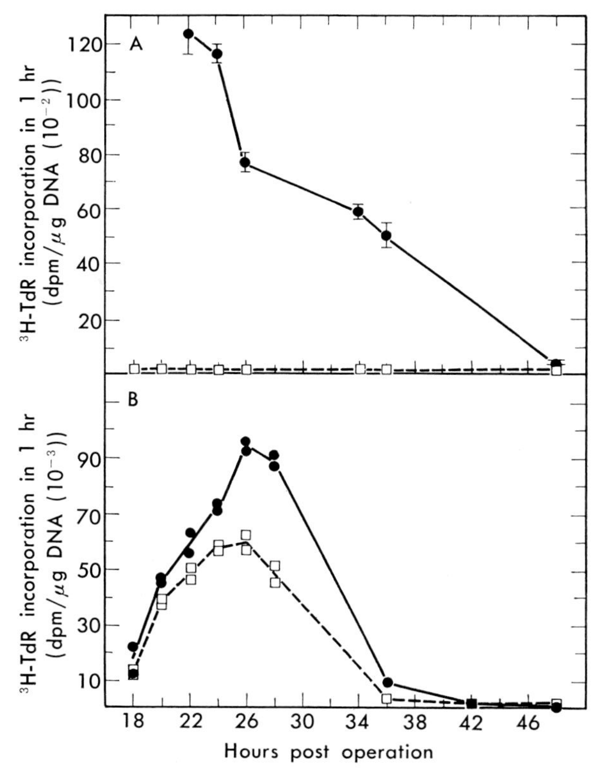

FIG. 7. Incorporation of ^{3}H-TdR into DNA by cells isolated (A) 12- and 18-hr post partial hepatectomy (similar results were obtained when cells were incubated in HI-WO/BA+13 mUnits/ml insulin) and (B) 15-hr post partial hepatectomy and cultured in medium with or without 5% rat serum as indicated. A: ● cells isolated 18-hr post operation (3 plates/point); □ cells isolated 12-hr post operation (2 plates/point). B: ● HI-WO/BA+insulin; □ HI-WO/BA+insulin +5% rat serum obtained 15-hr post operation.

tocol, cells were isolated 12-, 15-, and 18-hr post operation. The cells were cultured following isolation, and were given 1-hr pulses of ^{3}H-TdR at various times.

Figure 7A shows that cells isolated 18-hr post operation were incorporating peak levels of ^{3}H-TdR into DNA at the time of the first pulse, *i.e.*, 22-hr post operation. In contrast, cells isolated 12-hr post operation showed only baseline levels of ^{3}H-TdR incorporation into DNA. However, when cells were isolated 15-hr post operation, incorporation of ^{3}H-TdR into DNA increased from 18-hr post operation to a peak at 26-hr post operation, and then declined to very low levels during the next 10-hr (Fig. 7B). The second period of DNA synthesis, which peaks *in vivo* at 48-hr post operation (*13*), was not observed *in vitro*. Incubation of cells isolated 15-hr post operation with serum from rats sacrificed 15-hr post operation did not stimulate ^{3}H-TdR incorporation into DNA and may have inhibited the rate of incorporation (Fig. 7B).

Autoradiograms of cultures from these experiments showed that only 1–2% of the cells underwent DNA synthesis. This is considerably lower than that observed *in vivo*, where about 40% of the cells undergo cell division during the first period of cell proliferation (unpublished observations). We are currently studying methods to increase the number of cells synthesizing DNA in culture.

4. *Discussion and prospects*

The enzyme data presented in Figs. 4, 5, and 6 answer fewer questions than they pose. The level of ODC activity in cultured parenchymal cells was unresponsive to hormonal and nutritional induction during the first 6 hr following cell isolation and culturing (Fig. 6), but after this time HI-WO/BA alone, but not S-77 alone, induced ODC activity (Figs. 4 and 5). Adding insulin to HI-WO/BA resulted in an induction of ODC activity above that observed with HI-WO/BA alone, whereas adding insulin to S-77 resulted in no detectable ODC activity (Figs. 4 and 5). It is clear that insulin alone, in the absence of a proper nutritional environment, was not sufficient to support detectable levels of ODC activity in these cells. There are significant differences between HI-WO/BA medium and S-77 medium (*i.e.*, HI-WO/BA has higher levels of aspartate, glutamate, glutamine, leucine, glycine, histidine, lysine, methionine, phenylalanine, proline, serine, threonine, valine, tryptophan, and tyrosine, as well as oleate, albumin, and several co-factors not present in S-77), and we are presently exploring these differences in order to find which factor(s) missing from S-77, but present in HI-WO/BA, support(s) detectable levels of ODC activity in cultured liver parenchymal cells. It is of interest that Hogan *et al.* (*11*) reported that S-77 alone, in the absence of insulin, supported detectable levels of ODC activity in cultured hepatoma cells (HTC) under conditions in which the cells did not proliferate. The stimulation of ODC actvity appeared to be related, in part, to glutamine concentration. In cultured parenchymal cells, however, adding glutamine to 8 mM in S-77 medium failed to induce ODC activity (unpublished observations). These findings point out major differences in the regulation of ODC activity between HTC and normal adult liver cells cultured under conditions which we are using.

Mallette and Exton (*19*) recently reported a study on ODC induction in perfused rat livers. Insulin alone or with added casein hydrolysate gave 4–6-fold increases in ODC activity levels in 2 hr. In contrast, glucagon alone depressed ODC activity whereas glucagon+casein hydrolysate stimulated ODC activity in 2 hr. The insulin data appear to be in contrast to the findings reported in this paper, where insulin induction was dependent upon medium. However, there are 2 distinct differences between the perfused liver system used by Mallette and Exton (*19*) and the cultured hepatocyte technique described in the present paper: the perfused liver system involves a recirculating medium containing washed bovine red blood cells, and the perfused liver contains non-parenchymal as well as parenchymal cells. The latter point implies that one cannot be certain which cells or cell types are responding to the experimental variables, while the former pertains to the possibility that nutritive or unknown hormonal factors might be associated with the erythrocytes, and that such factors could affect the observed responses. In addition, Mortimore and his colleagues (*23*, *24*, *39*) have shown that considerable amounts of proteolysis, with release of free amino acids, occur during liver perfusion under conditions similar to those used by Mallette and Exton (*19*). In a recirculating system such a phenomenon would be expected to change the nutritional aspects of the perfusion medium during the course of the experiments.

The enzyme data presented in Figs. 4, 5, and 6 appear to augment animal studies reported by Panko and Kenney (*25*). These investigators found that both insulin and hydrocortisone induced increases in hepatic ODC activity in adrenalectomized rats. Figures 4, 5, and 6 of this report show that insulin induced ODC activity in adult liver parenchymal cell cultures as well. On the other hand, hydrocortisone treatment failed to affect ODC activity (Fig. 5B). However, treatment with hydrocortisone (*4*) or dexamethasone (Fig. 3) resulted in stimulation of TAT activity, showing that our cells retain receptors for these steroid hormones. The findings reported here imply that hydrocortisone, unlike insulin, does not act directly on the liver *in vivo* to stimulate ODC activity, and emphasize the usefulness of the cultured primary liver cell system in separating phenomena dependent upon interaction between liver cells and cells from other parts of the body from phenomena associated with liver cells alone.

As stated earlier, a major technical goal of our research is to find the metabolic "signals" which will initiate DNA synthesis and cell proliferation in cultured parenchymal cells, and then to halt cell division when those "signals" are removed or replaced with different "signals." To this end we are studying cells obtained from optimally entrained rats subjected to partial hepatectomy at a specified time (*13*) and killed at various time intervals thereafter. Initial studies indicated that cells incorporated ^{3}H-TdR into DNA on the *in vivo* schedule (Fig. 7), but the number of cells synthesizing DNA in culture represented only a fraction of the cells expected from *in vivo* studies (*13*; unpublished observations). We are currently investigating ways to increase the number of cells synthesizing DNA *in vitro*. Attempts to stimulate incorporation of ^{3}H-TdR into DNA of cultured adult hepatocytes from intact livers with known stimulators of *in vivo* DNA synthesis, *i.e.*, certain phorbol esters, have thus far been unsuccessful (unpublished observations). We are continuing studies along these lines.

Our ultimate research goal is to be able to mimic normal liver ontogeny *in vitro*, where complexities encountered in the intact animal can be controlled. The process of ontogeny encompasses the entire span of orderly metabolic events occurring as a dividing, fetal-like cell becomes a non-dividing, fully differentiated adult cell. The working hypothesis in our laboratory is that "oncogeny is blocked ontogeny." By this we mean that if the orderly process of ontogeny is interrupted or obstructed, the entire process may be diverted away from normal differentiation and towards an abnormal diseased state, which includes oncogenic transformation. However, the process of ontogeny in liver cells is poorly understood. The ability to mimic regenerative hyperplasia in culture would permit the study of ontogeny *in vitro*. A biochemical comparison of normal cells undergoing ontogeny with hepatoma cell lines should reveal regulatory differences which may prove highly significant to the phenomenon of oncogenic transformation.

A Comparison of Methods

Perhaps the best way to summarize the observations presented in this paper is to compare our methods and findings with those of Bissell *et al.* (*3*), the only other report to date rigorously dealing with short-term culturing of isolated adult hepatocytes. Table 1 shows such a comparison. As indicated, both methods deal with adult liver cells isolated from the same animal, the male Sprague-Dawley rat (180–350 g). Our animals are entrained to the "8+16" feeding schedule (*37*) at least 2 weeks prior to cell isolation, whereas Bissell *et al.* (*3*) do not mention the use of environmental controls for their animals. Cells used in the enzyme studies reported in the present paper were all isolated from intact adult liver. For studies involving incorporation of ^{3}H-TdR into DNA, cells were isolated 12, 15, or 18 hr after partial hepatectomy. Cells used by Bissell *et al.* (*3*) were routinely isolated 4 days after partial hepatectomy, a technique which may not result in the isolation and culturing of "completely" adult hepatocytes (*3*), especially in relation to glycolytic isoenzymes (*5*).

We began our studies with Ca^{2+}-free Hank's buffer as the perfusion medium (*2, 4, 6, 8*), but we are presently using Ca^{2+}-free Swim's S-77 medium with 4 mM glutamine. Both media are supplemented with albumin and insulin, and heparin (in the initial solution used to blanch the liver) or collagenase (in the digesting solution) and are aerated with a mixture of 95% oxygen+5% CO_2 (*2, 6, 8*). Our object is to supply cells with a spectrum of nutritional factors throughout the cell isolation procedure as well as insulin to encourage uptake of amino acids and glucose. This procedure may reduce the trauma associated with cell isolation, as suggested by Mortimore and his colleagues (*23, 24, 39*) who found that proteolytic enzymes in perfused liver were inhibited by amino acids and insulin. In addition, this procedure may shorten the recovery period needed for repair of lesions caused by the cell isolation procedure.

Bissell *et al.* (*3*) perfused with a balanced salts solution consisting of NaCl, KCl, Na_2HPO_4, and KH_2PO_4 supplemented with sodium succinate (to protect the microsomal enzyme *p*-nitroanisole O-demethylase) and collagenase (Table 1). The addition of 1 mM succinate to the perfusion medium may represent a useful

TABLE 1. Comparison of Methods and Parameters Studied in This Report and in Bissell *et al.* (*3*)

Parameter	This report	Bissell *et al.* (*3*)
Animal	Male Sprague-Dawley rats, 200–350 g	Male Sprague-Dawley rats, 180–300 g
Feeding and lighting schedule	Controlled "8+16" (*37*)	Not stated
Cell source	Isolated from normal intact adult liver or isolated according to controlled schedule following partial hepatectomy	Isolated 4 days after partial hepatectomy
Perfusion medium	Ca^{2+}-free Hank's buffer or Ca^{2+}-free Swim's S-77 medium with 4 mM glutamine and aerated with 95% oxygen+5% CO_2. Both media supplemented with 0.5% albumin, 13 mUnits insulin per ml(0.50 μg/ml), 8 μM streptomycin sulfate, and 100 units/ml penicillin. Additionally, 2 units/ml heparin in the first solution, or 0.05% collagenase in the second (digesting) solution.	0.8% NaCl, 0.04% KCl, 0.005% Na_2HPO_4, 0.006% KH_2PO_4, 1 mM sodium succinate, 0.05% collagenase, and 100 units/ml penicillin.
No. cells isolated from one liver	300–600 × 10^6 cells	100–200 × 10^6 cells
Incubation medium for cell attachment (first 24 hr in culture)	HI-WO/BA+insulin with collagen-coated plates, or Ham's F-12+FCS +insulin (plastic alone)	L-15 medium+1 mM succinate (no added hormones)
Inoculation	1,000 cells/mm²	3,000 cells/mm² (calculated from data)
Metabolic studies	1. TAT induction by hydrocortisone, dexamethasone 2. ODC induction by insulin, incubation medium 3. SDH activity[a] 4. Cholesterol synthesis[b] 5. DNA synthesis by cells isolated from partially hepatectomized rats	1. *p*-Nitroanisole O-demethylase activity 2. LDH activity 3. Glucose-6-phosphatase activity 4. Gluconeogenesis 5. Glycogen deposition induced by insulin 6. Glycogenolysis after exposure to glucagon 7. Stimulation of cyclic AMP levels by glucagon 8. Albumin synthesis 9. ATP levels in cultured cells

[a] Assayed by Dr. George Michalopoulos. [b] Assayed by Dr. Stanley Goldfarb.

innovation. Succinate is known to stimulate the Krebs citric acid cycle, and, as such, may help to protect cells against loss of ATP during isolation.

Using our perfusion technique, 300–600 × 10^6 cells are routinely isolated from one rat liver. The cells are attached in serum-free HI-WO/BA+insulin on collagen-coated plates, or in Ham's F-12+fetal calf serum+insulin on plastic alone. The cells are inoculated at about 1,000 cells/mm², a condition giving maximum cell attachment.

Bissell *et al.* (*3*) isolated 100–200 × 10^6 cells from one rat liver. The cells were cultured in Falcon plastic petri plates in serum- and glucose-free L-15 medium at initial densities of about 3,000 cells/mm², which favored formation of confluent monolayers.

Metabolic data reported in the present paper include induction studies on TAT and ODC, and DNA synthesis by cells isolated from rats given partial hepatectomies at a scheduled time under carefully controlled environmental conditions. In addition, serine dehydratase (SDH) activity and cholesterol biosynthesis have been demonstrated in our cells by members of Dr. Henry Pitot's research group (Table 1).

Bissell *et al.* (*3*) studied a different set of parameters. They investigated *p*-nitroanisole O-demethylase induction by benzo(a)pyrene, lactate dehydrogenase (LDH) activity, glucose-6-phosphatase activity, gluconeogenesis, glycogen deposition induced by insulin, glycogenolysis after exposure to glucagon, the stimulation of cyclic AMP levels by glucagon, albumin synthesis, and ATP levels in cultured cells.

Prospects for the method of studying isolated adult parenchymal liver cells in short-term monolayer culture appear very promising. The fact that widely differing perfusion and culture media can be used is particularly encouraging, since this implies that the conditions of cell isolation can be tailored to specific experimental protocols. It is likely that the primary adult liver parenchymal cell monolayer culture method will eventually lend itself to the study of essentially every aspect of liver cell development and metabolic regulation.

ACKNOWLEDGMENTS

Financial support was provided in part by Training Grant TO1-CA-5002 and Grant CA-07175 from the National Cancer Institute. The authors express their thanks to Dr. Harold Campbell for methodology employed in the assay of ornithine decarboxylase activity and to Mrs. Henryka Brania for technical assistance in cell culture.

REFERENCES

1. Bernhard, H. P., Darlington, G. J., and Ruddle, F. H. HEPA-1: a cell line expressing liver function *in vitro*. Abstr. Am. Soc. Cell Biol.-1971, 28, 1971.
2. Berry, M. N. and Friend, D. S. High-yield preparation of isolated rat liver parenchymal cells. A biochemical and fine structural study. J. Cell Biol., *43*: 506–520, 1969.
3. Bissell, M. D., Hammaker, L. E., and Meyer, U. A. Parenchymal cells from adult liver in non-proliferating monolayer culture. I. Functional studies. J. Cell Biol., *59*: 722–734, 1973.
4. Bonney, R. J., Becker, J. E., Walker, P. R., and Potter, V. R. Primary monolayer cultures of adult rat liver parenchymal cells suitable for the study of the regulation of enzyme synthesis. In Vitro, *9*: 399–413, 1974.
5. Bonney, R. J., Hopkins, H. A., Walker, P. R., and Potter, V. R. Glycolytic isoenzymes and glycogen metabolism in regenerating liver from rats on controlled feeding schedules. Biochem. J., *136*: 115–124, 1973.
6. Bonney, R. J., Walker, P. R., and Potter, V. R. Isoenzyme patterns in parenchymal and non-parenchymal cells isolated from regenerating and regenerated liver. Biochem. J., *136*: 947–954, 1973.
7. Butcher, F. R., Bushnell, D. E., Becker, J. E., and Potter, V. R. Effect of cordycepin on induction of tyrosine aminotransferase employing hepatoma cells in tissue culture. Exp. Cell Res., *74*: 115–123, 1972.

8. Crisp, D. M. and Pogson, C. I. Glycolytic and gluconeogenic enzyme activities in parenchymal and non-parenchymal cells from mouse liver. Biochem. J., *126*: 1009–1023, 1972.
9. Gerschenson, L. E., Anderson, M., Molson, J., and Okigaki, T. Tyrosine transaminase induction by dexamethasone in a new rat liver cell line. Science, *170*: 859–861, 1970.
10. Ham, R. G. Clonal growth of mammalian cells in a chemically defined synthetic medium. Proc. Natl. Acad. Sci. U.S., *53*: 288–293, 1965.
11. Hogan, B., Murden, S., and Blackledge, A. The effect of growth conditions on the synthesis and degradation of ornithine decarboxylase in cultured hepatoma cells. *In;* D. H. Russell (ed.), Polyamines in Normal and Neoplastic Growth, pp. 239–248, Raven Press, New York, 1973.
12. Hopkins, H. A., Bonney, R. J., Walker, P. R., Yager, J. D., Jr., and Potter, V. R. Food and light as separate entrainment signals for rat liver enzymes. Adv. Enzyme Regulation, *11*: 169–191, 1973.
13. Hopkins, H. A., Campbell, H. A., Barbiroli, B., and Potter, V. R. Thymidine kinase and DNA metabolism in growing and regenerating livers from rats on controlled feeding schedules. Biochem. J., *136*: 955–966, 1973.
14. Huang, Y. L. and Ebner, K. E. Induction of tyrosine aminotransferase in isolated liver cells. Biochim. Biophys. Acta, *191*: 161–163, 1969.
15. Iype, P. T. Cultures from adult rat liver cells. I. Establishment of monolayer cell-cultures from normal liver. J. Cell. Physiol., *78*: 281–288, 1971.
16. Johnson, M. E. M., Das, D. M., Butcher, F. R., and Fain, J. N. The regulation of gluconeogenesis in isolated rat liver cells by glucagon, insulin, dibutyryl cyclic adenosine monophosphates, and fatty acids. J. Biol. Chem., *247*: 3229–3235, 1971.
17. Lambiotte, M., Susor, W. A., and Cahn, R. D. Morphological and biochemical observations on mammalian liver cells in culture. Isolation of a clonal strain from rat liver. Biochimie, *54*: 1179–1187, 1972.
18. Lin, E. C. C. and Knox, W. E. Adaptation of the rat liver tyrosine-α-ketoglutarate transaminase. Biochim. Biophys. Acta, *26*: 85–88, 1957.
19. Mallette, L. E. and Exton, J. H. Stimulation by insulin and glucagon of ornithine decarboxylase activity in perfused rat livers. Endocrinology, *93*: 640–644, 1973.
20. Morley, C. G. D. The stimulation of liver ornithine decarboxylase by a factor from fetal calf serum. Biochem. Biophys. Res. Commun., *49*: 1530–1535, 1972.
21. Morrison, S. J. and Jenkin, H. M. Growth of *Chlamydia psittaci* strain meningopneumonitis in mouse L cells cultivated in a defined medium in spinner cultures. In Vitro, *8*: 94–100, 1972.
22. Morse, P. A., Jr. and Potter, V. R. Pyrimidine metabolism in tissue culture cells derived from rat hepatomas. I. Suspension cell cultures derived from Novikoff hepatoma. Cancer Res., *25*: 499–508, 1965.
23. Mortimore, G. E. and Mondon, C. E. Inhibition by insulin of valine turnover in liver. Evidence for a general control of proteolysis. J. Biol. Chem., *245*: 2375–2383, 1970.
24. Mortimore, G. E., Neely, A. N., Cox, J. R., and Guinivan, R. A. Proteolysis in homogenates of perfused rat liver: responses to insulin, glucagon and amino acids. Biochem. Biophys. Res. Commun., *54*: 89–95, 1973.
25. Panko, W. B. and Kenney, F. T. Hormonal stimulation of hepatic ornithine decarboxylase. Biochem. Biophys. Res. Commun., *43*: 346–350, 1971.
26. Perske, W. F., Parks, R. E., Jr., and Walker, D. L. Metabolic differences between

hepatic parenchymal cells and a cultured cell line from liver. Science, *125*: 1290–1291, 1957.

27. Pitot, H. C., Peraino, C., Morse, P. A., Jr., and Potter, V. R. Hepatomas in tissue culture compared with adapting liver *in vivo*. Natl. Cancer Inst. Monogr., *13*: 229–245, 1964.
28. Potter, V. R. Workshop on liver cell culture. Cancer Res., *32*: 1998–2000, 1972.
29. Potter, V. R., Baril, E. F., Watanabe, M., and Whittle, E. D. Systematic oscillations in metabolic functions in liver from rats adapted to controlled feeding schedules. Fed. Proc., *27*: 1238–1245, 1968.
30. Richardson, U. I., Tashjian, A. H., Jr., and Levine, L. Establishment of a clonal strain of hepatoma cells which secrete albumin. J. Cell Biol., *40*: 236–247, 1969.
31. Schreiber, G. and Schreiber, M. Review: the preparation of single cell suspensions from liver and their use for the study of protein synthesis. Subcellular Biochem., *2*: 321–367, 1973.
32. Snyder, S. H. and Russell, D. H. Polyamine synthesis in rapidly growing tissues. Fed. Proc., *29*: 1575–1582, 1972.
33. Thompson, E. B., Tomkins, G. M., and Curran, J. F. Induction of tyrosine α-ketoglutarate transaminase by steroid hormones in a newly established tissue culture line. Proc. Natl. Acad. Sci. U.S., *56*: 296–303, 1966.
34. Walker, P. R., Bonney, R. J., Becker, J. E., and Potter, V. R. Pyruvate kinase, hexokinase, and aldolase isozymes in rat liver cells in culture. In Vitro, *8*: 107–114, 1972.
35. Walker, P. R. and Potter, V. R. Allosteric properties of a pyruvate kinase isoenzyme from rat liver cells in culture. J. Biol. Chem., *248*: 4610–4616, 1973.
36. Walker, P. R. and Potter, V. R. Diurnal rhythms of hepatic enzymes from rats adapted to controlled feeding schedules. *In;* L. E. Scheving, F. Halberg, and J. E. Pauly (eds.), Chronobiology, pp. 17–22, Igaku Shoin, Tokyo, 1974.
37. Watanabe, M., Potter, V. R., and Pitot, H. C. Systematic oscillations in tyrosine transaminase and other metabolic functions in liver of normal and adrenalectomized rats on controlled feeding schedules. J. Nutrition, *95*: 207–227, 1968.
38. Waymouth, C. Rapid proliferation of sublines of NCTC clone 929 (strain L) mouse cells in a simple chemically defined medium (MB 752/1). J. Natl. Cancer Inst., *22*: 1003–1017, 1959.
39. Woodside, K. H. and Mortimore, G. E. Suppression of protein turnover by amino acids in the perfused rat liver. J. Biol. Chem., *247*: 6474–6481, 1972.

Discussion of Paper by Drs. Potter et al.

DR. WEBER: Do you have evidence that the cultures you use in the experiments reported still express the characteristic liver parenchymal cell functions, such as gluconeogenesis, *i.e.*, production of labeled glucose from labeled pyruvate or lactate?

DR. POTTER: Yes, there is good evidence for several of such functions. In systems very similar to our own, Dr. M. Bissell has a very relevant report in J. Cell Biol., *59*: 722-734, 1973.

DR. ICHIHARA: Have you tested effect of glucagon, cyclic AMP for induction of ornithine decarboxylase? We observed induction of the enzyme by these hormones using liver slices and dispersed cell suspension.

DR. POTTER: Such experiments will be done very soon.

DR. ICHIHARA: Have you noticed spontaneous induction of this enzyme, though small extent compared with that by hormones, in non-dividing cells?

DR. POTTER: Yes.

DR. ICHIHARA: In dividing cells there is large induction of this enzyme in late lag period without addition of any hormone. Do you think that these contradictory phenomena, such as induction by antagonizing hormones (insulin and glucagon), hormonal and non-hormonal induction, may be caused by the same mechanism?

DR. POTTER: I think induction can be stimulated at either transcriptional or translational level. That is, neither is rate limiting in an absolute sense.

DR. MURAMATSU: In your experiments with regenerating rat liver, it would be interesting to test shorter time intervals than 12 hr for the *in vitro* inoculation. Since DNA synthesis occurs as a consequence of a series of events beginning with early RNA synthesis, you may be able to determine the minimum *in vivo* period that is required for triggering these events.

Relationship between Cyclic AMP Level and Differentiation of Neuroblastoma Cells in Culture

Kedar N. Prasad, Shailendra K. Sahu, and Surendra Kumar

Department of Radiology, University of Colorado Medical Center, Denver, Colorado, U.S.A.

Abstract: Cyclic AMP induced irreversibly many differentiated functions in mouse neuroblastoma (NB) cells in culture. These include formation of long neurites, increase in size of soma and nucleus associated with an elevation of total RNA and protein contents, increase in levels of tyrosine hydroxylase (TH), choline acetyltransferase (ChA) and acetylcholinesterase (AChE), inhibition of cell division, increase in sensitivity of adenylate cyclase (AC) to catecholamines, and loss of tumorigenicity. Most of the "differentiated" cells accumulated in the G_1-phase of the cell cycle. Further studies using X-ray and serum-free medium (SFM) indicate that the expression of differentiated phenotype also occurred in the G_2-phase. The levels of AC, cyclic AMP, and cyclic AMP phosphodiesterase (PDE) and the sensitivity of AC to dopamine (DA) and norepinephrine (NE) increased during "differentiation" of mouse NB cells in culture; therefore, we suggest here that the reverse may be true during malignant transformation of nerve cells. Cyclic AMP also induced several differentiated functions in human NB cells in culture. In NB cells some of the differentiated functions are induced by agents which may or may not elevate cyclic AMP level. Based on our studies we conclude the following: (a) cyclic AMP induces many differentiated functions some of which, such as neurite formation and elevation of ChA and AChE, can be expressed without any change in cyclic AMP level; (b) the neurite formation can be expressed in the absence of an elevation of neural enzymes and of increased sensitivity of AC to catecholamines and *vice versa*; and (c) an elevation of cyclic AMP may not allow the expression of differentiated phenotype if the subsequent steps are not activated or if they are impaired. DA in the presence of an inhibitor PDE increased cyclic AMP level in intact NB cells. In human NB cells sodium butyrate induced cell lethality, neurite formation, and increased TH activity by elevating cyclic AMP level. Therefore a new experimental therapeutic model which involves the administration of sodium butyrate followed by an administration of L-dopa (precursor of DA) in the presence of an inhibitor of PDE is proposed in addition to existing modalities. NE and epinephrine (EP) which do not increase AC activity in homogenate of malignant NB cells markedly increased cyclic AMP level in mouse NB cells when the PDE activity was inhibited. Therefore we suggest that an elevation of cyclic AMP corrects the defective response of AC to NE and EP.

The factors which induce and regulate the expression of differentiated functions in nerve cells are poorly understood. Recently we have obtained some pertinent information on the above problem, using mouse neuroblastoma (NB) cell culture as an experimental model. Although the NB cell culture may not be a perfect model to study the neural differentiation, it has provided new information which appear to be pertinent at least in some aspects of neural differentiation. This is shown by the fact that normal nerve cell cultures after treatment with cyclic AMP show some responses similar to those observed with NB cells. We have shown that cyclic AMP induces irreversibly many differentiated functions in mouse NB cells in culture. These include formation of long neurites (*28*, *33*, *38*), increase in the size of soma and nucleus associated with an elevation of total RNA and protein contents (*42*), increase in the levels of tyrosine hydroxylase (TH) (*45*, *54*), choline acetyltransferase (ChA) (*36*), and acetylcholinesterase (AChE) (*9*, *39*), inhibition of cell division (*28*, *33*, *38*), loss of tumorigenicity (*29*), and increased sensitivity of adenylate cyclase (AC) to catecholamines (*32*). Some of the above differentiated functions are induced by agents which may or may not elevate cyclic AMP. For example, serum free medium (SFM) induces neurite formation (*50*), and increases AChE activity (*17*) but it does not increase TH (*17*, *54*), ChA (*36*), or AC activity (*32*). The sensitivity of AC to catecholamine also does not change in SFM-treated cell (*32*). 5-Bromodeoxyuridine (5-BrdU) also induces neurite formation (*49*) and increases (*31*) the activities of TH and ChA. SFM and 5-BrdU elevate the cyclic AMP level by about 2 fold (*40*). X-ray and 6-thioguanine do not change the cyclic AMP level (*40*), but induce neurite formation (*26*, *30*) and increase ChA activity (*30*, *36*): however, they do not increase TH activity (*30*, *54*). Sodium butyrate does not induce neurite formation (*33*) but increases the levels of cyclic AMP (*40*), and AC (*32*), without affecting the cyclic AMP phosphodiesterase (PDE) activity (*34*). In addition, the sensitivity of AC to catecholamines (*32*) and the activities of TH (*54*) and ChA (*36*) markedly increase in sodium butyrate-treated NB cells. Vinblastine sulfate and cytochalasin B which are known to block the assembly of microtubules and microfilaments, respectively, do not decrease (*52*) the prostaglandin (PG) E_1-stimulated cyclic AMP level; however, they completely block PGE_1-induced axon formation (*28*). Based on these observations the following conclusions have been reached: (a) cyclic AMP induces many differentiated functions some of which, such as neurite formation and elevation of ChA and AChE, can be expressed without any change in cyclic AMP level; (b) the neurite formation can be expressed in the absence of an elevation of neural enzymes and *vice versa*; (c) the neurite formation can also be expressed in the absence of increased sensitivity of AC to catecholamines and *vice versa*; and (d) an elevation of cyclic AMP may not allow the expression of differentiated phenotype if the subsequent steps are not activated or if they are impaired.

Features of NB Cells

The procedures for culturing and maintaining of mouse NB cells were previously described (*39*). These cells have a relatively high rate of glycolysis (*47*). They contain demonstrable levels of TH (*4, 43*), ChA (*4, 43*), AChE (*4, 43*), and catechol-*o*-methyltransferase (COMT) (*5, 37*), but lack tryptophan hydroxylase (*1, 43*). Mouse NB tumors have 5th band of muscle lactate dehydrogenase (*44*) which is absent from the brain. Four types of clone (*43*) have been isolated from mouse NB tumors. These include: (a) clone with TH but no ChA; (b) clone with ChA but no TH; (c) clone with neither TH nor ChA; and (d) clone with both TH and ChA. The first 3 types of clone (*1*) have also been isolated by other investigators. Cells from all of these clones have a doubling time of about 18–24 hr, show spontaneous morphological differentiation varying from 1 to 15% and produce malignant tumors when injected subcutaneously into male A/J mice.

Morphological Differentiation and Cyclic AMP

Analogs of cyclic AMP, PGE_1, and PDE inhibitors induced morphological differentiation as shown by the formation of long neurites and by an increase in the size of nucleus and soma (Fig. 1). Some cells which did not form long neurites increased in size of soma and nucleus. Table 1 shows the relative potency of various agents in causing morphological "differentiation." PGE_1, PDE inhibitors and 8-benzylthio cyclic AMP were more affective than $N^6,O^{2'}$-dibutyryl adenosine-3′,5′-cyclic monophosphate (dbcAMP). For a period of 4 days, no significant cell death occurred after the above treatments. The viability of attached cells as determined by the uptake of supravital stain (trypan blue in 1% saline) was similar to that of control (90–95%).

The number of morphologically differentiated cells after PGF_1-treatment is time and concentration dependent (*28*). A significant increase in the number of

TABLE 1. Effect of Various Cyclic AMP Agents on the Mouse NB Cells in Culture[a]

Treatment	"Differentiated" cells (% of total cells)
Control	9.0 ± 2.0[b]
Monobutyryl cyclic AMP (0.5 mM)	47.0 ± 5.0
dbcAMP (0.5 mM)	51.0 ± 4.8
8-Benzylthio cyclic AMP (400 μg/ml)	66.0 ± 4.5
PGE_1 (10 μg/ml)	72.0 ± 5.0
RO20-1724 (200 μg/ml)	71.0 ± 5.4
Papaverine (25 μg/ml)	79.0 ± 2.5

[a] Quantitative estimation of "differentiated" cells after treatment with various cyclic AMP agents. For quantitating the number of differentiated cells, cells (50,000) were plated in Falcon plastic dishes (60 mm) and treated with drug 24 hr after plating. The morphologically differentiated cells (cytoplasmic processes were greater than 50 μM long) were scored 3 days after treatment. At least 300 cells were counted and the number of "differentiated" cells were expressed as % of total cells. Each value represents an average of 6 to 8 samples. The data are summarized from a previous publication (*28, 33, 38*). [b] Standard deviation.

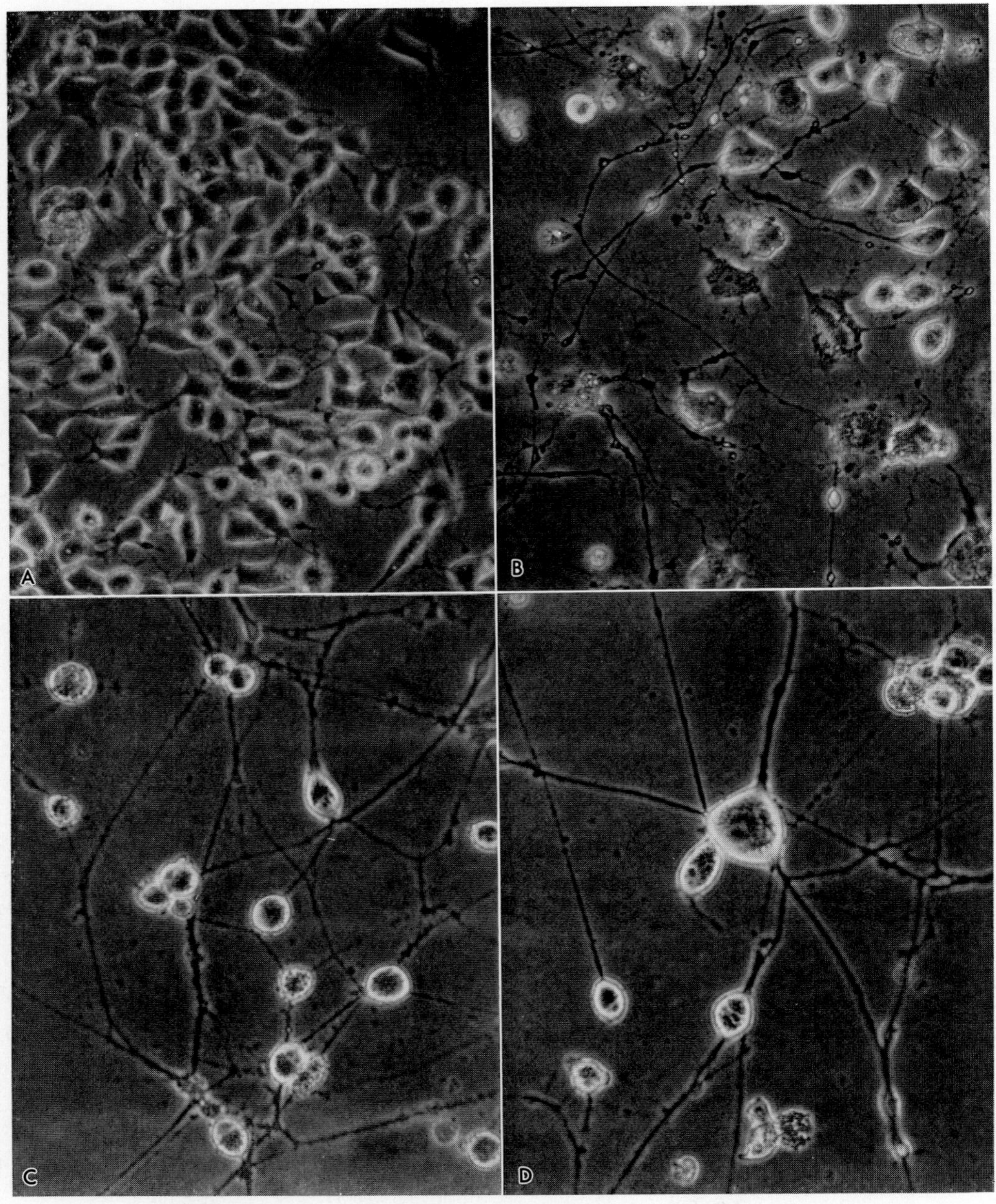

FIG. 1. Phase contrast micrographs of mouse NB cells ($NBA_{2(1)}$ clone) in culture. Cells (50,000) were plated in Falcon plastic dishes (60 mm) and PGE_1 (10 μg/ml) and RO20-1724 (200 μg/ml) were added separately 24 hr later. The medium and drug were changed at 3, 5, 8, and 11 days after treatment. Control culture (A) shows that cells grow in clumps and some of them have short cytoplasmic processes. PGE_1-treated culture 4 days after treatment (B) shows formation of long neurites. PGE_1-treated culture 14 days after treatment (C) shows that the remaining cells maintain their differentiated phenotype. RO20-1724-treated culture (D), in which the drug was removed 3 days after treatment and the micrograph was taken 8 days later, shows that many cells maintain their differentiated phenotype (*27*). ($\times$131)

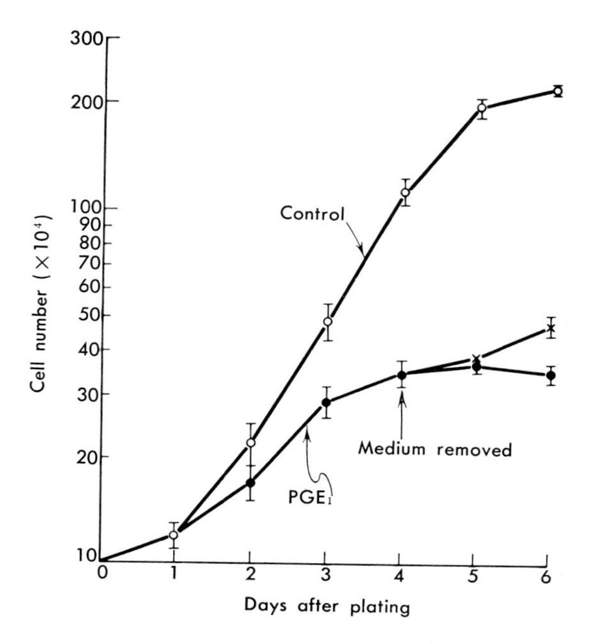

FIG. 2. Growth curve of NB cells *in vitro*. Cells were plated in Falcon plastic dishes (60 mm) and PGE_1 (10 μg/ml) was added 24 hr later. PGE_1 was removed from one group of dishes and fresh growth medium was added 3 days later. In another group of dishes fresh growth medium containing 10 μg/ml of PGE_1 was replaced at the same time. The growth medium in control dishes was also changed. The cell number was counted in a Coulter counter. Each point represents an average of 5 or 6 samples. Bar at each point shows standard deviations (*28*).

"differentiated" cells was noted 24 hr after treatment and cell division continued up to the third day (Fig. 2). This indicates that the inhibition of cell division temporally follows the induction of morphological differentiation and thus may be secondary to the neurite formation. The kinetics of morphological differentiation and growth after treatment with dbcAMP (*33*) or PDE inhibitors (*38*) were similar to those after PGE_1-treatment.

Cyclic AMP, 5′-AMP, theophylline, some 8-substituted analogs of cyclic AMP (8-hydroxyethylthio-, 8-amino-, 8-hydroxyethylamino-, and 8-methylamino-cyclic AMP), adenosine triphosphate, adenosine diphosphate, and sodium butyrate inhibited cell growth without causing morphological differentiation. These data indicate that the inhibition of cell division need not necessarily produce the expression of morphological differentiation; however, certain differentiated functions (*32, 36, 39, 54*) can be expressed after the inhibition of cell division.

Like cyclic AMP, guanosine-3′,5′-cyclic monophosphate (cyclic GMP) is also present in mammalian cells (*13*). Therefore, the effects of cyclic GMP on NB cell culture were examined. Both cyclic GMP and $N^2,O^{2'}$-dibutyryl cyclic GMP inhibited cell division without causing morphological differentiation.

Irreversibility of Growth Inhibition and Morphological Differentiation

The morphological differentiation and inhibition of growth induced by dbcAMP, PGE_1 or 4-(3-butoxy-4-methoxybenzyl)-2-imidazolidinone (RO20-1724) for the most part were irreversible (*28*, *33*, *38*), provided the drug was present in the medium for at least 3–4 days. When the "differentiated" cells, 3 days after treatment with RO20-1724 or PGE_1 were removed from dishes using Viokase solution and replated in separate dishes, cells attached and formed long neurites within 24 hr even though no drug was present during this period. The number of morphologically differentiated cells in the newly plated dishes was similar to those in which the drug was not removed. This indicates that the cellular factors which control the expression of differentiated phenotype remain functional after subculturing.

When the cultures were maintained in the presence of RO20-1724 or PGE_1 for 14 days, many dead cells were floating in the medium but most of the attached cells maintained their differentiated phenotype (Fig. 1C). At this time the treated cultures had 2–3 clones which appeared to be dividing in the presence of the drug. When RO20-1724 was removed 3 days after treatment and the culture was examined 8 days later, many cells maintained their differentiated phenotype (Fig. 1D) indicating further the irreversibility of neurite formation. dbcAMP also induces an irreversible neurite formation in human neuroblasts (*24*). The above findings are in contrast to the observations made on non-nerve cells in which cyclic AMP-effects are reversible at all times soon after the removal of the drug (*15*, *16*, *51*, *56*).

Requirements for the Expression of Differentiated Phenotype

Vinblastine sulfate (interfers with the assembly of microtubules), cytochalasin

TABLE 2. Effect of Metabolic Inhibitors on Neurite Formation and Cyclic AMP Levels[a]

Treatment	"Differentiated" cells (% of total cells)	Cyclic AMP level (pmoles/mg protein)
Control	4 ± 1	23 ± 4[b]
RO20-1724 (200 μg/ml)	71 ± 3	75 ± 10
RO20-1724 (200 μg/ml) +vinblastine sulfate (0.01 μg/ml)	0	62 ± 12
RO20-1724 (200 μg/ml) +cytochalasin B (0.1 μg/ml)	4 ± 1	73 ± 10
RO20-1724 (200 μg/ml) +cycloheximide (5 μg/ml)	4 ± 1	12 ± 3
RO20-1724 (200 μg/ml) +actinomycin D (5 μg/ml)	70 ± 3	82 ± 15

[a] 4-(3-Butoxy-4-methoxybenzyl)-2-imidazolidinone (RO20-1724), an inhibitor of cyclic AMP PDE, was added in NB cell culture 24 hr after plating either alone or in combination with a metabolic inhibitor. Actinomycin D was removed from the culture 1 hr after treatment and then fresh growth medium and RO20-1724 were added, but the remainder of the inhibitors remained in culture for 24 hr. The morphological differentiation and cyclic AMP level were determined 24 hr after treatment. Each value represents an average of at least 6 samples. [b] Standard deviation.

B (interfers with the assembly of microfilaments), and cycloheximide (inhibits protein synthesis) completely blocked the neurite formation induced by cyclic AMP, whereas actinomycin D (inhibits RNA synthesis) did not. Thus the expression of differentiated phenotype (*28*, *33*, *38*) requires at least the assembly of microtubules and microfilaments and the synthesis of new protein, but does not require new RNA synthesis. Although cytochalasin B and Vinblastine sulfate blocked the RO20-1724-induced neurite formation, but these agents failed to decrease the RO20-1724-stimulated increase in intracellular level of cyclic AMP (Table 2), indicating that an elevation of cyclic AMP may not allow the expression of differentiated phenotype if the subsequent steps involved in the formation of neurite are blocked. Cycloheximide decreased the cyclic AMP level, whereas actinomycin D did not.

Sensitivity of NB Clones to Cyclic AMP

Cells of most clones were sensitive to cyclic AMP in causing morphological differentiation. Also, some clones, irrespective of their neuronal cell type, were sensitive to PGE_1 but not to RO20-1724 and *vice versa* (*43*). The clone which was insensitive to RO20-1724 was also unresponsive to dbcAMP.

Tumorigenicity of "Differentiated" Cells

Control cells when injected subcutaneously produced tumors in all A/J mice, whereas the tumorigenicity of "differentiated" cells (4 days after treatment) was either partially or completely abolished (*29*). Uncloned cells were used since they were more nearly duplicate to *in vivo* condition. Since some cells were responsive to PGE_1 but not to PDE inhibitor and *vice versa*, PGE_1 was combined with RO20-1724 to maximize the effect on differentiation. Indeed, cells treated as above lost completely a prime feature of malignancy, their tumorigenicity (Table 3).

TABLE 3. Incidence of Tumors after Subcutaneous Injection of Control and "Differentiated" NB Cells[a)]

Treatment	Number of animals	Incidence of tumors (% of total)
Control cell treated with or without solvent	30	100
dbcAMP	15	50
RO20-1724	15	40
8-Benzylthio cyclic AMP	15	60
PGE_1	15	20
PGE_1+dbcAMP	15	0
PGE_1+8-benzylthio cyclic AMP	15	0
PGE_1+RO20-1724	16	0

a) Tumorigenicity of "differentiated" cells. Cells (10^5) were plated in Falcon plastic dishes (60 mm) and treated with drugs 24 hr later. dbcAMP (0.5 mM), RO20-1724 (200 μg/ml), 8-benzylthio cyclic AMP (400 μg/ml) or PGE_1 (10 μg/ml) were added individually or in combination with PGE_1 (10 μg/ml). After 4 days of incubation the control and differentiated cells (0.25×10^6) were injected subcutaneously into male A/J mice (6 to 8 weeks old). Cell viability in the control and drug-treated cultures was 90–95% (*29*). Animals were observed for 60 days.

Levels of TH, ChA, AChE, and COMT in "Differentiated" Cells

TH (*54*), a rate limiting enzyme in the biosynthesis of catecholamines was markedly increased (Table 4) by some analogs of cyclic AMP and papaverine. The morphologically differentiated NB cells induced by X-ray (*26*), SFM (*50*) and cytosine arabinoside (*17*) showed no change in TH activity (*17, 54*). Sodium butyrate which inhibits cell division without causing morphological differentiation (*33*) increased the TH activity (*54*) and cyclic AMP level (*40*). These data suggest that morphological differentiation and TH activity are independently regulated, and cyclic AMP may be involved in the regulation of TH activity. Our hypothesis that cyclic AMP may be involved in the regulation of TH activity has been confirmed by recent studies on the TH level of mouse adrenal gland (*12*) and mouse NB cells in culture (*45*).

TABLE 4. TH Activity and Differentiation of Mouse NB Cells in Culture[a]

Treatment	TH activity (pmoles product/30 min/10^6 cells)
Control, log phase	15.1±1.9[b]
Control, confluent phase	11.2±0.7
SFM	17.3±0.4
dbcAMP (0.25 mM)	473±17
8-Methylthio cyclic AMP (0.3 mM)	587±9
Papaverine (0.13 mM)	977±46
Sodium butyrate (0.5 mM)	300±12
X-irradiation (600 rads)	14±2
X-irradiation (600 rads)+sodium butyrate	764±90

[a] TH activity after treatment with cyclic AMP agents. NBP_2 clone was used in this study. This clone has both TH and ChA. NB (0.5×10^6) cells were plated in large Falcon plastic flasks (75 cm^2) and treated with drug or X-rays 24 hr later. The drug and medium were changed 2 days later and the enzyme was analyzed 3 days after treatment. Each value represents an average of at least 4 samples. The data are taken from a previous publication (*54*). [b] Standard deviation.

TABLE 5. Effect of Various Agents on ChA Activity of NB Cells[a]

Treatment	ChA activity (pmoles/15 min/10^6 cells)
Control (exponential)	260±35[b]
Control (confluent)	300±34
dbcAMP (0.5 mM)	1300±72
PGE_1 (10 μg/ml)	880±100
RO20-1724 (200 μg/ml)	1280±160
5′-AMP (0.25 mM)	1320±80
Sodium butyrate (0.5 mM)	760±100
X-irradiation (600 rads)	1640±144

[a] ChA activity after treatment with cyclic AMP agents. NB cells (0.5×10^6) were plated in the Falcon plastic flasks (75 cm^2) and each drug was added 24 hr later. Fresh growth medium and drug were added 2 days after drug treatment and the ChA was analyzed 3 days after treatment. Each value represents an average of 5–6 samples. The data are taken from a previous publication (*36*). [b] Standard deviation.

The ChA which synthesizes acetylcholine was markedly increased in "differentiated" cells induced by cyclic AMP and was also elevated after X-irradiation and after treatment with 5′-AMP and sodium butyrate (Table 5). The maximal increase in the ChA activity coincided with the cessation of cell division (*36*). These data indicate that the ChA activity and morphological differentiation are independently regulated, and that cyclic AMP is not necessarily involved in the regulation of ChA activity. Data on AChE (*39*) suggest a similar mode of regulation. Cyclic AMP is also not involved in the regulation of COMT activity, because the enzyme activity in cyclic AMP-induced "differentiated" cells does not change (*37*); however, it does increase in X-irradiated and sodium butyrate treated NB cells.

Changes in Nucleic Acid and Protein Contents in "Differentiated" Cells

Since the size of soma and nucleus increase during cyclic AMP-induced "differentiation" of NB cells, changes in total nucleic acid and protein contents were investigated 3 days after treatment. Table 6 shows that DNA content per cell among "differentiated" population markedly decreased indicating that most of cells accumulate in the G_1-phase of the cell cycle (*42*). The total RNA and protein contents increased by about 2 to 3 fold. This finding is consistent with observation made during differentiation of mammalian neurons.

It is generally presumed that the blocking of cells in the G_1-phase allows the expression of differentiated phenotype. This does not appear to be the case in mouse NB cells, because sodium butyrate-treated cells are blocked in the G_1-phase (*42*), but no expression of morphological differentiation occurs (*33*); however, the expression of several biochemical differentiated functions (*32*, *36*, *39*, *54*) do occur under above condition. In addition to cyclic AMP, other agents such as SFM (*50*), 6-thioguanine (*31*), X-ray (*26*, *36*), and 5-BrdU (*31*, *49*) also induce some differentiated functions; therefore the total nucleic acid and protein contents were measured. Table 7 shows that DNA content of SFM- and 6-thioguanine-treated cells did not significantly change indicating that the differentiated cells of may

TABLE 6. Total DNA, RNA, and Protein Contents in Cyclic AMP-induced "Differentiated" Mouse NB Cells in Culture[a]

Treatment	DNA (pg/cell)	RNA (pg/cell)	Protein (pg/cell)
Control	13.3±1.5[b]	15.3±1.0	500±29
dbcAMP	6.6±0.6	33.6±2.5	1580±122
PGE_1	6.0±1.6	24.4±1.9	870±47
RO20-1724	6.7±1.2	33±1.8	1016±54
Sodium butyrate	5.3±1.0	31.2±3.9	1479±111

[a] Nucleic acid and protein contents in cyclic AMP-induced "differentiated" cells. Cells (0.5×10^6) were plated in large Falcon plastic flasks (75 cm^2) and dbcAMP (0.5 mM), PGE_1 (10 μg/ml), RO20-1724 (200 μg/ml), and sodium butyrate (0.5 mM) were added separately 24 hr later. The total nucleic acid and protein contents were assayed 3 days after treatment. Each value represents an average of 4 to 6 samples. The data are presented from a previous publication (*42*). [b] Standard deviation.

TABLE 7. Total DNA, RNA, and Protein Contents in "Differentiated" Mouse NB Cells ($NBE_{(R)}^-$) in Culture Induced by Non-cyclic AMP Agents[a]

Treatment	DNA (pg/cell)	RNA (pg/cell)	Protein (pg/cell)
Control	19.8±2.7[b]	26±1.3	152±13
SFM	23.8±2.3	65±9.0	180±12
6-Thioguanine	25.4±3.5	79±12.0	346±7.0
5-BrdU	7±1.5	50±3.4	343±34
X-ray	62±13.8	153±27.0	768±39

[a] Cells ($0.5–1\times10^6$) were plated in large Falcon plastic flasks and SFM, 5-BrdU (5 μM), 6-thioguanine (0.5 μM), and X-rays (1,200 rads) were given separately 24 hr later. The total DNA, RNA, and protein contents were analyzed 3 days after treatment. Each value represents an average of 5 to 8 samples (*40*). [b] Standard deviation.

have been arrested in the G_1- and G_2-phase of the cell cycle. It has been generally presumed that cells accumulate in the G_1-phase of the cell cycle when SFM is added. This does not appear to be the case in this NB clone. 5-BrdU-treated cells had about one-third the DNA (*40*) of control cells, indicating that most of the 5-BrdU-treated cells accumulate in the G_1-phase. It should be pointed out that the DNA content per cell was less than that expected if the diploid cells were arrested in the G_1-phase; however, the NB cells are aneuploids (*1*), and therefore the DNA value in each phase of the cycle may not be comparable to diploid cells.

The X-ray induced "differentiated" cells had 3-fold higher DNA content than that of controls. This value was much higher than that expected if all cells were accumulated in the G_2-phase. Since the formation of polyploid cells is a well established response of irradiated mammalian cell culture, it is suggested that the expression of differentiated phenotype can occur in polyploid cells as well as in G_2-cells. Although most of mammalian neurons are diploid; some mammalian neurons such as Purkinje cells have tetraploid DNA contents (*22*).

Transport of Glucose in "Differentiated" Cells

It has been shown (*20*) that cyclic AMP reduces the transport of key macromolecules (glucose, uridine, and leucine) in malignant fibroblasts and thereby restores some of the characteristics of normal fibroblasts. However, cyclic AMP-induced differentiation of NB cells is not mediated by inhibition of glucose transport, because the uptake of ^{3}H-2-deoxyglucose was similar in control and cyclic AMP-induced "differentiated" cells (Table 8).

Changes in AC Activity and Increase in Sensitivity of AC to Catecholamines in "Differentiated" Cells

Figure 3 shows that dopamine (DA) and norepinephrine (NE) produced a maximal stimulation on AC activity to about the same level in both control and "differentiated" cells; but DA requires a much lower concentration than NE. However, in "differentiated" cells the effective concentrations of DA and NE needed

TABLE 8. Transport of ^{3}H-2-deoxyglucose in Cyclic AMP-induced "Differentiated" Mouse NB Cells in Culture[a)]

Treatment	% of control
Control (0.05 ml alcohol)	101±11[b)]
PGE_1 (10 μg/ml, 3 days)	73±9.4
RO20-1724 (200 μg/ml, 3 days)	85±9.0

[a)] NB cells (10^5) were plated in Falcon culture dish (60 cm^2) and RO20-1724 (200 μg/ml), an inhibitor of cyclic AMP PDE, and PGE_1 (10 μg/ml), was added 24 hr later. The drug and medium were changed 2 days after treatment and the glucose transport was performed 3 days after treatment using ^{3}H-2-deoxyglucose. The cells were washed twice with warm Earl's salt and then cells were incubated in Earl's salt containing ^{3}H-2-deoxyglucose (0.5 μCi/ml). After 10 min of incubation, cells were washed twice with cold Earl's salt and 2-ml reagent C (used in protein determination by Lowry method) was added to each dish. An aliquot was counted for radioactivity in Toulene-Triton mixture and an aliquot was taken for protein determination. The data were calculated as cpm/mg protein and then the values for treated cells were expressed as % of control. Each value represents an average of 6 samples. [b)] Standard deviation.

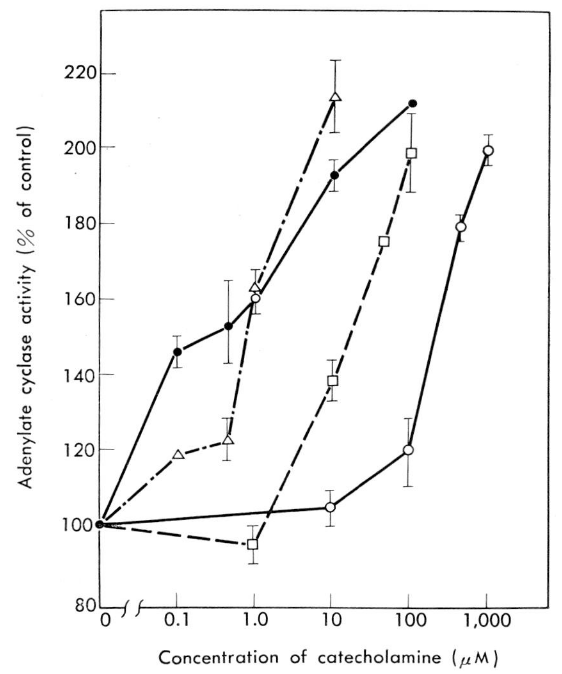

FIG. 3. Changes in AC activity in homogenates of control and "differentiated" mouse NB cells after treatment with DA and NE. RO20-1724, an inhibitor of cyclic AMP PDE, was used to induce "differentiation" in NB cells in culture. Cells (0.5×10^6) were plated in large Falcon plastic flasks (75 cm^2) and RO20-1724 (200 μg/ml) was added 24 hr later. Controls were treated with an equivalent volume of solvent. The drug and medium were changed 2 days after treatment and the AC activity in homogenate of NB cells was measured 3 days after treatment according to a modification (*25*) of the method of Krishna *et al.* (*21*). The basal levels of AC in control (15±1.4 pmoles/mg protein/min) and "differentiated" (21±1 pmoles/mg protein/min) cells were considered 100% control values and the AC values of treated cells were expressed as % of control. Each value represents an average of 8–12 samples. The bar at each point shows standard deviation (*32*). △ DA ("differentiated "cells); ● DA (control cells); □ NE ("differentiated" cells); ○ NE (control cells).

TABLE 9. Concentrations of Blocking Agents Which Produce 50% Inhibition of Catecholamine-stimulated Adenylate Cyclase Activity in Homogenate of NB Cells[a)]

Treatment	Concentrations (μM) which produce 50% inhibition		
	Haloperidol	Phentolamine	Propranolol
DA-stimulated AC activity	11	0.43	180
NE-stimulated AC activity	>100	2.3	1.7

a) Effect of haloperidol (blocks DA receptors), phentolamine (blocks α-adrenergic receptors), and propranolol (blocks β-adrenergic receptors) on DA and NE-stimulated AC activities in homogenates of "differentiated" mouse NB cells. RO20-1724, an inhibitor of cyclic AMP PDE, was used to induce "differentiation" in NB cells in culture. Cells (0.5×10^6) were plated in large Falcon plastic flasks (75 cm^2) and RO20-1724 (200 μg/ml) was added 24 hr later. Controls were treated with an equivalent volume of solvent. The drug and medium were changed 2 days after treatment and the AC activity in homogenate of NB cells was measured 3 days after treatment. The basal level of AC (21 ± 1 pmoles/mg protein/min) in "differentiated" cells was considered 100% control value and the AC values of treated cells were expressed as % of control. From the curves of each blocking agent the concentration which inhibited 50% of the catecholamine-stimulated AC activity was determined. Each value represents an average of 6 samples (*32*).

was about 10 times less than that required in control. Apomorphine (1 μM), which is known to mimic the effect of DA in the caudate nucleus also increased the AC activity. A low concentration of isoproterenol stimulated the AC activity in control cells, but in "differentiated" cells even a high concentration of the drug failed to do so. Acetylcholine did not stimulate the AC activity in control or "differentiated" cells.

Haloperidol, which specifically blocks the dopamine-"receptor", inhibited DA-stimulated AC activity; however, it did not significantly effect NE-stimulated enzyme activity (Table 9). A concentration of about 0.43 μM phentolamine (α-blocking agent) reduced DA-stimulated AC activity by 50%, whereas a similar amount of inhibition of NE-stimulated enzyme activity was achieved with about 2.3 μM phentolamine (Table 9). A concentration of about 180 μM propranolol (β-blocking agent) produced a 50% inhibition of DA-stimulated AC activity, whereas NE-stimulated enzyme activity required only 1.7 μM propranolol for a similar amount of inhibition (Table 9). Haliperidol and α- and β-adrenergic blocking agents did not affect the basal level of AC activity.

Our data demonstrate the presence of an AC, sensitive to very low concentration of DA, in homogenate of mouse NB cells. The occurence of DA-sensitive adenylate cyclase has been demonstrated by Kebabian *et al.* (*18*, *19*) in homogenates of mammalian superior cervical ganglia and basal ganglia, and by Brown and Makman (*6*) in mammalian retina. The AC has been suggested (*19*) to be the receptor for DA in mammalian brain. Our data indicate that the AC also may be linked to DA-"receptor" in NB cells. For example, DA stimulates AC activity and apomorphine, which mimics the actions of DA upon the dopamine-"receptor" of the caudate nucleus (*3*, *53*), also increases AC activity of NB cells. Haloperidol, which antagonizes the DA-"receptors" within the caudate nucleus (*7*) also blocks DA-stimulated AC activity in NB cells. It has been shown that the DA-"receptors" of the caudate nucleus is not blocked by β-adrenergic blocking agents, but is weakly antagonized by α-blocking agents (*2*). The AC of

TABLE 10. Effect of Prostaglandin E_1 and Catecholamines on AC Activity in Homogenate of "Differentiated" NB Cells[a)]

Treatment	AC activity (pmoles/mg protein/min)
Basal level	21±1[b)]
DA (100 μM)	36±4
NE (100 μM)	42±5.2
PGE_1 (10 μM)	41±1
DA+NE	69±3
NE+PGE_1	74±2
DA+PGE_1	38±1
PGE_1+propranolol (10 μM)	40±2
PGE_1+phentolamine (10 μM)	20±1

[a)] Effect of DA and PGE_1 alone or in combination with NE on the AC activity of "differentiated" mouse NB cells. RO20-1724, an inhibitor of cyclic AMP PDE, was used to induce "differentiation" in NB cells in culture. Cells (0.5×10^6) were plated in large Falcon plastic flasks (75 cm^2) and RO20-1724 (200 μg/ml) was added 24 hr later. Controls were treated with an equivalent volume of solvent. The drug and medium were changed 2 days after treatment and the AC activity in homogenate was measured 3 days after treatment. Each value represents an average of 6 samples (*32*). [b)] Standard deviation.

caudate nucleus shows a similar response with respect to β- and α- blocking agents (*19*). Our results show that the DA-stimulated AC activity of NB cells is not affected by a low concentration of β-blocking agent, but is markedly inhibited by a low concentration of α-blocking agent. Thus the AC of NB cells and caudate nucleus shows a quantitatively different response with respect to α-blocking agent.

The mouse NB cells also appear to contain NE-sensitive AC which has pharmacological properties different from those of DA-sensitive AC. NE-stimulated AC activity is inhibited by a low concentration of β-blocking agent, but affected little by a low concentration of α-blocking agent and haloperidol, whereas DA-stimulated AC activity is inhibited by a low concentration of α-blocking agent and haloperidol, but not by β-blocking agent. The fact that DA and NE produces an additive stimulatory effect on AC activity of "differentiated" NB cells (Table 10) suggests that DA and NE interact at different receptor sites. This suggestion is further supported by the observation that the combination of PGE_1 and NE produces an additive stimulatory effect on AC activity, whereas, the combination of DA and PGE_1 does not (Table 10). The observation that the effects of DA and PGE_1 are not additive coupled with the observation that a low concentration of phentolamine blocks the effect of PGE_1 (Table 10) suggests the interesting possibility that these 2 agents may interact at a common site.

The present study shows that the regulation of AC in malignant NB cells in culture is different from that in malignant glial cells in culture. This is demonstrated by the fact that the AC activity of malignant NB cells is stimulated by a low concentration of DA, but not by a low concentration of NE; whereas, the AC activity of malignant glial cells (*8*, *11*) is stimulated by a low concentration of NE, but not by a low concentration of DA.

Table 11 shows that the AC activity in SFM-treated cells (3 days after treat-

TABLE 11. Effect of DA and NE on AC Activity in Homogenates of SFM and Sodium Butyrate-treated NB Cells[a]

Treatment	AC activity (pmoles/mg protein/min)		
	Basal	DA (100 μM)	NE (100 μM)
Control	15±1.4[b]	32±1	18±2.2
Sodium butyrate (3 days)	35±1.4	60±1.8	68±1.5
SFM (3 days)	16±1.5	19±1.2	17±1

[a] Effect of DA and NE on sodium butyrate- and SFM-treated cells. Cells (0.5×10^6) were plated in large Falcon plastic flasks (75 cm^2) and SFM and sodium butyrate (0.5 mM) were given separately 24 hr later. The drug and medium were changed 2 days after treatment and the AC activity in homogenate was measured 3 days after treatment. Each value represents an average of 6 samples (*32*).
[b] Standard deviation.

ment) was similar to that in control cells. In addition, the AC activity was not significantly stimulated either by DA or NE. SFM induces neurite formation (*50*). In sodium butyrate treated-cells (3 days after treatment) the basal level of AC was increased by about 230% of control, and DA and NE increased further the enzyme activity by about 2-fold. Sodium butyrate does not induce neurite formation (*33*). Therefore, we suggest that the neurites can be formed in the absence of an increased sensitivity of AC to catecholamines and *vice versa*.

Changes in Cyclic AMP Level by Dopamine

Although DA stimulates AC activity in broken cell preparation (*32*), it did not increase the intracellular level of cyclic AMP in intact cells (Table 12). This may be due to a rapid degradation of cyclic AMP by PDE. Indeed, when the PDE activity was blocked prior to DA-treatment, DA markedly increased the intracellular level of cyclic AMP (Table 12). Therefore it is not expected that the addition of ex-

TABLE 12. Effect of Catecholamines on Cyclic AMP Level in NB Cells in Culture[a]

Treatment	Cyclic AMP level (pmoles/mg protein/min)
Control	24±4[b]
DA	23±5
NE or EP	27±6
RO20-1724 (200 μg/ml, 1 hr	144±27
RO20-1724+DA (10^{-4} M, 15 min)	587±101
RO20-1724+NE (10^{-4} M, 15 min)	545±65
RO20-1724+EP (10^{-4} M, 15 min)	588±45

[a] NB cells (0.5×10^6) of clone $NBA_{2(1)}$ were plated in large Falcon plastic flasks (75 cm^2). The medium was changed first 2 days after plating and every day thereafter. On 4th day, the fresh medium was changed and RO20-1724 (an inhibitor of cyclic AMP PDE) was added. After 1 hr of incubation DA, NE, or EP was added to culture containing RO20-1724 and the cells were further incubated for 15 min. After incubation, cells were quickly washed once with phosphate buffer solution, pH 7 and 2 ml of 5% cold TCA was added. The cells were moved, homogenized, and cyclic AMP was determined according to Gilman's method (*10*) and protein was determined according to the method of Lowry *et al.* (*23*).
[b] Standard deviation.

ogenous dopamine would mimic the effect of cyclic AMP. We have shown (*27*) that DA reversibly inhibits the cell division when NB cells in culture are exposed to DA for an hour, whereas, NE and epinephrine (EP) under a similar experimental condition has no such effect. DA is auto-oxidized in solution, therefore, the removal of the drug was necessary to avoid the effect of oxidative products.

Correction of a Defective AC Response to Catecholamines by Cyclic AMP

NE and EP at low concentrations do not stimulate AC activity in malignant neuroblastoma cell homogenate (*32*); however, they stimulated the intracellular level of cyclic AMP to the same extent as that produced by DA, when the PDE activity was inhibited prior to catecholamine-treatment (Table 12). Thus it appears that an elevation of cyclic AMP level corrects the defective response of AC to NE and EP. If this is true for human cell systems, a new therapeutic model can be developed for those human diseases which result from a defective AC-response to hormones.

TABLE 13. Levels of AC, Cyclic AMP, and Cyclic AMP PDE in "Differentiated" NB Cells

Treatment	Cyclic AMP (pmoles/mg protein/hr)	AC (pmoles/mg protein/min)	Cyclic AMP PDE (pmoles/mg protein/min)
Control	12 ±1.5[a]	15±1.4	36± 2.0[b]
"Differentiated" (RO20-1724)	42.3±4.4	21±1	103±12
"Differentiated" (PGE_1)	47.1±5.3	22±1.5	88± 5.2

The data were summarized from previous publications (*32, 34, 35*).
[a] Standard deviation. [b] A higher value in a previous publication (*34*) was due to a calculation error.

Working Hypothesis for Malignancy of Nerve Cells

Our data indicate (*32, 34, 35*) that the levels of AC, cyclic AMP, PDE, and the sensitivity of AC to DA and NE increase during cyclic AMP-induced "differentiation" of mouse NB cells in culture (Table 13); therefore, the reverse may be true during malignant transformation of nerve cells. However, the PDE activity in NB cells remains unusually high. The fact that DA which stimulates AC activity can not increase cyclic AMP level until the PDE activity is inhibited and coupled with the observation that NE and EP which do not increase AC activity increases cyclic AMP level when the PDE activity is inhibited indicates that the persistence of unusually high level of PDE activity may account for the low levels of cyclic AMP and AC, and for the insensitivity of AC to catecholamines. Therefore, we propose a working hypothesis that one of the primary lesions during malignant transformation of nerve cells is an abnormal regulation of PDE activity which allows the expression of high amounts of this enzyme. The high PDE activity then leads to low levels of cyclic AMP and AC, and to the insensitivity of AC to catecholamines. It is interesting to note that the AC and PDE levels in adult hamster cerebrum increases 2 fold in comparison to newborn animals (*55*) and the brain AC undergoes an age-dependent increase in its sensitivity to NE (*48*).

Differentiation of Human NB Cell Culture and Cyclic AMP

Like mouse NB tumors (*1*, *43*) human tumors contain more than one neuronal cell type (*41*). Human NB cells in culture also show morphological differentiation after treatment with dbcAMP (2.0 mM), papaverine (1–10 μg/ml) (Fig. 4B) and PGE_1 (10 μg/ml). The time of expression and extent of differentiation are concentration dependent. The combination of PGE_1 (10 μg/ml) and papaverine (2.5 μg/ml) allows the expression of differentiated phenotype much earlier and to a much greater extent than that produced by each individual agent. Among all cyclic AMP agents, papaverine was most potent in causing morphological differentiation. Sodium butyrate, a degradative product of dbcAMP in solution, induced neurite formation in a dose-dependent fashion (Fig. 4C). This is in contrast to mouse cells (*33*) in which sodium butyrate reversibly inhibits the cell division without causing morphological differentiation. SFM (Fig. 4D), 5-BrdU (Fig. 4E), and X-ray also induced neurite formation similar to that observed in mouse cells. Our recent study (unpublished observation) shows that the sodium butyrate increased the cyclic AMP level by about 2 fold: but SFM, X-ray, and 5-BrdU did not. Thus like in mouse cells, the neurites in human NB cells can be induced by agents which do not change the intracellular level of cyclic AMP. The control culture had extremely low level of TH (5 ± 2.0 pmoles/mg protein) but dbcAMP (0.5 mM) and sodium butyrate (0.5 mM) treated cells had TH activity about 550 ± 102 and 72 ± 19 pmoles/mg protein, respectively. Thus many of responses of human NB cells to various differentiating agents are similar to those observed in mouse NB cells.

Suggestion of a New Experimental Therapeutic Model for NB Tumor

Based on our studies we suggest a new experimental therapeutic model for the treatment of NB tumor in addition to existing modalities. This involves an administration of sodium butyrate followed by an injection of L-dopa, a precursor of dopamine, in the presence of an inhibitor of cyclic AMP phosphodiesterase. The following experimental data support the above model: (a) an elevation of cyclic AMP induces irreversibly many differentiated functions in human and mouse NB cells in culture; (b) DA increases the intracellular level of cyclic AMP in the presence of an inhibitor of PDE by stimulating the AC activity; (c) in human NB cells, sodium butyrate causes cell lethality, neurite formation, and an elevation of tyrosine hydroxylase activity; however, sodium butyrate has very little or no effect on the growth of several other mammalian non-nerve cells in culture (*15*, *16*, *51*, *56*). In addition, the AC activity of sodium-butyrate treated cells is much higher than control cells and is further stimulated by catecholamines (*32*).

CONCLUSION

The fact that dbcAMP also induces neurite formation in human neuroblasts (*24*), chick embryo dorsal root ganglion (*46*) and mouse embryo sensory ganglion (*14*) indicates that mouse NB cell culture may be a pertinent model to study the

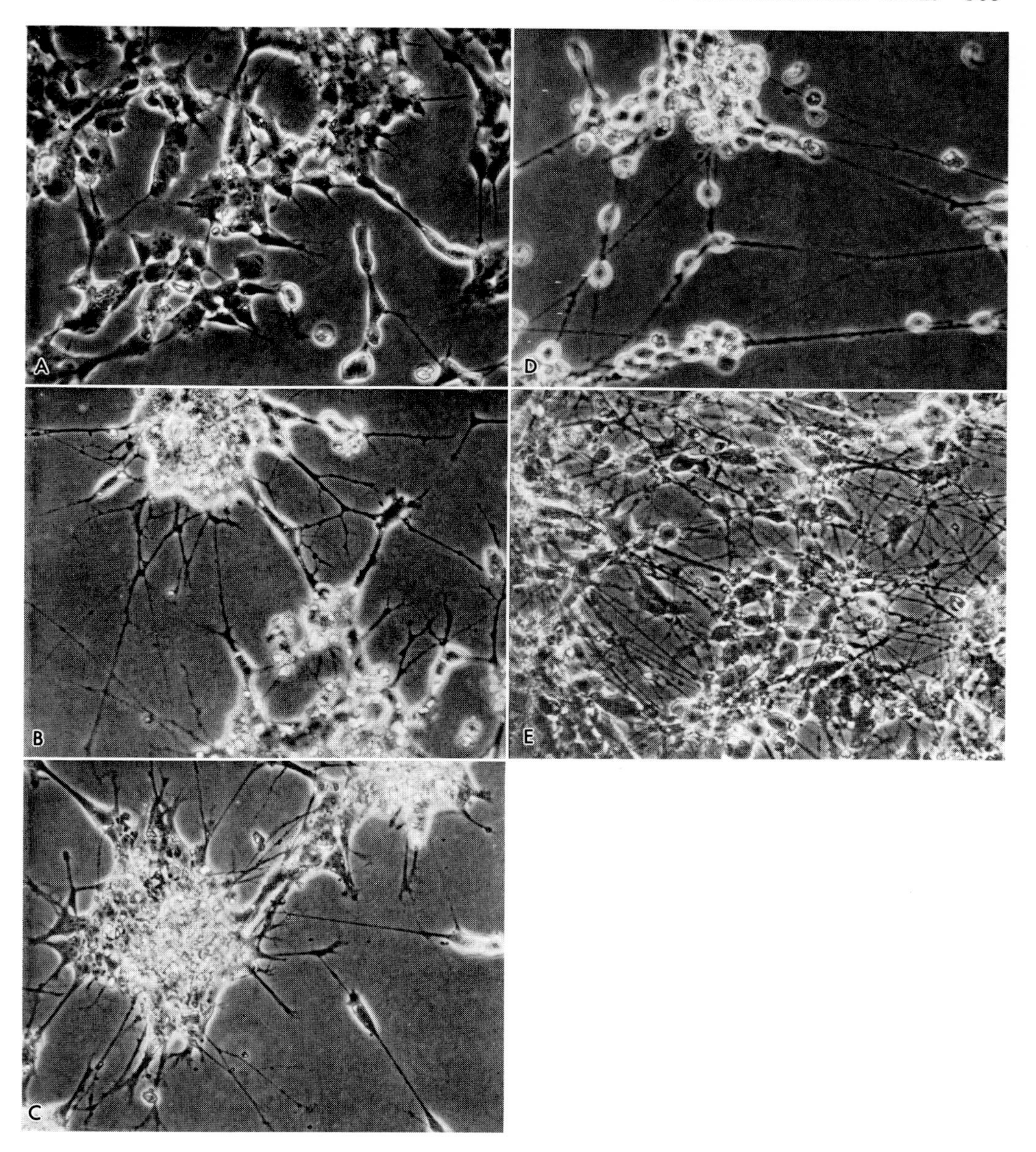

FIG. 4. Phase contrast micrographs of human NB cells in culture (IMR-32 clone). Cells were plated in Falcon plastic dishes (60 mm) and papaverine (2.5 μg/ml), sodium butyrate (0.5 mM), SFM, and 5-BrdU (2.5 μM) were added individually 4 days after plating. The drug and medium were changed every 2–3 days and the cultures were maintained for 10–13 days. Control culture (A) shows that cells grow in clumps and exhibit no spontaneous morphological differentiation (cytoplasmic processes greater than 50 μM in length). Papaverine-treated culture 10 days after treatment (B) shows the formation of extensive neurites. Many cell deaths occurred during this period. Sodium butyrate-treated culture 10 days after treatment (C) also shows the formation of extensive neurites. Some cell death occurred during this period. SFM-treated culture 3 days after treatment (D) and 5-BrdU-treated cultures 10 days after treatment (E) show an extensive neurite formation (*27*).

problem of differentiation. Among various agents which are known to induce some differentiated functions in mouse NB cells in culture: cyclic AMP is the only agent which induces at least eight differentiated functions, some of which can be induced by agents other than cyclic AMP. The neurites can be expressed in the absence of any of parameters of biochemical differentiation. The mechanism of cyclic AMP-induced "differentiation" of NB cells remains to be elucidated; however, data published thus far indicates that cyclic AMP activates both the translational and transcriptional control of gene expression and thereby regulates the expression of many differentiated functions.

ACKNOWLEDGMENT

This work was supported by US PHS NS-09230, CA-12247, and DRG-1182 from the Damon Runyon Cancer Research Fund. We thank Drs. H. Sheppard of Hoffmann-La Roche and J. E. Pike of UpJohn Company for supplying RO20-1724 and prostaglandin, respectively. We thank Miss Laurel Millhouse, Miss Katrina Gilmer, and Mrs. Marianne Gaschler for their technical help.

REFERENCES

1. Amano, T., Richelson, E., and Nirenberg, M. Neurotransmitter synthesis by neuroblastoma clones. Proc. Natl. Acad. Sci. U.S., *69*: 258–263, 1972.
2. Andén, N. E., Dahlsröm, A., Fuxe, K., and Larsson, K. Functional role of the nigroneostriatal dopamine neurons. Acta Pharmacol. Toxicol., *24*: 263–274, 1966.
3. Andén, N. E., Rubenson, A., Fuxe, K., and Hökfelt, T. Evidence for dopamine receptor stimulation by apomorphine. J. Pharm. Pharmacol., *19*: 627–629, 1967.
4. Augusti-Tocco, G. and Sato, G. Establishment of functional clonal lines of neurons from mouse neuroblastoma. Proc. Natl. Acad. Sci. U.S., *64*: 311–315, 1969.
5. Blume, A., Gilbert, F., Wilson, S., Farber, J., Rosenberg, R., and Nirenberg, M. W. Regulation of acetylcholinesterase in neuroblastoma cells. Proc. Natl. Acad. Sci. U.S., *67*: 786–792, 1970.
6. Brown, J. H. and Makman, M. H. Stimulation by dopamine of adenylate cyclase in retinal homogenates and of adenosine 3′,5′cyclic monophosphate formation in intact retina. Proc. Natl. Acad. Sci. U.S., *69*: 539–543, 1972.
7. Carlsson, A. and Lindquist, M. Effect of chlorpromazine or haloperidol on formation of 3-methoxytyramine and normetanephrine in mouse brain. Acta Pharmacol. Toxicol., *20*: 140–144, 1963.
8. Clark, R. B. and Perkins, J. P. Regulation of adenosine 3′,5′ cyclic monophsphate concentration in cultured human astrocytoma cells by catecholamines and histamine. Proc. Natl. Acad. Sci. U.S., *68*: 2757–2760, 1971.
9. Furmanski, P., Silverman, D. J., and Lubin, M. Expression of differentiated functions in mouse neuroblastoma mediated by dibutyryl cyclic adenosine monophosphate. Nature, *233*: 413–415, 1971.
10. Gilman, A. G. A protein binding assay for adenosine 3′,5′-cyclic monophosphate. Proc. Natl. Acad. Sci. U.S., *67*: 305–312, 1970.
11. Gilman, A. G. and Nirenberg, M. Effect of catecholamines on the adenosine 3′,5′-cyclic monophosphate concentrations of clonal satellite cells of neurons. Proc. Natl. Acad. Sci. U.S., *68*: 2165–2168, 1971.

12. Guidotti, A. and Costa, E. Involvement of adenosine 3′,5′ cyclic-monophosphate in the activation of tyrosine hydroxylase elicited by drugs. Science, *179*: 902–904, 1973.
13. Hardman, J. G., Robison, G. A., and Sutherland, E. W. Cyclic nucleotides. Ann. Rev. Physiol., *33*: 311–386, 1971.
14. Hass, D. C., Hier, D. B., Aranson, B. G. W., and Young, M. On a possible relationship of cyclic AMP to the mechanism of action of nerve growth factor. Proc. Soc. Exp. Biol. Med., *140*: 45–47, 1972.
15. Hsie, A. W. and Puck, T. T. Morphological transformation of Chinese hamster cells by dibutyryl adenosine cyclic 3′,5′-monophosphate and testosterone. Proc. Natl. Acad. Sci. U.S., *68*: 358–361, 1971.
16. Johnson, G. S., Friedman, R. M., and Pastan, J. Restoration of several morphological characteristics of normal fibroblasts in sarcoma cells treated with adenosine 3′,5′-cyclic monophosphate and its derivatives. Proc. Natl. Acad. Sci. U.S., *68*: 425–429, 1971.
17. Kates, J. R., Winterton, R., and Schlessinger, K. Induction of acetylcholinesterase activity in mouse neuroblastoma tissue culture cells. Nature, *229*: 345–346, 1971.
18. Kebabian, J. W. and Greengard, P. Dopamine-sensitive adenylate-cyclase: Possible role in synaptic transmission. Science, *174*: 1346–1349, 1971.
19. Kebabian, J. W., Petzold, G. L., and Greengard, P. Dopamine-sensitive adenylate cyclase in caudate nucleus of rat brain and its similarity to the "dopamine-receptor." Proc. Natl. Acad. Sci. U.S., *69*: 2145–2149, 1972.
20. Kram, R. and Tomkins, G. M. Pleiotypic control by cyclic AMP: Interaction with cyclic GMP and possible role of microtubules. Proc. Natl. Acad. Sci. U.S., *70*: 1659–1663, 1973.
21. Krishna, G., Weiss, B., and Brodie, B. B. A simple sensitive method for the assay of adenyl cyclase. J. Pharm. Exp. Therap., *163*: 379–385, 1968.
22. Lapham, L. W. Tetraploid DNA content of purkinje neurons of human cerebellar cortex. Science, *159*: 310–312, 1968.
23. Lowry, O. H., Rosebrough, N. J., Farr, A. L., and Randall, R. J. Protein measurement with the folin phenol reagent. J. Biol. Chem., *193*: 265–275, 1951.
24. MacIntyre, E. H., Perkins, J. P., Wintersgill, C. J., and Vatter, A. W. The responses in culture of human tumor astrocytes and neuroblasts to N^6,$O^{2'}$-dibutyryl adenosine 3′,5′-cyclic monophosphoric acid. J. Cell Sci., *11*: 639–667, 1971.
25. Perkins, J. P. and Moore, M. M. Adenylate cyclase of rat cerebral cortex. J. Biol. Chem., *246*: 62–68, 1971.
26. Prasad, K. N. X-ray-induced morphologic differentiation of mouse neuroblastoma cells *in vitro*. Nature, *234*: 471–474, 1971.
27. Prasad, K. N. Effect of dopamine and 6-hydroxydopamine on the mouse neuroblastoma cells *in vitro*. Cancer Res., *31*: 1457–1460, 1971.
28. Prasad, K. N. Morphological differentiation induced by prostaglandin in mouse neuroblastoma cells in culture. Nature New Biol., *236*: 49–52, 1972.
29. Prasad, K. N. Cyclic AMP-induced differentiation mouse neuroblastoma cells lose tumourigenic characteristics. Cytobios, *6*: 163–166, 1972.
30. Prasad, K. N. Role of cyclic AMP in the differentiation of neuroblastoma cell culture. *In;* H. Gratzner and J. Schults (eds.), The Role of Cyclic Nucleotides in Carcinogenesis (Miami Winter Symposia), vol. 6, pp. 207–237, New York, 1973.
31. Prasad, K. N. Differentiation of neuroblastoma cells induced in culture by 6-thioguanine. Int. J. Cancer, *12*: 631–636, 1973.

32. Prasad, K. N. and Gilmer, K. N. Demonstration of dopamine-sensitive adenylate cyclase in malignant neuroblastoma cells and change in sensitivity of adenylate cyclase to catecholamines in "differentiated" cells. Proc. Natl. Acad. Sci. U.S., *71*: 2525–2529, 1974.
33. Prasad, K. N. and Hsie, A. W. Morphological differentiation of mouse neuroblastoma cells induced *in vitro* by dibutyryl adenosine 3′: 5′-cyclic monophosphate. Nature New Biol., *233*: 141–142, 1971.
34. Prasad, K. N. and Kumar, S. Cyclic 3′,5′-AMP phosphodiesterase activity during cyclic AMP-induced differentiation of neuroblastoma cells in culture. Proc. Soc. Exp. Biol. Med., *142*: 406–409, 1973.
35. Prasad, K. N. and Kumar, S. Cyclic AMP and the differentiation of neuroblastoma cells. *In;* B. Clarkson and R. Baserga (eds.), Control of Proliferation in Animal Cells, pp. 581–594, Cold Spring Harbor Laboratory, New York, 1974.
36. Prasad, K. N. and Mandal, B. Choline acetyltransferase level in cyclic AMP and X-ray induced morphologically differentiated neuroblastoma cells in culture. Cytobios, *8*: 75–80, 1973.
37. Prasad, K. N. and Mandal, B. Catechol-O-methyltransferase activity in dibutyryl cyclic AMP, prostaglandin and X-ray induced differentiated neuroblastoma cell culture. Exp. Cell Res., *74*: 532–535, 1972.
38. Prasad, K. N. and Sheppard, J. R. Inhibitors of cyclic nucleotide phosphodiesterase induced morphological differentiation of mouse neuroblastoma cell culture. Exp. Cell Res., *73*: 436–440, 1972.
39. Prasad, K. N. and Vernadakis, A. Morphologic and biochemical study in X-ray and dibutyryl cyclic AMP-induced differentiated neuroblastoma cells. Exp. Cell Res., *70*: 27–32, 1972.
40. Prasad, K. N., Gilmer, K., and Kumar, S. Morphologically "differentiated" mouse neuroblastoma cells induced by noncyclic AMP agents: levels of cyclic AMP, nucleic acid and protein. Proc. Soc. Exp. Biol. Med., *143*: 1168–1171, 1973.
41. Prasad, K. N., Mandal, B., and Kumar, S. Demonstration of cholinergic cells in human neuroblastoma and ganglioneuroma. J. Pediat., *82*: 677–679, 1973.
42. Prasad, K. N., Kumar, S., Gilmer, K., and Vernadakis, A. Cyclic AMP-induced differentiated neuroblastoma cells: changes in total nucleic acid and protein contents. Biochem. Biophys. Res. Commun., *50*: 973–977, 1973.
43. Prasad, K. N., Mandal, B., Waymire, J. C., Lees, G. J., Vernadakis, A., and Weiner, N. Basal level of neurotransmitter synthesizing enzymes and effect of cyclic AMP agents on the morphological differentiation of isolated neuroblastoma clones. Nature New Biol., *241*: 117–119, 1973.
44. Prasad, R., Prasad, N., and Prasad, K. N. Esterase, malate and lactate dehydrogenesis activity in murine neuroblastoma. Science, *181*: 450–451, 1973.
45. Richelson, E. Stimulation of tyrosine hydroxylase activity in an adrenergic clone of mouse neuroblastoma by dibutyryl cyclic AMP. Nature New Biol., *242*: 175–177, 1973.
46. Roisen, F. J., Murphy, R. A., Pichichero, M. E., and Braden, W. G. Cyclic adenosine monophosphate stimulation of axonal elongation. Science, *175*: 73–74, 1972.
47. Sakamoto, A. and Prasad, K. N. Effect of DL-glyceraldehyde on mouse neuroblastoma cell culture. Cancer Res., *32*: 532–534, 1972.
48. Schmidt, M. J. and Robison, G. A. Cyclic AMP, adenyl cyclase and the effect of norepinephrine in the developing rat brain. Fed. Proc., *29*: 479, 1970.

49. Schubert, D. and Jacob, F. 5-Bromodeoxyuridine-induced differentiation of a neuroblastoma. Proc. Natl. Acad. Sci. U.S., *67*: 247–254, 1970.
50. Seeds, N. W., Gilman, A. G., Amano, T., and Nirenberg, M. W. Regulation of axon formation by clonal lines of a neural tumor. Proc. Natl. Acad. Sci. U.S., *66*: 160–167, 1970.
51. Sheppard, J. R. Restoration of contact-inhibited growth to transformed cells by dibutyryl 3′,5′-monophosphate. Proc. Natl. Acad. Sci. U.S., *68*: 1316–1320, 1971.
52. Sheppard, J. R. and Prasad, K. N. Cyclic AMP levels and morphological differentiation in mouse neuroblastoma cells. Life Science, *12*: 431–439, 1973.
53. Ungerstedt, U., Butcher, L. L., Butcher, S. G., Andén, N. E., and Fuxe, K. Direct chemical stimulation of dopaminergic mechanisms in the neostriatum of the rat. Brain Res., *14*: 461–471, 1969.
54. Waymire, J. C., Weiner, N., and Prasad, K. N. Regulation of tyrosine hydroxylase activity in cultured mouse neuroblastoma cells. Elevations induced by analogs of adenosine 3′,5′-cyclic monophosphate. Proc. Natl. Acad. Sci. U.S., *69*: 2241–2242, 1972.
55. Weiss, B., Shein, H. M., and Snyder, R. Adenylate cyclase and phosphodiesterase activity of normal and SV_{40} virus transformed hamster astrocytes in cell culture. Life Science, *10*: 1253–1260, 1971.
56. Wijk, R. V., Wicks, W. D., and Clay, K. Effect of derivatives of cyclic 3′,5′-adenosine monophosphate on the growth, morphology and gene expression of hepatoma cells in culture. Cancer Res., *32*: 1905–1911, 1972.

Discussion of Paper by Drs. Prasad et al.

DR. JOHNSON: We have studied the effects of N^6-substituted adenine derivatives on the morphology of L-929 fibroblasts. We find that they induce cell elongation very rapidly and dramatically. Interestingly they do not significantly alter the intracellular cyclic AMP levels. The mechanism of action of these compounds is unknown at the present time but they should be useful in elucidating the actions of cyclic AMP on these cells.

DR. PRASAD: This is interesting. We have also found that certain differentiated functions in NB cells can be expressed in the absence of any change in cyclic AMP level.

DR. OHNO: The dogma is one neuron, one neural transmitter, yet you had clones which apparently synthesize 2 transmitters. Now you make this type of clone differentiate by whatever means, do these in a differentiated state still synthesize both transmitters?

DR. PRASAD: Yes, the TH is increased by 50-fold, and ChA is increased 3-fold. Therefore our data indicate that both neurotransmitters can be synthesized in the same neuron. I don't know whether such neurons exist in the mammalian nervous system or they reflect an abnormal regulation of neural enzymes in malignant nerve cells. Since the basal level of TH in such clone is very low, the same investigators have ignored the existence of clone having both neurotransmitter synthesizing enzymes. However, we have been able to stimulate TH activity in such clone by 50-fold and therefore we feel that both neurotransmitters can be synthesized in the same clone of mouse NB cells.

DR. WEBER: At the January 1973 Cyclic AMP Conference in Miami my report was perhaps the only one to demonstrate that Na butyrate had marked, dose-dependent action on hepatomas in tissue culture. Thus, you may wish to modify your present claim that as you said, "as you know Na butyrate has no effect on growth rate of any other cultured cells and thus appears specific to NB." As you recall my lecture, published several months ago, Na butyrate inhibited in a dose-dependent manner and even killed some of the hepatoma cell lines. I also reported on certain of the biochemical effects of Na butyrate which involves inhibition of key glycolytic enzymes.

DR. PRASAD: In the same meeting we also reported the effect of Na butyrate on several parameters of mouse NB cell in cultures. I did not mean to imply that Na butyrate has no effect on any other mammalian cells in culture. We know from work of other investigators that Na butyrate has no effect on the growth rate of Chinese hamster ovary cells, fibroblasts, and at least one clone of hepatoma used by Dr. Wieks of University of Colorado Medical Center.

DR. ONO: Have you ever tested fatty acid other than butyrate, for example, propionic acid or a longer chain? In our cell system, mouse mammary carcinoma cell, butyrate is the best inducer of alkaline phosphatase, but other fatty acids also exhibit induction.

DR. PRASAD: I am glad you have butyrate-effect on alkaline phosphatase. We have tried the effect of γ-aminobutyric acid and β-hydroxybutyric acid, and found no effect on growth rate or enzyme activity. We have not tried other fatty acid. We will try in the future.

DR. HOLTZER: Is there any evidence of a mature nerve cell being transformed? My understanding is that such cells cannot be transformed. In brief it is likely that in your cultures you really have a mixture of phenotypes and that the cell that propagates the tumor may not contain any of the enzymes that are unique to nerve cells. This point strikes me as of some importance simply because it shows that the presumptive nerve cell is very different from its daughter nerve cells.

DR. PRASAD: There is no evidence that mature nerve cells ever transform to malignant form. In my opinion, the malignant transformation may occur before the maturation of the nerve cell. Our NB clones have neural specific enzymes, therefore they indeed are nerve cells. There are some clones which have no specific neural enzyme. We do not know if they represent presumptive nerve cells or they are like sensory neurons which do not have either TH or ChA. I agree with you that the presumptive nerve cells are different from daughter nerve cells.

Dr. Prasad: There is no evidence that mature nerve cells ever transform [illegible]. In my opinion [illegible] contamination may occur [illegible] of the [illegible] clones [illegible] enzymes, therefore [illegible] which have no specific neural [illegible]. We do not [illegible] which do not have [illegible]. I agree with you that the [illegible] nerve cells are different from [illegible] nerve cells.

Variants of Yoshida Ascites Sarcoma and α-Fetoprotein

Hidehiko Isaka,[*1,*2] Hiroshi Satoh,[*1] and Hidematsu Hirai[*3]

Sasaki Institute, Tokyo, Japan[*1] *and Department of Biochemistry, School of Medicine, Hokkaido University, Sapporo, Japan*[*3]

Abstract: Yoshida sarcoma cells change their biological characteristics, when transplanted in immunologically conditioned Donryu rats, giving rise to variant cancer cell populations. The variants are characterized by decreased growth rate, epithelial character of the cells and altered chromosome patterns. The variant cells showed a production of α-fetoprotein (AFP), although the original tumor cells did not.

Population analysis of the Yoshida sarcoma was performed, as regards AFP production. The results were suggestive of the presence of AFP-producing cells in the original Yoshida sarcoma, with variable production of AFP. The production of AFP by cultured Yoshida sarcoma cells was accelerated by cyclic AMP treatment.

LY-variant tumors of the so-called Yoshida ascites sarcoma were first observed by Satoh (*13*) in 1961. The variants are characterized by decreased growth rate and epithelial character of the cells. Since then, many LY variants of this kind have been produced from Yoshida sarcoma (*14*) and more than 10 transfer lines are maintained in this laboratory at the present. The main problems raised by these variants are concerned with the origin of Yoshida sarcoma and the differentiation of cancer cells. The question of the hepatic cell origin of Yoshida tumor seems to have been clarified by morphological studies of the variants (*20*, *21*). Production of α-fetoprotein (AFP) by the variant cells may provide further support for the liver cell origin of the tumor (*18*). Biochemical studies on the hexokinase pattern of LY-variant cells showed some significant findings which may shed light on cancer differentiation (*17*). However, there still remain some problems, for example, the real cause(s) or factor(s) inducing such variants and the exact mechanism responsible for the decreased growth rate of these variants.

---

*2 Present address: Department ot Pathology, Kagoshima University, School of Medicine, Kagoshima, Japan.

This paper describes the development and characteristics of LY variants and their AFP production, as well as some related findings.

Development of LY Variants

Eight LY variants of the Yoshida sarcoma were developed under the following conditions.

LY-336: Twelve Donryu rats were transplanted intraperitoneally with 0.3 ml of mouse leukemia SN-36 (*11*). The leukemic cells proliferated well in the animals for the first week but disappeared later. Eighty-four days after heterotransplantation, 10^3 cells of Yoshida sarcoma were injected intraperitoneally in the same animals. The cells failed to grow and the animals survived. This means that a kind of cross-immunity in transplantation was established between the SN-36 mouse leukemia and the Yoshida sarcoma of the rat. In order to determine the limit of this transplantation resistance, further repeated challenges were made with increasing numbers of Yoshida sarcoma cells (Table 1). The 5th challenge resulted in the growth of Yoshida sarcoma, but the ascites picture was not that of the original tumor. The original Yoshida sarcoma was composed of individually isolated tumor cells in ascites (Fig. 1). However, the successfully growing tumors after the 5th challenge with Yoshida sarcoma showed cell clusters or "island" formation in ascites, just like ascites hepatoma (Fig. 2). This transformation of the ascites picture was observed in 5 out of the 12 series of rats (Table 2). These new types of Yoshida sarcoma were designated as LY variants or LY tumors and 3 of them were maintained by animal passage in normal Donryu rats. An unusually prolonged survival

TABLE 1. Unsuccessful Transplantation of Tumors in Donryu Rats[a]

Transplantation			
1st	2nd	3rd	4th
SN-36[b] (0.3 ml)	YS[c] (10^3)	YS[c] (10^4)	YS[c] (10^7)

[a] Twelve animals per group were treated in the same way. [b] SN-36: mouse leukemia with ascitic growth (amount of ascites transplanted). [c] YS: Yoshida sarcoma (No. of cells transplanted).

TABLE 2. Successful Transplantation of Yoshida Sarcoma in Pretreated Rats[a]

Cell lines	Rats, sex (g)	Days after 4th pretreatment	No. of cells	Latent days	Survived days[b]
LY-306	f 90	21	10^5	12	84
LY-336	f 80	55	10^5	3	145
LY-346	f 80	55	10^6	3	289
LY-320	f 100	21	10^6	2	142
LY-323	f 88	55	10^5	3	245

[a] Successful transplantation occurred in 5 out of 12 pretreated rats. Island formation was observed in all successfully growing tumors. [b] These rats survived a remarkably long time as compared with the 7-day average survival of Yoshida sarcoma rats.

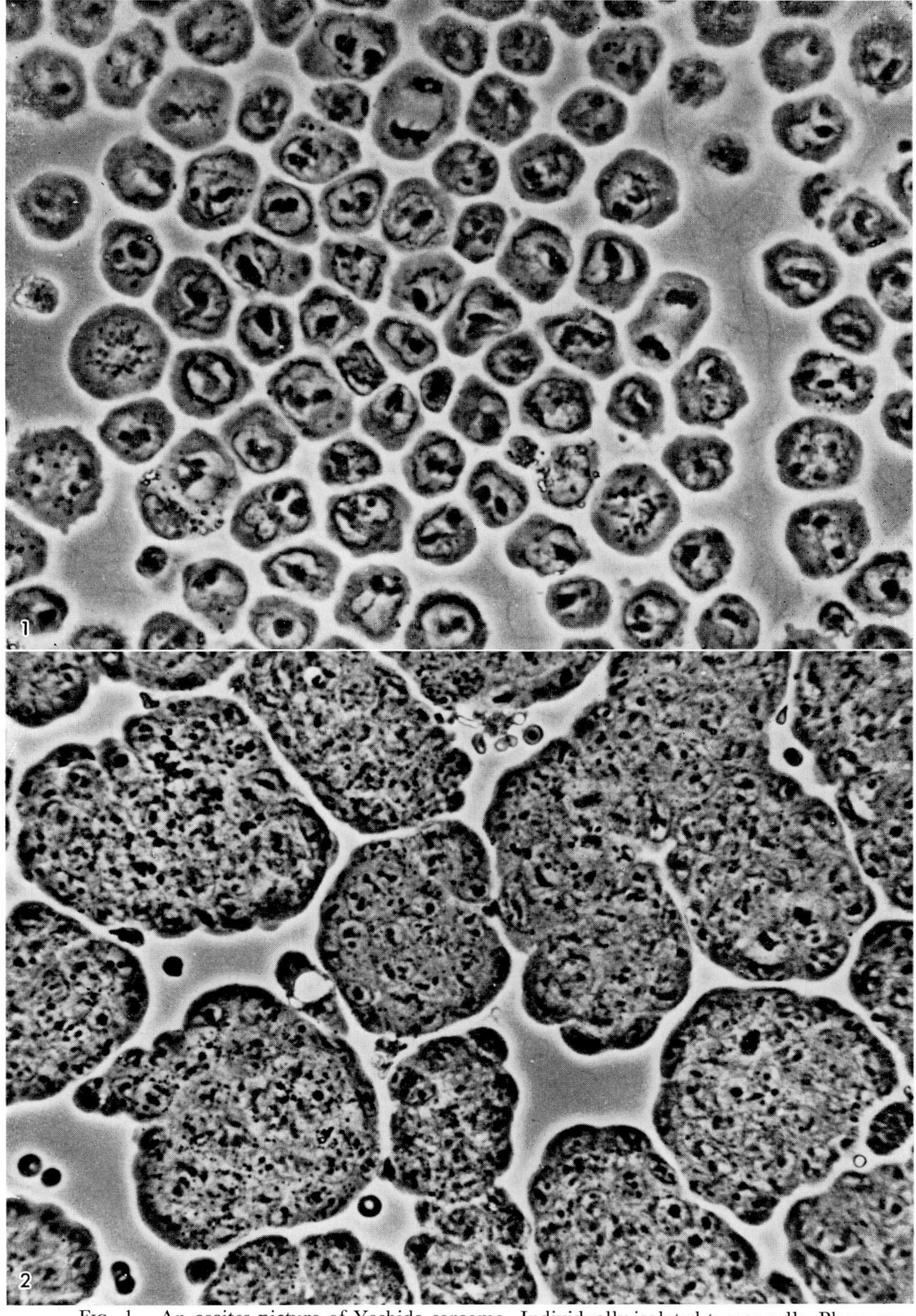

FIG. 1. An ascites picture of Yoshida sarcoma. Individually isolated tumor cells. Phase contrast.

FIG. 2. An ascites picture of LY-7. Huge tumor-cell islands in ascites. Phase contrast.

TABLE 3. Survival Days of Rats Bearing LY Tumors[a]

Cell lines	Transfer generations									
	1	2	3	4	5	6	7	8	9	10 →
LY-306	144	101	148	60	51	41	82	115	126	276 →
LY-336	195	135	113	177	51	186	—[b]	305	168	266 →
LY-346	58	98	226	95	275	88	123	31	58	43 →

[a] In every transfer generation of every cell line, the tumor was transplanted only in one rat. [b] Complete regression of the tumor occurred after transplantation to the next recipient.

of rats in which this transformation had taken place was seen (Table 2). Autopsy findings of the animals revealed no sign of spontaneous tumor incidence in any peritoneal organs and tissues. Table 3 shows the unusually decreased growth rate of the 3 LY variants for the first 10 transfer generations. Of the 3 lines, LY-336 has been serially maintained up to now. The development of LY-336 has been reported in detail (*20*).

LY-7: A polyploid subline of the Yoshida sarcoma was transplanted in a Donryu rat intraperitoneally. The polyploid tumor cells proliferated for several days but failed to grow later. In this rat, 10^7 cells of Yoshida sarcoma were transplanted intraperitoneally 83 days after the inoculation of polyploid Yoshida sarcoma. Thirty days after the last transplantation, island-forming tumor cells were seen in accumulating ascites in the rat. Details have been published elsewhere (*20*).

LY-52, LY-54: The LY transformation occurred from Yoshida sarcoma in the host rat to which immunological conditioning was applied. Based on this finding, Satoh pretreated normal Donryu rats with an intraperitoneal injection of rat materials as follows: the circulating blood was taken from Donryu rats in which Yoshida sarcoma had failed to proliferate. The blood was injected intraperitoneally into normal rats in a group, 1 ml per rat. Then, each of them received an intraperitoneal inoculation of Yoshida sarcoma (10^7 tumor cells) 4 weeks after the blood injection. One of the rats showed abundant island-forming tumor cells in ascites 6 days after the inoculation and died of ascites tumor 11 days later. Autopsy of this animal revealed about 10 ml of dense, hemorrhagic ascites containing numerous islands and their growth in the mesentery and hilar portions of the liver, as well as retroperitoneal lymph nodes. In another rat of the same group similar island-forming cells were detected in accumulating ascites, with similar autopsy findings. The tumor ascites of these rats were transplanted successfully and designated as LY-52 and LY-54, respectively.

LY-80: A large amount of physiological saline was injected, under high pressure, into the portal vein of a normal Donryu rat in order to obtain liver containing the minimum amount of blood. Then the liver was extirpated and homogenized. One milliliter of the liver homogenate was injected into the abdominal cavity of normal Donryu rats and 10^5 Yoshida sarcoma cells were inoculated intraperitoneally into each of the same animals 4 weeks later. One rat in this group showed a slightly extended abdomen due to ascites, including numerous island cells, 6 days after the inoculation of Yoshida sarcoma. Autopsy disclosed no sign of spontaneous tumor development except for the ascites tumor of island tumor cells.

LY-5, LY-6, LY-1: These LY variants started from 3 different animals in the same experimental group. The group consisted of Donryu rats in which the ascites hepatoma of Wistar rat, AH-34, regressed 2 weeks after the intraperitoneal transplantation of 10^6 AH-34 cells and disappeared entirely from the ascitic fluid. Ten million cells of Yoshida sarcoma were injected intraperitoneally into these rats at various times after the regression of AH-34. Island-forming cells, LY-5, appeared in the ascitic fluid of a rat 93 days after the disappearance of AH-34 cells. The animal survived for 201 days with a markedly distended abdomen due to accumulated ascites. Two other variants of the island type developed in rats into which Yoshida sarcoma cells were injected 55 and 72 days after the regression of AH-34 cells, respectively. These animals died from ascites of island-forming cells 159 and 98 days later, respectively. Variant cells were also subjected to serial transplantations in normal Donryu rats.

Reproducibilities of such LY variants from Yoshida sarcoma are very low, even if the above procedures are applied to Donryu rats. It is evident that immunological conditioning had been given to the rats in which LY transformation occurred. However, the immunologically conditioned state of rats is difficult to analyse. It may be noteworthy that the original Yoshida sarcoma, from which the present LY variants developed, is a clonal cell line derived from a single Yoshida sarcoma cell inoculated in a rat. Analogues of such variants were described by Ishidate and Isaka (*9*).

Characteristics of LY Variants

One of the most characteristic findings of LY variants is the decreased growth rate. Table 4 shows the mean survival time of host rats bearing 8 LY variants and of the original Yoshida sarcoma during serial transfers. The LY-variant animals survived for a remarkably long time compared with the 7-day average survival of Yoshida-sarcoma animals. The generation time *in vitro* of LY variant cells was also prolonged markedly (*10*).

TABLE 4. Growth of Yoshida Sarcoma and Its LY Variants

Cell lines	Takes (%)	Mean survival (days)	Ascitic picture[a]
LY-336	95	48	Island type
LY-1	82	45	Island type
LY-5	96	44	Mixed type
LY-7	100	19	Island type
LY-6	30	40	Island type
LY-52	100	12	Mixed type
LY-54	100	13	Mixed type
LY-80	100	13	Mixed type
Yoshida sarcoma	100	7	Free cell type

[a] Island type: almost all the cells form islands. Free cell type: individually isolated cells. Mixed type: a mixture of islands and free cells.

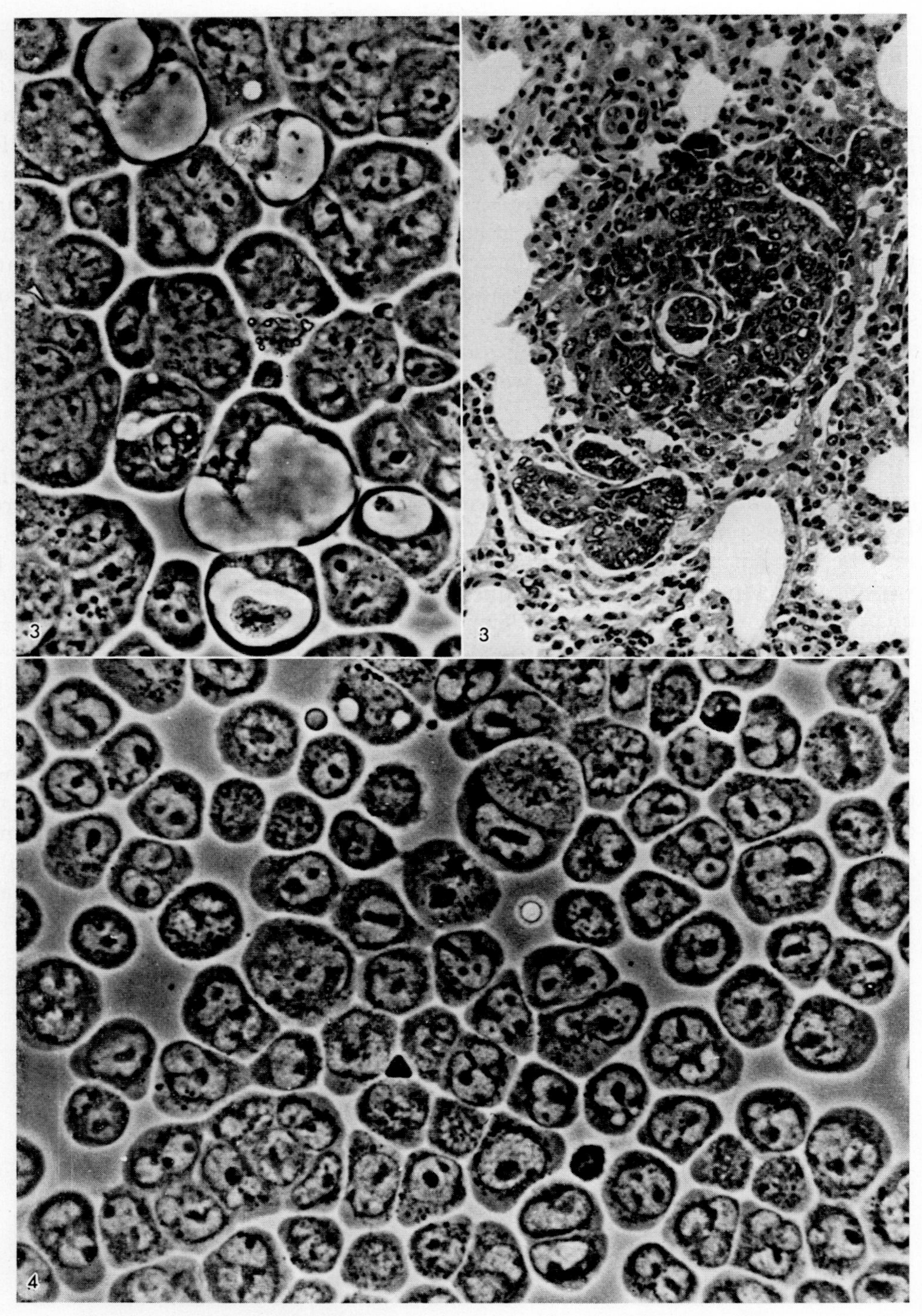

FIG. 3. Tumor-cell islands with lumen in ascites of LY-336 (left, phase contrast) and an adenocarcinomatous pattern of LY-336 cells in the lung metastasis (right, Hematoxylin-Eosin (H-E) stain).

FIG. 4. Small islands and free tumor cells in ascites of LY-54. Phase contrast.

The ascitic picture of LY variants differs considerably from that of Yoshida sarcoma. As stated above, the variant cells make islands just like the ascites hepatoma. According to the size and frequency of islands in ascites, 2 categories could be seen, *i.e.*, island type and mixed type, as shown in Table 4. The island-type LY variant, in which almost all the cells form numerous and large islands in ascites, shows a typical adenocarcinomatous pattern of cells pathologically in invasive and metastatic lesions (Fig. 3). The mixed type is a mixture of abundant free tumor cells and small-sized islands of lower frequency (Fig. 4). In contrast, the Yoshida sarcoma is a free cell-type tumor in which all the tumor cells are individually isolated. Electron microscopy of the variants revealed the development of terminal bar desmosomes in each of neighboring cells constituting the same island (*16*). These findings indicate that the variant cells have authentic epthelial characteristics. Characteristics of the ascites picture have been maintained up to now in each variant cell line.

Karyological examinations of the 8 LY variants (*3*) showed that the modal chromosome number was 42 in LY-52, LY-80, and LY-54; 54 in LY-5; 55 in LY-6; 58 in LY-336; 59 in LY-7; and 66 in LY-1; whereas it was 40 in the original Yoshida sarcoma (Fig. 5). The variant lines had larger chromosomes in the mode. Typical

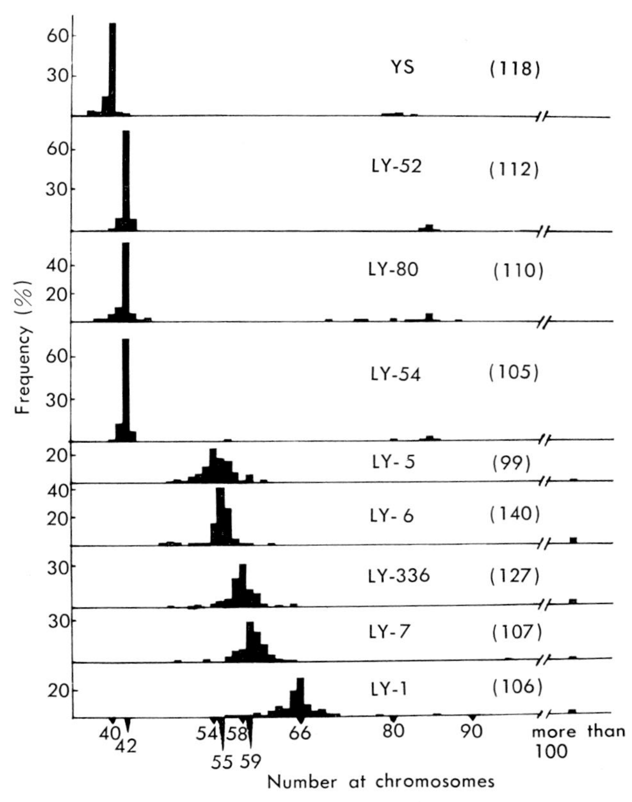

FIG. 5. Distribution of chromosome number in Yoshida sarcoma (YS) and LY variants. The number of cells examined is shown in parentheses.

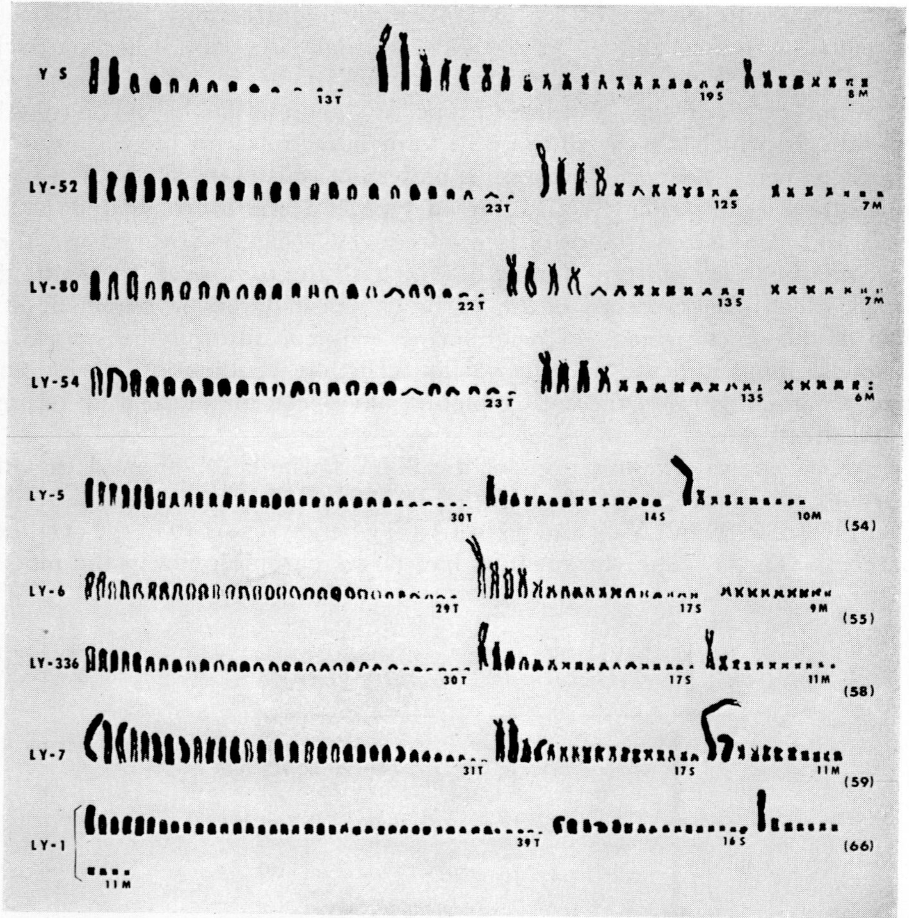

FIG. 6. Metaphase chromosomes of Yoshida sarcoma (YS) and LY variants. Arranged from left to right in groups of the same type; telocentrics (T), submeta- and subtelocentrics (S), and metacentrics (M), in order of decreasing size. The chromosome number of each group is shown.

metaphase chromosomes in the modal region of each variant line are indicated in Fig. 6. Chromosomes were classified into 3 types, telocentrics (T), metacentrics (M), and submeta- and subtelocentrics (S) in the figure. Karyotypes of LY variants differed from each other and from the Yoshida sarcoma. There was, however, a similarity in the chromosome patterns of 3 LY variants, LY-52, LY-80, and LY-54, while there were considerable dissimilarities in those of the remaining variants. The most conspicuous differences are seen in numbers of larger S and M chromosomes with different shape and size. These findings may imply that all the LY variants are different mutants of Yoshida sarcoma, considering that karyotype alteration of tumor cells is thought to be a sign of mutational changes of tumor cells.

Sensitivity of LY-variant cells to nitrogen mustard N-oxide was determined *in vivo* (*15*). Minimum effective dose of the drug (*8*) is a criterion for the drug sensitivity of cells. It was demonstrated that the LY transformation of Yoshida

TABLE 5. Sensitivity to Nitrogen Mustard N-oxide in Yoshida Sarcoma and LY Variants

Cell lines	MED[a] (mg/kg)	MTD[b] (mg/kg)
Yoshida sarcoma	1	50
LY-336	5	50
LY-5	25	50
LY-7	25	50
LY-1	50	50

[a] The minimum amount of the drug required to induce nitrogen mustard effects on half the tumor cells. [b] The maximum tolerance dose of the drug.

sarcoma cells resulted in decreased sensitivity to nitrogen mustard N-oxide as indicated in Table 5.

Production of AFP by the LY variants was studied for the first time by Watabe and Hirai (*18*). Ouchterlony's precipitation reaction of LY tumor materials showed "positive" in 4 out of 7 lines, and ^{125}I radioimmunoassay by the double-antibody method of Nishi and Hirai (*12*), was positive in all 7 cases (*19*). However, the Yoshida sarcoma as a whole cell population has never shown any sign of AFP production.

These results suggested that the epithelial-cell nature of Yoshida sarcoma, which had been masked for a long time, was unmasked by a presently unknown effect and that the original epithelial character was manifested, with accompanying chromosomal alteration and slowdown of growth rate. Based on these facts, Yoshida (*6*, *20*, *21*) assumed that the Yoshida sarcoma was not a sarcoma but a free cell tumor of hepatic cell origin, when considered together with the developmental history of the tumor, data on the ascitic conversion of azo dyeinduced hepatomas into free cell-type tumors and development of island-type variants from free cell-type ascites hepatomas. It was considered that AFP production by LY-variant cells might be a problem of gene expression for the AFP production of Yoshida tumor cells (*2*). AFP is a product of hepatic cells of embryonal, neonatal and cancerous states, as well as of some regenerating liver, at least in the rat (*1*). Hence the producibility of AFP in Yoshida tumor cells and their free cell-type variants was studied.

Production of AFP by Free Cell-type Variants of Yoshida Tumor

Isaka *et al.* (*4*) established 3 further variant lines of Yoshida sarcoma, characterized by lowered growth and production of AFP, in 1971. They cultivated both cloned and noncloned Yoshida tumor cells *in vitro* for almost one year. Cultures consisted of subcultures treated with carcinogenic substances including 4-nitroquinoline 1-oxide. Details of this experiment have been reported elsewhere (*4*). At various times after the beginning of culture, cloned cells were transplanted to normal Donryu rats intraperitoneally or intralienally. Thus 17 cell lines of Yoshida tumor were established by animal passage and 3 of them were AFP-producing. Table 6 indicates the doubling time measured *in vitro* and the survival of host rats

TABLE 6. Growth of AFP-producing Variant Tumor Cells of Yoshida Ascites Sarcoma

Variant cell lines	Doubling time *in vitro* (hr)	Survival days of host rats
SB-7-2	28.2	12
UP	14.7	11
UO	44.1	14
Yoshida sarcoma (control)	20.9	8

TABLE 7. Concentration of AFP in Rats and Culture Medium of Variant Cells of Yoshida Sarcoma

	SB-7-2	UP	UO
Serum[a]	0.4	0.24	0.1
Ascites[a]	0.2	0.1	0.06
Culture Medium[b]	1.2×10^{-9} (0.23)[c]	6.3×10^{-10} (0.12)[c]	3.1×10^{-10} (0.058)[c]

[a] Collected from the host rats 7 days after transplantation. Inoculum, 10^7 cells. Numbers indicate mg AFP in 100 ml serum or ascites. [b] Collected 7 days after cultivation of cells in Eagle's minimum essential medium supplemented with 10% calf serum. Inoculum, 5×10^4 cells/ml. Numbers indicate mg AFP produced per cell per day. [c] Numbers indicate mg AFP in 100 ml culture medium.

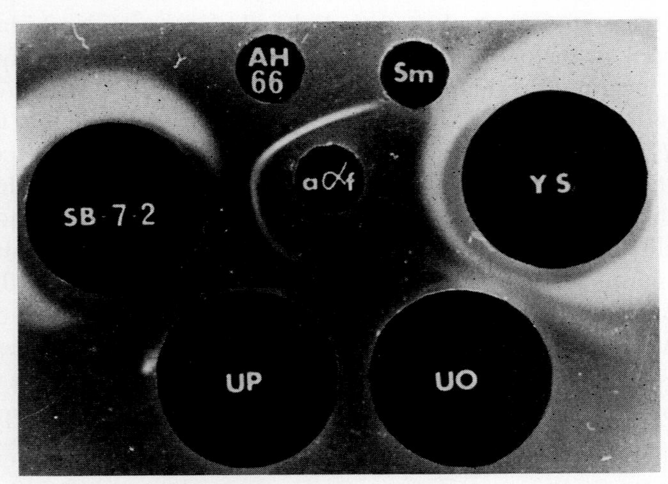

FIG. 7. Double-immunodiffusion pattern of the sera of rats bearing ascites hepatoma AH-66 (AH 66), 3 free-cell variants of Yoshida sarcoma (SB-7-2, UP, and UO), Yoshida sarcoma (YS), and normal adult rat serum (Sm). Anti-AFP horse serum is shown by $a\alpha_f$. The fused precipitation lines represent antigenic identity and absence of reaction between the antiserum and Yoshida sarcoma or normal rat serum.

of the 3 cell lines. Marked slowdown of growth was noted, compared to the original Yoshida sarcoma. Variant cells of the 3 lines grow in ascitic fluid as isolated cells and they have never shown any island formation. They are different, cytogenetically altered cells. Production of AFP by these variant cells is shown in Table 7 and Fig. 7. No correlation between island formation and AFP production of cells was demonstrated in the Yoshida ascites tumor (*6*).

Isolation of AFP-producing Cells from Yoshida Ascites Tumor

Population analysis of the original Yoshida sarcoma was made as regards the producibility of AFP, in order to learn whether any AFP-producing cells were present originally. A soft agar culture of Yoshida tumor cells was made for this purpose and examined for the development of clonal colonies from individually isolated seed cells. The plating efficiency of Yoshida cells in the soft agar culture plate was 1.14%. Fifteen clonal colonies were isolated by means of a pipette and cultivated serially using TD-40 flasks. The medium was Eagle's minimum essential medium, supplemented with 20% calf serum and 0.1% Bactopeptone. Cells of each clone grew floating in the medium as free tumor cells. The doubling time of these clones ranged from 13.5 to 49.7 in hour.

Detection of AFP was done by ^{125}I radioimmunoassay, with 4-day culture medium of each clone containing about $5\text{–}8 \times 10^5$ cells/ml, which were freed by centrifugation before use (*5*). Two out of 10 clones *in vitro* showed a distinct production of AFP, as indicated in Table 8. This means that both the clones have synthesized AFP and excreted it into the media. The doubling time of 2 AFP-producing clones was 13.5 and 18.4 hr, respectively, and was not slower than that of all the remaining clones. It is now evident that the original Yoshida sarcoma

TABLE 8. Concentration of AFP in Culture Media of 10 Clones of Yoshida Sarcoma *in Vitro*

Clones	Conc. (ng/ml)	Clones	Conc. (ng/ml)
UYSCL-31	51	UYSCL-38	62
UYSCL-32	(—)	UYSCL-39	(—)
UYSCL-33	(—)	UYSCL-40	(—)
UYSCL-34	(—)	UYSCL-42	(—)
UYSCL-35	(—)	UYSCL-43	(—)

The concentration was determined by ^{125}I radioimmunoassay, by the double-antibody method, using cell-free fluids prepared from the culture medium containing $5\text{–}8 \times 10^5$ tumor cells/ml. (—) represents values below 20 ng/ml.

TABLE 9. Concentration of AFP in Ascites and Serum of Rats with 13 Subclones of Yoshida Sarcoma Clone UYSCL-38

Subclones	Conc. (ng/ml)	Subclones	Conc. (ng/ml)
UYSCL-38-1	317	UYSCL-38-23 (serum)	(—)
UYSCL-38-2	(—)	UYSCL-38-24	(—)
UYSCL-38-4	(—)	UYSCL-38-28	300
UYSCL-38-6	900	UYSCL-38-36	245
UYSCL-38-17	600	UYSCL-38-37	300
UYSCL-38-19	(—)	UYSCL-38-40	65
UYSCL-38-22 (serum)	65		
UYSCL-38 (control)	252		

The concentration was determined by ^{125}I radioimmunoassay. (—) represents values below 20 ng/ml. Cell-free ascitic fluids prepared from the tumor ascites containing $2\text{–}10 \times 10^7$ tumor cells/ml were used except for the two marked (serum).

TABLE 10. Concentration of AFP in 4-Day Culture Media of Cyclic AMP- and Theophylline-treated Yoshida Sarcoma Clone UYSCL-38

Treatment	Conc. (ng/cell)
Cyclic AMP 2 mM	7.1×10^{-4}
Cyclic AMP 1	3.1×10^{-4}
Cyclic AMP 0.5	1.5×10^{-4}
Cyclic AMP 0.5 and theophylline 0.1	3.3×10^{-4}
Control	8.1×10^{-5}

The concentration was measured by ^{125}I radioimmunoassay with cell-free culture media and indicated by values per cell after counting the number of cells in the medium.

contains cells producing AFP. It is supposed, however, that the production of AFP *in vivo* by them is so poor that the AFP concentration in host animals would have been insufficient to detect.

One cell was picked up by means of a micropipette under a microscope from an AFP-producing clone and was transplanted intraperitoneally into a normal rat. Out of 50 recipient rats, 13 animals showed takes of subclonal tumors. Table 9 indicates the concentration of AFP in the ascites or serum from the rats bearing subclonal tumors. Variable production of AFP by the subclonal cells *in vivo* was suggested.

It is said that cyclic AMP could have some basic regulatory action on the growth of cultured cells and cause cell specialization. Extracellular addition of cyclic AMP to the medium of AFP-producing clone, 5, 2, 1, or 0.5 mM/ml, induced a retardation of cell growth of the clone (*7*). The production of AFP in the growth-depressed cells is accelerated as indicated in Table 10, although this study is now in progress and the results remain subject to confirmation.

The production of AFP by Yoshida ascites tumor cells and their variants may support, but not absolutely, the hypothesis of the liver-cell origin of Yoshida tumor. AFP may present a clue to the change or differentiation of hepatocarcinoma cells.

ACKNOWLEDGMENTS

This work was supported partially by a grant from the Princess Takamatsu Cancer Research Fund, a grant from the Tokyo Club and a Grant-in-Aid for Cancer Research from the Ministry of Education of Japan.

REFERENCES

1. Abelev, G. I. Production of embryonal serum α-globulin by hepatomas: reviews of experimental and clinical data. Cancer Res., *28*: 1344–1350, 1968.
2. Grabar, P., Stanislawski-birencwajg, M., Oisgold, S., and Uriel, J. Immunochemical and enzymatic studies on chemically induced rat-liver tumors. UICC Monograph, *2*: 20–31, 1967.
3. Isaka, H. Chromosomal features and DNA content of Yoshida ascites sarcoma and its slow-growing variants characterized by island formation in ascites. Gann, *59*: 327–339, 1968.

4. Isaka, H., Umehara, S., Hirai, H., and Tsukada, Y. Development of variant tumor cells of Yoshida ascites sarcoma producing α-fetoprotein. Gann, *63*: 63–71, 1972
5. Isaka, H., Umehara, S., Hirai, H., Tsukada, Y., and Watabe, H. Isolation of α-fetoprotein-producing cells from Yoshida ascites sarcoma and its clone. Gann, *64*: 133–138, 1973.
6. Isaka, H. and Tsukada, Y. A comment on α-fetoprotein-producing variants of Yoshida ascites sarcoma, with reference to the hypothesis of hepatic-cell origin of the ascites tumor. GANN Monograph on Cancer Research, *14*: 289–299, 1973.
7. Isaka, H., Umehara, S., Umeda, M., Hirai, H., Tsukada, Y., and Watabe, H. Increased production of α-fetoprotein by cyclic 3′,5′-adenosine monophosphate-treated Yoshida ascites sarcoma cells *in vitro*. Gann, *65*: 79–83, 1974.
8. Ishidate, M., Kobayashi, K., Sakurai, Y., Satoh, H., and Yoshida, T. Experimental studies on chemotherapy of malignant growth employing Yoshida sarcoma animals. (II) The effect of N-oxide derivatives of nitrogen mustard. Proc. Japan Acad., *27*: 493–500, 1951.
9. Ishidate, M., Jr. and Isaka, H. Studies on two variant strains of Yoshida sarcoma characterized by island formation in ascites. Gann, *57*: 413–426, 1966.
10. Kuroki, T., Sato, M., and Satoh, H. The growth of LY variants of Yoshida sarcoma. Nippon Rinsho (Jap. J. Clin. Med.), *24*: 19–21, 1966 (in Japanese).
11. Nakamura, K. Transplantable leukemia as an ascites tumor: its histogenesis and progress of the disease. Natl. Cancer Inst. Monogr., *16*: 149–206, 1964.
12. Nishi, S. and Hirai, H. Radioimmunoassay of α-fetoprotein in hepatoma, other liver diseases, and pregnancy. GANN Monograph on Cancer Research., *14*: 79–87, 1973.
13. Satoh, H. Induction of resistance in animals to tumor transplantation. Kenkyu Hokoku Shuroku, (Annu. Rep. Co-oper. Res. Min. Educ. Cancer), 84–94, 1964 (in Japanese).
14. Satoh, H. and Ichimura, H. Variant sublines of ascites hepatomas developed by prolonged animal passage or cold storage at −80°C. Nippon Rinsho (Jap. J. Clin. Med.), *24*: 10–12, 1966 (in Japanese).
15. Satoh, H. Personal communication.
16. Sugano, H., Ikawa, Y., and Satoh, H. Ultrafine structures of LY variants of Yoshida sarcoma. Nippon Rinsho (Jap. J. Clin. Med.), *24*: 14–18, 1966 (in Japanese).
17. Sugimura, T. Decarcinogenesis, a newer concept arising from our understanding of the cancer phenotype. *In;* W. Nakahara (ed.), Chemical Tumor Problems, pp. 269–284, Japan Society for Promotion of Science, Tokyo, 1970.
18. Watabe, H. and Hirai, H. Production of α-fetoprotein by transplantable rat hepatoma. GANN Monograph on Cancer Research, *14*: 279–287, 1973.
19. Watabe, H. Personal communication.
20. Yoshida, T., Isaka, H., and Satoh, H. Problems of the origin of Yoshida sarcoma. Arzneimittelforschung, *14*: 735–741, 1964.
21. Yoshida, T. Comparative studies of ascites hepatomas. Methods Cancer Res., *6*: 97–157, 1971.

Discussion of Paper by Drs. Isaka et al.

DR. SILAGI: Have you any explanation for the protection against the rat Yoshida ascites sarcoma in animals pretreated with the mouse leukemia cells? Have you checked for any antigens common to the 2 cell types? Have you checked for virus production in both cell types?

DR. ISAKA: We have never checked immunologically for any antigens common to the 2 cell types. No checking for virus production in either cell type was done.

DR. F. SCHAPIRA: It is said that AFP is not absolutely specific for hepatoma, for example, it may be found in teratoblastoma.

DR. ISAKA: Yes, indeed! AFP is just one item of evidence for the theory of hepatocellular origin of Yoshida ascites tumor.

DR. PRASAD: Could you induce AFP in a variant which normally does not express AFP?

DR. ISAKA: LY variants show the production of AFP normally.

DR. IKAWA: In the study of ^{125}I radioimmunoassay, if the background is as high as 20 ng/ml, I wonder whether the low level in the original Yoshida tumor cells has any significance.

DR. ISAKA: The ^{125}I radioimmunoassay of AFP of normal materials shows various values below 20 ng/ml. In healthy men, the concentration of AFP in serum is about 5 ng/ml. Four-day culture of original Yoshida sarcoma, about 800,000 cells/ml, shows a zero value by this method. Significant values were determined to be more than 20 ng/ml by this method. Clones of Yoshida sarcoma, UYSCL-31 and UYSCL-38, have significant values of AFP.

DR. WEBER: What was the histological diagnosis of the original Yoshida sarcoma?

DR. ISAKA: The original histological picture of the so-called Yoshida sarcoma was sarcomatous.

Control of Alkaline Phosphatase Activity in Cultured Mammalian Cells: Induction by 5-Bromodeoxyuridine, Cyclic AMP, and Sodium Butyrate

Hideki Koyama and Tetsuo Ono
Department of Biochemistry, Cancer Institute, Tokyo, Japan

Abstract: Treatment of a somatic hybrid cell line, B-6, in culture with the thymidine analog 5-bromodeoxyuridine (BrdU) results in a marked induction of alkaline phosphatase activity. This induction is reversible, is antagonized by the addition of thymidine, and requires continued synthesis of RNA and protein following incorporation of the analog into the DNA. The hybrid line can produce hyaluronic acid; this is inhibited by treatment with BrdU. The above and other patterns of induction and inhibition effects are quite similar, suggesting the existence of similar mechanisms, irrespective of their inverse relationship. We may propose that the induction is caused by an increase in the transcription rate of the gene(s) for alkaline phosphatase, or by a decrease in the transcription rate of a presumed inactivating enzyme(s) with rapid turnover, which might inactivate the phosphatase itself or its mRNA. Also, alkaline phosphatase in the cells is more markedly induced by cyclic AMP and sodium butyrate. This induction requires both new RNA and protein synthesis.

BrdU inhibits several differentiated functions, while cyclic AMP induces them. However, with regard to alkaline phosphatase, both chemicals not only induce the enzyme, but also act on it synergistically.

When the induction in several other cell lines was compared, some, but not all, responded to 1, 2, or all of the above 3 inducers. A similar marked difference in response was observed even between clones that were isolated from mouse mammary carcinoma (FM3A) line by cloning in soft agar medium. Some clones failed to yield alkaline phosphatase induction in response to 1 or 2 drugs. In addition, utilizing a selection method specific for the enzyme, clones having higher levels of alkaline phosphatase were isolated successfully from the FM3A cells. The development of such inducers and of mutants concerning the expression of alkaline phosphatase activity may provide a unique and useful system for studying the control mechanism of gene expression and differentiation in mammalian cells.

Alkaline phosphatase is an enzyme that is widely distributed in all tissues *in vivo*, but higher levels of activity are found in some tissues, such as small intestine, kidney, placenta, and mammary gland (*8*). This enzyme shows a tissue-specific pattern of isoenzymes, which have different enzymic properties (*39*). Although its biological function has not yet been elucidated completely, alkaline phosphatase seems likely to play a direct role in the absorption process of nutrients as a differentiated function in these tissues (*8*).

So far, tissue culture methods have provided a useful tool for the study of the role of enzymes in various differentiated functions at the cellular level. It has been shown that many mammalian cell lines in culture have alkaline phosphatase activity, the levels of which are quite different in lines derived from different species, organs, and tissues (*5*). However, there is no work dealing with the exact origin, role in cell multiplication and metabolism *in vitro*, or the *in vivo* function, of alkaline phosphatase. With respect to regulation *in vivo*, on the other hand, intestinal alkaline phosphatase is induced by fat or fatty acid ingestion in the rat (*10*), or by adrenal glucocorticoid treatment in developing amphibian, chicken, and mouse (*4*, *31*). In some cultured cells, phosphatase is induced by cultivation in the presence of prednisolone (*6*), substrate (*7*), or under unusual growth conditions such as hyperosmolarity (*32*) and low temperature (*13*).

The present work concerns the cellular regulation of alkaline phosphatase activity in cultured mammalian cells. Recently, we found marked induction of the enzyme on treatment with 3 chemicals: 5-bromodeoxyuridine (BrdU), cyclic AMP, and sodium butyrate, and studied their induction mechanisms. Many recent studies have shown that BrdU suppresses differentiation, while cyclic AMP induces it. Therefore it was very interesting to see how the mechanisms of alkaline phosphatase induction by BrdU and cyclic AMP are related to such induction and inhibition effects of the chemicals on differentiation. Sodium butyrate is a kind of fatty acid, and infusion of fatty acid *via* the portal vein results in a rise in rat-lymph alkaline phosphatase (*10*). Thus, phosphatase induction in sodium butyrate-treated cells is an *in vitro* manifestation of the phenomenon found *in vivo*.

In addition, with mouse mammary carcinoma cell lines, we tried to isolate control mutants which failed to respond to inducers, or which had very high levels of the enzyme activity compared with the original line. The development and use of these mutants should facilitate an understanding of the mechanisms of induction and of the biological function of alkaline phosphatase *in vivo*. Furthermore, this system should provide a useful model for studying gene expression and differentiation in mammalian cells.

Induction of Alkaline Phosphatase Activity by BrdU

The cells mainly used here were of the hybrid B-6 line, which had been derived from somatic-cell hybridization of mouse mammary carcinoma and Chinese hamster lung cells *in vitro* in our laboratory (*26*). The phenomenon of alkaline phosphatase induction by BrdU was first found in this line. Figure 1 shows the morphology of cells growing in suspension with a doubling time of about 16 hr. This line is charac-

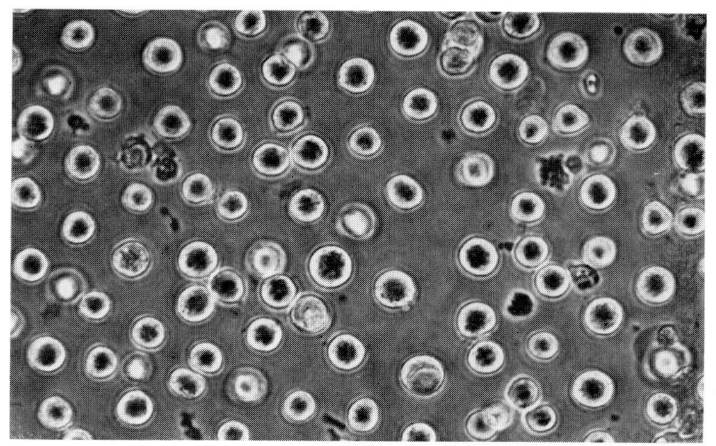

FIG. 1. Living B-6 cells. Phase contrast. (×330)

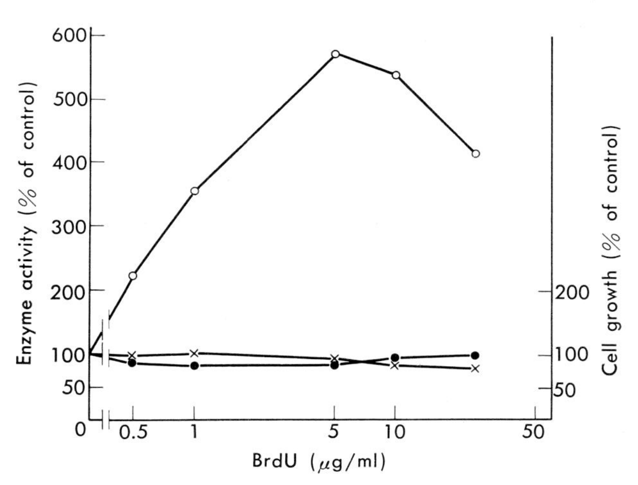

FIG. 2. Effect of BrdU on cell growth and activities of alkaline and acid phosphatase in B-6 cells as a function of concentration. B-6 cells (5×10^4 cells/ml) were inoculated into a modified Eagle's minimal essential medium supplemented with 5% calf serum, Bactopeptone, and various concentrations of BrdU and cultured for 3 days in a 5% CO_2 incubator at 37°C. Then, an aliquot of the culture was taken out to assay cell protein and the reminder was used to assay enzyme activity. They were collected by low-speed centrifugation, washed twice with ice-cold saline and stored at −76°C. Enzyme extracts were made by sonicating thawed cells suspended in 1–2 ml of 0.25M sucrose for 1–2 min, and alkaline and acid phosphatase assays were performed by the method of Lowry (*28*). Cell protein and protein of these extracts were determined by the procedure of Oyama and Eagle (*33*) using serum albumin as a standard. Unless otherwise cited, the enzyme activity was shown in specific activity units, defined as nmoles *p*-nitrophenol liberated from *p*-nitrophenol-phosphate per hr per mg protein. In this figure, the results are represented percent of control. In control cultures, the cell protein and specific activities of alkaline and acid phosphatases were 220 μg/ml, 59 and 1,220 nmoles/hr/mg protein, respectively. ○ alkaline phosphatase; ● acid phosphatase; × cell protein.

terized by the ability to produce hyaluronic acid (*21*), and this differentiated function was reversibly inhibited by treatment of the cells with BrdU, following a reduction in hyaluronic acid synthetase activity (*22*). However, alkaline phosphatase was induced under the same conditions (*23*, *24*).

B-6 cells were grown in various concentrations of BrdU for 3 days, and alkaline and acid phosphatase activities were assayed in cell-free extracts (Fig. 2). Alkaline phosphatase activity in cells treated with 5–10 μg/ml of the analog was markedly induced, whereas there was no significant change in acid phosphatase activity. Cell growth, as estimated by the protein content of those cultures, was little affected. We have found a small decline in cell number and viability in treated cultures, as reported previously (*22*).

The time course of the induction is shown in Fig. 3, where B-6 cells were cultured in the presence of BrdU for either 72 hr or 15 days and assayed for alkaline phosphatase levels at the indicated times (*24*). In Fig. 3a, the induction at 5 μg/ml proceeded rapidly and linearly after a lag period of 24 hr and did not level off even after 72 hr. The cultures exposed to 1 μg/ml showed a 3-fold greater activity than control cultures at 48 hr, but there was no elevation in the enzyme activity thereafter. The levels in control cultures did not change significantly throughout the period studied. In Fig. 3b, where B-6 cells were serially subcultured in the analog over 3 days, it is evident that it takes more than 6 days (2 transfer generations) for the induction to become maximal and that the plateau levels depend on the levels of BrdU added to the medium.

The BrdU effect was completely reversible (Fig. 4) (*24*). When cells that had

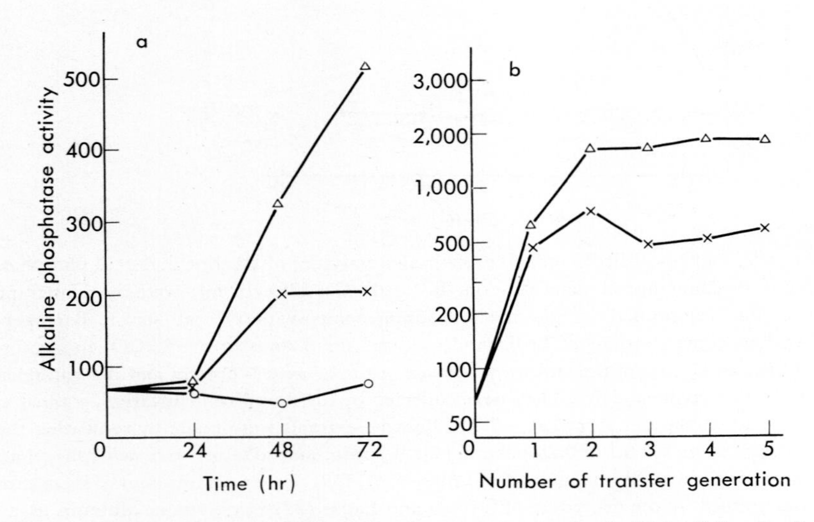

FIG. 3. Time course of alkaline phosphatase induction by BrdU in B-6 cells. B-6 cells were cultured for 72 hr (a) or 15 days (b) in the presence or absence of BrdU, and alkaline phosphatase activity was assayed at various times as indicated, as in Fig. 1. ○ control; × 1 μg/ml of BrdU; △ 5 μg/ml of BrdU. Alkaline phosphatase activity: nmoles/hr/mg protein.

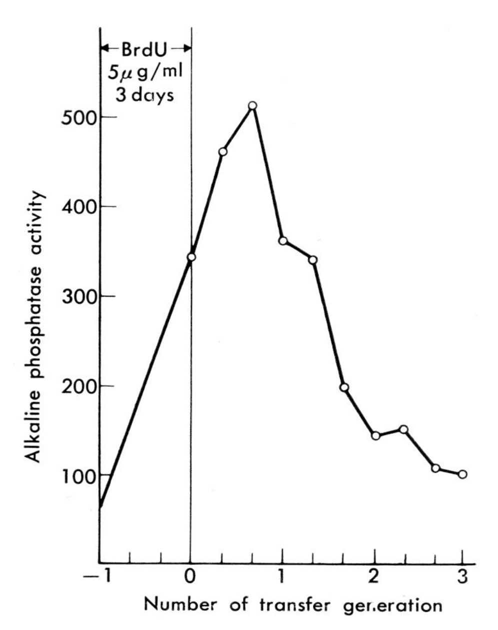

FIG. 4. Decrease of alkaline phosphatase activity induced by BrdU after its removal from the medium. B-6 cells were treated with 5 μg/ml of BrdU for 3 days as described in Fig. 2. At zero time, they were washed, transferred into the normal medium and subcultured serially every 3 days. Alkaline phosphatase was assayed at various times as indicated. Alkaline phosphatase activity: nmoles/hr/mg protein.

been grown with the analog for 3 days were transferred to normal medium and cultured, the elevated alkaline phosphatase activity at zero time continued rising for another 2 days even after transfer of the treated cells to the analog-free medium, then began to drop and returned to base level in 8 to 9 days.

Generally, enzyme induction by hormones or substrates stops immediately on the removal of these inducers and begins to disappear rapidly due to the decay of the induced enzyme proteins caused by turnover and their dilution due to cell multiplication. In this study, however, the fact that further induction continues for 2 days even after the removal of BrdU suggests that the inducer (BrdU) may be tightly fixed to the cells in some way. BrdU is well-known to be incorporated into cellular DNA when the cell grows in it. This was also demonstrated for the B-6 line by density gradient centrifugation in cesium chloride (*37*) and by autoradiography using ^{3}H-BrdU (not shown). Therefore, it seems likely that BrdU acts on the cells producing alkaline phosphatase by virtue of incorporation into the DNA. This idea is also supported by the finding that the effect of BrdU does not appear in the presence of thymidine. Table 1 shows that alkaline phosphatase activity was not increased when the analog was added with 2 to 4 times the amount of thymidine.

In order to demonstrate the necessity of DNA synthesis during BrdU treatment, namely, the necessity for analog incorporation into the cellular DNA, we first determined the shortest exposure time of cells to BrdU necessary to produce a

TABLE 1. Suppression of Alkaline Phosphatase Induction by Simultaneous Addition of Thymidine with BrdU in B-6 Cells

BrdU (μg/ml)	Thymidine (μg/ml)	Alkaline phosphatase activity (nmoles/hr/mg protein)
—	—	95
5	—	370
5	5	276
5	10	179
5	20	119
5	40	102
5	60	91
—	60	83

B-6 cells were grown fcr 3 days in media containing either BrdU alone, thymidine alone, or both, at the indicated concentrations, and alkaline phosphatase was assayed as described in Fig. 1.

TABLE 2. Induction of Alkaline Phosphatase Activity by Different Times of Treatment of B-6 Cells with BrdU

Time treated with BrdU (hr)	Alkaline phosphatase activity (nmoles/hr/mg protein)
0	44
9	148
18	259
34	267
48	267

Actively growing B-6 cells 24 hr after inoculation were treated with BrdU (5 μg/ml) for various lengths of time as indicated, then were washed and transferred into normal medium. At 72 hr, all cultures were harvested and assayed for alkaline phosphatase activity.

detectable rise in the specific activity of alkaline phosphatase. As indicated in Table 2, actively growing cells were incubated with BrdU for various lengths of time, transferred to normal medium, harvested at 72 hr and then assayed for enzyme activity. The data show that even 9 hr of treatment is sufficient to induce the enzyme activity.

We checked whether or not induction occurs on the simultaneous addition of 2 DNA synthesis inhibitors with BrdU. Hydroxyurea or cytosine arabinoside blocked ^{3}H-thymidine incorporation into the B-6 cell DNA more than 96% 1 hr later when employed at a concentration of 40 μg/ml or 2 μg/ml, respectively. Cells were treated with BrdU for 9 hr under such conditions following addition of those inhibitors, as shown in Table 3. Then all cultures were washed and incubated for an additional 38 hr in complete normal medium. Table 3 shows that in the cultures treated with either inhibitor, a slight increase in specific activity occurred. However, in the cultures treated with a combination of BrdU and either inhibitor, the activity was just the same as that found in those treated with inhibitor alone, and it was somewhat lower than the activity in the culture grown with only BrdU. This result demonstrates that incorporation of BrdU into DNA is necessary for its

TABLE 3. Requirement of DNA Synthesis for Alkaline Phosphatase Induction by BrdU in B-6 Cells

Hydroxyurea	Additions		Alkaline phosphatase activity (nmoles/hr/mg protein)
	Cytosine arabinoside	BrdU	
−	−	−	51
+	−	−	75
+	−	+	76
−	+	−	97
−	+	+	96
−	−	+	138

Either hydroxyurea (40 μg/ml) or cytosine arabinoside (2 μg/ml) was added to actively growing B-6 cells 24 hr after inoculation. One hour later, BrdU (5 μg/ml) was further added to some cultures, followed by incubation for 9 hr. Then, all cultures were washed, transferred into normal medium and incubated for an additional 38 hr. They were harvested and assayed for alkaline phosphatase activity.

effect on alkaline phosphatase, suggesting that the analog acts at the gene level.

BrdU is a potent mutagen in microbial cells (*40*). However, there is no evidence of such an effect in mammalian cells. The present effect of the analog is unlikely to be due to its mutagenicity, because a very large increase in enzyme activity appears after a short lag time (24 hr) as compared to the generation time (16 hr) and the induction is completely reversible.

Cultured mammalian cells in general have an S period of 6 to 10 hr during which DNA is replicated semiconservatively. Since the B-6 line doubles in 16 hr and the BrdU effect occurred upon treatment of the cells for 9 hr, it can be further concluded that BrdU incorporation into one strand of DNA is sufficient to cause the induction.

The induction also requires continued synthesis of both RNA and protein. When cells grown in BrdU for 18 hr were transferred to analog-free medium with either actinomycin or cycloheximide, there was no induction, as shown in Table 4. This result is compatible with the need for BrdU incorporation into DNA. Prob-

TABLE 4. Requirement of RNA and Protein Synthesis for Alkaline Phosphatase Induction by BrdU in B-6 Cells

BrdU	Additions		Alkaline phosphatase activity (nmoles/hr/mg protein)
	Actinomycin S3	Cycloheximide	
−	−	−	47
−	+	−	25
−	−	+	10
+	−	−	214
+	+	−	39
+	−	+	43

BrdU (5 μg/ml) was added to actively growing B-6 cells 24 hr after inoculation. These cultures were incubated for 42 hr, washed to remove the analog, transferred into fresh medium containing either actinomycin S3 (0.5 μg/ml) or cycloheximide (1 μg/ml) and collected at 72 hr for alkaline phosphatase assay. Before this experiment, the levels of antibiotics used here had been checked to block RNA and protein synthesis by over 90% 16 hr after addition to cultures by means of incorporation of ^{3}H-uridine and ^{14}C-phenylalanine in the acid-insoluble fraction of B-6 cells.

ably the primary site of the analog action is at the transcriptional level of mRNA which is directly or indirectly responsible for the expression of alkaline phosphatase to maintain its level within the cells. At the present stage, we may propose that, after incorporation into DNA, BrdU causes alkaline phosphatase induction by increasing the transcription rate of the gene(s) for the enzyme, or decreasing the transcription rate of a presumed enzyme(s) with rapid turnover, which might inactivate alkaline phosphatase itself or its mRNA in the cells.

BrdU suppresses the expression of several differentiated functions (*1*, *35*, *36*, *38*) as well as hyaluronic acid production in the hybrid line used here (*22*). Thus, it was of interest to study what relationship existed between the 2 rather different actions of BrdU: induction of alkaline phosphatase and inhibition of mucopolysaccharide synthesis. A detailed discussion has already been published (*24*) and will also be presented in this symposium by Ono *et al.* Anyway, we found a great similarity between the 2 effects in several criteria which we took for comparison. This strongly suggests that the 2 reactions, irrespective of their dissimilar nature, are caused by similar or common mechanisms during the growth of B-6 cells with BrdU. It is therefore probable that the suppression of differentiation by BrdU may be understood more easily and precisely through analysis of the mechanism of alkaline phosphatase induction.

Recently we knew a similar work done by Martin *et al.*, who reported that BrdU treatment of human cells increases the frequency of alkaline phosphatase positive cells which can be stained histochemically (*30*).

Induction of Alkaline Phosphatase Activity by Cyclic AMP and Sodium Butyrate

The alkaline phosphatase in the B-6 line has also been discovered to be induced

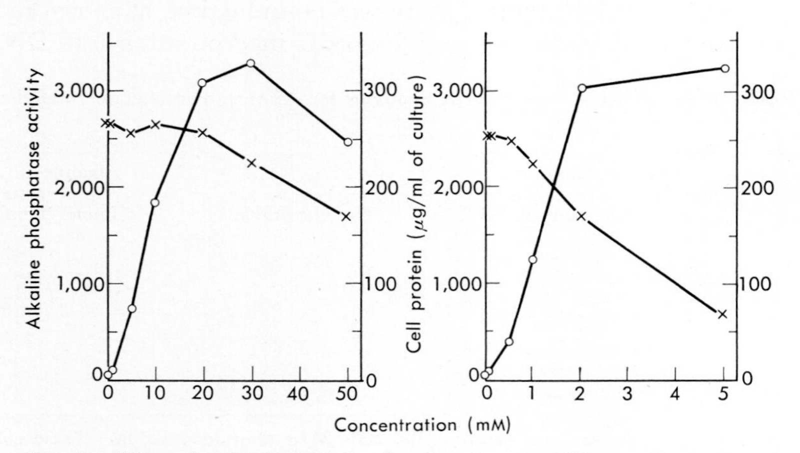

FIG. 5. Effect of cyclic AMP (a) and sodium butyrate (b) on cell growth and alkaline phosphatase activity in B-6 cells as a function of concentration. B-6 cells were treated in the same way as described in Fig. 2, except that various concentrations of either cyclic AMP or sodium butyrate were added to the cells in place of BrdU. ○ alkaline phosphatase activity (nmoles/hr/mg protein); × cell protein.

by cyclic AMP and sodium butyrate. Figure 5 illustrates the dose-response curve of this induction. B-6 cells grown in both chemicals at optimal concentrations (20–30 mM for cyclic AMP or 2 mM for sodium butyrate) for 3 days produced a more than 60-fold increase in the enzyme activity. At this time, this concentration of cyclic AMP inhibited cell growth by less than 10%, while sodium butyrate at 2 mM reduced it to 65%. In earlier studies (*20*), we used a cyclic AMP derivative, dibutyryl cyclic AMP, which is much more active than the naturally occurring nucleotide. Recently, it was reported that this derivative is so unstable in culture conditions that it rapidly degrades into monobutyryl cyclic AMP, cyclic AMP, and butyric acid (*19*). The reason for the greater effectiveness of dibutyryl cyclic AMP may be some synergistic action of those degradation products within the cells, since we observed that they very much potentiated their mutual effects on alkaline phosphatase. The levels of cyclic AMP employed here are rather high. It would be better to use cyclic AMP itself, not the derivative, in order to investigate the induction mechanism of the phosphatase.

Adenosine, AMP, ADP, and ATP were not effective. Theophylline, which is an inhibitor of phosphodiesterase, not only caused induction, but also potentiated the cyclic AMP effect. Prostaglandin E, which activates adenylcyclase on the cell membrane to increase cyclic AMP levels in the cells, was also effective, but other hormones that behave like prostaglandins were ineffective.

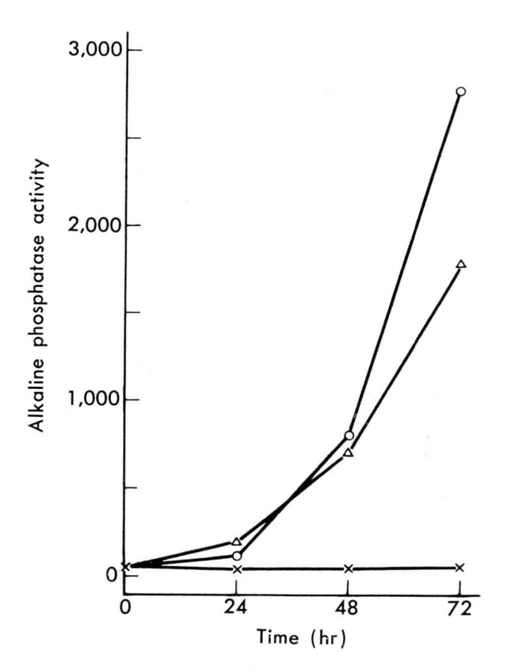

FIG. 6. Time course of alkaline phosphatase induction by cyclic AMP and sodium butyrate. B-6 cells were grown in the presence or absence of cyclic AMP (20 mM), or sodium butyrate (2 mM). At 24, 48, and 72 hr, parts of the cultures were harvested and assayed for alkaline phosphatase. × control; △ cyclic AMP; ○ sodium butyrate. Alkaline phosphatase activity: nmoles/hr/mg protein.

Addition of cyclic AMP or sodium butyrate to B-6 cultures led to a rapid and exponential rise of alkaline phosphatase activity after a lag period of 24 hr and this remained active over 72 hr (Fig. 6). Thus, the kinetics of this enzyme induction are characterized by a long lag period prior to a detectable elevation in specific activity after addition of any of the above 3 inducers, suggesting a long half-life of the phosphatase. In fact, we have seen that the induced activity decayed with a half-life longer than 24 hr, when B-6 cells grown with cyclic AMP or sodium butyrate for 72 hr were transferred to normal medium and cultured.

Similarly to the induction by BrdU, neither direct activation of the enzyme with both chemicals, nor the presence of inhibitors or activators were found to be involved in the induction. No induction occurred in the presence of actinomycin (0.5 μg/ml) or cycloheximide (1 μg/ml) with those inducers. It seems likely that the induction requires continued synthesis of both RNA and protein.

Treatment with cyclic AMP converts transformed fibroblasts in culture to normal ones as regards cell morphology, growth properties relating to restoration of contact inhibition, and membrane properties (*16*, *17*). In addition, the cyclic nucleotide induces glutamine synthetase (*3*), tyrosine aminotransferase (*2*), collagen synthesis (*15*), mucopolysaccharide synthesis (*11*, *25*), melanin synthesis (*18*), and differentiation of neuroblastoma (*34*). These effects are clearly directed to differentiation from less-differentiated or undifferentiated states of cells, and this process may be related to the reversion of malignant cells to normal. From this point of view, alkaline phosphatase induction by cyclic AMP may be regarded as a manifestation of differentiation. On the other hand, BrdU blocks differentiation, as discussed above, so BrdU has just the opposite effect to cyclic AMP. Yet this analog not only induced alkaline phosphatase, but also potentiated the cyclic AMP effect in B-6 cells. This contradiction remains unresolved.

In relation to the butyrate effect, some short-chain fatty acids such as valeric, caproic, and caprylic acids, analogous to it in chemical structure, showed similar inducing ability. These fatty acids are normal components in milk, and in the mammary gland during lactation high levels of alkaline phosphatase activity are maintained (*9*). Thus, in this tissue, we could expect any relationship between the existence of these acids and high levels of the phosphatase to appear. In addition, octanoic (caprylic) acid infusion induced alkaline phosphatase in rat lymph (*10*). We may suggest from these data that the fatty acids act for alkaline phosphatase as internal inducers or stabilizers in these tissues. Also, a similar role might be shared by cyclic AMP, because it is more ubiquitous than short-chain fatty acids.

Induction of Alkaline Phosphatase Activity in Various Cell Lines

In order to test the generality of the present phenomenon, several other cell lines in culture were exposed to BrdU, cyclic AMP, and sodium butyrate for 3 days under the same conditions as B-6 cells. Table 5 summarizes the results. Induction was judged as positive (+) when the specific activity increased more than 2-fold in the treated cells. The B-6 line and its parental line FC-1 or FM3A of mouse mammary carcinoma origin responded to all these inducers, but other cell lines,

TABLE 5. Induction of Alkaline Phosphatase in Various Cell Lines

Cell line	Alkaline phosphatase induction		
	BrdU	Cyclic AMP	Sodium butyrate
B-6	+	+	+
FM3A	+	+	+
FC-1	+	+	+
CHL	−	+	+
L-929	−	−	+
7288CtC	−	−	−
Sen	+		+
HT-4	+	−	+

Cells were seeded at a density of 4×10^4 to 10^5 cells/ml into medium containing either BrdU (5 μg/ml), cyclic AMP (20 mM), or sodium butyrate (2 mM) and cultured for 3 days. Cell harvesting and alkaline phosphatase assay were performed as described in the legend to Fig. 1. The results are expressed as positive (+) when alkaline phosphatase specific activity in the treated cultures increased more than 2-fold compared with those in the controls. FM3A, C3H mouse mammary carcinoma cells; FC-1, 8-azaguanine-resistant cells derived from FM3A cells; CHL, Chinese hamster lung cells; L-929, C3H mouse fibroblast cells; 7288CtC, Morris rat hepatoma 7288C cells; Sen, Buffalo rat fibroblast cells; HT-4, golden hamster brain cells transformed by adenovirus type 12.

including the other parental line, CHL (Chinese hamster lung origin), mainly derived from species other than mouse, did not necessarily respond to all inducers. In the case of Sen cells established from rat fibroblasts, we noticed a significant decrease in the phosphatase level following treatment with BrdU. The pattern and extent of the response was quite different among these individual cell lines. These results indicate that alkaline phosphatase in some, but not all, cell lines is induced by treatment with 1, 2, or all 3 chemicals, and, furthermore, the inducibility of cells seems unlikely to be correlated with other cellular properties such as cell species and tissue origin, growth rate, morphology, tumorigenicity, *etc.*

New System for the Study of the Regulation of Alkaline Phosphatase

Up to now we have used an artificially prepared interspecies hybrid line, but extensive work with it might raise problems in the future because of its genetic complexity and instability. Hence we changed to FM3A cells, because the latter cells had a low constitutive activity of alkaline phosphatase with all 3 inducers, as shown in Table 5, grew rapidly in a floating fashion and easily formed single colonies with a high plating efficiency in soft agar medium. The cells were cloned by the method of Goto and Sato (*14*), and the clones were cultured to obtain a pure line highly responsive to inducers.

We compared the inducibility of alkaline phosphatase in the 48 clones isolated. Table 6 shows the data obtained for 12 clones. Severalfold differences were found in the basal activity among individual clones. However, no clone was found to be completely deficient in alkaline phosphatase activity. It should be noted that there was a greater difference in the pattern and extent of response to inducers among these clones. This difference was as large as that found between cell lines of other

TABLE 6. Induction of Alkaline Phosphatase in Clones Derived from Mouse Mammary Carcinoma (FM3A) Cells

Clone	Alkaline phosphatase activity (nmoles/hr/mg protein)			
	None	BrdU	Cyclic AMP	Sodium butyrate
C21	66	537	2,760	2,000
C22	190	800	2,190	1,280
C23	79	364	1,180	797
C24	103	862	1,600	1,720
C25	61	340	880	873
C26	86	718	1,230	651
C27	37	55	36	264
C28	45	510	2,070	1,320
C32	36	432	1,950	1,540
C33	47	353	768	825
C35	32	163	1,040	983
C36	40	37	106	85

Clones were seeded at a density of 8×10^4 cells/ml into the medium containing either BrdU (5 μg/ml), cyclic AMP (20 mM), or sodium butyrate (2 mM) and cultured for 2 days. Cell harvesting and alkaline phosphatase assay were performed as described in Fig. 1.

species origin (Table 4). In many clones, the degree of induction by one chemical was parallel to that by the other two, but in some clones such as C27, C36, or others (not shown here), alkaline phosphatase was hardly induced by 1 or 2 of these inducers. This result indicates the extreme heterogenity of FM3A cells, since they had never been cloned. Also, it shows that one can easily isolate control mutants for the expression of alkaline phosphatase activity which fail to respond to inducers. Comparison of inducible clones and these uninducible mutants would make it easier to analyze the induction mechanisms. Recently, Levinsohn and Thompson isolated a tyrosine aminotransferase induction regulation variant from rat hepatoma cultures (*27*). This variant had normal basal levels of enzyme but failed to respond to the usual inducing steroids.

Using a selection technique specific for alkaline phosphatase, it is possible to select clones with higher enzyme activity from a cell population with low activity. This method was first developed by Maio and De Carli (*29*) with human cells and was then modified by Goto *et al.* (*12*) to apply it to clones growing in agar. We used the latter method after a minor modification. The original FM3A cells, having a very low activity (50 nmoles/hr/mg protein average specific activity), were cloned in soft agar medium and 14 days later, visible clones appearing were overlayed with Tris buffer (pH 9.4) containing *p*-nitrophenol phosphate plus 1% agar and incubated for 15–30 min at 37°C. The frequency of clones stained yellow by hydrolysis of the substrate was counted, and then they were picked up and transferred to normal medium. All clones isolated in this way were healthy enough to grow thereafter. Table 7 shows the isolation process of one such experiment. During the 3 cycles of cloning, the frequency of yellow clones increased gradually; the mean specific activities, though considerably different even between clones in each step, also rose gradually. As a consequence, we were able to obtain a clone (Fh3-4-4) showing

TABLE 7. Isolation Process of Clones with High Levels of Alkaline Phosphatase Activity from Mouse Mammary Carcinoma (FM3A) Cells

	1st cloning		2nd cloning		3rd cloning
	–Fh1, 270		–Fh3-1, 508		–Fh3-4-1, 2,040
	–Fh2, 98		–Fh3-2, 906		–Fh3-4-2, 1,730
	–Fh3, 319		–Fh3-3, 617		–Fh3-4-3, 4,460
	–Fh5, 291		–Fh3-4, 5,500		–Fh3-4-4, 6,870
FM3A, 50 ⟶	–Fh6, 111	⟶	–Fh3-5, 398	⟶	–Fh3-4-5, 2,770
	–Fh7, 284		–Fh3-6, 557		–Fh3-4-6, 1,310
	–Fh8, 83		–Fh3-7, 254		–Fh3-4-7, 1,700
	–Fh9, 151		–Fh3-8, 140		–Fh3-4-8, 2,330
	(0.20)		(5.5)		(90)

Fh1, ..., Fh3-1, ..., and Fh3-4-1, ... show clones isolated by the selection method specific for alkaline phosphatase (see text). The values on the right of the original FM3A cells or their clones are the specific activities of alkaline phosphatase (nmoles/hr/mg protein). The activity of the FM3A cells is the average of the cumulative data, while that of the clones was assayed within 4–15 days after isolation. The values in parentheses are the percentages of clones stained yellow.

more than a 130-fold greater enzyme activity than the initial FM3A line. We have already isolated several similar clones and are studying the stability of their high levels of alkaline phosphatase during prolonged cultivation.

One can ask many questions about these mutant clones. Why or by what mechanism do they possess such high levels of alkaline phosphatase? Are their phosphatases induced further upon treatment with BrdU, cyclic AMP, or sodium butyrate? Are their phosphatases the same as that in the FM3A cells, or in cells induced by the above chemicals? Are their high levels correlated with any other cellular properties? As, in particular, the original FM3A line is of mammary carcinoma origin and the mammary gland has a high phosphatase activity, are any characteristics involved in differentiation or function of the tissue reexpressed in these mutants? All these questions, however, remain to be answered. Whatever the results, they will provide useful information about the biological function as well as the regulation of alkaline phosphatase.

REFERENCES

1. Abbott, J. and Holtzer, H. The loss of phenotypic traits by differentiated cells. V. The effect of 5-bromodeoxyuridine on cloned chondrocytes. Proc. Natl. Acad. Sci. U.S., *59*: 1144–1151, 1968.
2. Butcher, F. R., Becker, J. E., and Potter, V. R. Induction of tyrosine aminotransferase by dibutyryl cyclic AMP employing hepatoma cells in tissue culture. Exp. Cell Res., *66*: 321–328, 1971.
3. Chader, G. J. Hormonal effects of the neural retina: Induction of glutamine synthetase by cyclic-3′,5′-AMP. Biochem. Biophys. Res. Commun., *43*: 1102–1105, 1971.
4. Chieffi, G. and Carfagna, M. The alkaline phosphatase of intestinal epithelium of *Bufo vulgaris* tadpoles during metamorphosis. The influence of hydrocortisone on the epithelial phosphatase *in vitro*. Acta Embryol. Morph. Exp., *3*: 213–217, 1960.

5. Cox, R. P. and MacLeod, C. M. Alkaline phosphatase content and the effects of prednisolone on mammalian cells in culture. J. Gen. Physiol., *45*: 439–485, 1962.
6. Cox, R. P. and MacLeod, C. M. Hormonal induction of alkaline phosphatase in human cells in tissue culture. Nature, *190*: 85–87, 1961.
7. Cox, R. P. and Pontecorvo, G. Induction of alkaline phosphatase by substrates in established cultures of cells from individual human donors. Proc. Natl. Acad. Sci. U.S., *47*: 839–845, 1961.
8. Ferney, H. N. Mammalian alkaline phosphatase. *In;* P. D. Boyer (ed.), The Enzymes, vol. 4, pp. 417–447, Academic Press, New York and London, 1971.
9. Folley, S. J. and Greenbaum, A. L. Changes in the arginase and alkaline phosphatase contents of the mammary gland and liver of the rat during pregnancy, lactation and mammary involution. Biochem. J., *41*: 261–269, 1947.
10. Glickman, R. M., Alpers, D. H., Drummey, G. D., and Isselbacher, K. J. Increased lymph alkaline phosphatase after fat feeding: Effects of medium-chain triglycerides and inhibition of protein synthesis. Biochim. Biophys. Acta, *201*: 226–235, 1970.
11. Goggins, J. F., Johnson, G. S., and Pastan, I. The effect of dibutyryl cyclic adenosine monophosphate on synthesis of sulfated acid mucopolysaccharides by transformed fibroblasts. J. Biol. Chem., *247*: 5759–5764, 1972.
12. Goto, M., Kuroki, T., and Sato, H. Cytogenetical analysis of alkaline phosphatase of FM3A cells by the agar culture method. *In;* Symposium for Cell Biology (Japan Society for Cell Biology), vol. 20, pp. 205–214, 1969 (in Japanese).
13. Goto, M., Kuroki, T., and Sato, H. Temperature effect on alkaline phosphatase activity in cultured FM3A cells. Proc. Jap. Cancer Assoc., 28th Annu. Meet., p. 39, 1969 (in Japanese).
14. Goto, M. and Sato, H. Studies on tissue culture of ascites tumors. III. Colony formation of Yoshida sarcoma cells in agar medium. Sci. Rep. Res. Inst. Tohoku Univ. -C, *12*: 319, 1965.
15. Hsie, A. W., Jones, C., and Puck, T. T. Further changes in differentiation state accompanying the conversion of Chinese hamster cells to fibroblastic form by dibutyryl adenosine cyclic 3′,5′-monophosphate and hormones. Proc. Natl. Acad. Sci. U.S., *68*: 1648–1652, 1971.
16. Hsie, A. W. and Puck, T. T. Morphological transformation of Chinese hamster cells by dibutyryl adenosine cyclic 3′,5′-monophosphate and testosterone. Proc. Natl. Acad. Sci. U.S., *68*: 358–361, 1971.
17. Johnson, G. S., Friedman, R. M., and Pastan, I. Restoration of several morphological characteristics of normal fibroblasts in sarcoma cells treated with adenosine-3′,5′-cyclic monophosphate and its derivatives. Proc. Natl. Acad. Sci. U.S., *68*: 425–429, 1971.
18. Johnson, G. S. and Pastan, I. $N^6,O^{2'}$-Dibutyryl adenosine 3′,5′-monophosphate induces pigment production in melanoma cells. Nature New Biol., *237*: 267–268, 1972.
19. Kaukel, E., Mundhenk, K., and Hilz, H. N^6-Monobutyryladenosine 3′,5′-monophosphate as the biologically active derivative of dibutyryladenosine 3′,5′-monophosphate in HeLa S3 cells. Eur. J. Biochem., *27*: 197–200, 1972.
20. Koyama, H., Kato, R., and Ono, T. Induction of alkaline phosphatase by cyclic AMP or its dibutyryl derivative in a hybrid line between mouse and Chinese hamster in culture. Biochem. Biophys. Res. Commun., *46*: 305–311, 1972.
21. Koyama, H. and Ono, T. Initiation of a differentiated function (hyaluronic acid synthesis) by hybrid formation in culture. Biochim. Biophys. Acta, *217*: 477–487, 1970.

22. Koyama, H. and Ono, T. Effect of 5-bromodeoxyuridine on hyaluronic acid synthesis of a clonal hybrid line of mouse and Chinese hamster in culture. J. Cell Physiol., *78*: 265–271, 1971.
23. Koyama, H. and Ono, T. Induction of alkaline phosphatase by 5-bromodeoxyuridine in a hybrid line between mouse and Chinese hamster in culture. Exp. Cell Res., *69*: 468–470, 1971.
24. Koyama, H. and Ono, T. Further studies on the induction of alkaline phosphatase by 5-bromodeoxyuridine in a hybrid line between mouse and Chinese hamster in culture. Biochim. Biophys. Acta, *264*: 497–507, 1972.
25. Koyama, H., Tomida, M., and Ono, T. Unpublished.
26. Koyama, H., Yatabe, I., and Ono, T. Isolation and characterization of hybrids between mouse and Chinese hamster cell lines. Exp. Cell Res., *62*: 455–463, 1970.
27. Levisohn, S. R. and Thompson, E. B. Tyrosine aminotransferase induction-regulation variant in tissue culture. Nature New Biol., *235*: 102–104, 1972.
28. Lowry, O. H. Micromethods for the assay of enzymes. Methods Enzymol., *4*: 366–381, 1957.
29. Maio, J. J. and De Carli, L. L. Distribution of alkaline phosphatase variants in a heteroploid strain of human cell in tissue culture. Nature, *196*: 600–601, 1962.
30. Martin, G. M., Derr, M. A., and Sprague, C. A. Alkaline phosphatase constitutivity: A marker for the estimation of somatic cell "mutation" in man. Ann. N.Y. Acad. Sci., *166*: 433–446, 1969.
31. Moog, F. The adaptations of alkaline and acid phosphatase in development. *In;* D. Rudnick (ed.), Cells, Organism and Milieu, pp. 121–155, Roland Press Co., New York, 1959.
32. Nitowsky, H. M., Herz, F., and Geller, S. Induction of alkaline phosphatase in dispersed cell cultures by changes in osmolarity. Biochem. Biophys. Res. Commun., *12*: 293–299, 1963.
33. Oyama, V. I. and Eagle, H. Measurement of cell growth in tissue culture with a phenol reagent (folin-ciocalteau). Proc. Soc. Exp. Biol. Med., *91*: 305–307, 1956.
34. Prasad, K. N. and Hsie, A. Morphologic differentiation of mouse neuroblastoma cells induced *in vitro* by dibutyryl adenosine 3′,5′-cyclic monophosphate. Nature New Biol., *233*: 141–142, 1971.
35. Silagi, S. and Bruce, S. A. Suppression of malignancy and differentiation in melanolic melanoma cells. Proc. Natl. Acad. Sci. U.S., *66*: 72–78, 1970.
36. Stockdale, F., Okazaki, K., Nameroff, M., and Holtzer, H. 5-Bromodeoxyuridine: Effect on myogenesis *in vitro*. Science, *146*: 533–535, 1964.
37. Tomida, M., Koyama, H., and Ono, T. Hyaluronic acid synthetase in cultured mammalian cells producing hyaluronic acid: Oscillatory change during the growth phase and repression by 5-bromodeoxyuridine. Biochim. Biophys. Acta, *338*: 352–363, 1974.
38. Wessels, N. K. DNA synthesis, mitosis and differentiation in pancreatic acinar cells *in vitro*. J. Cell Biol. *20*: 415–433, 1964.
39. Wilkinson, J. H. Phosphatase isoenzymes. *In;* Isoenzymes, pp. 239–278, Champan and Hall, London, 1970.
40. Zamenhof, S., de Giovanni, R., and Greer, S. Induced gene unstabilization. Nature, *181*: 827–829, 1958.

Discussion of Paper by Drs. Koyama and Ono

DR. OHNO: It is wonderful to have *constitutive* clones. Have you hybridized *constitutive* clones with *inducible* clones to see the phenotype of the hybrids? This is pertinent to the question of *negative versus positive* control.

DR. KOYAMA: No, I have not hybridized them. I am interested in doing so.

DR. PRASAD: Have you measured cyclic AMP after sodium butyrate treatment?

DR. KOYAMA: No, I have not. I want to measure the cyclic AMP level in sodium butyrate-treated cells as well as in the BrdU-treated cells.

DR. JOHNSON: The concentration of cyclic AMP (20 mM) which you used is very large. The contaminating butyric acid in your dibutyryl cyclic AMP preparation is indeed a problem in the interpretation of your results with this compound. However, it is easily removed by ether extraction at pH 2.5–3.0. Other cyclic AMP derivatives could be used, such as N^6-monobutyryl cyclic AMP or 8-Br-cyclic AMP. These derivatives are relatively stable and can be used at much lower concentrations than 20 mM. They could be used to substantiate the functions of cyclic AMP in your system.

DR. KOYAMA: Thank you for your comment.

DR. RUTTER: In support of your view that the stimulation by BrdU of alkaline phosphatase is probably mechanistically related to the inhibition by BrdU of differentiated functions in other systems, Dr. S. Githens in our laboratory has found that BrdU, which blocks the accumulation of cell-specific exocrine proteins and insulin, stimulates the alkaline phosphatase levels with a similar or identical concentration dependence. This effect coincides with the presence of "bubbles" formed by the accumulation of fluid in the ductules of the pancreas cells. The alkaline phosphatase activity appears to be identical with that found in high concentrations in pancreas duct cells. Thus in this system BrdU may either increase the proportion of duct-like cells in the culture, or increase the alkaline phosphatase level in individual cells. Attempts to resolve this question histologically have given equivocal results.

Dr. Thompson: I would like to mention some work Dr. D. Aviv and I have done which may pertain to phenotypic variation in tumor cells (Aviv and Thompson, Science, *177*: 1201, 1972). We cloned HTC cells without any (known) selective pressures and tested the clones for induction of tyrosine aminotransferase by dexamethasone. Clones from wild-type HTC cells displayed a range of inducibility. We selected a low- and a high-inducing clone and recloned each; then we tested the subclones for inducibility. Each clone gave rise to a broad range of subclones. The average inducibility of the subclones still showed a distinction between low and high, with some overlap.

We do not know the reason for this rapid development of variants, but we think it occurs too fast to be readily explained by classical mutation rates. It suggests an epigenetic phenomenon. In any case, workers attempting to assess changes in phenotype in cultured tumor lines (and perhaps *in vivo* as well) should be aware that this can occur.

Dr. Paul: Dr. Hicky in my laboratory has performed an experiment similar to the one described by Dr. Thompson but investigating the inducibility of hemoglobin synthesis by dimethylsulfoxide in individual subclones of the Friend cell clone 707. He has obtained similar results. This cell is not euploid but has a rather stable karyotype. I, therefore, agree with Dr. Thompson that this may represent an epigenetic phenomenon. This could be due to segregation of cytoplasmic factors, not necessarily stable mRNA but, mathematically, the number of factors per cell must be quite small, of the order of 10 or less.

Control of Enzyme Induction and Growth in Tumor Cells and Cell Hybrids

E. Brad Thompson, Marc E. Lippman, and Linda B. Lyons

Laboratory of Biochemistry, National Cancer Institute, National Institutes of Health, Bethesda, Maryland, U.S.A.

Abstract: Hepatoma tissue culture (HTC) cells, a line of rat hepatoma cells in culture, provide a valuable model system for the study of certain differentiated functions in tissue culture. They differ functionally in several ways from L cells, and these contrasting functions have been studied in somatic cell hybrids. In HTC cells but not in L cells glucocorticoids induce the enzyme tyrosine aminotransferase, while L cells but not HTC cells display inhibition of macromolecular synthesis and of glucose utilization in response to such steroids. Both cell types contain specific cytoplasmic steroid receptors. In HTC×L cell hybrids, the aminotransferase is not expressed while the inhibitory responses of the L cell are still present. The hybrids appear to have both L cell and HTC cell steroid receptors.

Although neither parental cell type shows contact inhibited growth, many hybrids between them do so. This restoration of "normal" growth by fusing a pair of transformed parental lines is an intriguing result, the pursuit of which may shed light on problems of growth control.

Malignant cells in tissue culture may be particularly helpful in studying the regulation of gene expression. Their virtues are that they are easy to grow in a controlled environment; they grow relatively rapidly; and they sometimes express functions characteristic of the differentiated state. We have been studying one such line for several years. These hepatoma tissue culture (HTC) cells were placed in culture from the ascites form of Morris hepatoma 7288c, which had been originated and carried in the inbred Buffalo rat (*24*). HTC cells possess the characteristic hepatic enzyme, tyrosine aminotransferase (L-tyrosine: 2-oxoglutarate aminotransferase, EC 2.6.1.5). Upon exposure to physiologic levels of hydrocortisone or other analogous steroids, HTC cell tyrosine aminotransferase is induced from its baseline level to a new level several times higher (*32*), a response much like that of liver *in vivo*. The mechanism by which this induction takes place is under study in several laboratories in the hope that it will shed light on the general problem of gene control in animal cells. Several reviews recently have outlined the state of information with respect to this and other presumably analogous enzyme inductions (*1, 29, 31, 33*). Therefore we will not outline the physiology of the response in this paper. Only 1 or 2 points must be stressed. First, the steroid-provoked induction of the transaminase appears to require RNA and protein synthesis and can be correlated with the accumulation of transaminase-synthesizing polysomes in the cytoplasm (*5, 23, 25, 32*). Second, the induction is thought to require the interaction of steroid with a specific receptor protein in the cytosol,* followed by the movement of the steroid-receptor complex to the nucleus, where it presumably binds to specific DNA-containing sites, (*2, 3*). Binding of steroid to cytosol receptor requires no metabolic energy and occurs at 0°C, but the subsequent steps are temperature-sensitive, complex, and not fully understood.

We have been studying these phenomena on several levels. This paper will outline our results from experiments utilizing the technique of somatic cell hybridization. A brief general review of the use of cell hybrids to study differentiated cell functions has recently appeared (*8*). In the work we will discuss 2 types of cells have been employed: the HTC cells described above, and L cells, a well-known transformed mouse fibroblast line (*9*). L cells lack both basal and inducible hepatic tyrosine aminotransferase but respond to glucocorticoid hormones by showing inhibition of macromolecular synthesis and glucose uptake. These responses occur in only a few tissues *in vivo* and might then be analogous to differentiated functions. The L cell × HTC hybrids can be used, therefore, to ask whether there is dominant or recessive expression of reciprocal, contrasting, steroid-specific responses.

Besides their responses to steroids, the 2 parent lines have other functions which we have utilized in these hybrids as will be seen below. These functions fall into 4 groups, listed in Table 1.

TABLE 1. Some Characteristics of L and HTC Cell Phenotypes

	L	HTC
Group 1		
Thymidine kinase	Absent	Present
Hepatic tyrosine aminotransferase	Absent	Present
Induction of tyrosine aminotransferase by glucocorticoids	Absent	Present
Production of C′2	Absent	Present
Group 2		
Hypoxanthine-guanine phosphoribosyl transferase	Present	Absent
Cyclic AMP		
Content	Higher	Lower
Increase after prostaglandin E	Yes	No
Inhibitory responses to glucocorticoids	Present	Absent
Group 3		
Glucocorticoid receptor in cytoplasm	Present	Present
Group 4		
Contact inhibited growth	No	No

* We here will use the term cytosol to refer to the non-particulate fraction of the cytoplasm, operationally defined as the supernatant fraction after 100,000 × g centrifugation for 30 min.

The first group consists of HTC cell-specific functions, which include thymidine kinase, production of the second component of rat complement (C′ 2) and tyrosine aminotransferase, both basal and steroid-inducible.

The second group of functions are L cell-specific. First is hypoxanthine-guanine phosphoribosyl transferase. The 2 lines of HTC cells used in the present experiments, HTC AR1 and HTC H1, both lack hypoxanthine-guanine phosphoribosyl transferase, having been selected for growth in up to 10^{-4}M 6-mercaptopurine. Other L cell-specific markers are a high content of cyclic AMP relative to that in the HTC cell, known for its extremely low level of this compound (*11*, *22*), the ability of prostaglandin to stimulate increased cyclic AMP levels, and a number of specific inhibitory responses of glucocorticoids (*14*, *26*). These latter responses include a reduction in glucose transport and reduced incorporation of precursors into DNA, RNA, and protein. These responses appear to require some sort of inductive events by the glucocorticoids; *e.g.*, it is believed in L cells that glucocorticoids induce some function(s) responsible for the inhibitory events. The L cell line used in these experiments (LB82) lacks thymidine kinase, having been selected for growth in bromodeoxy uridine (BrdU) (*21*).

The third and fourth groups of functions are possessed by both HTC and L cells. Both types of cell contain specific cytoplasmic glucocorticoid receptors, believed to be essential for expression of each cell's characteristic response to the hormones (*1–3*, *29*). For L cells, this idea has recently been reinforced by the demonstration that steroid-resistent L cells can be isolated and that such cells lack the receptor (*17*). The final phenotype common to both L and HTC cells is that of non-contact inhibited growth. This term is used here simply to mean a lack of density-dependent inhibition of replication under the culture conditions employed in these experiments.

Previous experiments have shown that in heterokaryons and long term hybrids between 2 distinct lines of tyrosine aminotransferase inducible cells of rat origin and non-inducible cells of rat, mouse, or human origin, the inducible enzyme is not expressed (*6*, *7*, *15*, *27*, *28*, *34*). In some cases, induction has been recovered after partial segregation of chromosomes from the hybrid progeny (*7*, *34*). The mechanism for this loss of inducibility in hybrids is presently unknown. One possibility might be that the cytoplasmic steroid receptor is somehow lost or destroyed in the hybrid cell. Loss or alteration of receptor has been correlated with loss of glucocorticoid specific response in several systems (*4*, *17*, *19*, *20*).

In the present studies, therefore, we wished to ask several questions. Would there be extinction of tyrosine aminotransferase in these hybrids? Would the L cell-specific responses to glucocorticoids also be lost? What would happen to the cytoplasmic steroid receptors in the hybrids? In addition, we were interested in expression of the other cell-specific biochemical markers mentioned above. The growth characteristics and cell morphology of the hybrids were also to be observed.

Expression of HTC and L Cell Characteristics in HTC × L Hybrids

The initial hybridization was carried out simply by co-cultivating the HTC

TABLE 2. Biochemical Markers in Hybrid Clone HL5 (HTC AR1×LB82)

	Parental cells		Hybrid
	HTC AR1	LB82	HL5 hybrid
C′2 production (molecules/cell/hr)	65	0	33
Cyclic AMP content (pmole/mg protein)			
Basal	1	5	5
+ Epinephrine[a]	3	5	5–7
+ Prostaglandin E	1	25–50	25–50

[a] Epinephrine and prostaglandin were added to the culture medium at various doses and for various times. The results shown are examples from the optimum dose and time.

TABLE 3. Response of Tyrosine Aminotransferase to Glucocorticoid in HTC Cells, L Cells, and Hybrid Clone HL5

	Parental cells		Hybrid
	HTC AR1	LB82	HL5
Tyrosine aminotransferase[a]			
Control	6.7	1.8	1.6
$+10^{-6}$M dexamethasone PO_4	16.3	1.6	1.4
% immunoreactive[b] (hepatic)	≥90%	0–10%	0–10%
Inhibitory responses[c]			
Glucose utilization: % inhibition at 2 hr[d]	1.3	14	22
Thymidine incorporation control	11,200 cpm	14,400 cpm	11,000 cpm
+ Dexamethasone PO_4	9,500	7,600	3,900
% inhibition	15%	47%	65%

[a] Enzyme activity expressed as nmole product formed/min/mg cell protein. [b] Enzyme inactivated by exposure to antibody known to be non-specific to hepatic tyrosine aminotransferase (*12*). [c] See Refs. *16* and *17*. In each case cells were exposed to 10^{-7}–10^{-6}M dexamethasone phosphate for a period of time appropriate to evoke the maximum response, and the values given here are those for the full response. [d] Assayed by method of Kattwinkel and Munck (*13*). Percent inhibition was calculated relatively to untreated controls.

subline HTC AR1 with LB82 cells in selective medium (*15*). Several colonies of hybrid cells were picked and the remaining colonies pooled. One hybrid colony, HL5, was selected for detailed studies. described below. The constitutive HTC cell marker, C′2 production, continued to be expressed in the hybrids (Table 2). The table also shows that cyclic AMP levels in the hybrid resembled those of the L cell parent, both in basal amount and in their response to selected inducers. These functions therefore represent reciprocal biochemical markers indicating the presence of functional HTC and L cell genes.

As might be predicted, tyrosine aminotransferase was not expressed in the hybrids, nor was it induced by glucocorticoids (Table 3). Furthermore, treatment with actinomycin D up to 5 μg/ml did not evoke the aminotransferase (*30*). The trace amount of apparent tyrosine aminotransferase in L cells is probably not true hepatic enzyme, since it is not inactivated by antiserum prepared against that enzyme. The activity of both basal and induced HTC cells, on the other hand, is virtually all irreversibly lost upon exposure to such antiserum. None of the hybrid

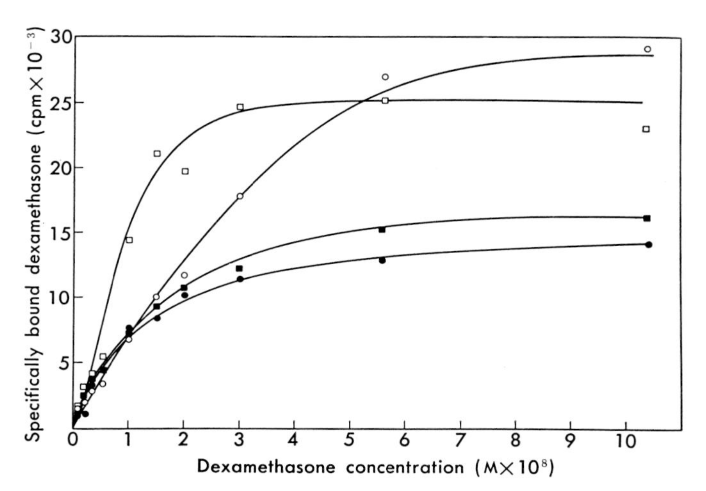

Fig. 1. Saturation curves of specific cytoplasmic glucocorticoid receptors in HTC-H1 (●), L (■), and HL5 (□) cells and in a mixture of HTC and L cell cytosols (○). Binding studies carried out with ^{3}H-dexamethasone as described in Ref. *17*. Figure is reproduced with permission from the publisher (*17*).

enzyme activity present in trace amounts in the hybrid was antibody-sensitive. Once again, this differentiated hepatic function had proved to be recessive.

The L cell group of steroid-specific inhibitory responses, however, continued to be expressed in the hybrids (Table 3). Since the presence of these responses implied that cytoplasmic glucocorticoid receptor is functional in the HL5 hybrid clone, we examined cytosol preparations for receptor sites. As Fig. 1 shows, saturable glucocorticoid-specific sites exist in plenty in the hybrid cells as well as in either parent. These hybrid-cell receptors saturate at dexamethasone concentrations similar to their parents and have similar affinity constants for the steroid.

Distinguishing the receptors from the 2 cell types chemically or immunologically is difficult since neither has yet been purified. However, on the basis of heat inactivation and *in vitro* nuclear binding studies we believe that the HTC and L cell glucocorticoid binding proteins can be distinguished and that both types can be detected in the hybrid (*16*, *17*).

In summary, these studies show a non-reciprocal inhibition of distinctive glucocorticoid responses in HTC × L cell hybrids. The HTC cell-specific induction of tyrosine aminotransferase is lost, while the L cell-specific inhibitory responses remain. Glucocorticoid specific receptor sites exist in the cytoplasm of the HL hybrids. These results led us to wonder whether there might be a difference in L cell and HTC cell glucocorticoid receptors. It might be possible, for instance, that all the HTC-specific receptor was missing and only the L cell type was present in the hybrid, explaining the phenotype of the hybrid cell.

We examined the receptors by allowing them to interact with nuclei *in vitro*. Nuclei were prepared from whole cells and allowed to interact with free steroid or with cytosol plus steroid, preincubated at 20°C so that active steroid-receptor complexes were present. We found that virtually no free steroid bound to nuclei,

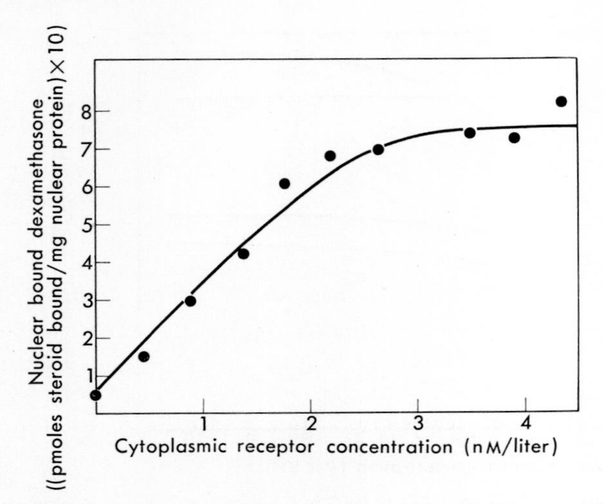

FIG. 2. Example of nuclear binding curve showing saturation of nuclear sites with radioactive dexamethasone (^{3}H) in the presence of glucocorticoid receptor (cytosol). In this case both cytosol and nuclei were from the HL5 hybrid. Experimental technique can be found in Ref. *16*. Briefly, nuclei and cytosol were prepared and separated by homogenization and centrifugation. Then half the nuclei and half the cytosol were incubated for 2 hr at 0°C with radioactive dexamethasone; the other half of each with radioactive plus a 100-fold excess of non-radioactive dexamethasone. Then the nuclei and cytosols were mixed in varying proportions, combining those fractions preincubated with radioactive steroid only in one series and those fractions preincubated with radioactive plus excess non-radioactive steroid in a second series. All were then incubated at 20°C, and at completion of binding, specific nuclear bound steroid calculated by the difference between total nuclear bound counts and nuclear bound counts with excess non-radioactive steroid present. Cytosol receptor concentration was estimated independently as in Fig. 1.

whereas in the presence of cytosol containing receptors, radioactive steroid would bind to the nuclei. This binding was a temperature-sensitive process, because no nuclear binding was seen at 0°C. At 37°C there was brief binding but the system became inactive within a few minutes. At 20°C, the temperature used in all these experiments, binding proceeded stably for several hours. Using this binding reaction, we could study the interaction of various receptor-containing cytosols with specific, saturable nuclear sites (*16*). A typical saturation curve is shown in Fig. 2. Each cell type tested, HTC, L, and HL5 hybrid showed a similar curve to the one displayed when its own cytosol and nuclei were combined.

We next carried out mixing experiments. From curves like that in Fig. 2, the saturation level of steroid-cytosol complex in each type of nucleus with each type of cytosol was determined. Then to the nuclei of a given cell type saturating levels of homologous receptor-steroid complex were added along with additional cytosol containing receptor-steroid complex from another cell type (*17*). We observed that additional binding occurred when L cell cytosol plus radioactive steroid was incubated with HTC nuclei along with saturating amounts of homologous (HTC) cytosol. For example, line one, Part I of Table 4 shows the saturation level of nuclear binding of steroid in HTC nuclei incubated with HTC cytosol. An increment of L cell cytosol "ΔLA9" is additive to the HTC-saturated nuclei (compare lines 3 and 4 with line 1). Furthermore, cytoplasm from the HL5 hybrid is also additive to nuclei already saturated with steroid-receptor from HTC cytoplasm (compare lines 1 and 5). Conversely, L cell nuclei saturated with homologous (L) steroid-receptor complex could bind additional radioactive steroid from HTC or HL5 cytosol (Table 4, Part II). Again it should be emphasized that in the presence of HL5 hybrid cytosol along with homologous cytosol, more radioactive steroid could bind to either nuclear type, than could bind with saturating levels of its own receptor alone. Similar results were obtained when HL5 nuclei were used.

In contrast, when any of the 3 types of nuclei were first saturated with HL5 cytosol and its steroid-receptor complex, no additional radioactive steroid could bind when either L cell or HTC cytosol was added. An example of this is shown, using HL5 nuclei, in Table 4, Part III. These results suggest that nuclear binding can be used to distinguish L cell from HTC glucocorticoid receptors. It seems that these receptors differ from one another with respect to their nuclear binding *in vitro*, and that the hybrid cell, HL5, contains material capable of saturating both L cell and HTC nuclear receptor sites. These results are expressed in pseudo-equation form below, where the expressions HTC_{cyt}, L_{cyt}, or $HL5_{cyt}$ refer to the appropriate cytosol receptor carrying steroid in an activated complex.

HTC_{cyt} + nuclei $\rightarrow$ nuclei-HTC_{cyt} (saturated for HTC receptor)
Nuclei-HTC_{cyt} + L_{cyt} $\rightarrow$ nuclei-HTC_{cyt}-L_{cyt} (L receptor increment bound)

or

L_{cyt} + nuclei $\rightarrow$ nuclei-L_{cyt} (saturated for L receptor)
Nuclei-L_{cyt} + HTC_{cyt} $\rightarrow$ nuclei-L_{cyt}-HTC_{cyt} (HTC receptor increment bound)

or

$$\left.\begin{array}{c}HTC_{cyt}\\ \text{or}\\ L_{cyt}\end{array}\right] + \text{nuclei} \rightarrow \left[\begin{array}{c}\text{nuclei-}HTC_{cyt}\\ \text{or}\\ \text{nuclei-}L_{cyt}\end{array}\right. \quad \text{(saturated for HTC or L receptor)}$$

$$\left.\begin{array}{c}\text{Nuclei-}HTC_{cyt}\\ \text{or}\\ \text{Nuclei-}L_{cyt}\end{array}\right] + HL5 \rightarrow \left[\begin{array}{c}\text{nuclei-}HTC_{cyt}\text{-}HL5_{cyt}\\ \text{or}\\ \text{nuclei-}L_{cyt}\text{-}HL5_{cyt}\end{array}\right. \quad \text{(HL5 hybrid receptor bound)}$$

but

$HL5_{cyt}$ + nuclei $\rightarrow$ nuclei-$HL5_{cyt}$ (saturated for HL5 receptor)
Nuclei-$HL5_{cyt}$ + HTC_{cyt} $\not\rightarrow$ no HTC receptor increment can be bound

or

Nuclei-$HL5_{cyt}$ + L_{cyt} $\not\rightarrow$ no L receptor increment can be bound

The fact that saturable nuclear binding occurs *in vitro* does not show that the sites we observe are physiologic; it does, however, provide a way of distinguishing various receptors. Figure 3 shows a model depicting separate nuclear receptor sites for L cell and HTC cell receptor. Although the figure depicts the sites as mutually exclusive, we cannot be certain at this point whether these nuclear sites are unique for each of the 2 receptors or whether there is some degree of overlap. The failure of the hybrids to express tyrosine aminotransferase therefore cannot be ascribed to loss of the HTC receptor in the hybrid. It also appears that L cell receptor cannot substitute for HTC receptor, since the hybrid contains both types but does not express the aminotransferase.

TABLE 4. Nuclear Binding of Glucocorticoid-cytosol Receptor Complex

	Cytosol	Radioactivity bound in nuclei specified (pmoles ^{3}H steroid bound/mg nuclear protein)
		HTC nuclei
I.	HTC[a]	11 (9–12)[b]
	LA9	7 (6–7.5)
	HTC+ΔLA9	13.5 (13–14)
	ΔLA9	2 (1.8–2.2)
	HTC+ΔHL5	18 (17–19)
		LA9 nuclei
II.	LA9	5.3 (4.7–5.9)
	HTC	12 (11–13)
	LA9+ΔHTC	8 (6–9.5)
	ΔHTC	5 (4.5–6)
	LA9+ΔHL5	10 (9–10.5)
		HL5 nuclei
III.	HL5	21 (18–24)
	LA9	7 (6.5–8.5)
	HTC	11.5 (10–13)
	HL5+ΔHTC	19.5 (16.5–22)
	HL5+ΔLA9	18 (17–19)
	ΔLA9+ΔHTC	20 (17.5–22.5)

[a] Source of cytosol. Saturating levels used unless preceded by Δ, which signifies subsaturating level. Thus for example, HL5 + ΔHTC indicates a saturating amount of HL5 cytosol + an additional subsaturating amount of HTC cytosol. [b] Data given as an average of 4–8 assays with range are shown in parentheses.

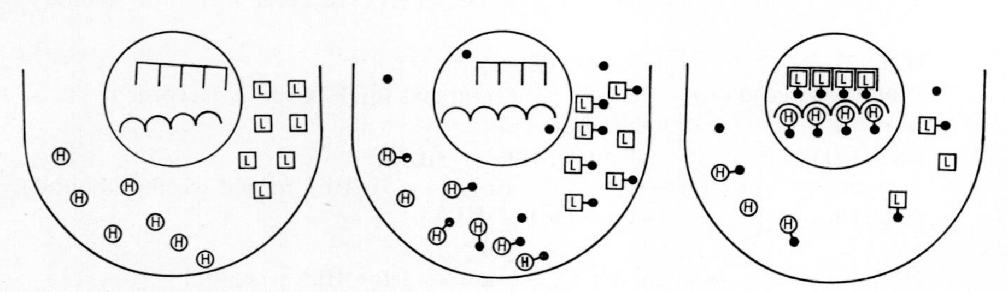

FIG. 3. Model of cytosol and nuclear steroid receptors in HTC×L cell hybrids. Left, hybrid cell in absence of steroid. Cytosol contains 2 classes of receptor, L-cell (L) and HTC-cell (H). Nucleus has unoccupied sites specific for each. Center, steroid (●) is added and binds to cytosol receptors. Right, activated receptor-steroid complexes have entered nucleus and saturated their respective nuclear sites.

It is possible that the tyrosine aminotransferase-bearing chromosomes have been lost from all these hybrids. We feel that this is unlikely since this result requires that the aminotransferase-chromosome be lost in preference to others at a very early time after hybrid formation. No evidence is available to support this; and as we have mentioned, the aminotransferase is not measurable in heterokaryons when both nuclei are present and also has been recovered in hybrids after considerable chromosomal loss (*7*, *28*, *34*). In other experiments, we also found that HTC receptor cannot substitute for L cell receptor. By fusing receptor-containing HTC cells with a receptor-lacking L cell variant line, we obtained hybrids which contained receptor, but which nevertheless failed to demonstrate the typical L cell responses to glucocorticoids (*17*). Furthermore, mere presence of receptor, even of the homologous cell type, does not guarantee steroid responsiveness. We have described several lines of cells with full levels of receptor which are nevertheless steroid unresponsive (*18*).

Growth Control in HTC×L Cell Hybrids

Another very interesting phenotypic aspect of these hybrids was their mode of growth. Both the L cell and HTC cell parents continued to grow even after reaching confluence. The hybrids, on the other hand, did not do so. This became apparent to us as we watched their behavior from day to day, and we became interested in applying to these cells some of the available standards for evaluating the nature of growth. In addition, we wished to see what happened to the growth characteristics of these hybrids as they gradually segregated out more and more chromosomes.

All the initial HTC×L cell hybrids seemed to show this new mode of growth (*15*). Of this group, we have studied the HL5 clone, used for the steroid receptor experiments described above, for over 2 years. After about a year's growth in nonselective medium, it altered its style of growth, becoming less regular in morphology, piling up, and continuing to grow even after becoming confluent. At this time its chromosome content had been reduced from the initial 106 to about 80 chromosomes. Since then, the hybrid has continued to segregate chromosomes and presently contains about 55. Its current morphology and growth pattern seen by phase-contrast light microscopy resemble those of its L cell parent. Some of the characteristics of these cell lines are summarized in Table 5.

To study these phenomena further, we have made a series of new HTC×L cell hybrids (Lyons and Thompson, manuscript in preparation). For these we have used a newly developed HTC parent, HTC H1, which lacks the phosphoribosyl transferase and is inducible for tyrosine aminotransferase. We fused these with the thymidine kinase-free LB82, assisting the fusion with β-propiolactone-inactivated

TABLE 5. Growth Characteristics of HTC and L Cells Compared with Hybrid Clone HL5 before and after Partial Segregation

Cell line	Chromosome mode	Thymidine incorporation at confluence	Growth	Shape	Nucleus / cytoplasm	Plasma membrane
HTC AR1	62	High	Not contact inhibited	Resemble epitheloid tumor	High	Birefringent
LB82	50	High	Not contact inhibited	Fibroblastic	High	Birefringent
HL5 (original clone)	106	Low	Contact inhibited	Resemble epithelial monolayer	Low	Only slightly birefringent
HL5 (later)	80	High	Not contact inhibited	Resemble L cells	High	Birefringent

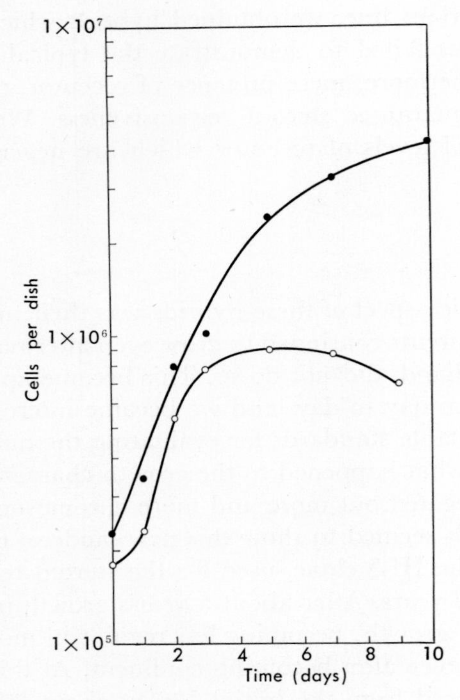

FIG. 4. Comparison of growth of a typical freshly isolated contact inhibited HTC×L cell hybrid clone (○) with the non-contact inhibited HL5 late segregant (●). Equal numbers of cells were plated into a series of tissue culture dishes on day zero and fed every other day. At each time point shown, dishes of cells were removed for trypsinization and counting.

Sendai virus. In this case, we tried to estimate the frequency with which hybrids showing "contact inhibited" morphology were obtained. By observing the initial colonies of hybrid cells arising in selective medium designed to eliminate the parental cells, it was clear that in this Sendai promoted fusion at least, not all hybrid colonies display "contact inhibited" type of growth. However, a significant fraction of the hybrid colonies in each fusion experiment do appear to do so. We estimate this fraction to represent perhaps one-quarter to one-half of all the hybrids.

From these new fusions we have again selected several hybrid colonies, representing independent fusion events, and have begun to examine them for their growth characteristics. Figure 4 compares the growth curve of a typical hybrid with that of a non-contact inhibited cell line. The non-contact inhibited cell shown is the HL5 segregant after it lost contact inhibited growth, but the growth curve shown by L cells or HTC cells is virtually identical. As can be seen, the contact inhibited hybrid reaches maximum cell density and stops growing at a much lower cell number than the non-contact inhibited line.

The morphology of the new "contact inhibited" hybrids is variable, but different from either parent. Some clones consist of flat epithelial cells; others resemble 3T3 cells, a more fibroblastic morphology. In a typical contact inhibited hybrid clone, some of the cells at the margin of the colony are dividing and all appear to

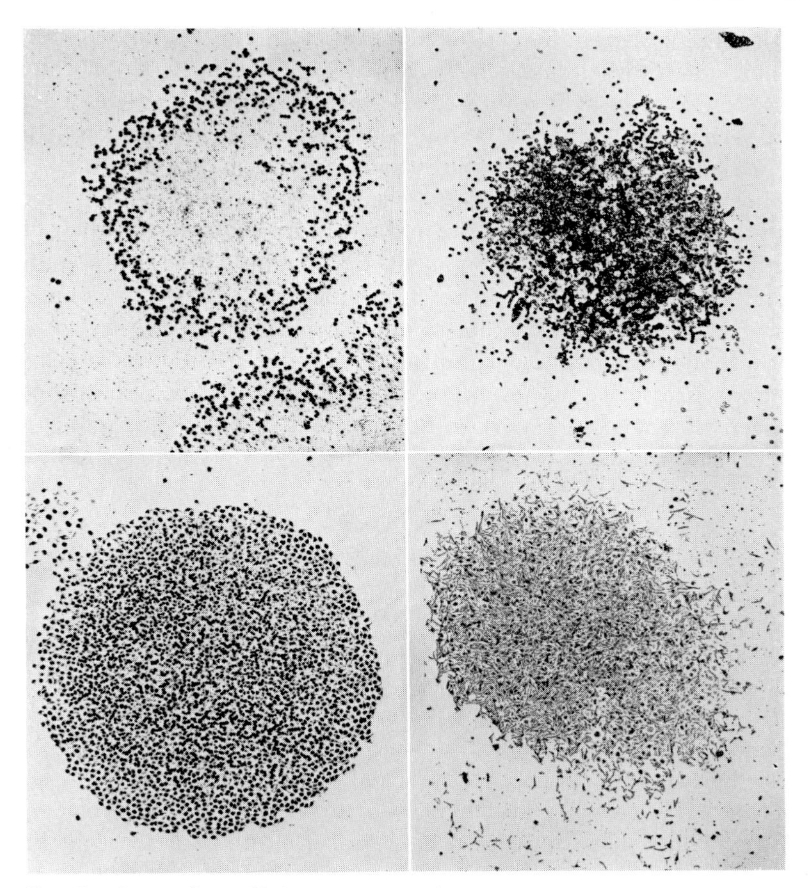

FIG. 5. Comparison of ^{3}H-thymidine uptake by colonies of various cells and hybrids. Colonies were grown from single cells to appropriate size and then exposed several hours to the radioactive nucleoside. After labelling, unincorporated radioactivity was removed, the cells washed and fixed, and radioautographs prepared by methods previously described (*10, 15*). Upper right, HTC-H1; lower right, LA9; upper left, fresh unsegregated HTC×L cell hybrid; lower left, HL5 segregant which has lost density dependent growth control.

have motile membranes, while the cells deep in the colony are dividing rarely and have quiescent membranes. This behavior can be verified biochemically by growing up colonies, exposing them to ^{3}H-thymidine, washing to remove unincorporated thymidine, and preparing autoradiographs. Under proper conditions, a contact inhibited colony will evidence its nature by incorporating the label only in the dividing peripheral cells, producing an annular labelling pattern (*10*). In contrast, non-contact inhibited cells will show uptake of label randomly through each colony, with little or no concentration of uptake at the margins. Figure 5 compares labelling patterns exemplified by HTC cells, L cells, contact inhibited HL hybrids, and the non-contact inhibited HL5 revertant. Since the LB82 parent lacks thymidine kinase and cannot incorporate thymidine into DNA, the labelling pattern of a clone of LA9 cells is shown. It can be seen that the parental lines (upper right and lower right) and the HL5 segregant (lower left) label in the center as well as at the margins of their colonies, while a relatively unsegregated fresh H×L colony (upper left) labels predominantly at its margins. We are currently assessing a series of HL hybrids by a series of parameters commonly used to evaluate growth characters. In addition, we are beginning to explore the biochemistry of these cells.

To conclude, it seems that fusion of the rat hepatoma line HTC with chemically transformed mouse L cells can produce hybrids which display growth characteristics different from either parent. The parents do not show density dependent inhibition of growth, while the hybrids do so. It is as if the parental lines have provided complimentary elements necessary for contact inhibition to the hybrids. If this is indeed so, these complementary function(s) may depend on the chromosomal make-up of the hybrids, because after considerable segregation of chromosomes, one hybrid has been observed to revert to non-contact inhibited growth. Of course it may be that the chromosome loss and loss of density dependent growth are merely coincident.

CONCLUSIONS

HTC cells provide a model system convenient for the study of certain differentiated functions in tissue culture. One approach, discussed here, has been the use of somatic cell hybrids to investigate such functions. In short-term heterokaryons and long-term hybrids between HTC cells containing inducible tyrosine aminotransferase and L cells (or other cell types) lacking the enzyme, the inducible enzyme is lost. On the other hand, L cell-specific responses to steroids are inhibitory in nature. These responses are still expressed in HTC×L cell hybrids. Glucocorticoid-specific receptors are found in the cytosol fraction of HTC cells and L cells; HTC×L cell hybrids also contain glucocorticoid receptors. Nuclear binding studies suggest that L cell and HTC receptors differ and that both types exist in the hybrids. Thus, the failure of tyrosine aminotransferase to be expressed in hybrids cannot be due to loss of HTC-type receptor. Also L cell receptor cannot substitute for that of the HTC cells.

These results suggest that the negative control of tyrosine aminotransferase expression in hybrids is a relatively specific event. One attractive model to explain these results is that there are specific gene sites required to interact with the HTC cytosol receptor-steroid complex in order to allow expression of the aminotransferase. Such sites may be masked in the hybrid, perhaps due to the dominant effect of some protein(s) present in the non-inducible cell. Of course, many other possibilities exist and cannot be ruled out by the existing data. The fact that heterokaryons also fail to express this enzyme, however, suggests that the negative element in the cell lacking the aminotransferase must have access to the cytoplasm.

Although neither HTC nor L cells are contact inhibited, hybrids between the 2 are contact inhibited. It is as though each parent has lost contact inhibited growth by a different mechanism, and the hybrids restore these functions in a complementary fashion. The molecular mechanisms for these phenomena remain to be elucidated, but offer promising areas for exploring gene control in animal cells.

REFERENCES

1. Baxter, J. D. and Forsham, P. H. Tissue effects of glucocorticoids. Am. J. Med., *53*: 573–589, 1972.
2. Baxter, J. D. and Tomkins, G. M. The relationship between glucocorticoid binding and tyrosine aminotransferase induction in hepatoma tissue culture cells. Proc. Natl. Acad. Sci. U.S., *65*: 709–715, 1970.
3. Baxter, J. D. and Tomkins, G. M. Role of DNA and specific cytoplasmic receptors in glucocorticoid action. Proc. Natl. Acad. Sci. U.S., *69*: 1892–1896, 1972.
4. Baxter, J. D., Harris, A. W., Tomkins, G. M., and Cohn, M. Glucocorticoid receptors in lymphoma cells in culture: relationship to glucocorticoid killing activity. Science, *171*: 189–191, 1971.
5. Beck, J. P., Beck, G., Wong, K. Y., and Tomkins, G. M. Synthesis of inducible tyrosine aminotransferase in a cell-free extract from cultured hepatoma cells. Proc. Natl. Acad. Sci. U.S., *69*: 3615–3619, 1972.
6. Benedict, W. F., Nebert, D. W., and Thompson, E. B. Expression of aryl hydrocarbon hydroxylase induction and suppression of tyrosine aminotransferase induction in somatic-cell hybrids. Proc. Natl. Acad. Sci. U.S., *69*: 2179–2183, 1972.
7. Croce, C. C., Litwack, G., and Koprowski, H. Human regulatory gene for inducible tyrosine aminotransferase in rat-human hybrids. Proc. Natl. Acad. Sci. U.S., *70*: 1268–1272, 1973.
8. Davis, F. M. and Adelberg, E. A. Use of somatic cell hybrids for analysis of the differentiated state. Bacteriol. Rev., *37*: 197–214, 1973.
9. Earle, W. R. Production of malignancy *in vitro*. IV. The mouse fibroblast cultures and changes seen in the living cells. J. Natl. Cancer Inst., *4*: 165–212, 1943.
10. Fisher, H. W. and Yeh, J. Contact inhibition in colony formation. Science, *155*: 581–582, 1967.
11. Granner, D., Chase, L. R., Aurbach, G. D., and Tomkins, G. M. Tyrosine aminotransferase: enzyme induction independent of adenosine 3′,5′-monophosphate. Science, *162*: 1018–1020, 1968.
12. Hayashi, S., Granner, D. K., and Tomkins, G. M. Tyrosine aminotransferase. Purification and characterization. J. Biol. Chem., *242*: 3998–4006, 1967.
13. Kattwinkel, J. and Munck, A. Activities *in vitro* of glucocorticoids and related steroids on glucose uptake by rat thymus cell suspensions. Endocrinology, *79*: 387–390, 1966.
14. Kemper, B. W., Pratt, W. B., and Aronow, L. Nucleic acid synthesis in intact nuclei isolated from mouse fibroblasts: characterization of the system and effects of glucocorticoids. Mol. Pharmacol., *5*: 507–531, 1969.
15. Levisohn, S. R. and Thompson, E. B. Contact inhibition and gene expression in HTC/L cell hybrid lines. J. Cell Physiol., *81*: 225–232, 1973.
16. Lippman, M. E. and Thompson, E. B. Differences between cytoplasmic glucocorticoid binding proteins shown by heterogeneity of nuclear acceptor sites. Nature, *246*: 352–355, 1973.
17. Lippman, M. E. and Thompson, E. B. Steroid receptors and the mechanism of the specificity of glucocorticoid responsiveness of somatic cell hybrids between hepatoma tissue culture cells and mouse fibroblasts. J. Biol. Chem., *249*: 2483–2488, 1974.
18. Lippman, M. E., Perry, S., and Thompson, E. B. Cytoplasmic glucocorticoid bind-

ing proteins in glucocorticoid unresponsive human and mouse leukemic cell lines. Cancer Res., *34*: 1572–1576, 1974.

19. Lippman, M. E., Halterman, R. H., Leventhal, B. G., Perry, S., and Thompson, E. B. Glucocorticoid-binding proteins in human acute lymphoblastic leukemic blast cells. J. Clin. Invest., *52*: 1715–1725, 1973.
20. Lippman, M., Halterman, R., Perry, S., Leventhal, B., and Thompson, E. B. Glucocorticoid binding proteins in human leukemic lymphoblasts. Nature New Biol., *242*: 157–158, 1973.
21. Littlefield, J. W. The use of drug-resistant markers to study the hybridization of mouse fibroblasts. Exp. Cell Res., *41*: 190–196, 1966.
22. Makman, M. H. Conditions leading to enhanced response to glucagon, epinephrine, or prostaglandins by adenylate cyclase of normal and malignant cultured cells. Proc. Natl. Acad. Sci. U.S., *68*: 2127–2130, 1971.
23. Miller, J. V., Jr., Cuatrecasas, P., and Thompson, E. B. Partial purification by affinity chromatography of tyrosine aminotransferase-synthesizing ribosomes from hepatoma tissue culture cells. Proc. Natl. Acad. Sci. U.S., *68*: 1014–1018, 1971.
24. Odashima, S. and Morris, H. P. Studies on the conversion of Morris transplantable hepatomas into ascitic form. GANN Monograph, *1*: 55–64, 1966.
25. Peterkofsky, B. and Tomkins, G. M. Effect of inhibitors of nucleic acid synthesis on steroid-mediated induction of tyrosine aminotransferase in hepatoma cell cultures. J. Mol. Biol., *30*: 49–61, 1967.
26. Pratt, W. B. and Aronow, L. The effect of glucocorticoids on protein and nucleic acid synthesis in mouse fibroblasts growing *in vitro*. J. Biol. Chem., *241*: 5244–5250, 1966.
27. Schneider, J. A. and Weiss, M. C. Expression of differentiated functions in hepatoma cell hybrids. I. Tyrosine aminotransferase in hepatoma-fibroblast hybrids. Proc. Natl. Acad. Sci. U.S., *68*: 127–131, 1971.
28. Thompson, E. B. and Gelehrter, T. D. Expression of tyrosine aminotransferase activity in somatic-cell heterokaryons: evidence for negative control of enzyme expression. Proc. Natl. Acad. Sci. U.S., *68*: 2589–2593, 1971.
29. Thompson, E. B. and Lippman, M. E. Mechanism of action of glucocorticoids. Metabolism, *23*: 159–202, 1974.
30. Thompson, E. B., Granner, D. K., and Tomkins, G. M. Superinduction of tyrosine aminotransferase by actinomycin D in rat hepatoma (HTC) cells. J. Mol. Biol., *54*: 159–175, 1970.
31. Thompson, E. B., Levisohn, S. R., and Miller, J. V., Jr. Steroid control of tyrosine aminotransferase in hepatoma tissue culture (HTC) cells. *In;* V. H. T. James and L. Martini (eds.). Hormonal Steroids (Proc. 3rd Int. Congr. on Hormonal Steroids), pp. 463–471, Excerpta Medica, Amsterdam, 1971.
32. Thompson, E. B., Tomkins, G. M., and Curran, J. F. Induction of tyrosine α-ketoglutarate transaminase by steroid hormones in a newly established tissue culture cell line. Proc. Natl. Acad. Sci. U.S., *56*: 296–303, 1966.
33. Tomkins, G. M., Gelehrter, T. D., Granner, D., Martin, D., Jr., Samuels, H. H., and Thompson, E. B. Control of specific gene expression in higher organisms. Science, *166*: 1474–1480, 1969.
34. Weiss, M. C. and Chaplain, M. Expression of differentiated functions in hepatoma cell hybrids: reappearance of tyrosine aminotransferase inducibility after the loss of chromosomes. Proc. Natl. Acad. Sci. U.S., *68*: 3026–3030, 1971.

Discussion of Paper of Drs. Thompson et al.

DR. PAUL: The HTC×L cell hybrids involve an interspecific cross between rat and mouse cells. Is there a specific loss of rat chromosomes from these hybrids?

DR. THOMPSON: An interesting question. We are trying to establish differential chromosome strains to answer it. From gross morphology we see no preferential loss of either rat or mouse chromosomes, but this is a weak criterion. We have noted that our Sendai-fused cells seem to be segregating more quickly than the original spontaneously fused ones. I wonder whether anyone else has noted this phenomenon?

DR. WEBER: It may well be too early perhaps to ask you whether your results support the conclusions of Dr. H. Harris and Dr. G. Klein which indicate that after fusion with the progressive loss of chromosomes malignancy reappears? Your data showing the reappearance of loss of contact inhibition might point in this direction.

DR. THOMPSON: I would rather not comment yet about whether our results pertain to the question of suppression of malignancy in hybrids, since contact inhibited growth sometimes is not a true indicator of tumorigenicity (or lack of it). We have just begun experiments testing the tumorigenicity of our hybrids in nude mice, and I prefer to await the results of those experiments.

DR. OHNO: What did you think of Dr. H. Koprowski's experiment? In human-rat hepatoma cell hybrids, tyrosine aminotransferase is said to remain *non-inducible* so long as the human X is kept by hybrids.

DR. THOMPSON: I think those are very interesting experiments which deserve an effort at confirmation. We are now attempting to do so. Of course, finding repression of tyrosine aminotransferase to be syntonic with the human X chromosome in an interspecific hybrid, although of great interest, is still not quite the ideal experiment. That would be to locate the putative repressor gene in an intraspecific cross.

DR. IKAWA: Have you ever tried isolation of flat type revertants out of the original HTC cells?

Dr. Thompson: We have tried many times to isolate for flat or contact inhibited revertants of HTC or L cells, but with no success. We have not tried colcemid to enhance the appearance of flat cells, however.

Dr. Ikawa: I have isolated a flat type revertant from Kirsten MSV non-producer of BALB/3T3 origin. The cells of the revertant produced tumors in the syngeneic hosts with longer latency. Surprisingly enough, the explants of the tumor also showed flat type colonies and showed favorable contact inhibition.

CONTROL OF DIFFERENTIATION

SUPPRESSION OF DIFFERENTIATION

Reversible Suppression of Differentiation and Malignancy in Bromodeoxyuridine-treated Melanoma Cells

Selma Silagi and Jean R. Wrathall

Laboratory of Cell Genetics, Department of Obstetrics and Gynecology, Cornell University Medical College, New York, N.Y., U.S.A.

Abstract: Culture of melanotic melanoma cells (clone $B_5$59), derived from the B16 mouse melanoma, with 1–3 μg/ml of 5-bromodeoxyuridine (BrdU) results in the following major effects: (1) alteration in cell morphology and growth pattern; (2) suppression of melanogenesis; (3) suppression of tumorigenic potential; and (4) greatly increased production of C-type virus. These effects appear to be mediated by incorporation of BrdU into DNA as an analog of thymidine, but are reversible upon growth of BrdU-treated cells in medium without the drug. During a 7-day period of growth with BrdU, the highly melanotic cells which usually grow as piled foci of spindle-shaped or dendritic cells become amelanotic and flat and exhibit "contact inhibition." Melanogenesis is suppressed through a coordinated reduction of tyrosinase activity and premelanosome formation. A dialyzable, heat-stable inhibitor of tyrosinase activity appears in BrdU-treated cells, but is quantitatively insufficient to account for the observed reduction of tyrosinase activity. A single cell cycle (30 hr) with BrdU (3 μg/ml) is sufficient to completely suppress tyrosinase activity by 7 days, and the kinetics of reduction are similar to those from cultures grown with BrdU for this entire period. Cells grown continuously in BrdU are unpigmented, fibroblastic, and exhibit no tyrosinase activity or evidence of melanosomes or premelanosomes. The reduction of tumorigenicity by BrdU is time and dose dependent. No tumors are formed when cells grown for 14 days with BrdU (3 μg/ml) are injected into syngeneic adult mice. Such treated cells produce tumors in many neonates and immunosuppressed adult mice, indicating an immunological component in the reduction of tumorigenicity by BrdU. Further, injections of non-tumorigenic BrdU-treated melanoma cells protect mice against tumor formation by untreated melanoma cells. Production of C-type virus is greatly increased and virus-associated Gross cell surface antigen becomes detectable on BrdU-treated cells. The number of virus particles produced by these cells appears to be positively correlated with their ability to protect mice against untreated melanoma cells. The use of BrdU with these melanoma cells offers a probe to dissect cellular control of both differentiated function and malignant potential.

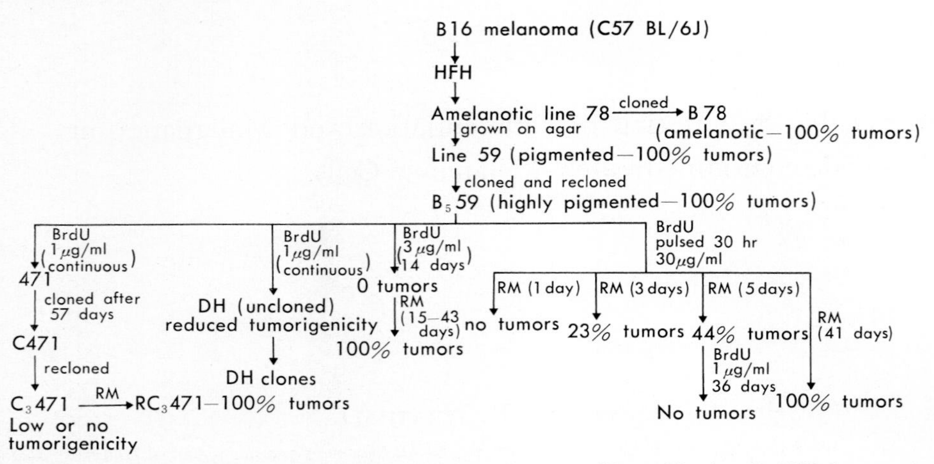

FIG. 1. Lineage and treatment of melanoma cell clones. (From Silagi *et al.*, 1972, by permission from the National Academy of Sciences.) RM: normal medium without BrdU.

The use of cultured mouse melanoma cells provides a model system in which the control of a complex differentiated function, melanogenesis, and the ability to form lethal tumors in syngeneic animals, *i.e.* malignancy, may be studied. We have been using cells derived (*11*) from the B16 spontaneous mouse melanoma (C57BL/6) for such studies. The lineage and treatment of these cells are shown in Fig. 1. The thymidine analogue, 5-bromodeoxyuridine (BrdU), has allowed experimental manipulation of both differentiated function and tumorigenicity in these cells. BrdU has been shown to suppress differentiation in a variety of other cell types, both embryonic and adult (*2, 4, 8, 12, 16, 32, 35–37*). The use of BrdU in the melanoma cell system is providing a powerful probe to dissect cellular control of both differentiated function and malignant potential (*25, 28–31, 38*).

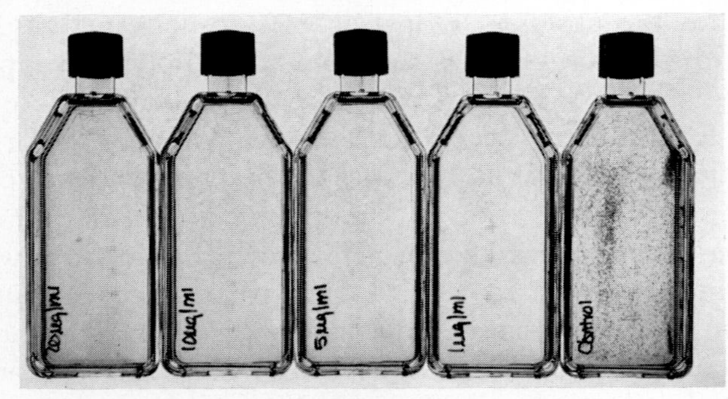

FIG. 2. Effect of BrdU on pigment production in embryonic iris epithelium. Only pigmented colonies are visible. Unpigmented cells cover almost the entire surface of the flasks containing BrdU. Pigmented iris epithelium was from 15-day chick embryos. Flasks of 6th subculture photographed 9 days after beginning treatment with BrdU at the labeled concentrations. Each flask was seeded with 5×10^5 pigmented cells and allowed to grow in Eagle's minimal medium supplemented with 10% fetal calf serum and 1% chick embryo extract for 4 days before treatment began. (From Silagi and Bruce, 1970, by permission from the National Academy of Sciences.)

1. Reversible Suppression of Melanogenesis

A. Pigmented embryonic tissues

Primary cultures of pigmented iris epithelium from 15-day-old chick embryos divided in culture and remained pigmented for at least 9 subcultures. When such cultures were grown in the presense of BrdU (1–20 μg/ml) for 9 days the cells continued to divide but became unpigmented (Fig. 2) (*30*). Similar results have been reported for primary cultures of pigmented retinal cells from 8-day-old chick embryos (*8*) when grown with BrdU (3–30 μg/ml). The suppression was reversible and

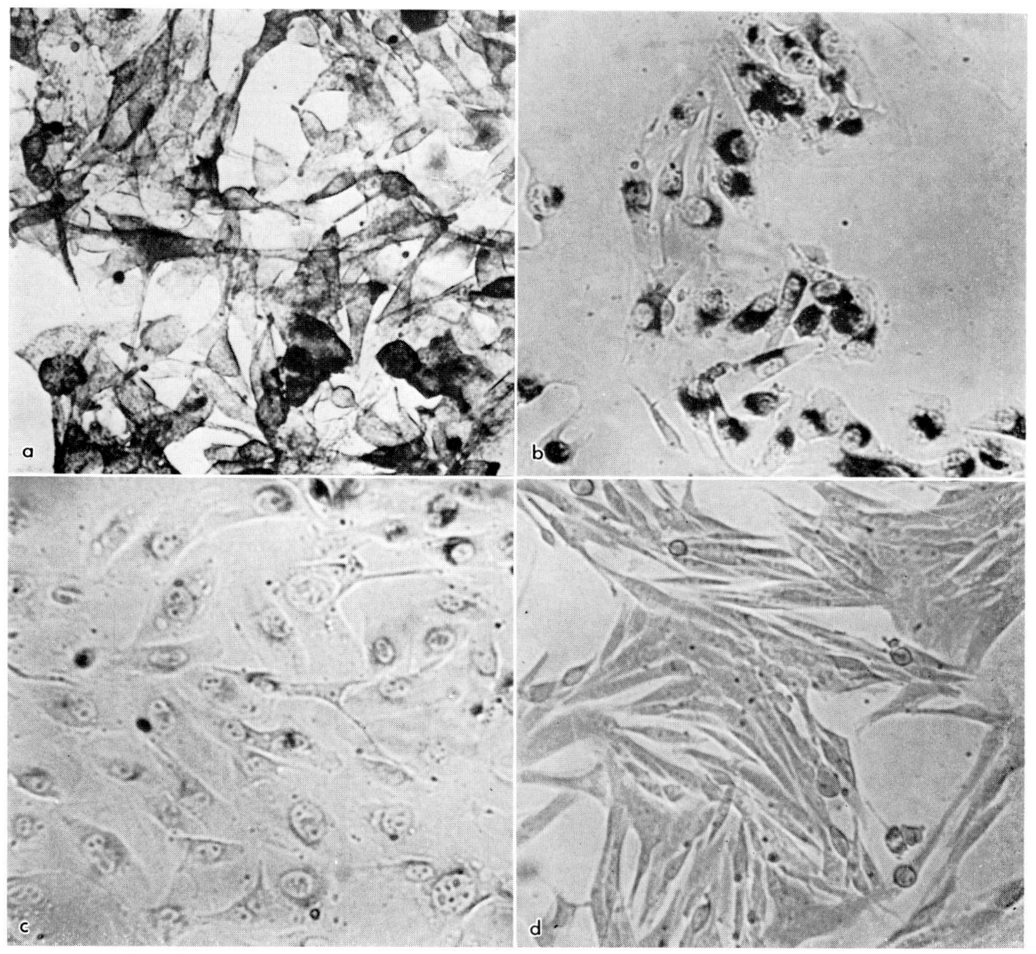

FIG. 3. a. $B_5$59 melanoma cells grown in regular medium. Living cells growing in a piled reticulated fashion. (×180). b. $B_5$59 cells grown for 3 days with 3μg of BrdU/ml. Living cells which have become disassociated and whose melanin granules are concentrated in a juxtanuclear position. (×180). c. $B_5$59 cells grown for 7 days with 3 μg of BrdU/ml. Living cells are nearly amelanotic. (×180). d. $C_3$471 cells derived from long-term culture of $B_5$59 cells with 1 μg of BrdU/ml. Living cells. Completely amelanotic cells grow in a fibroblastic manner. (×180). (From Wrathall *et al.*, 1973, by permission of the Rockefeller University Press.)

pigmented cells reappeared upon culture of the treated cells in medium without BrdU.

B. Morphological effects in melanoma cells

Whenever melanotic melanoma cells (clone $B_5 59$) are grown in medium (Eagle's minimal medium with 10% fetal calf serum) containing BrdU (1–3 μg/ml) for 48 hr to 1 week, there is little or no effect upon growth of the cells, but there is a marked effect on pigment production and cell morphology (*30, 38*). Untreated $B_5 59$ cells grow in a reticulated, piled manner, and are generally spindle shaped or dendritic, with melanin granules dispersed throughout the cytoplasm (Fig. 3a). Growth in BrdU-containing medium causes alterations in morphology which are discernible by 2 days and obvious by 3 days (Fig. 3b). Cells disengage from their piles, flatten and enlarge. Melanin granules become concentrated near the nucleus, with concomitant loss of pigmentation from cytoplasmic processes, the latter becoming almost invisible in unstained specimens, except with phase microscopy. During this period, cell division is virtually unaffected by BrdU concentrations of less than 5 μg/ml (Fig. 4). When grown for longer periods with BrdU the cells become more flattened, exhibit "contact-inhibition" of growth and increased adhesiveness to the surface on which they are growing. By 7 days of treatment many cells

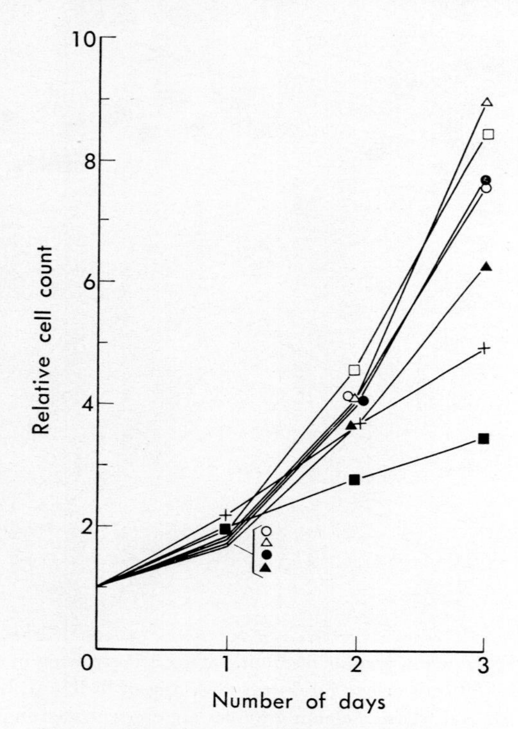

Fig. 4. Growth curves of melanotic melanoma cells grown in medium supplemented by concentrations of BrdU ranging from 0 to 20 μg/ml. □ control; △ 5; ● 3; ○ 1; ▲ 8; + 10; ■ 20 μg/ml. (From Silagi and Bruce, 1970, by permission from the National Academy of Sciences.)

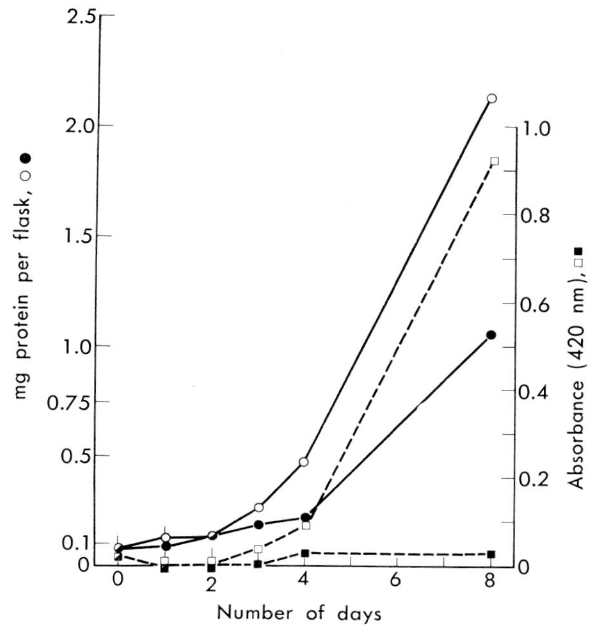

FIG. 5. Comparison between growth (as determined by protein concentration) and melanin concentration, in melanotic melanoma cells grown in normal culture medium (○, □) and in medium supplemented by 3.3×10^{-6}M BrdU (1 μg/ml) (●, ■). (From Silagi *et al.*, 1972, by permission from the National Academy of Sciences.)

appear amelanotic by light microscopy (Fig. 3c). These observations are supported by measurements of changes in melanin concentration during culture of cells with BrdU (Fig. 5).

All of these effects can be prevented by addition of equimolar or greater quantities of thymidine with the BrdU. None of them appear if during BrdU treatment the cells are prevented from synthesizing DNA by sublethal concentrations (0.5–1 μg/ml) 1-β-D arabinofuranosylcytosine (Ara-C). Cells in medium to which both Ara-C and BrdU (3 μg/ml) are added appear indistinguishable for at least 48 hr from cells to which Ara-C only or no additive is added, after which toxic effects of Ara-C predominate. No effects of BrdU can be discerned at any time in such non-dividing cultures.

Continuous growth in 1–3 μg of BrdU/ml produces unpigmented lines, and single-cell clones have been derived from these lines. Clones grown in 1 μg/ml of BrdU tend to grow in parallel rows and appear similar to fibroblasts. Cells of one of these, clone $C_3$471, which has been used extensively in studies of the BrdU effects, are shown in Fig. 3d.

C. *Effects on tyrosinase activity*

Tyrosinase (*o*-diphenol: oxygen oxidoreductase) catalyses the conversion of tyrosine to melanin (*15, 22*). Vertebrate melanin synthesis occurs *in vivo* within specialized tyrosinase-containing organelles, the melanosomes. Tyrosinase appears

TABLE 1. Distribution of Tyrosinase Activity in Subcellular Fractions

Cellular fraction	Tyrosinase specific activity (units/mg protein)	Percent control[a)]
Control[b)]		
Nuclear[c)]	0.0093	—
Large particulate	0.0129	—
Supernatant	0.0154	—
Total	0.0134[d)]	
BrdU-treated[b)]		
Nuclear[c)]	0.0006	6.5
Large particulate	0.0022	17.1
Supernatant	0.0014	9.1
Total	0.0011[d)]	

[a)] (Specific activity of BrdU treated/specific activity of control) × 100. [b)] Replicate cultures of $B_5$59 cells were grown for 3 days in regular medium (control) or medium containing 3 μg of BrdU/ml (BrdU treated). [c)] Contained melanin granules sedimenting with the nuclei. [d)] Total units/total mg protein.

to be synthesized by the ribosomes and transported through endoplasmic reticulum to the Golgi area (*27*). Here tyrosinase-containing new premelanosomes appear to be formed from dilated cisternae of smooth endoplasmic reticulum (*19*). These develop an ordered matrix which is then progressively obscured by the synthesis of melanin. Cell homogenates, subjected to differential centrifugation, contain tyrosinase activity in the large particulate fraction which contains premelanosomes and melanosomes. Enzyme activity is also found in microsomal and soluble fractions of such homogenates (*27*).

A progressive decrease in tyrosinase activity occurs during culture of melanotic melanoma cells (clone $B_5$59) with BrdU (3 μg/ml) (*38*). Both soluble and particulate tyrosinase activity is decreased (Table 1). The decrease in soluble enzyme

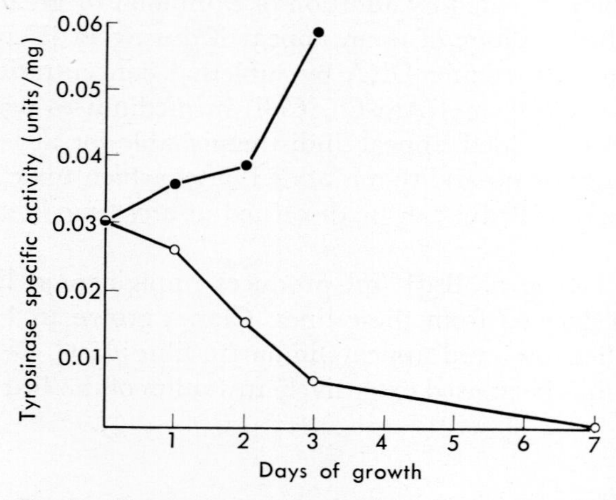

FIG. 6. Changes in tyrosinase specific activity of the supernatant fraction of $B_5$59 cells grown in regular medium (●) and replicate cultures grown in medium containing 3 μg of BrdU/ml (○). Data from one time course experiment. (From Wrathall *et al.*, 1973, by permission of the Rockefeller University Press.)

activity, assayed spectrophotometrically by following the oxidation of dihydroxyphenylalanine (L-DOPA) to dopachrome at 475 nm (*21*), is detectable by 24 hr of culture in the presence of BrdU. By 7 days of treatment tyrosinase activity usually reaches undetectable levels (Fig. 6). This decrease in enzyme activity is accompanied by the appearance of an inhibitor of tyrosinase in extracts of BrdU-treated cells which causes a lag period in enzyme assays. Dialysis of these extracts eliminates this lag period and results in a significant increase in detectable units of tyrosinase activity. Extracts of $C_3$471 cells that had been grown continuously in medium containing BrdU (1 μg/ml) contain no detectable tyrosinase activity before or after dialysis although they contain dialyzable heat-stable material capable of inhibiting tyrosinase activity (Table 2). The inhibitor in BrdU-treated cells is similar to in-

TABLE 2. Inhibition of Tyrosinase Activity by Supernatant Fraction of $C_3$471 Cells

Treatment	$B_5$59 (25 μl)	$C_3$471 (200 μl)	Mixtures[a] (50 μl)	Mixtures[a] (100 μl)	Mixtures[a] (200 μl)
Untreated	100[b]	0	75.4	51.1	31.5
Heated 10 min at 100°C	0	0	—	46.7	36.8
Dialyzed 0.05 M barbital, pH 8.6	104.1	0	—	100.9	—

[a] $C_3$471 supernatant fraction (volume and treatment as indicated) was added to 25 μl untreated $B_5$59 supernatant fraction immediately before assay. [b] Activity expressed as percent of units detected in untreated $B_5$59 supernatant fraction.

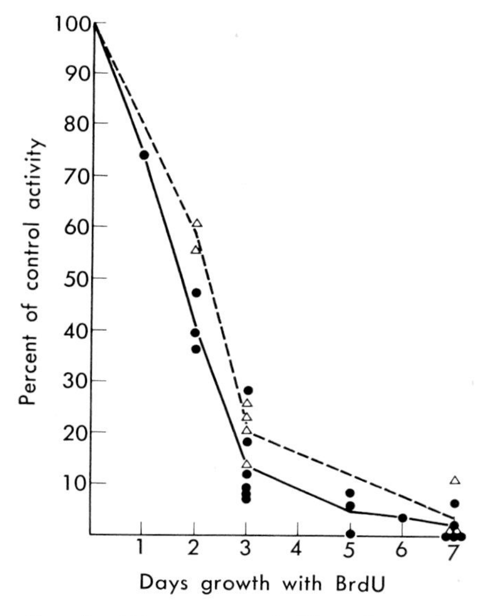

FIG. 7. Time course of effect of growth with BrdU (3 μg/ml) on tyrosinase specific activity. Summary of 10 experiments. Values for the cultures grown with BrdU are expressed as a percentage of the specific activity of a replicate control culture grown in regular medium for the same period. Activities determined on supernatant fractions before (●) and usually also after dialysis (△) overnight at 4°C against 0.05M Barbital buffer (pH 8.6). Lines are drawn between the arithmetic means of multiple experimental points. (From Wrathall *et al.*, 1973, by permission of the Rockefeller University Press.)

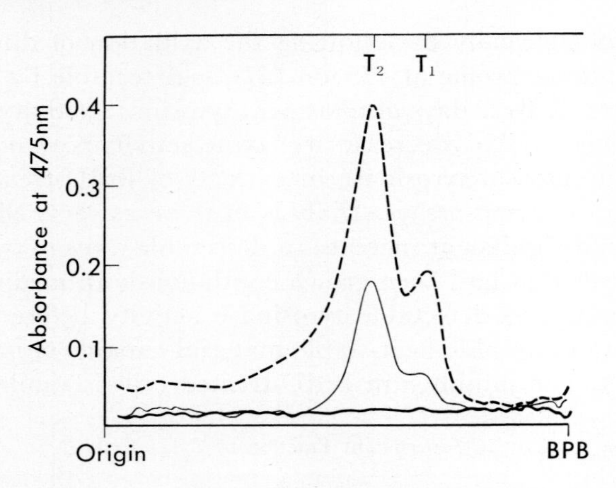

FIG. 8. Effect of growth with BrdU (3 μg/ml) on soluble tyrosinase (T_1 and T_2) pattern, as shown by acrylamide gel electrophoresis. Dialyzed supernatant fraction containing ~188 μg protein was applied to each gel. Densitometric tracings of gels scanned at 475 nm with a Gilford recording spectrophotometer equipped with a linear transport attachment; migration of bromphenol blue dye marker (BPB). ---- control; —— 3 days BrdU; —— 7 days BrdU. (From Wrathall *et al.*, 1973, by permission of the Rockefeller University Press.)

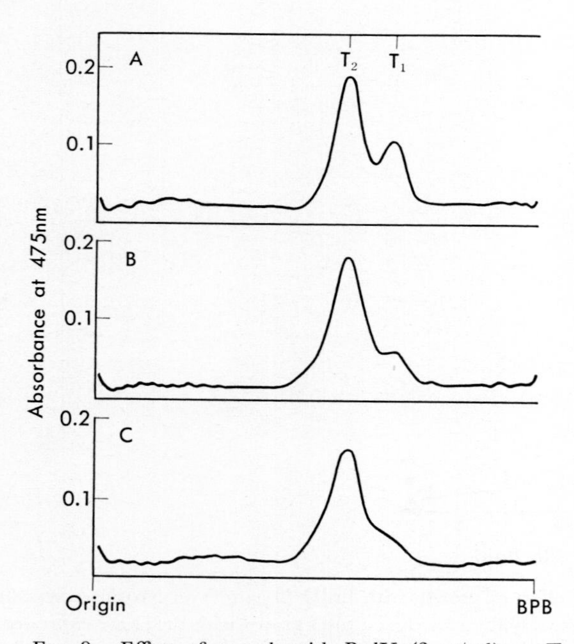

FIG. 9. Effect of growth with BrdU (3 μg/ml) on T_1 and T_2 forms of soluble tyrosinase. Dialyzed supernatant fraction containing approximately 0.0006 U of tyrosinase activity was applied to each gel. A, control $B_5$59 cells grown for 3 days on regular medium; B, cells of replicate culture grown for 3 days with BrdU; C, cells grown for 7 days with BrdU. (From Wrathall *et al.*, 1973, by permission of the Rockefeller University Press.)

hibitors prepared from both hamster melanoma (*24*) and an amelanotic mouse melanoma (*6*) and could facilitate the suppression of melanogenesis and/or the maintenance of the amelanotic state. However, only a quantitatively minor part of the reduction in tyrosinase activity during BrdU treatment can be attributed to this dialyzable inhibitor (Fig. 7).

Soluble tyrosinase has been separated by chromatographic (*20*) and electrophoretic (*5, 9*) means into multiple molecular forms. Both the T_1 and T_2 forms are present in extracts of control $B_5$59 cells and are reduced during BrdU-treatment (Fig. 8). However, there is an earlier and preferential reduction in the T_1 form of soluble tyrosinase (Fig. 9). The C locus in the mouse apparently contains the structural gene for tyrosinase (*7*) and the multiple soluble forms probably result from secondary modifications of the product of this locus (*10*). Tyrosinase solubilized from smooth membrane (*26*) and melanosomal (*17*) fractions of mouse melanoma has been reported to be exclusively of the T_1 form. In light of these reports, the earlier reduction of T_1 during BrdU treatment infers that T_1 is a precusor form from which T_2 may form, perhaps by degradative processes.

D. Effects on premelanosome formation

The reduction of tyrosinase activity in melanoma cells treated with BrdU is coordinated with a suppression of premelanosome formation (*38*). Extensive premelanosome formation in control $B_5$59 cells is indicated by the presence of large numbers of premelanosomes in all stages of development and the histochemical detection of tyrosinase reaction product in Golgi saccules, Golgi-associated vesicles and smooth-surfaced tubules (Fig. 10a, b). Cells treated for 3 days with BrdU (3 μg/ml) have a decreased number of premelanosomes but these are concentrated in the juxtanuclear area. Reduced premelanosome formation is indicated by the general absence of tyrosinase reaction product in Golgi saccules and Golgi-associated smooth-surfaced tubules of these cells (Fig. 10c). Very few melanosomes are observed in 7-day-treated cells, and neither tyrosinase reaction product nor early forms of premelanosomes are present (Fig. 10d). Long-term BrdU-grown $C_3$471 cells contain no melanosomes or premelanosomes, and there is no histochemical evidence of tyrosinase activity. Thus BrdU treatment appears to suppress melanogenesis completely in melanoma cells. In light of the complex genetic control of this differentiated function (*18*), this infers a coordinated effect by BrdU on a number of gene products.

E. Reversibility

The suppression of melanoma cell pigmentation by BrdU is reversible upon culture of the treated cells in medium without BrdU (*28, 30*). Figure 11 shows the experimental manipulation of pigmentation possible with BrdU treatment and reversal. As already noted (section 1, A), suppression of pigmentation in embryonic pigmented iris epithelium (*30*) and pigmented retina cells (*8*) is also reversible. These results indicate that the suppression of pigmentation does not occur through BrdU-induced mutations. The demonstration of reversible suppression of pigmentation both in mass population and in single-cell-derived clones isolated in BrdU (*30*)

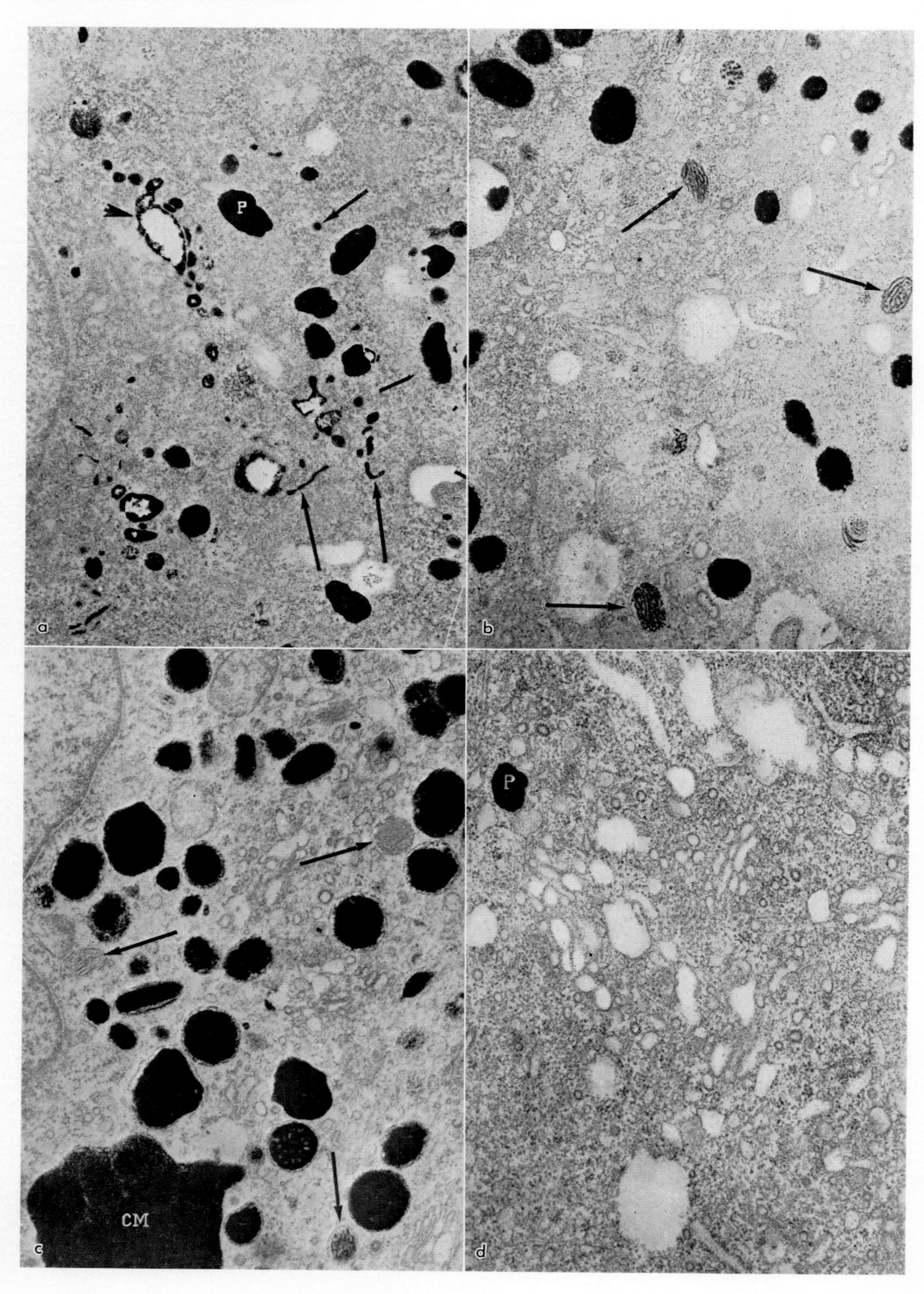
P
a
b
P
c
CM
d

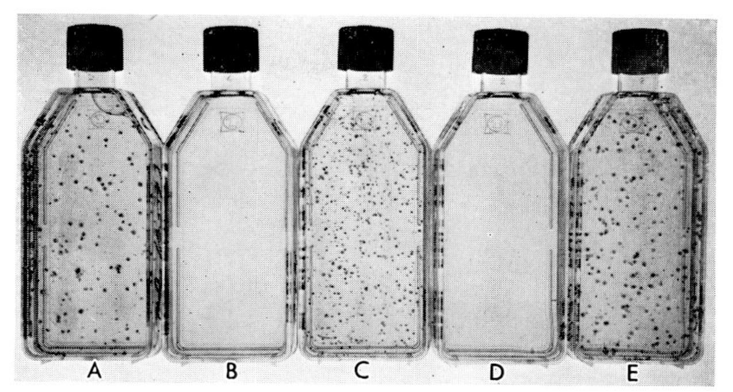

FIG. 11. Back and forth reversibility of BrdU effect on pigment production. The Falcon flasks were seeded with approximately 10,000 cells 1 week before this photograph was taken. A, untreated $B_5$59 melanoma cells (control); B, clone C471 after long term growth in 1 μg per ml of BrdU; C, clone C471 allowed to grow in normal medium and to revert to pigmented, piled up morphology; D, loss of pigment and piled up morphology by reverted repigmented clone C471 (as in C) after return to BrdU (1 μg/ml); E, rereversal to the pigmented state of cells in group D, after growth in normal medium. Replicate flasks trypsinized for cell counting showed that each of the flasks contained approximately an equal number of cells, indicating that the growth rate was unaffected by the treatments. (From Silagi, 1971, by permission from the Tissue Culture Association.)

indicates further that BrdU is not acting through selection of aberrent cells in the control cell population. The suppression of melanogenesis appears to result from an effect of BrdU on regulation of this complex differentiated function.

F. *Kinetics*

Continuous growth in BrdU is not required for suppression of tyrosinase activity in melanoma cells. Additional experiments (Wrathall, unpublished data) have shown that a single cell cycle in the presence of BrdU is sufficient to suppress enzyme activity completely. Figure 12 shows the time course of reduction in enzyme activity after a 30-hr pulse with BrdU (3 μg/ml). Enzyme activity was determined on the supernatant fraction as previously described (*38*). Tyrosinase activity was undetectable 6 days after the "pulse" and the kinetics of reduction mirrored those seen with continued BrdU treatment (compare Fig. 12 with Fig. 6). The long period required for complete suppression of enzyme activity therefore does not appear to be due to a requirement for continued incorporation of the drug, but would be consistent

←FIG. 10. Thin sections of cells incubated for tyrosinase activity. a. $B_5$59 cell grown in regular medium. Tyrosinase reaction product is localized in the inner Golgi saccule (arrowhead), Golgi-associated smooth surfaced tubules (long arrows) and vesicles (short arrow), and in premelanosomes (P). (×20,000). b. $B_5$59 cell grown in regular medium. Treated before incubation with a tyrosinase inhibitor, DECA. The absence of reaction product in intermediate premelanosomes (arrows) and associated structures indicates the specificity of the tyrosinase reaction. (×20,000). c. $B_5$59 cell grown for 3 days with 3 μg of BrdU/ml. Tyrosinase reaction product is localized primarily in the compound melanosome (CM). Many premelanosomes (arrows) contain no reaction product. (×19,000). d. $B_5$59 cell grown for 7 days with 3 μg of BrdU/ml. No reaction product is present in the Golgi saccules or any associated structure. P, premelanosome. (×20,000). (From Wrathall *et al.*, 1973, by permission of the Rockefeller University Press.)

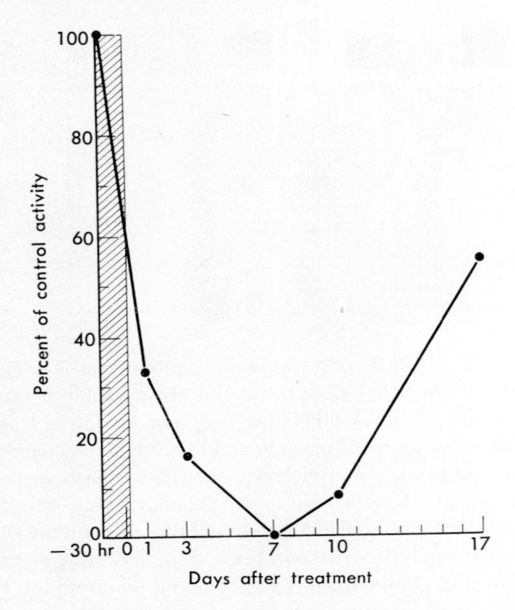

FIG. 12. Time course of effect on tyrosinase specific activity of BrdU (3 μg/ml) for 30 hr and return to normal medium. Values are expressed as a percentage of the specific activity of replicate control cultures grown continuously in medium without BrdU. ▨ BrdU (3 μg/ml).

with a lengthy half life for tyrosinase or its mRNA. The reappearance of tyrosinase activity by the 11th day of the experiment demonstrated a gradual reversal of BrdU-induced suppression. Microscopic examination of such reversing cultures revealed that reversal to pigmentation did not occur as synchronously as its original suppression, making it difficult to draw inferences from the kinetics of reversal in mass populations.

G. Correlations with BrdU effects in other systems

A similar reversible suppression of cell-specific cytodifferentiation has been described for several embryonic cell types such as chondrocytes (*2*, *8*), and for other continuous cell lines expressing differentiated functions, such as hepatoma cells (*32*). The initiation of differentiation in embryonic precursor cells such as myoblasts, (*8*, *35*) and precursor erythroblasts (*16*, *37*), is also sensitive to inhibition by BrdU. In all these systems cell division and total cell macromolecular synthesis appear minimally affected by BrdU concentrations which profoundly reduce cell-specific differentiated products. This preferential effect of BrdU on differentiated function is also seen in the melanoma cells. BrdU thus provides a tool for the study of regulation of differentiation in systems employing either embryonic cells or established cell lines. In the melanoma cell system BrdU treatment also affects the malignant potential of the cells.

2. *Reversible Suppression of Tumorigenicity by BrdU*

A. *Time and dose dependence*

Control $B_5$59 cells cause rapidly growing tumors and death in every adult C57BL/6 mouse injected subcutaneously with 2×10^5 or more viable cells (trypan blue exclusion test). The TD50 (number of melanoma cells needed to produce tumors in 50% of inoculated mice) is 6×10^4 cells (Fig. 13). BrdU-grown cells exhibit decreased or absent tumorigenic potential depending upon concentration and length of time grown in BrdU. Figure 14 shows the effect on tumorigenicity of growth in 3 μg/ml of BrdU for up to 14 days. Each mouse was inoculated with 10^6 cells, a 20-fold increase over the TD50 of control cells. No tumors developed in 60 animals inoculated with 10^6 viable cells grown for 14 days in 3 μg/ml of BrdU. Animals were observed for more than 100 days after inoculation.

Cells grown in 1 μg/ml of BrdU show greater variability in tumorigenicity. Again the standard test has been 10^6 viable cells/mouse inoculated subcutaneously (s.c.). Uncloned cells tested after 29 days of growth in BrdU formed tumors in 20/26 mice, but tumorigenicity of single cell clones derived from the same cells was from 10–70% (*30*). When clone C471 which initially formed tumors in 1/10 mice was recloned, subclone $C_3$471 tested in early passages formed tumors in 0/40 mice (*29*). In an independent series, uncloned cells grown in 1 μg/ml for 92 days formed tumors in 2/5 mice. When these cells were cloned and tested soon after cloning the results shown in Table 3 were obtained. Again variability among the clones was exhibited. Figure 13 shows the results obtained in a dose-response curve for clone $C_3$471 in an early passage, showing that inoculations 100-fold higher than the TD50 for control cells produced no tumors.

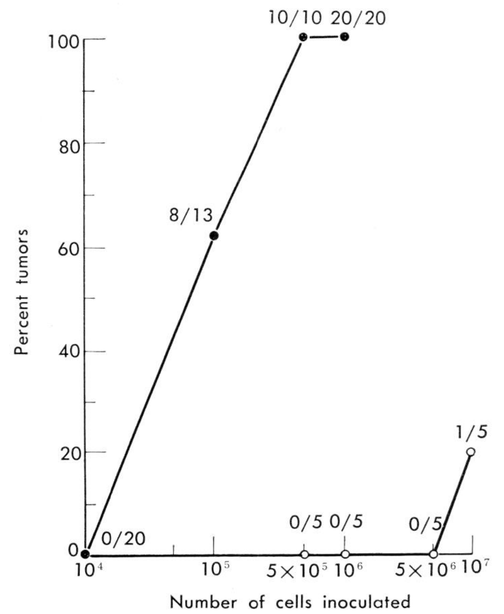

FIG. 13. Dose response curves for control malignant melanoma cells, clone $B_5$59, ●, and for clone $C_3$471 (BrdU, 1 μg/ml) in an early passage, ○.

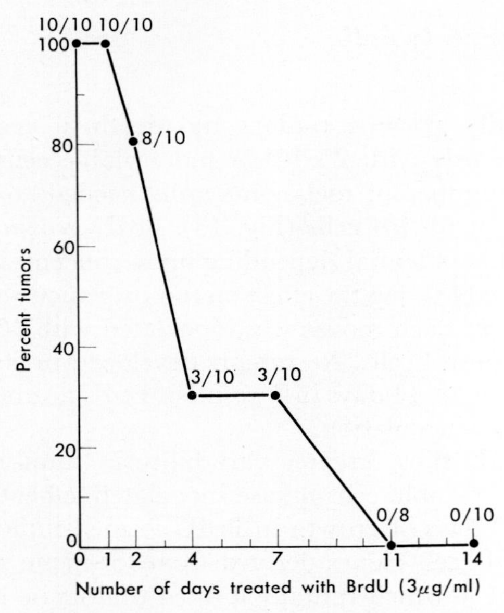

FIG. 14. Effect on tumorigenicity of growth of B_559 melanoma cells for different lengths of time in medium containing BrdU (3 μg/ml). The fractions next to each point show number of tumor-bearing animals/total number of animals inoculated with 10^6 viable cells.

TABLE 3. Variability among Clones Derived from Cells Grown in BrdU (1 μg/ml)[a),b)]

Clone	Tumor incidence
DH-H1	0/10
DH-A12	2/10
DH-F6	3/9
DH-B3	3/10
A471	7/10
B471	1/10
C471	1/15
C_1471	10/10
C_2471	0/4
C_3471	1/93

a) Cells were injected subcutaneously at 10^6 viable cells/mouse. DH clones derived from DH line. 471 clones derived from 471 line. All were derived from B_559 cells grown continuously in medium containing 1 μg of BrdU/ml. b) Clones were injected between passage 2–7 after cloning. C_1, C_2, and C_3 were subclones of C471, passage 7. Later passages became more tumorigenic (Ref. *29*).

A single cell cycle in BrdU (30 μg/ml for 30 hr) produced a marked effect on tumorigenicity. Tumorigenicity was reduced to zero after 1 day in normal medium subsequent to the pulse, and rose slightly during the following 5 days. When the cells were tested after 41 days in normal medium, all mice injected developed tumors (Fig. 15) (*29*). The initial decrease and subsequent increase in tumorigenicity are similar to effects on tyrosinase activity of a single cell cycle in 3 μg/ml of BrdU (section 1, F).

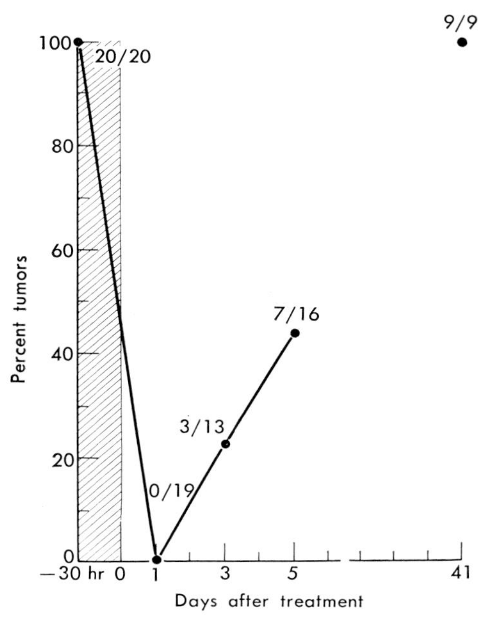

FIG. 15. Time course of effect on tumorigenicity of treatment with BrdU (30 μg/ml) for 30 hr and return to normal medium. See Fig. 14 for explanation of fractions next to each point. ▨ BrdU (30 μg/ml).

B. Reversibility of BrdU-grown cells to tumorigenicity after growth in normal medium

Populations of cells that became non-tumorigenic or reduced in tumorigenicity after continuous growth in 1–3 μg/ml of BrdU or in 30 μg/ml for 30 hr returned to full tumorigenic potential after a period of growth in medium devoid of BrdU (*28,*

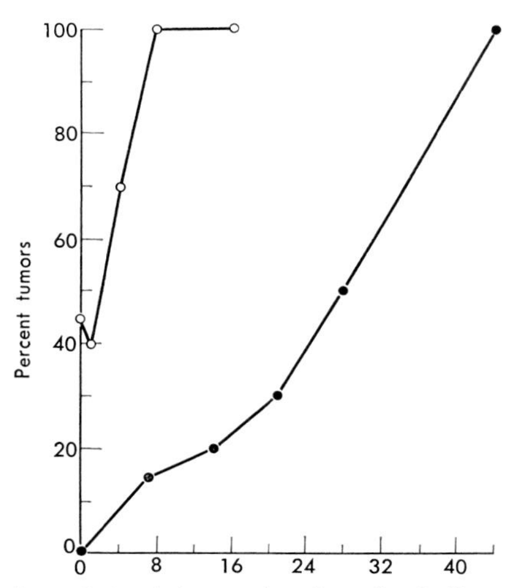

FIG. 16. Time course of reversibility of $B_5$59 cells after growth in BrdU for 7 days (2.6 μg/ml), ○, and for 14 days (3 μg/ml), ●, and subsequent return to normal medium.

29). Growth for 7 days in 2.6 μg/ml of BrdU-reduced tumorigenicity to about 40% of controls. After 7 days of growth in normal medium tumorigenicity returned to control level (Fig. 16). Growth for 14 days in 3 μg/ml of BrdU obliterated tumorigenic potential. In 4 reversal experiments, 42/44 mice injected with cells treated for 14 days in BrdU and then grown for 15–43 days in normal culture medium developed tumors. One such experiment is shown in Fig. 16. One non-tumorigenic clone grown for 138 days in 3 μg/ml of BrdU caused tumors in 7/8 animals when tested after growth for 47 days in normal medium (*28*). Similar results were obtained with cells grown for 133–201 days in 1 μg/ml of BrdU and then for a similar length of time in non-BrdU-containing medium. All tests were made with 10^6 cells per mouse.

C. *Reduction of tumorigenicity in other mouse tumor lines*

The effect of growth in BrdU on the tumor forming ability of 4 other mouse strains was tested. These were neuroblastoma C1300, adrenal cortex tumor, amelanotic melanoma clone B78, and a cloned hybrid cell line derived by fusion of B16 melanoma with L cells. Their sensitivity to the analog varied, both with respect to toxicity and concentration of BrdU and period of treatment required to reduce tumorigenicity. All showed a marked decrease in tumorigenicity after growth in BrdU (*28*) (Table 4). Reversibility was not tested.

TABLE 4. Effect of BrdU on Various Mouse Tumor Lines

Cell line	BrdU concentration (μg/ml)	Number of days grown	Tumor incidence
Amelanotic melanoma (C57BL/6)	10	14	0/10
Adrenal cortex (LAF_1)	1	14	2/7
Neuroblastoma C1300 (A/J)	1	50	6/10
Hybrid clone A946 (C57BL × C_3H)	3	23	3/10

D. *Tumorigenicity, replication, and metabolism of BrdU-grown cells*

Tumor-forming capacity drastically decreased in cells grown in 1 μg of BrdU/ml (Table 3) whereas their doubling time (24 hr) and plating efficiency (46–80%) were close to that of the parental line. Viability of parental and BrdU lines was usually about 90%. The effects of growth in 3 μg of BrdU/ml on replicative and tumor-forming capacity are shown in Table 5. The decrease in plating efficiency and increase in doubling time are insufficient to explain the complete inability of cells grown for 14 days in BrdU to form tumors in normal adult C57BL/6 mice. Based on a TD50 of 6×10^4 for untreated melanoma cells, one would expect about 50% of mice inoculated with 10^6 cells of 5% plating efficiency (3 μg/ml for 14 days) to produce tumors, but none did. These cells are flat, enlarged, completely "contact-inhibited," and adhere strongly to the surface of the culture vessel. The change from the control cells' focus-forming mode of growth would indicate changes in properties of surface membranes.

We have found amino acid pool differences between B_559 cells and C_3471 cells grown continuously in 1 μg of BrdU/ml, (*25*). The BrdU-grown cells show sig-

TABLE 5. Comparison between Control and BrdU-grown Melanoma Cells[a)]

	Control	BrdU grown (3 μg/ml)	
		7 days	14 days
Percent tumors	100	40–50	0
Latent period (days)	8–20	20–100	—
Plating efficiency (percent)	67–92	40–80	5–25
Population doubling time (hr)	24	24–40	~48

a) See text (section 2, D) and Table 3 for cells grown in 1 μg/ml BrdU.

nificant increases in intracellular free amino acid concentrations per unit of cell protein for 10 out of 13 amino acids quantitated. The concentrations of proline and glu (glutamine+glutamate) are significantly higher in these cells than in the malignant control cells, whether calculated per cell or per unit cell protein. The molar percentage of 5 amino acids in the protein of BrdU-treated cells differs significantly from that of untreated melanoma cells. These alterations in amino acid pools and protein amino acids may reflect metabolic changes characteristic of expression or suppression of malignancy, and therefore merit further investigation.

E. Virus production by BrdU-grown melanoma cells

Melanoma cells grown in BrdU, at differing concentrations and for differing periods of time, significantly increase production of virus with the morphology of murine leukemia virus (*29*) (Fig. 17). This is similar to the reports of Lowy *et al.* (*14*) and Aaronson *et al.* (*1*) that BrdU can induce previously unexpressed endogenous virus in various lines, including embryonic mouse cell lines. Budding or extracellular virus particles are very rare in the parental B_559 line, detectable only by electron microscopy. Tests of controls for virus by XC-plaque assay (*23*) and for group specific antigens 1 and 3 were negative, but were positive in C_3471 cells. These cells usually give from $1–4 \times 10^6$ plaque-forming units (p.f.u.) and DH clones (Fig. 1) about $0.7–1 \times 10^6$ p.f.u. per ml of 100-fold concentrated culture medium in which cells grew to a full monolayer.

The etiological role of the virus is unknown. We have injected close to 100 neonates with the virus (subcutaneously, intraperitoneally, intrathymically, and intracranially). The mice were observed for a minimum of 100 days, but none developed leukemia or melanoma.

Tests for Gross cell surface antigen (GCSA) of C_3471 cells were positive, although the control B_559 cells proved negative (*31*). GCSA represents a substance produced by cells at the surface, presumably in response to information coded for by the virus genome. It is the same antigen found on lymphocytes of mice with naturally occurring (Gross) leukemia.

We documented another change at the cell surface—an increase of at least 100-fold in cytotoxicity titers (50% kill) with $H\text{-}2^b$ typing antiserum of C_3471 cells over B_559 cells. The significance of the latter change is unclear, but changes in 2 known cell surface antigens reinforce the visual observation of cell membrane alteration, and may help explain the changes in immunogenic potential of the BrdU-grown cells (section 2, F).

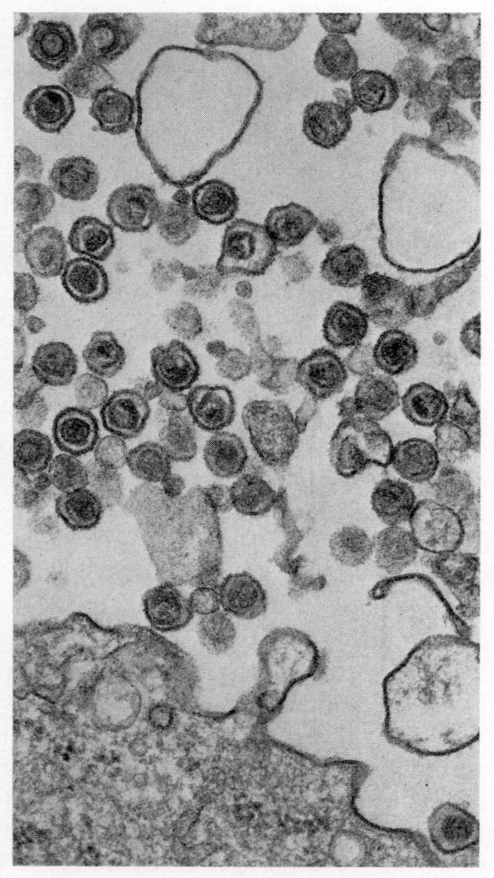

FIG. 17. Typical accumulation of virus particles having the morphology of murine leukemia virus (enveloped A and type C) located in close contact with the surface of a C_3471 cell treated with 1 μg BrdU per ml for 248 days. (×50,240). (From Silagi *et al.*, 1972, by permission of the National Academy of Sciences.)

F. Immunogenicity of BrdU-grown cells

Weekly injections of melanoma cells grown for 4–12 months in 1 μg of BrdU/ml protected adult C57BL/6 animals against challenge with a usually lethal dose of 10^6 untreated melanoma cells (*28*). Protection against this challenge was proportional to the number of preinjections, ranging from 7% with 1 inoculation to 90% with 4 inoculations with C471 cells (Fig. 18). Other early experiments showed that cells grown for 14 days in 3 μg/ml of BrdU protected 17% of the mice, whereas untreated B_559 cells did not protect any against the same challenge dose. The degree of protection in adult animals was also roughly proportional to the number of virus particles being produced by the "immunizing" cells (*29*).

We found, however, that when C_3471 cells were injected into adult C57BL/6 mice treated with antithymocyte serum (ATS) or into neonates, tumors grew and killed all the mice. When melanoma cells which had been grown for 14 days in

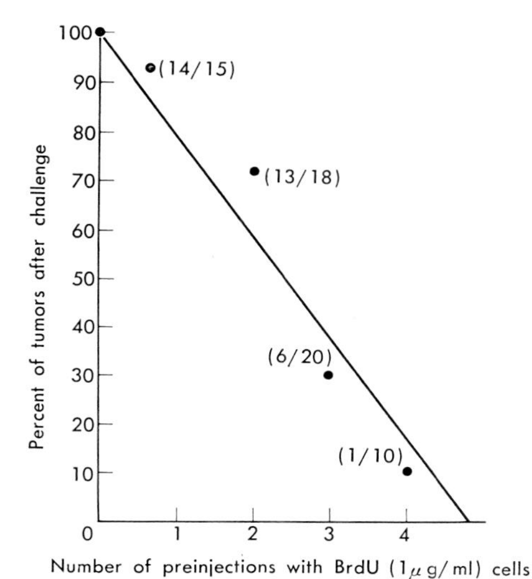

FIG. 18. Effect of increasing numbers of preinjections with BrdU-treated cells on resistance to challenge with untreated melanoma cells (10^6 per mouse). (From Silagi, 1971, by permission of the Tissue Culture Association.)

3 μg of BrdU/ml and which were non-tumorigenic in adult mice (10^6/mouse) were inoculated into ATS-treated adults and into neonates, they formed tumors and killed 72% of the former and 21% of the latter (Tables 6 and 7) (*31*). Latent periods were increased over those for control $B_5$59 cells. This may be due to the fact that these cells have a plating efficiency reduced from control levels. Nonetheless, their ability to grow *in vivo* is proved by tumor formation in neonates and ATS-treated adult mice. Thus, the ability of BrdU-grown cells to form tumors in immunologically compromised mice, their ability to protect normal adults against melanoma, their greatly increased production of C-type virus and their increased expression of 2 cell surface antigens (section 2, B), all make it likely that one component of the loss of tumorigenicity of these cells is a change in their antigenicity.

Similar results to ours were obtained by Barbieri *et al.* (*3*) who found that C-

TABLE 6. Tumorigenicity of BrdU-grown Cells in Mice Immunosuppressed with ATS

BrdU (μg/ml)	Control mice			ATS-treated mice	
	Number of tumors				
	Saline or untreated	NRS treated	MLP days	Number of tumors	MLP days
3	0/10 (100)[a]	0/7 (131)	—	13/18 (130)	65±10
1	5/15 (36)	4/19 (36)	27±7	20/20 (36)	15±2

All mice were injected with 10^6 cells. Cells were grown in 3 μg/ml of BrdU for 14 days, or continuously in 1 μg/ml of BrdU for almost 1 year, as indicated. ATS, antithymocyte serum. NRS, normal rabbit serum. MLP, mean latent period ± standard deviation; $P < 0.02$. [a] Number of mice with tumors/total number of mice. Numbers in parentheses represent number of days when surviving animals were observed.

TABLE 7. Tumorigenicity in Neonatal Mice of BrdU-grown Cells (3 μg/ml)

Number of cells	Control cells (untreated)	4–5 days in BrdU	7–8 days in BrdU	13–16 days in BrdU
10^2	4/16	—	—	—
10^3	9/9	—	—	—
10^4	15/15	3/4	5/11	2/19
10^5	6/6	4/5	9/18	4/11[a)]
10^5	6/6	6/6	3/10	4/19[b)]

Number of mice with tumors/total number of mice. All animals with tumors died. Surviving animals were kept under observation for 96–117 days. [a)] 1–2$\times 10^5$ cells/mouse combined. [b)] 0.5–1$\times 10^6$ cells/mouse combined.

type particles, Gross cell surface antigen and decreased tumorigenicity occurred in an *in vitro* "spontaneously" transformed C57BL/6 lung cell line after prolonged culture with methylcholanthrene. Untreated transformed cultures were negative for C-type particles and antigen, and retained high tumor-producing capacity. Stephenson and Aaronson (*34*) also found an increase in cell surface antigens and immunogeniciy in mouse sarcoma virus-transformed lines associated with virus production after superinfection with leukemia virus. Lieber and Todaro (*13*) have hypothesized that transformed cells that express endogenous C-type virus may be more antigenic than those that do not, thus providing a natural means of controlling the growth of such neoplastic cells.

G. Working hypothesis of BrdU suppression of tumorigenicity

Our working hypothesis based on the accumulated data is that there are several components to the suppression of malignancy by growth in BrdU. (a) BrdU enhances production of C-type virus which confers new properties on the cells, including an increase to measurable levels of already existing but undetectable cell surface antigens. The antigens would elicit an immune response in injected animals, which in turn would reject the BrdU-grown cells before they had an opportunity to grow into a tumor. Although not detectable in control cells, these antigens may nonetheless be related to the rare viral particles observed in $B_5$59 cells by electron microscopy. This could explain the ability of animals, immunized with BrdU-grown cells, to recognize and reject untreated melanoma cells. (b) Reduced replicative capacity is probably a component in reduced tumor-forming ability of cells treated with at least 3 μg/ml of BrdU. Plating efficiency is, however, not reduced sufficiently (see section 2, D) to provide more than a fraction of the loss of tumorigenic potential. (c) The change in the cells' surface properties related to cell-cell contact, their flattening and firmer adherence to the glass or plastic on which they grow, makes it probable that growth in BrdU is having many effects on the plasma membrane. These effects may be related to the intrinsic change that distinguishes a normal from a malignant cell. Further studies of the kinetics of change of cell membranes of melanoma cells during growth in BrdU may help elucidate this problem.

3. Mechanism of BrdU Effects

Since BrdU is an analog of thymidine the probability is high that its effects are mediated by incorporation into DNA. This hypothesis is supported by several lines of evidence from BrdU-treated melanoma cells: (1) the morphological effects of BrdU (see section 1, B) are completely prevented by inhibition of DNA synthesis with 1-β-D-arabinosylcytosine (*28*); (2) tritiated BrdU is demonstrable in the nuclei of treated cells by autoradiography, and is incorporated into acid-insoluble material, as determined by scintillation counting (*30*); and (3) the presence in the medium of equimolar concentrations of thymidine along with BrdU results in cells morphologically indistinguishable from controls (*30*). The "pulse" experiment (section 1, F) infers that the suppression of tyrosinase activity can result from incorporation of BrdU into only a single strand of DNA.

The incorporation of BrdU in place of thymidine has been documented in other systems where BrdU suppresses differentiated functions, by isolation of the substituted DNA followed by base analysis and/or density gradient analysis (*32*, *36*, *37*). A maximum of 2 cell cycles in the presence of BrdU is sufficient to prevent the initiation of hemoglobin synthesis in erythroblasts (*37*). Incorporation of BrdU during a single DNA synthetic cycle is sufficient to reduce the synthesis of tyrosine aminotransferase by hepatoma cells to less than 50% of controls (*32*).

Incorporation of BrdU results in a decreased rate of *synthesis* of cell-specific products without equivalent effects on total cell protein, DNA or RNA synthesis (*32*, *36*, *37*). This has led to the hypothesis that BrdU exerts an effect at the transcriptional level. Evidence supporting this hypothesis has been reported by Turkington *et al.* (*36*). They detected reduction of prolactin-induced polysome formation in mammary gland cultures when prolactin-induced casein synthesis was inhibited by BrdU. The preferential effect of BrdU on synthesis of cell-specific products could reflect a selective effect on gene transcription. However, since BrdU appears to be extensively and uniformly incorporated into the genome (*37*), this selective effect of BrdU does not have an obvious structural basis. Preferential effects could result from: (1) generalized reduction in the rate of transcription but differential labilities of certain mRNA's and their proteins (*33*); (2) selective inhibition of transcription perhaps due to substitution of BrdU in pyrimidine-rich initiator sites (*36*); or (3) temporal sensitivity in which only the initiation of new "programs" of gene transcription is inhibited by BrdU (*37*). The actual molecular basis of the preferential inhibition of differentiated functions by BrdU is still open to investigation. A critical question in the melanoma cell system is whether there is an interrelationship between the control mechanisms governing the effects of BrdU on melanogenesis and those affecting tumorigenicity.

CONCLUSIONS

The scheme of the effects of BrdU on melanoma cells depicted in Fig. 19 may be useful in visualizing the complexity of effects requiring a molecular understanding in this system. This scheme does not attempt to designate intermediate products,

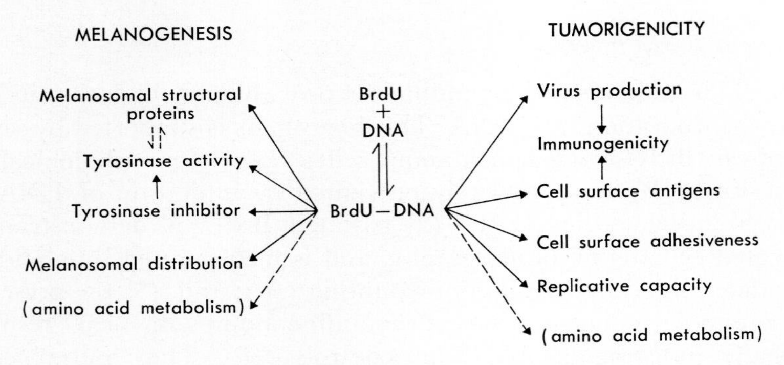

FIG. 19. Scheme showing multiple effects of growth of melanoma cells in BrdU that pertain to melanogenesis and tumor formation. No attempt is made to designate intermediate products precise causal relationships or pleiotropic effects.

exact causal relationships or possible pleiotropic effects. Further study may reveal an even more complex array of changes during BrdU treatment.

Our investigations show that suppression of melanogenesis by BrdU occurs through a coordinated effect on the structural and enzymic proteins required for melanin synthesis. These results are consistent with cessation of synthesis of tyrosinase and melanosomal structural proteins in BrdU-treated cells, and indicate that there may be a "program" of gene activity for melanogenesis which is regulated as a unit. Whether the inhibitor of tyrosinase activity detected in BrdU-treated cells plays a role in regulating this "program" may be determined by further study. Also, the possible regulatory interrelationships between tyrosinase and other melanosomal structural proteins warrants further investigation. Melanogenesis is a complex form of differentiated function which can be studied in these melanoma cells with far greater ease than in normal melanocytes. The understanding obtained from such study can be compared and hopefully related to basic control mechanisms during normal embryonic differentiation.

Investigations on the suppression of tumorigenicity by BrdU are revealing separable components involved in the tumorigenic potential of melanoma cells. Both reduction in the intrinsic ability of the cells to grow *in vivo*, and changes evoking host defense mechanisms appear to be operating in the reduction of tumorigenicity by BrdU treatment. Host response may depend on antigenic changes, at least some of which appear to be related to the induction of C-type virus, in BrdU-treated cells. The untreated melanoma cells grow in a manner suggestive of "transformed" cells. BrdU treatment produces flattened, more adhesive cells which express "contact inhibition" and are therefore more similar to "normal" cells in culture. These changes must involve alterations in the cell membranes which could produce antigenic differences eliciting host response, or which could, in a direct way, reduce the cells' capacity to form tumors *in vivo*. Continuing investigations with this system offer promise of revealing basic components of tumorigenic potential in malignant cells in general.

By such studies further information may be obtained on the normal regulation of gene activity in eukaryotic cells and the perturbations of this regulation which produce malignant cells.

ACKNOWLEDGMENTS

This work was supported in part by U.S. Public Health Service Grant CA 10095 from the National Cancer Institute, Grant DRG 1095 from the Damon Runyon-Walter Winchell Cancer Fund, and a grant from the Esther and Morris Grossman Foundation. S. S. is a recipient of Faculty Research Award PRA-77 from the American Cancer Society and J. R. W. was a Damon Runyon Cancer Research Fellow during this investigation.

REFERENCES

1. Aaronson, S. A., Todaro, G. J., and Scolnick, E. M. Induction of murine C-type viruses from clonal line of virus-free BALB/3T3 cells. Science, *174*: 157–159, 1971.
2. Abbott, J. and Holtzer, H. The loss of phenotypic traits by differentiated cells. V. The effect of 5-bromodeoxyuridine on cloned chondrocytes. Proc. Natl. Acad. Sci. U.S., *59*: 1144–1151, 1968.
3. Barbieri, D., Belehradek, J., Jr., and Barski, G. Decrease of tumor-producing capacity of mouse cell lines following infection with mouse leukemia viruses. Int. J. Cancer, *7*: 364–371, 1971.
4. Bischoff, R. and Holtzer, H. Inhibition of myoblast fusion after one round of DNA synthesis in 5-bromodeoxyuridine. J. Cell Biol., *44*: 134–150, 1970.
5. Burnett, J. B., Seiler, H., and Brown, I. V. Separation and characterization of multiple forms of tyrosinase from mouse melanoma. Cancer Res., *27*: 880–889, 1967.
6. Chian, L. T. Y. and Wilgram, G. F. Tyrosinase inhibition: its role in suntanning and in albinism. Science, *155*: 198–200, 1966.
7. Coleman, D. L. Effect of genic substitution on the incorporation of tyrosine into the melanin of mouse skin. Arch. Biochem. Biophys., *96*: 562–568, 1962.
8. Coleman, A. W., Coleman, J. R., Kankel, D., and Werner, I. The reversible control of animal cell differentiation by the thymidine analog, 5-bromodeoxyuridine. Exp. Cell Res., *59*: 319–328, 1970.
9. Holstein, T. J., Burnett, J. B., and Quevedo, W. C., Jr. Genetic regulation of multiple forms of tyrosinase in mice: acion of a and b loci. Proc. Soc. Exp. Biol. Med., *126*: 415–418, 1967.
10. Holstein, T. J., Quevedo, W. C., Jr., and Burnett, J. B. Multiple forms of tyrosinase in rodents and lagomorphs with special reference to their genetic control in mice. J. Exp. Zool., *177*: 173–184, 1971.
11. Hu, F. and Lesney, P. F. The isolation and cytology of two pigmented cell strains from B16 mouse melanoma. Cancer Res., *24*: 1634–1643, 1964.
12. Lasher, R. and Cahn, R. D. The effects of 5-bromodeoxyuridine on the differentiation of chondrocytes *in vitro*. Develop. Biol., *19*: 415–435, 1969.
13. Lieber, M. M. and Todaro, G. J. Spontaneous and induced production of endogenous type-C RNA virus from a clonal line of spontaneously transformed BALB/3T3. Int. J. Cancer, *11*: 616–627, 1973.
14. Lowy, D. R., Rowe, W. P., Teich, N., and Hartley, J. W. Murine leukemia virus: High frequency activation *in vitro* by 5-iododeoxyuridine and 5-bromodeoxyuridine. Science, *174*: 155–156, 1971.

15. Mason, H. S. The chemistry of melanin. III. Mechanism of the oxidation of dihydroxyphenylalanine by tyrosinase. J. Biol. Chem., *172*: 83–99, 1948.
16. Miura, Y. and Wilt, F. H. The effects of 5-bromodeoxyuridine on yolk sac erythropoiesis in the chick embryo. J. Cell Biol., *48*: 523–532, 1971.
17. Miyazaki, K. and Seiji, M. Tyrosinase isolated from mouse melanoma melanosome. J. Invest. Derm., *57*: 81–86, 1971.
18. Moyer, F. H. Genetic effects on melanosome fine structure and ontogeny in normal and malignant cells. Ann. N.Y. Acad. Sci., *100*: 584–606, 1963.
19. Novikoff, A. B., Albala, A., and Biempica, L. Ultrastructural and cytochemical observations on B-16 and Harding-Passey mouse melanomas. The origin of premelanosomes and compound melanosomes. J. Histochem. Cytochem., *16*: 299–319, 1968.
20. Pomerantz, S. H. Separation, purification, and properties of two tyrosinases from hamster melanoma. J. Biol. Chem., *238*: 2351–2357, 1963.
21. Pomerantz, S. H. and Li, J. P. Tyrosinases (hamster melanoma). *In;* H. Tabor and C. W. Tabor (eds.), Methods in Enzymology, vol. 17, Part A, pp. 622–626, Academic Press, New York, 1970.
22. Raper, H. S. The aerobic oxidases. Physiol. Rev., *8*: 245–282, 1928.
23. Rowe, W. P., Pugh, W. P., and Hartley, J. W. Plaque assay techniques for murine leukemia viruses. Virology, *42*: 1136–1139, 1970.
24. Satoh, G. J. and Mishima, Y. Tyrosinase inhibitor in Fortner's amelanotic and melanotic malignant melanoma. J. Invest. Derm., *48*: 301–303, 1967.
25. Schulman, J. D., Wrathall, J. R., Silagi, S., and Doores, L. Altered amino acid concentrations accompanying suppression of malignancy of mouse melanoma cells by 5-bromodeoxyuridine. J. Natl. Cancer Inst., *52*: 275–277, 1974.
26. Seiji, M., Itakura, H., and Irimajiri, T. Tyrosinase in the membrane system of mouse melanoma. *In;* V. Riley (ed.), Pigmentation—Its Genesis and Control, pp. 525–542, Appleton-Century-Crofts, New York, 1972.
27. Seiji, M., Shimao, K., Birbeck, M. S. C., and Fitzpatrick, T. B. Subcellular localization of melanin biosynthesis. Ann. N.Y. Acad. Sci., *100*: 497–533, 1963.
28. Silagi, S. Modification of malignancy by 5-bromodeoxyuridine. Studies of reversibility and immunological effects. In Vitro, *7*: 105–114, 1971.
29. Silagi, S., Beju, D., Wrathall, J., and deHarven, E. Tumorigenicity, immunogenicity and virus production in mouse melanoma cells treated with 5-bromodeoxyuridine. Proc. Natl. Acad. Sci. U.S., *69*: 3443–3447, 1972.
30. Silagi, S. and Bruce, S. A. Suppression of malignancy and differentiation in melanotic melanoma cells. Proc. Natl. Acad. Sci. U.S., *66*: 72–78, 1970.
31. Silagi, S., Newcomb, E. W., and Weksler, M. E. Relationship of antigenicity of melanoma cells grown in 5-bromodeoxyuridine to reduced tumorigenicity. Cancer Res., *34*: 100–104, 1974.
32. Stellwagen, R. H. and Tomkins, G. M. Preferential inhibition by 5-bromodeoxyuridine of the synthesis of tyrosine aminotransferase in hepatoma cell cultures. J. Mol. Biol., *56*: 167–182, 1971.
33. Stellwagen, R. H. and Tomkins, G. M. Differential effect of 5-bromodeoxyuridine on the concentrations of specific enzymes in hepatoma cells in culture. Proc. Natl. Acad. Sci. U.S., *68*: 1147–1150, 1971.
34. Stephenson, J. R. and Aaronson, S. A. Antigenic properties of murine sarcoma virus-transformed BALB/3T3 nonproducer cells. J. Exp. Med., *135*: 503–515, 1972.

35. Stockdale, F., Okazaki, K., Nameroff, M., and Holtzer, H. 5-Bromodeoxyuridine: effect on myogenesis *in vitro*. Science, *146*: 533–535, 1964.
36. Turkington, R. W., Majumder, G. C., and Riddle, M. Inhibition of mammary gland differentiation *in vitro* by 5-bromo-2′-deoxyuridine. J. Biol. Chem., *246*: 1814–1819, 1971.
37. Weintraub, H., Campbell, G., Le, M., and Holtzer, H. Identification of a developmental program using bromodeoxyuridine. J. Mol. Biol., *70*: 337–350, 1972.
38. Wrathall, J. R., Oliver, C., Silagi, S., and Essner, E. Suppression of pigmentation in mouse melanoma cells by 5-bromodeoxyuridine. Effects on tyrosinase activity and melanosome formation. J. Cell Biol., *57*: 406–423, 1973.

Discussion of Paper of Drs. Silagi and Wrathall

DR. SUGIMURA: Do tumors other than melanoma lose the tumorigenicity with BrdU treatment?

DR. SILAGI: Table 4 gives the answer. Also see the answer to Dr. Thompson's question.

DR. THOMPSON: You mentioned that there are several examples of BrdU causing reduced tumorigenicity in mouse cell lines. Would you comment further on the universality (or lack of it) of this effect for all mouse lines and for other spiecies as well? We have tried treating HTC cells, a rat cell line with BrdU and have been unable to reduce tumorigenicity or rescue any C-particle type viruses.

DR. SILAGI: Table 4 in this paper shows that the lines we tested with BrdU were all from mice: neuroblastoma C1300, clone 2A, adrenal cortical tumor, amelanotic melanoma and a cloned hybrid cell line derived by fusion of B16 melanoma and L cells. The concentration of BrdU and length of time of growth in BrdU varied for different lines. Dr. Marquardt (personal communicated) obtained reduction of tumorigenicity in a chemical carcinogen induced mouse prostate tumor line with very low concentration of BrdU. Dr. Klement (Nature New Biol., *234*: 12–14, 1971) has reported induction of C-type particles in rat cells with BrdU. As to the universality of these effects, if more investigators would try a range of concentrations and times in the treatment of malignant cells with BrdU, and inject a range of cell concentrations into appropriate hosts, we might get the answer. In some cases, it may be necessary to clone as the treated cells, as we did with some of our lines continuously grown in 1 μg/ml of BrdU, to obtain non-tumorigenic strains.

DR. SUGIMURA: How many kinds of structural proteins are identified so far? Please give us any information if available.

DR. SILAGI: The only structural proteins we have identified as being suppressed with growth in BrdU are those involved in melanosome formation.

DR. JOHNSON: We tested the effect of BrdU on the pigmentation of a Cloudman melanoma which I obtained from American Type Culture Collection. We were unable to detect any increase or decrease in pigmentation in 4 days growth with 10 μg/ml BrdU. Have you tested this cell line?

DR. SILAGI: We never tested the Cloudman melanoma. I am surprised at the result you obtained. If *the cells were dividing* and not infected with mycoplasma I would expect them to stop making melanin. Mycoplasma being rich in thymidine would preferentially remove the BrdU from the medium. Other investigators have confirmed our results as to loss of pigment in other pigmented strains of B16 melanoma. Also pigmented embryonic iris epithelium and retina lose pigment on growth in BrdU.

DR. HOZUMI: I am interested in the membrane changes of the BrdU-treated melanoma cells in connection with reduction of the tumor. Have you ever checked the surface changes besides surface antigens, such as changes of the receptor of phytohemagglutinins?

DR. SILAGI: I have not yet checked other surface changes in the BrdU-treated melanoma cells.

Quantal Cell Cycles, Normal Cell Lineages, and Tumorigenesis

H. Holtzer, S. Dientsman, S. Holtzer, and J. Biehl

Department of Anatomy, School of Medicine, University of Pennsylvania, Philadelphia, Pennsylvania, U.S.A.

Abstract: Thus far the technique of cell cloning has been directed toward securing a relatively homogeneous population of cells. The experiments to be described were designed to isolate a single progenitor cell that could *in vitro* generate a phenotypically diverse progeny. These experiments demonstrate that the leg muscles of advanced chick embryos harbor goodly numbers of mononucleated cells that consistently give rise to cells that establish myogenic and fibrogenic lineages.

Cultures of normal 8-day leg muscles were trypsinized and single cells from these cultures plated into separate wells of a Microtest Plate. Approximately 30% of these single cells gave rise to "myogenic clones." Each myogenic clone contained at least 2 multinucleated myotubes, plus varying numbers of mononucleated cells. When such myogenic clones were trypsinized and sub-cultured they invariably yielded large numbers of mononucleated cells. When these replicating mononucleated cells were grown in ^{3}H-proline, ^{3}H-glucosamine, or ^{35}S-sulfate they were shown to synthesize α $(I)_2$ $\alpha 2$ collagen chains, hyaluronic acid and varying amounts of chondroitin sulfate. In brief, mononucleated cells derived from myogenic clones were microscopically and biochemically indistinguishable from authentic fibroblasts. It is worth stressing that the progenitor mesenchymal cell that yielded daughter cells constituting the myogenic and fibrogenic lineages, did not yield cells constituting the chondrogenic lineage.

These results will be discussed in terms of (1) mechanisms that "move" cells into, and through the various compartments of, the myogenic, fibrogenic, and chondrogenic lineages, and (2) suggesting that while there may be "multipotential systems," there are in fact no "multipotential cells."

Cell Physiology and Cell Differentiation

The basic assumption of this brief review is that the genetic mechanisms that "move" differentiating cells from compartment to compartment in emerging myogenic and erythrogenic lineages (*9, 10, 13, 14*) are also the mechanisms that move normal cells into their respective neoplastic compartments. By way of establishing perspective, particularly for the molecularly-oriented biologists and pathologists, it is imperative to distinguish between (1) the genetic events responsible for expressing the physiological repertoire of a given cell at a given time, and (2) the genetic events that *initially* establish the metabolic repertoire of that cell type. The first issue deals with the fact that only myoblasts (My) and only erythroblasts have the option to synthesize such cell-specific luxury molecules as skeletal myosin and hemoglobin. The second issue deals with the cumulative genetic events that occurred in early cells of the myogenic or erythrogenic lineages which determine the limited synthetic capacities of My and erythroblasts. The second issue must also account for the observation that My and erythroblasts do not have the option to synthesize such inappropriate molecules as melanin, insulin, adrenocoricotropic hormone (ACTH), or albumin, even though the necessary information is stored in their DNA. A variety of non-specific exogenous molecules may trigger a given My or erythroblast to assemble myofibrils or synthesize hemoglobin, but no exogenous molecule will induce (1) a zygote, (2) a blastula cell, (3) a gastrula cell, (4) a presumptive neuroblast, (5) a chondroblast (Cb), (6) a hepatoma cell, or (7) a lymphoma cell to transform into a My or erythroblast. In contrast, it is probable that many non-specific exogenous agents will induce hemoglobin synthesis in Friend leukemic cells-tumors of cells already in the early compartments of the erythroid lineage (*5, 18*).

The major implication of this view is that exogenous molecules such as hormones, embryonic "inducers," tissue "factors," cyclic AMP, or chemical carcino-

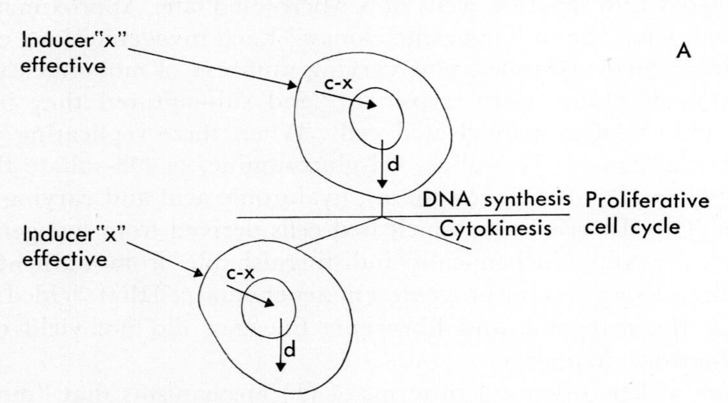

FIG. 1A. Cell physiology. This is a simple diagram of interactions involved in cell physiology as distinct from cell differentiation. Component "c," a cytoplasmic substance found is all cells of this compartment of this lineage, forms a complex with inducer "x." The complex is able to react with the nucleus in such a manner that a new luxury molecule "d" is formed. In the absence of inducer "x," luxury molecule "d" is not produced because although component "c" is present in the cytoplasm, it is unable to interact with the nucleus by itself. Cells in this compartment are able to undergo proliferative mitoses, producing daughter cells which also synthesize component "c" and therefore in the presence of inducer "x" are able to synthesize luxury molecule "d." It is also possible for cells in this compartment to undergo a quantal cell cycle, after which the resulting daughter cells are able to synthesize a new luxury molecule. The study of quantal cell cycles is a concern of cell differentiation.

gens introduce no new information into the responding cell; at most they activate or alter pre-existing sets of metabolic options. Cataloguing the responses of a given cell to a host of exogenous agents will extend our knowledge of cell physiology, but it will not in itself further understanding of cell differentiation.

This fundamental but often over-looked distinction between cell physiology and cell differentiation is illustrated in Fig. 1. If exogenous molecules induce the same response in the mother cell that they induce in the daughter cell, then the cellular

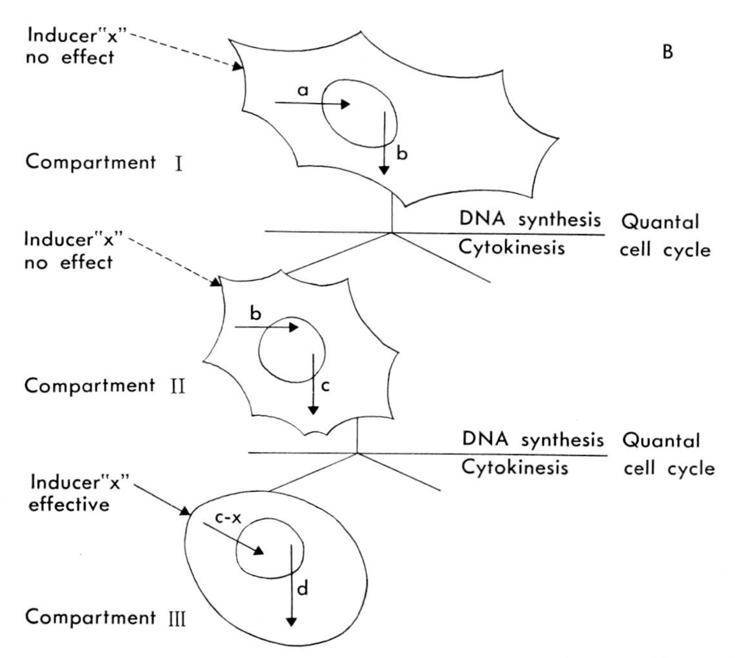

Fig. 1B. Cell differentiation. This is an equally simple diagram of interactions involved in cell differentiation. A cell in compartment I of this lineage contains cytoplasmic component "a," which reacts with the nucleus in such a way that luxury molecule "b" is formed. If inducer " x" be added to such cells, there is no effect. A cell in compartment I may undergo a quantal cell cycle. This type of cell cycle permits a reprogramming of synthetic activities. As the result of such a cell cycle, the daughter cells, in compartment II behave differently from their mothers in compartment I. Cytoplasmic component "b," which was synthesized in compartment I cells, is now able to react with the nucleus in such a way that a new luxury molecule "c" is produced. If inducer "x" be added to compartment II cells, it will have no effect. At this stage luxury molecule "c" is present in the cytoplasm, but is unable to react with inducer "x" under any circumstances. In order for such a reaction to take place, cells in compartment II must undergo a quantal cell cycle. The compartment II cells, emerging from that cell cycle, now contain component "c" in a form that is able to react with the nucleus. Whether it does, and so is instrumental in the synthesis of luxury molecule "d," depends upon the presence of inducer "x" and the formation of a c-x complex. Note that this set of reactions is the same as that illustrated in Fig. 1. There are many variations possible on this scheme. Cells in any compartment may undergo determined or indefinite numbers of proliferative cell cycles before going through the next quantal cell cycle. Quantal cell cycles may be asymmetrical, one daughter moving into the next compartment, the other remaining behind as a stem cell. It is also possible for both daughters to differ from each other as well as from the mother cell.

circuitry being probed relates to cell physiology. To study differentiation requires studying a system that involves, at a minimum, one quantal cell cycle, *i.e.*, one cell division that produces a daughter cell with metabolic options that were not available to the mother cell (*8–10, 13, 14*).

Presumptive Myoblast (PreMy)

Consider, for example, the PreMy and the My, cells in the penultimate and ultimate compartments of the myogenic lineage. My are post-mitotic cells which in permissive environments fuse to form multinucleated myotubes and are able to exercise their option to synthesize and organize cell-specific contractile proteins into striated myofibrils. This aspect of myogenesis is amenable to analysis in terms of how sets of structural genes for skeletal myosin, tropomyosin, and actin are transcribed, how such transcripts are translated, and how the resulting monomers might assemble into homo- and heteropolymers (*15*). More difficult to contend with experimentally is the way in which replicating PreMy with their unique metabolic program were derived from their particular ancestral cells, and how they in turn, by way of DNA synthesis and nuclear division, give rise to My (*12*). The PreMy must be viewed as a covertly differentiated cell with as unique and as limited a set of metabolic options as is found in any terminally differentiated cell, for only the daughters of PreMy acquire the option to differentiate into My. PreMy, though they cannot as yet be distinguished biochemically from other covertly differentiated embryonic cells, have few options in terms of the luxury molecules they can synthesize; they cannot be induced, for example, to synthesize hemoglobin, albumin, ACTH or even to deposit metachromatic chondroitin sulfate. PreMy do not even have the option to synthesize and organize contractile proteins into striated myofibrils, though their daughter cells do. This means that the differentiated PreMy must be the unique division product of a still earlier differentiated cell in the myogenic lineage and so on back to the uniquely differentiated zygote. Each compartment of the myogenic lineage then consists of cells with unique, but limited, metabolic repertoires. The cells in earlier compartments are obligatory precursors to those in later compartments. This view of lineages renders untenable the notion that any embryonic cell or any tumor cell is in fact "undifferentiated" or "multipotential" (*1, 10*). It also raises the question of what kind of mechanism "moves" cells within a lineage so that the succeeding compartment consists of progeny with a different but limited and predictable set of new metabolic options.

Mutipotential Primitive Cell

The following experiments were designed to test the hypothesis that there is a "multipotential" primitive mesenchyme cell that can form My, or fibroblasts (Fb), or chondroblasts (Cb) (*2, 19, 20*). More specifically the experiments were designed to determine: (1) whether the primitive mesenchyme (Ms) cell has the option to differentiate itself into either a terminal My, or Fb, or Cb; (2) whether its immediate daughters can yield any 1, 2, or 3 of these terminal phenotypes; or (3) whether there is an obligatory requirement for 2 or 3 division cycles before the progeny of a given primitive Ms cell can produce these 3 terminal mesodermal cells. Mononucleated cells from the leg muscles of chick embryos were cultured for

TABLE 1. Primary Clones from Mass Cultures of 8-Day Muscle

Exp. No.	Total number of single cells cloned	2 or more cells after 1 week	50 or more cells after 3 weeks	
			With myotubes	No myotubes
M6	96	50	17	3
M7	91	55	10	8
M8	96	51	29	5
M11	88	38	13	12

All clones were initiated by introducing a single cell into the well of a Microtest Plate. The variability in numbers of clones scored as not having myotubes is due to: (1) myotubes often degenerated leaving only mononucleated cells; (2) often the small myotubes were obscured by overgrowth of mononucleated cells; (3) different batches of medium promoted fusion to different degrees; (4) mechanical damage associated with trypsinization affects the viability of the cloned cells. These unknowns render a more rigorous quantitation of this kind of material dubious. Some myogenic clones displayed a few, others over 100 multinucleated myotubes, plus goodly numbers of mononucleated cells. Only myogenic clones with sizeable numbers of both mononucleated cells and myotubes were used to initiate secondary monolayer cultures.

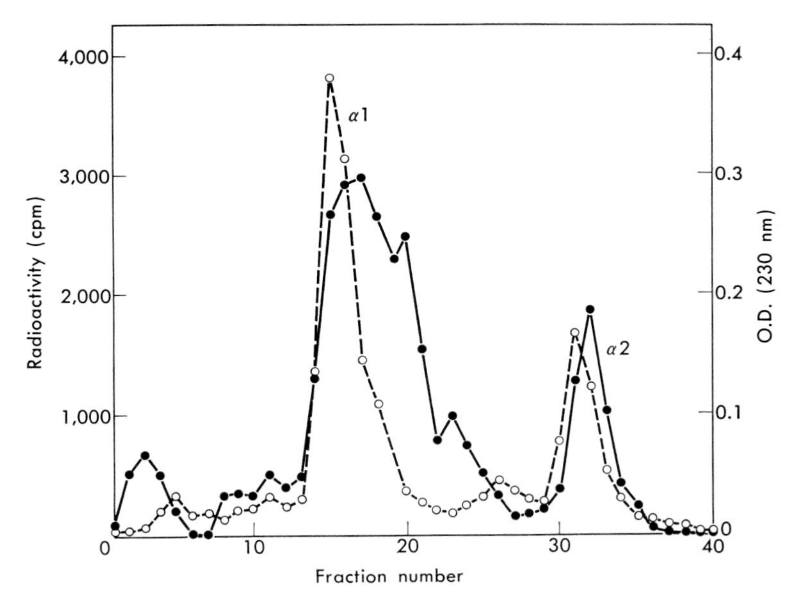

FIG. 2. Subunit structure of newly synthesized collagen from a confluent secondary monolayer culture. The culture was incubated for 24 hr with ^{3}H-proline together with ascorbic acid (50 μg/ml) and B-aminoproprionitrile (125 μg/ml) and after extraction mixed with lathyritic chick skin collagen and analyzed by chromatography on a carboxymethyl cellulose column by the procedures described in Schiltz *et al.* (1973). ● O.D. 230 nm of carrier lathyritic collagen; ○ cpm of ^{3}H-proline incorporation. Two peaks of radioactivity can be observed corresponding to the $\alpha 1$ and $\alpha 2$ peaks of the carrier lathyritic collagen. In this experiment the ratio of radioactivity in the $\alpha 1/\alpha 2$ peaks was 2.1.

4 days. The cultures were trypsinized and single cells were taken up into a micropipette under the microscope and introduced into collagen-coated wells of Microtest Plates (*1, 16*). After a month the wells were scored, and as shown in Table 1, it was found that approximately 50% of the single cells put into the wells failed to replicate and approximately 20% gave rise only to clones of less than 10 cells. Many of these small clones could not be reliably classified as either "myogenic" or "fibrogenic." The great majority of the remaining 30% of the wells contained myogenic clones consisting of multinucleated myotubes with cross-striated myofibrils and many mononucleated cells. The larger myogenic clones were trypsinized and all the resulting mononucleated cells from one clone plated in one 100 mm Petri dish and grown for varying periods with or without further sub-culturing. In this way large numbers of cells, all derived from a single cell that yielded a myogenic clone were obtained. The major finding of these experiments from the viewpoint of lineages was that the replicating cells from the original myogenic clones formed a stable line of cells, which were microscopically indistinguishable from Fb. Fur-

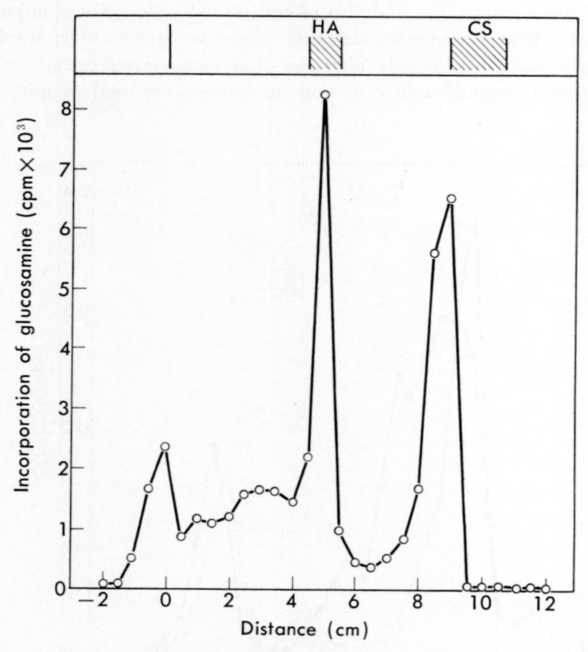

FIG. 3. Glycosaminoglycan synthesis in a confluent secondary monolayer culture. ³H-glucosamine was added for 24 hr, the glycosaminoglycans isolated and fractionated by high voltage electrophoresis as described in Mayne *et al.* (1971). The position of the hyaluronic acid (HA) and chondroitin sulfate (CS) standards are indicated at the top of the figure. That the first peak is hyaluronic acid is demonstrated by: (1) hydrolysis of the peak in 0.05N HCl and separation of glycosamine from galactosamine showed 92% of the labelled hexosamine to be glucosamine; (2) on treatment with testicular hyaluronidase 81% of the peak was destroyed. That the second peak is chondroitin sulfate is shown by: (1) hydrolysis of the peak showed 94% of the labelled hexosamine to be galactosamine; (2) the peak is completely sensitive to chondroitinase ABC and AC. Similar digestion experiments with similar results were performed with ³⁵S-sulfate labeled glycosaminoglycans.

thermore, as shown in Figs. 2 and 3, the descendents of cells from a myogenic clone synthesize $\alpha(I)_2\alpha 2$ collagen chains and the same kinds of glycosaminoglycans that are synthesized by definitive muscle Fb. These fibroblastic cells did not fuse to form myotubes and did not detectably change their properties even after 4 sub-cultures. The most simple interpretation of these findings is that in embryonic leg muscles a surprisingly large number of cells are not yet committed either to the myogenic or fibrogenic lineage, but are members of a common precursor compartment. These cells we visualize as presumptive-myogenic-fibrogenic cells (PreMyFb cells).

Question: what is the relationship between PreMyFb cells and cartilage cells? Are PreMyFb cells the putative multipotential primitive Ms cell? To answer these questions the original cells from which the myogenic clones were established, or the cloned cells, or the progeny of the muscle clones, were grown under conditions known to permit presumptive chondroblasts (PreCb) to move from the penultimate into the ultimate compartment of the chondrogenic lineage (*3, 11*). None of these cells differentiated into Cb. From this it was concluded that PreMyFb cells are not the equivalent of cells in the primitive Ms compartment, for if they were, they should have also been able to yield chondrogenic cells.

From these results, and those reported using stage 17–18 limb bud cells (*4*), we have proposed the lineage shown in Fig. 4 to account for the derivation of My, Fb, and Cb from a common ancestral cell, the primitive Ms cell. If lineages are to be useful, their nodal points must be translated into specific generations of cells and

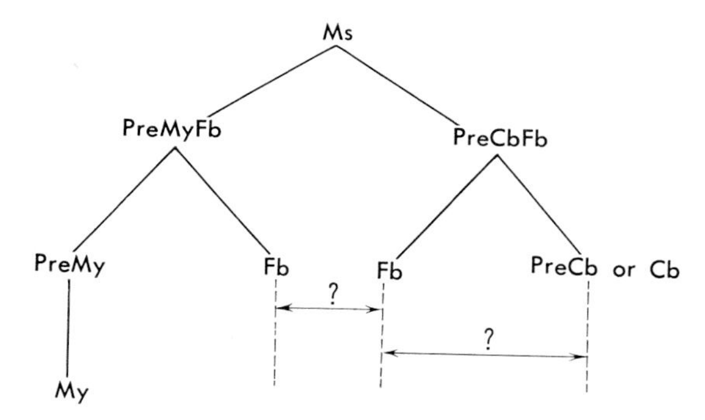

Fig. 4. A proposed lineage based on binary decisions beginning with cells in the "presumptive mesenchyme" (PreMs) compartment that leads to My, to Fb, and to Cb. PreMy compartment consists of cells yielding post-mitotic, My cells. Ellipses indicate that cells within that compartment are replicating and capable of either increasing the population within that compartment or of yielding progeny for the next compartment. As some terminally differentiated Fb cells emerge very early in development and readily replicate it may be there is no presumptive fibroblast compartment. Currently there is no way of identifying PreCb. While the small clones described in this paper may have been derived from cells within the PreMy compartment, the larger clones are presumed to be from the PreMyFb compartment. It is also suggested that the PreMyFb compartment is virtually depleted in day 18 muscles, whereas the Ms compartment must be depleted before day 8. Indeed from the data on limb buds and on somites there is nothing to rule out the possibility that the Ms compartment is depleted even in 3 day embryos, and that by stage 17–18 all mesenchymal cells have at least entered the PreMyFb or PreCbFb compartments. The arrow and question marks indicate that the progeny of Cb will form Fb, and that the progeny of Fb will form Cb, Fb, however, do not form My.

the restricted metabolic options open to cells in any one compartment delineated. It is also important to determine, with respect to Fig. 4, where cell division is dispensible, facultative, or obligatory in moving cells from one compartment to the next. For example, if PreMy cells are blocked at the G1-S interface by either fluorodeoxyuridine (FUdR) or cytosinarabinoside (Ara-C), they will not assemble striated myofibrils, though this option is promptly expressed by their daughter cells which are produced when the block is removed. These findings have been interpreted as demonstrating a requirement for DNA synthesis and nuclear division in moving cells from one compartment to the next in a lineage (*12*, *13*, *15*).

Little is actually known about this proposed lineage. It is by no means clear whether the Ms cell is obliged to divide asymmetrically, alwasy yielding one Pre MyFb and one presumptive-chondrogenic-fibrogenic (PreCbFb) cell, or whether the event is a probabilistic one, each division being symmetrical, but cell divisions, in response to exogenous fluctuations, lead to either 2 PreMyFb cells or 2 PreCbFb cells. It is also possible that there is no PreMy compartment as such but following a finite number of cell divisions the PreMyFb cells throw off fibrogenic cells with a greater frequency than myogenic cells; this would be a programmed form of transdetermination (*6*, *7*). Irrespective of whether Fb arise only from the PreMyFb compartment or also from the PreMy compartment, the genetic shunt into the Fb compartment involves a relatively irreversible genetic decision, since Fb cannot be induced to differentiate into muscle cells. Another intriguing possibility predicted by Fig. 4 is that Fb may have multiple origins, not only along with myogenic and chondrogenic cells, but possibly along with cells in other lineages such as corneal cells, gut cells, *etc.*

The major theoretical implication of Fig. 4 postulates the existence of 2 kinds of cell cycles, the proliferative and the quantal (*9*, *10*, *13*). Proliferative cell cycles *increase* the numbers of cells *within* the Ms, or PreMyFb, or PreCbFb compartments, but quantal cell cycles are required to move cells *into* or *out* of these compartments. While the population of cells depicted in Fig. 4 has the properties of a "multipotential" *system*, any one cell in any one compartment is at the most only "bipotential." A given Ms cell does not have the option itself of directly changing into, or within one generation of giving rise to, My, or Cb, or Fb. Similarly, the only option, in terms of phenotypic diversification, open to a given Ms cell is to yield PreMyFb and/or PreCbFb cells. Our prediction is that multipotential cells, in the sense that a single cell has *many* options of itself diversifying into *many* phenotypes, do not exist in eukaryotes, and that step-wise diversification is the sum of many binary decisions made during quantal cell cycles.

Tumorigenesis

Now what have these notions of cell lineages, quantal cell cycles, and the emergence of terminally differentiated cells to do with tumorigenesis? We begin by emphasizing that early tumor cells derived from different types of normal cells have more in common with their normal progenitor cells than they have with each other. Just as there is no "general, undifferentiated" normal cell, only normal myogenic, normal chondrogenic, or normal liver cells, so there is no "general, malignant" cell, but only malignant myogenic, malignant chondrogenic, or malignant liver cells. Malignant cells of even the most anaplastic kind retain to varying degrees some metabolic options that characterized their normal progenitors. No single macromolecule shared by all tumor cells demarks them as a class from all normal cells. We suggest that the "first" tumor cell to arise in a given population of normal cells behaves in a manner analogous to the division product of a quantal cell cycle in normal development. It possesses some of the limited metabolic options of its normal progenitor, but also displays some new ones not found in the mother. The properties of single aberrent cells reflect both the lineage and the particular compartment within that lineage in which the normal mother was located. For example, the following sequence might account for the emergence of a chondrosarcoma: a replicating Cb undergoes an "unscheduled" quantal cell cycle yielding "transformed" daughter cells. These daughters retain the options to synthesize chondroitin sulfates and cartilage collagen chains, but in addition they have acquired the option to synthesize glycoprotein that renders their cell surfaces insensitive to a normal mitotic inhibitor and so these cells escape the usual constraints on cell replication. This transformed cell increases in number by repeated proliferative cell cycles, leading to a homogeneous tumor population. Additional quantal cell cycles among these tumor cells leads to a sub-population of chondrosarcoma cells with still fewer properties of a normal Cb, and so on. This scheme accounts for (1) partial deviation tumors and (2) progression in a tumor. It states that the kind of genetic changes which "move" cells from compartment to compartment in normal development can also lead to the emergence of malignant cells. Tumor cells are then considered to be an alternative phenotypic expression within the cells lineage from which they arose.

With this in mind we can speculate further that the original transformed cell in the Friend leukemic tumor was a cell in the early compartment of the erythroid lineage (*5, 18*); the original cell in a rhabdomyosarcoma was a cell in the middle compartments of the myogenic lineage (*17*); whereas the testicular teratocarcinoma was derived from a transformed germinal cell that could still serve as a common precursor to many distinct lineages. According to this view, the transformed teratocarcinoma cell has to undergo the same sequence of quantal cell cycles to produce a recognizable striated muscle or a cartilage nodule as does its normal germ-cell counterpart.

ACKNOWLEDGMENT

This work was supported in part by grants from the American Cancer Society (VC 45A), the National Institutes of Health, Grant No. HL 15835 to the Pennsylvania Muscle Institute, and the National Institute of Child Health and Human Development (HD-00189).

REFERENCES

1. Abbott, J., Schiltz, J., and Holtzer, H. Phenotypic complexity of myogenic clones. PNAS, U.S.A., *71*: 1506–1510, 1974.

2. Caplan, A. The teratogenic action of nicotinamide analogues 3-acetylpyridine and 6-aminonicotinamide on developing chick embryos. J. Exp. Zool., *178*: 351–358, 1971.
3. Chacko, S., Abbott, J., Holtzer, S., and Holtzer, H. The loss of phenotypic traits by differentiated cells. VI. Behavior of the progeny of a single chondrocyte. J. Exp. Med., *130*: 417–441, 1969.
4. Dientsman, S., Biehl, J., Holtzer, S., and Holtzer, H. Myogenic and chondrogenic lineages in developing chick embryo limb buds. Develop. Biol., 1974, in press.
5. Friend, C., Scher, W., Holland, J., and Sato, T. Hemoglobin synthesis in murine virus-induced lukemic cells *in vitro*: stimulation of erythroid differentiation by dimethyl sulfoxide, PNAS, U.S.A., *68*: 378–382, 1971.
6. Gehring, W. *In;* H. Ursprung and R. Nothiger (eds.), The Biology of Imaginal Disks, pp. 35–58, Springer-Verlag, Berlin, 1972.
7. Hadorn, E. Dynamic of determination. *In;* M. Locke (ed.), Major Problems in Developmental Biology, pp. 85–104, Academic Press, New York, 1967.
8. Holtzer, H. Control of chondrogenesis in the embryo. Biophys. J., *4*: 239–251, 1964.
9. Holtzer, H. *In;* H. Padykula (ed.), Control Mechanisms in Tissue Cells (ISCB Symposium), vol. 6, pp. 69–87, Academic Press, New York, 1970.
10. Holtzer, H. Myogenesis. *In;* O. Schjeide and J. de Vellis (eds.), Cell Differentiation, pp. 476–503, Van Nostrand Reinhold Co., New York, 1970.
11. Holtzer, H. and Abbott, J. Oscillations of the chondrogenic phenotype *in vitro*. *In;* H. Ursprung (ed.), The Stability of the Differentiated State, pp. 1–10, Springer Verlag, Heidelberg, 1968.
12. Holtzer, H. and Sanger, J. Myogenesis: old views rethought. *In;* B. Banker, R. Pryzblyski, and J. van der Meulen (eds.), Research in Muscle and the Muscle Spindle, pp. 120–131, Excerpta Medica, Amsterdam, 1972.
13. Holtzer, H., Sanger, J., Ishikawa, H., and Strahs, K. Selected topics in skeletal myogenesis. Cold Spring Harbor Symp. Quant. Biol., *37*: 549–566, 1972.
14. Holtzer, H., Weintraub, H., Mayne, R., and Mochan, B. The cell cylce, cell lineage and cell differentiation. *In;* A. Moscona and A. Monroy (eds.), Current Topics in Developmental Biology, vol. 6, pp 229–256, Academic Press, New York, 1973.
15. Holtzer, H., Mayne, R., Weintraub, H., and Campbell, G. *In;* J. Pollak and J. Lee (eds.), The Biochemistry of Gene Expression in Higher Organisms, pp. 287–304, Australia and New Zealand Book Co., Sydney, 1973.
16. Holtzer, H., Rubinstein, N., Chi, J., Dientsman, S., and Biehl, J. *In;* A. Milhorat (ed.), Exploratory Concepts in Muscular Dystrophy and Related Disorders, Excerpta Medica, Amsterdam, 1974, in press.
17. Nameroff, M., Reznik, M., Anderson, P., and Hansen, J. Differentiation and control of mitosis in a skeletal muscle tumor. Cancer Res., *30*: 596–600, 1970.
18. Ostertag, W., Crozier, T., Kluge, N., Melderis, H., and Dube, S. Action of 5-bromodeoxyuridine on the induction of haemoglobin synthesis in mouse leukaemia cells resistant to 5-BUDR. Nature New Biol., *243*: 203–205, 1973.
19. Searls, R. The role of cell migration in the development of embryonic chick limb bud. J. Exp. Zool., *166*: 39–50, 1967.
20. Zwilling, E. *In;* M. Locke (ed.), 27th Symposium on Developmental Biology, pp. 184–207, Academic Press, New York, 1968.

Discussion of Paper by Drs. Holtzer et al.

DR. OSTERTAG: Do you get any evidence for "transdetermination" in your system? Is it possible to "wipe out" an instituted program of a cell by 5-bromodeoxyuridine (BrdU) and then to get "transdetermination"?

DR. HOLTZER: We have looked, with no success, for what would be obvious and dramatic instances of "transdetermination." For example, somites, early heart cells, or precursor Fb or muscle cells treated with BrdU do not yield erythroblasts, pigment cells, obvious nerve cells, or even epithelial cells. On the other hand BrdU-suppressed chondrogenic cells will eventually give rise to cells that operationally are indistinguishable from Fb in terms of their collagen chains and glycosaminoglycans. However, it must be stressed that *in vitro* even in the absence of BrdU chondrogenic cells eventually give rise to fibrogenic cells (Schiltz, Mayne, and Holtzer, Differentiation, *1:* 97, 1973).

DR. RUTTER: The model which you proposed, based upon binary decisions, assumes that the differentiation event occurs in one cycle thus conferring different properties on a daughter cell. If this is not the case, then it is possible that there is a "metastable" stage which is developmentally plastic, and the number of options may be different. Thus it seems to me that the binary model you propose, though attractive, is a model. There is little argument that cells have different developmental potential, the question which remains open is the general pattern of restricting potential.

DR. HOLTZER: I too like the concept of metastable state (Holtzer *et al.*, 1973). But the whole thrust at our argument is missed when you say we "assume the differentiation event occurs in one cycle." That is precisely what we say does not occur. There is no "one differentiating event," no "single inductive interaction" that lifts a cell from a biochemically virginal state into a differentiated state. A given quantal cell cycle, at most, yields daughter cells of the *next*, and only next, compartment in a given lineage. The terminal differentiated state is the integral of several ancestral quantal cell cycles.

At no transition point in a lineage, we argue, does any given cell have the option to enter more than 2 succeeding compartments, that is the crucial issue. Again I stress I know of no experiment demonstrating that a given type cell could be induced to transform, or to give rise to daughter cells that could differentiate into more than 2 cell types. If a given cell type is to give rise to 3 cell types, than a minimum of 2 quantal cycles is required, whereas 5 cell types, I predict, would require a minimum of 3 quantal cell cycles.

Lastly, I would disagree that the basic question is the restricting of potential. I would turn your statement upside down and ask what mechanisms step-wise open the potential, or make available in daughter cells information not available in the mother or grandmother cell? There are several compartments separating the zygote from the My, erythroblast, or pancreatic cell, and the issue is not how cells lose something but how they "open" the pertinent parts of the genome.

DR. PAUL: Hadorn's work on determination and transdetermination does not to be in complete agreement with your conclusions. Perhaps we do not have adequate criteria to define at what stage of development imaginal disc cells is arrested but do not the transdetermination results suggest the possibility that these cells have multipotentiality rather than precisely bipotentiality?

DR. HOLTZER: There is nothing in Hadorn's work, as I understand it, that demonstrates multipotentiality. There is no evidence when a given cell in an imaginal disc enters a particular lineage that excludes entry into another lineage. Nor is anything known of how synchronized the passage of cells in a given imaginal disc may be within a single lineages or between lineages. Lastly, nothing is known of how many cells in a disc undergo "transdetermination." If only a small number of cells are involved initially, they may well be the "least differentiated" and so have not entered the later stages of any one particular lineage.

Regulation of Hyaluronic Acid Synthesis in Cultured Mammalian Cells: Its Relation to Cell Growth and 5-Bromodeoxyuridine Effect

Tetsuo Ono, Hideki Koyama, and Mikio Tomida

Department of Biochemistry, Cancer Institute, Tokyo, Japan

Abstract: A hybrid cell line (B-6), able to produce hyaluronic acid, was induced by somatic cell hybridization of mouse mammary carcinoma and Chinese hamster lung cells, and has been maintained *in vitro*. Studies on the expression of this function demonstrated that (1) morphological differentiation is not always a prerequisite for functional differentiation, (2) hyaluronic acid synthesis occurs only during the exponential growth phase, and (3) the expression of this function is mainly regulated at the level of hyaluronic acid synthetase activity with a short half-life and oscillatory change during the growth phase.

The inhibitory effect of 5-bromodeoxyuridine (BrdU) on the differentiation was found both in the content of hyaluronic acid secreted and in the synthetase level. The inhibition is not all-or-none, is completely reversible, and requires the incorporation of the analog into the cell DNA, in agreement with the inhibition of differentiation by BrdU in other systems.

Considering these 2 facts, the mechanism of the BrdU effect may be interpreted as a preferential reduction in the transcription rate of gene(s) for hyaluronic acid synthetase with a high turnover rate of the enzyme.

The evidence accumulated so far in many laboratories has indicated that anomalies of gene expression are general in cancer cells, and moreover it was suggested that cancer is a disease of differentiation (*15*), or the ontogeny is blocked ontogeny (*19*). In these circumstances, studies on the regulation mechanism of gene expression or differentiation in mammalian cells are most relevant and urgent for the elucidation of the essential changes in cancer. Conversely, cancer cells with many anomalies of gene expression have offered superb experimental systems to study the regulation of gene expression.

The cell hybridization technique developed originally by Barski *et al.* (*1*) introduced a new methodology for studying gene expression in mammalian cells. By the use of this technique, regulation in the synthesis of macromolecular substances as well as the expression of malignancy has been pursued in many different systems (*1*, *5*, *7*, *8*). To explore another system, we tried to isolate hybrids between cultured mouse mammary carcinoma and Chinese hamster cell lines by using UV-irradiated parainfluenza virus (HVJ). Among the hybrids thus obtained, we found one clone that produced hyaluronic acid, which is considered to be a differentiated function of fibroblast cells (*13*). Concerning the regulation mechanism of hyaluronic acid synthesis in this cell line, it was demonstrated that hyaluronic acid synthesis is regulated by hyaluronic acid synthetase and not by UDP-glucose dehydrogenase (*24*). Thus, this system seems to be very useful for elucidating the control mechanism of expression of differentiated functions in mammalian cells.

Recently, the expression of differentiated functions in many cell-culture systems has been shown to be inhibited by relatively low concentrations of 5-bromodeoxyuridine (BrdU), which does not interfere with cell multiplication (*4*, *21*). Also, the tumorigenicity of melanoma cells has been shown to decrease on treatment with the analog (*21*). In our hybrid cell, it was demonstrated that BrdU has the same inhibitory effect on hyaluronic acid synthesis, although our cell is an artificial, interspecific hybrid line (*10*). In this paper we will review our experiments.

Initiation of Hyaluronic Acid Synthesis by Hybrid Formation in Culture

Somatic cell hybridization is a useful technique for elucidating the regulatory mechanism of expression of both differentiated functions and other properties in mammalian cells, as already reported by some workers (*8*). For the same purpose, we isolated hybrids between mouse and Chinese hamster lung-cell lines, examined their characteristics, and compared them with those of the parent lines. During the course of our investigation, an unexpected and interesting finding was that the culture medium of one clonal hybrid line, B-6, became very viscous as the cells grew. Subsequent observation that the viscosity disappeared rapidly on mild treatment with bovine testicular hyaluronidase and that the addition of acetic acid to B-6 medium gave rise to a precipitated "mucin clot," which is known to be a specific and unique reaction for high-molecular-weight hyaluronic acid in the presence of proteins (*9*), suggested the production of hyaluronic acid by the cells.

The mouse cell line used here was 8-azaguanine-resistant FC-1 line derived from C3H mouse mammary carcinoma cell, FM3A established by Nakano (*16*). The other cell line was CHL, which was established from the lung tissue of a newborn female Chinese hamster in our laboratory (*12*). Methods for cell fusion were the same as those of Okada (*17*) using UV-irradiated HVJ, as described previously (*13*).

Chromosome analysis gives us the most direct and convincing evidence to determine whether hybrids have been produced from a cross between 2 distinct parent lines, because the karyotype of FC-1 line is quite different from that of CHL line. Table 1 shows the chromosome constitutions of parent and 4 clonal hybrid lines examined in detail among 11 lines. These data for hybrids were collected within 50 days after isolation by cloning. FC-1 cells had 43 chromosomes as a mode, 3 of which were biarmed. These biarmed elements, which never exist in normal mouse cells,

TABLE 1. Chromosome Constitution of Parent and Clonal Hybrid Lines

Cell line	No. of total chromosomes (T)	No. of biarmed chromosomes (B)	B/T (%)
FC-1	41–45 (43)	2–4 (3)	7.0
CHL	21–23 (22)	21–23 (22)	100
A-5	99–105 (105)	20–23 (22)	21.0
B-1	91–97 (95)	17–25 (22)	23.2
B-4	114–123 (121)	15–20 (17)	14.1
B-6	87–96 (90)	11–17 (12)	13.3

Fifty metaphases of each line were counted. The numbers in parentheses represent the modal chromosome number of cells analyzed.

consisted of 2 relatively large metacentrics (or submetacentrics) and one small subtelocentric. On the other hand, CHL cells maintained exactly normal, female chromosome complements of 22, which were all biarmed. As shown in Table 1, B-1 and B-6 cells among the hybrids listed were characterized by the presence of somewhat smaller numbers of chromosomes, 95 and 90, respectively, in which biarmed complements were 22 and 12. On the basis of these chromosome numbers and of karyotypes in the parent line, both hybrids are also likely to have been derived from a cross between two FC-1 and one CHL cells, though some fractions of their chromosome complements must have been deleted.

Table 2 indicates the yield of cells and the amount of mucopolysaccharide secreted per 10^6 cells of each line at various culture times. B-6 cells accumulated 18–20 μg in the medium per 10^6 cells at 3–5 days. In contrast, only one-fifth to one-tenth of that amount was detected in the media of other clonal hybrids and both parent lines proliferated until their maximum level of growth. However, it is doubtful whether these latter values are in fact due to hyaluronic acid, because the optical densities obtained from the estimation of glucuronic acid in the media were too low

TABLE 2. Comparison of the Ability of Parent and 4 Clonal Hybrid Lines to Produce Hyaluronic Acid

Cell line	Day	No. of total cells ($\times 10^{-6}$ per dish)	Hyaluronic acid produced (μg/10^6 total cells)	Mucin clot test
FC-1	4	7.3	0.38	–
	5	6.5	1.24	–
CHL	5	5.2	1.27	–
	6	5.6	2.08	–
A-5	4	3.8	2.97	–
	5	3.7	1.44	–
B-1	3	3.8	2.70	–
	5	4.6	2.78	–
B-4	4	2.4	4.18	–
B-6	3	3.8	21.0	+
	4	4.2	19.2	+
	5	4.8	18.0	+

for significant determination and the mucin clot tests gave all negative results except in the case of B-6 cells.

B-6 cells are round and grow singly in a floating form. Although, in general, hyaluronic acid synthesis is said to occur only in fibroblasts, the present observations with B-6 cells imply that morphological differentiation is not always a prerequisite for functional differentiation.

We have not yet identified from which parent genome the hyaluronic acid synthesis was induced in B-6 cells, but since only the B-6 clone produces hyaluronic acid among the hybrids, this induction is not due to the complementation of imperfect phenotypes of both parent cells. The activation of dormant gene by cell hybridization has also been reported by the group of Dr. Weiss in the case of serum albumin synthesis by mouse fibroblasts (*18*). They explained the mechanism in terms of gene dosage. We would like to propose a new term "hybriduction" for these phenomena, that is, the induction of a phenotype not expressed in either parent cell by hybridization.

Hyaluronic Acid Synthesis and Cell Growth

Generally differentiation is not coupled with growth, but this is not the case in hyaluronic acid synthesis. It has already been shown that the rate of hyaluronic acid production in some fibroblasts continues undiminished during the actively growing phase with DNA synthesis, but that the rate falls off as DNA synthesis or cell multiplication ceases in the stationary growth phase (*16*). These facts indicate that there is some kind of shut-off mechanism for mucopolysaccharide synthesis operating with the cessation of growth or DNA synthesis in cultured fibroblasts. Thus, it seemed important to test for the existence of a similar control mechanism in the artificial interspecific hybrid B-6 line. The results are shown in Fig. 1. The

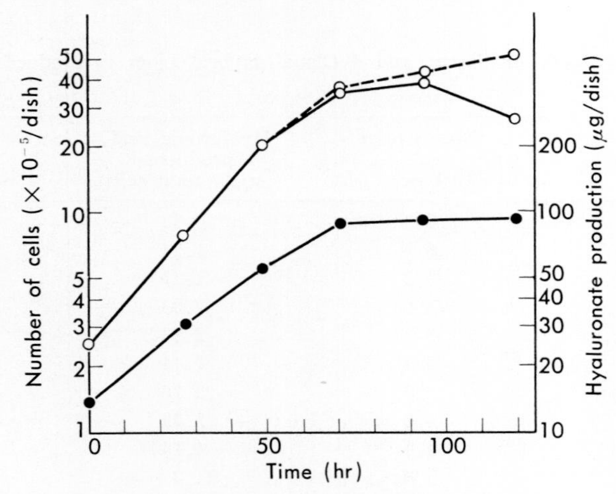

FIG. 1. Kinetics of cell growth and hyaluronic acid synthesis of B-6 line. O- -O total number of cells; O—O number of viable cells; ●—● hyaluronic acid content in the medium.

growth curve of B-6 cells is closely correlated with the curves for accumulation of hyaluronic acid in the medium. This shows that even in the present hybrid cells, hyaluronic acid synthesis takes place under the same control mechanism as in fibroblasts (*6*).

Hyaluronic acid is synthesized from glucose, which is converted to UDP-N-acethylglucosamine and UDP-glucuronic acid *via* several enzyme reactions, and finally the 2 intermediates are condensed to form hyaluronic acid by the synthetase. The processes from glucose to hyaluronic acid and the enzyme catalysing each process are illustrated in Fig. 2. In this process, the key enzyme for this differentiation is considered to be UDP-glucose dehydrogenase (Enzyme 9 in Fig. 2) and hyaluronic acid synthetase (Enzyme 10 in Fig. 2). Accordingly, we examined the enzyme levels in the cells in relation to their growth phase. These results are illustrated in Fig. 3. As shown clearly here, the enzyme activity increased without any lag period after inoculation into the fresh medium, and the increase lasted during the early and middle logarithmic growth phases, but decreased rapidly thereafter. This al-

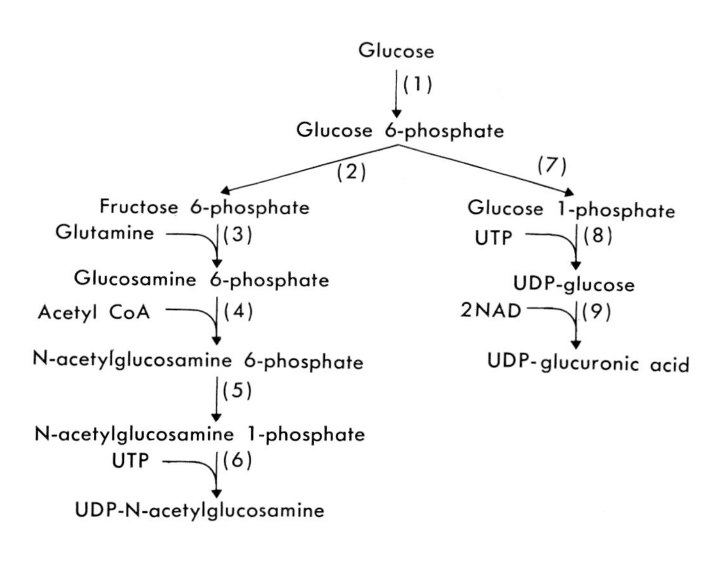

nUDP-glucuronic acid + nUDP-N-acetylglucosamine
(10) → Hyaluronic acid + 2nUDP

Enzymes:
(1) Hexokinase
(2) Phosphohexoisomerase
(3) Fructose 6-phosphate aminotransferase
(4) Glucosamine 6-phosphate N-acetyltransferase
(5) Phosphoacetylglucosamine mutase
(6) UDP-N-acetylglucosamine pyrophosphorylase
(7) Phosphoglucomutase
(8) UDP-glucose pyrophosphorylase
(9) UDP-glucose dehydrogenase
(10) Hyaluronic acid synthetase

FIG. 2. Pathway of hyaluronic acid biosynthesis.

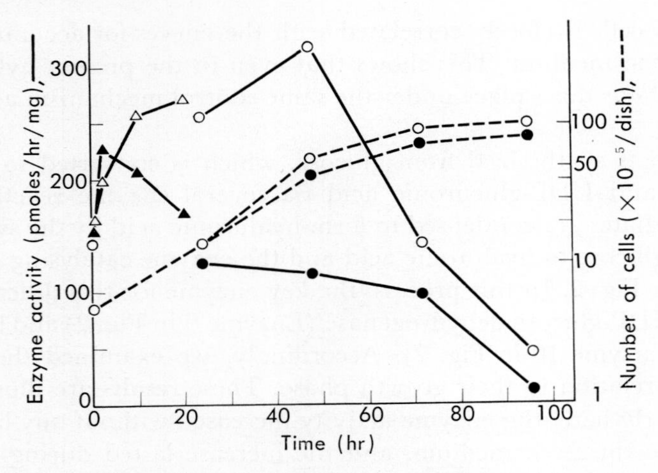

FIG. 3. Fluctuation of hyaluronic acid synthetase activity during the growth phase and time course of BrdU effect. Cells (5×10^5) of a 3-day-old culture were inoculated into 10 ml of fresh medium with or without 5 μg/ml of the analog. A portion of culture was harvested daily, and enzyme activity and number of cells were measured. Hyaluronic acid synthetase activity: ●—● BrdU-treated culture; ○—○ untreated cultures. Number of cells: ●- -● BrdU-treated cultures; ○- -○ untreated cultures. A larger inoculum size than that in the above experiment ($1–2.5 \times 10^6$ cells) was employed to examine the early change in enzyme activity during the first 20 hr after inoculation, and only enzyme activity was measured. ▲ BrdU-treated cultures; △ untreated cultures.

teration of enzyme level is intimately correlated with the accumulation curve of hyaluronic acid shown in Fig. 1. On the other hand, UDP-glucose dehydrogenase did not change significantly throughout the growth phase (data not shown). Therefore, it can be concluded that hyaluronic acid synthesis is mainly regulated at the level of its synthetase in the cells. A similar conclusion can be drawn from our recent work using rat fibroblast cells which synthesize hyaluronic acid (unpublished data). We will come back later to the effect of BrdU on hyaluronic acid synthetase activity during the growth phase.

Suppression of Hyaluronic Acid Synthesis by BrdU

From the results described above, the expression of this differentiated function is found to be closely related to active cell growth. However, if any chemical can be found which separates the 2 processes, hyaluronic acid synthesis and cell growth, one may obtain more information on the regulation of this differentiation. Among many chemicals screened for this purpose, the thymidine analogs, BrdU and 5-iododeoxyuridine (IdU) inhibited hyaluronic acid synthesis in this cell, with less effect on cell growth.

First, the effect of BrdU on growth and hyaluronic acid synthesis of B-6 cells was studied by treatment of the cells with various concentrations of the drug for 3 days. The relative growth rates of treated cultures and the relative amounts of hyaluronic acid produced by them were expressed as percentages of those in control

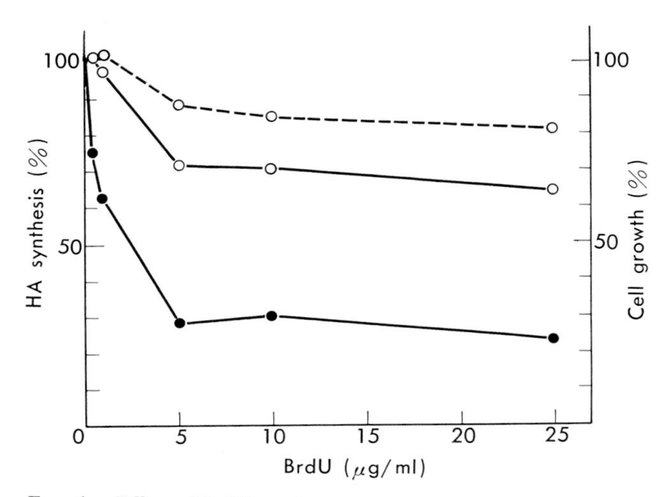

Fig. 4. Effect of BrdU on the growth and hyaluronic acid synthesis of B-6 cells as a function of concentration. Cells (2.5×10^5) were grown for 3 days in the standard medium containing each concentration of BrdU. The results are expressed as relative percentages of the control values. Untreated control cultures yielded an average of 3.4×10^6 total cells per dish after 3 days, their viability as checked by the Trypan Blue exclusion test was more than 95%, and the mean amount of hyaluronic acid produced per 10^6 total cells was 23.9 μg. ● hyaluronic acid synthesis; ○—○ viable; ○- -○ total.

cultures. Figure 4 shows that the presence of BrdU at concentrations of 0.5 to 1 μg/ml had no effect on the multiplication of the cells, but that at higher concentrations some inhibitory effect appeared. Treatment with concentrations ranging from 5 to 25 μg/ml resulted in a constant decrease of approximately 15 % in the total number of cells or 30 % in that of viable cells. In contrast, the effect of the analog on the production of hyaluronic acid differed significantly from its effect on growth. The amount of hyaluronic acid produced by the cells treated with BrdU at 0.5 or 1 μg/ml was reduced to about 75 or 60 %, respectively, of the control level. This reduction was more pronounced at the higher concentrations. The cultures grown in 5 to 25 μg/ml of BrdU were able to secrete only 27 % hyaluronic acid on the average, as compared with untreated cultures, so that the BrdU treatment diminished the amount of mucopolysaccharide synthesized per cell to about one-third of the level of the control. These results clearly indicate that BrdU inhibits hyaluronic acid synthesis either more strongly than or in preference to the multiplication of B-6 cells, in agreement with the data reported previously by Bischoff and Holtzer (*3*) for chick-embryo amnion cells.

Next, we studied the inhibition of hyaluronic acid synthetase in cells treated with BrdU. Figure 5 shows the dose-response relationship of this inhibition; during a 2-day treatment with BrdU the total cell protein was unchanged and UDP-glucose dehydrogenase increased slightly. Conversely, hyaluronic acid synthetase activity decreased greatly. This pattern is quite similar to that of the decrease of hyaluronic acid content secreted in the medium. In Fig. 3, the effect of BrdU at 5 μg/ml on hyaluronic acid synthetase activity of B-6 cells during the course of each

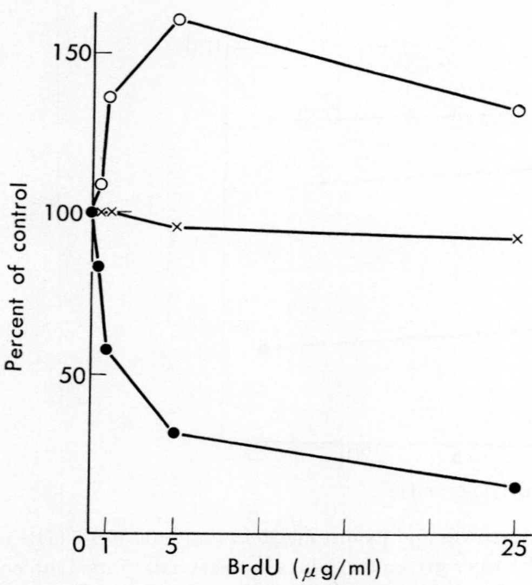

FIG. 5. Effects of BrdU on the activities of hyaluronic acid synthetase and UDP-glucose dehydrogenase and content of total protein in B-6 cells as a function of concentration. Cells (5×10^5) of a 3-day-old culture were inoculated into 10 ml of medium containing each concentration of BrdU and cultured for 48 hr. The results are expressed as percentages relative to the control values. ● hyaluronic acid synthetase activity; ○ UDP-glucose dehydrogenase activity; × content of total cell protein.

growth phase is included along with the change of hyaluronic acid synthetase activity of a control culture. This activity in the BrdU-treated cells increased similarly to that in control cells for approximately the first 2 hr, after which it began to decrease. The maximum relative suppression was 60% at 48 hr. The data suggest that growth in the analog does not cause a complete loss of hyaluronic acid synthetase activity. On the other hand, the activity of UDP-glucose dehydrogenase remained nearly constant throughout the growth phase (not shown).

Inhibition of hyaluronic acid synthesis by BrdU is completely reversible, as shown in Fig. 6. When B-6 cells treated with BrdU for 3 days were transferred to normal medium and cultured, the reduced level of production of hyaluronic acid recovered completely to the control level in 3 transfer generations, after a lag period of one generation (Fig. 6). Also, the reduced activity of hyaluronic acid synthetase in the cells treated with BrdU for 16 hr, that is, one generation, returned to the normal level of activity after removal of the analog and 3 to 4 subsequent generations of culture in normal medium (Fig. 7).

The effect of BrdU is overcome by the addition of 2 to 4 times the amount of thymidine with respect to the analog to the medium. The relatively prolonged effect of BrdU even after its removal and the effect of thymidine suggest that BrdU acts on the cells by being incorporated into the cellular DNA, and that the incorporation of BrdU in one strand of DNA is effective in suppressing this function. A result which supports this view was obtained in the following experiment, which is shown in Table 3. In the presence of hydroxyurea, BrdU is not taken up into DNA

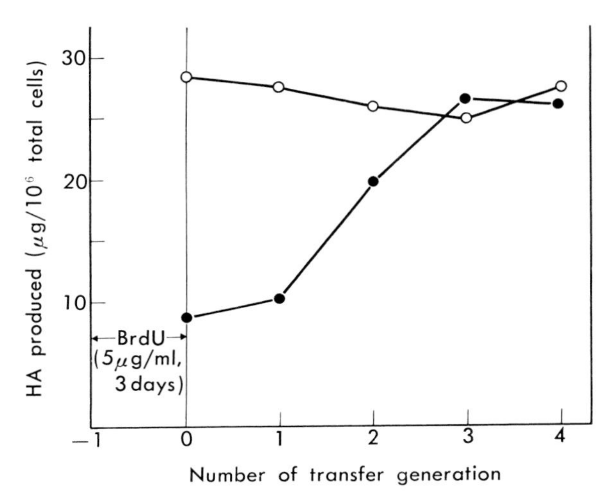

Fig. 6. Recovery from the inhibited state of hyaluronic acid synthesis produced by BrdU, following transfer into analog-free medium. Two sets of cultures containing 2.5×10^5 cells/dish were incubated in the standard medium with and without 5 μg BrdU/ml for 3 days. The BrdU-treated cells were washed twice with agent-free medium, transferred into the standard medium and subcultured in the same manner as untreated cells. In each of the transfer generations, cell count and determination of the hyaluronic acid content were carried out. ○ control cultures; ● BrdU-treated cultures.

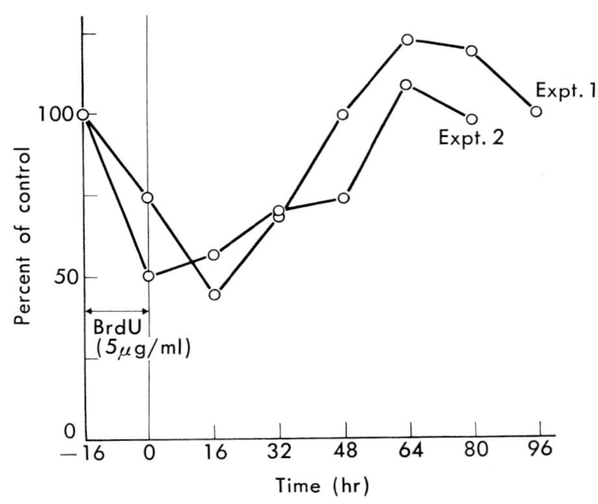

Fig. 7. Reversibility of the effect of BrdU on hyaluronic acid synthetase. After one generation in medium containing 5 μg/ml of the analog, B-6 cells were washed with the analog-free medium, transferred into fresh medium and diluted periodically to keep the cell concentration in the range of 1×10^5 to 5×10^5 cells/ml. Values for the analog-treated cells were expressed as percentages of the values for control cultures. The results of 2 independent experiments are shown.

of the cells, as confirmed by cesium chloride density gradient centrifugation of the DNA. After treatment of the cells with BrdU (5 μg/ml) for 16 hr, the hyaluronic acid synthetase activity decreased to 60%, and no recovery of activity was observed

TABLE 3. Prevention of the BrdU Effect on Hyaluronic Acid Synthetase by the Simultaneous Addition of Hydroxyurea

Additions	Final concentration (μg/ml)	Hyaluronic acid synthetase activity: Treatment (16 hr) Specific activity[a]	Treatment (16 hr) Relative activity	Treatment (16 hr) → Removal (20 hr) Specific activity[a]	Treatment (16 hr) → Removal (20 hr) Relative activity
None	—	268	100	212	100
BrdU	5	169	63.1	142	66.7
Hydroxyurea	50	107	39.9	235	111
BrdU + hydroxyurea	5 50	119	44.4	210	99.0
Thymidine	480	270	101	—	—
FUdR	1	247	92.2	—	—

Cells of a 3-day-old culture were inoculated into fresh medium containing each reagent. After incubation for 16 hr, one portion of the cultures was harvested, and the remainder was transferred into a reagent-free medium and incubated for another 20 hr.

[a] pmoles of incorporated ^{14}C-glucuronic acid/hr per mg protein.

20 hr after the removal of BrdU. On treatment of cells with BrdU (5 μg/ml) in the presence of hydroxyurea, the synthetase activity was reduced to 44%. This reduction was mainly attributable to the action of hydroxyurea, since hydroxyurea alone reduced the synthetase activity, probably by the inhibition of protein synthesis, and no decrease of synthetase activity was observed by other inhibitors of DNA synthesis, such as excess thymidine or fluorodeoxyuridine (FUdR). When the cells treated with BrdU in the presence of hydroxyurea were transferred to a drug-free medium and incubated for another 20 hr, the enzyme activity returned to the con-

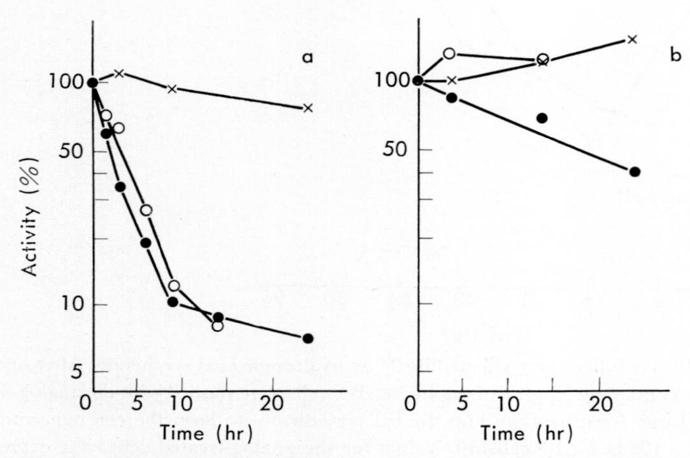

FIG. 8. Decay of activity of hyaluronic acid synthetase (a) and UDP-glucose dehydrogenase (b) after treatment with cycloheximide or actinomycin D. Cells ($1-2.5\times10^6$) of a 2-day-old culture were inoculated to 10 ml of medium containing either 10 μg/ml of cycloheximide (●) or 5 μg/ml of actinomycin D (○), or to normal control medium (×). A portion of each of the cultures was harvested at several intervals thereafter and enzyme activities were assayed. The activities are expressed as percent of the initial specific activity.

trol level, while that of BrdU-treated cells still remained suppressed. These results seem to prove the requirement for incorporation of BrdU into DNA for the inhibitory effect of BrdU on hyaluronic acid synthesis to become apparent.

Turnover Rates of Hyaluronic Acid Synthetase and Its mRNA

When B-6 cells were incubated in the presence of a relatively large dose of actinomycin D (5 μg/ml) or cycloheximide (10 μg/ml), the enzyme synthetase activity decayed with a half-life of about 3.5 and 2 hr, respectively (Fig. 8a). On the other hand, UDP-glucose dehydrogenase activity had a long life and remained stable for over 15 hr (Fig. 8b). These data further support the above view that hyaluronic acid production is regulated by the level of hyaluronic acid synthetase and that BrdU inhibits mucopolysaccharide synthesis by decreasing the enzyme level in the treated cells. High turnover rates of messenger RNA and enzyme protein of hyaluronic acid synthetase might account for the selective inhibition of this enzyme by BrdU treatment, as suggested by Stellwagen and Tomkins (*22*).

DISCUSSION

Using the same hybrid cell line B-6 as reported in this paper, Koyama, from our laboratory has described in this symposium the induction of alkaline phosphatase by BrdU(*11*). The induction of alkaline phosphatase was found in experiments to determine the effect of BrdU on enzymes other than hyaluronic acid synthetase, which was suppressed by the analog. As expected, acid phosphatase activity in this cell was absolutely refractory to BrdU treatment. The induction of enzyme activity by BrdU was quite unexpected. However, there have been reported many cases of RNA tumor virus induction by BrdU from non-producing cells (*20, 26*). The RNA of oncogenic RNA viruses is considered to be reversely transcribed and integrated into host-cell DNA. Consequently the mechanism of induction is quite comparable to that of the induction of messenger RNA.

The fairly intimate relationship between the suppression of hyaluronic acid synthetase and induction of alkaline phosphatase by treatment with BrdU was very remarkable, although the 2 actions of BrdU are opposite in direction, *i.e.*, induction and suppression. The similarities between the 2 actions of BrdU are as follows:
1) among the pyrimidines and their ribo- and deoxyribonucleosides, BrdU, IdU, and bromodeoxycytidine are effective, and the action of these analogs is prevented by thymidine addition,
2) in both cases, 5 μg/ml of BrdU is the minimum dose to obtain maximum effects,
3) both actions follow similar kinetics, such as lag period (one generation time), reversibility and even the recovery curve after removal of BrdU. Those intimate relationships between the 2 actions are so remarkable that they seem unlikely to be fortuitous.

In both actions, it was shown that the incorporation of BrdU into at least one strand of DNA duplex was essential. However, we don't know why one enzyme is induced and another is suppressed, while most other enzymes, such as those for cell growth, are not sensitive to BrdU treatment. In this regard, we can mention some models proposed so far. Stellwagen and Tomkins (*23*) suggested 2 possibilities as mechanisms by which BrdU could preferentially inhibit transcription from selected genes. One possibility is that the sensitivity of a transcription unit to BrdU might increase with size. The other is that the promotors for some genes are more sensitive than others, because of high thymine content or because of some specific sequence. The later model seems most appealing, because Riggs *et al.* (*14*) suggested that the *lac* operator of *Escherichia coli* has a high content of thymine and moreover they demonstrated that BrdU substitution in the *lac* operator enhances the binding of *lac* repressor to it. Riggs also speculated that the altered binding of regulatory protein caused by the incorporation of BrdU into the DNA involved with regulation of virus production results in the induction of C-type virus.

Also, in a review on the action of BrdU on differentiation, Wilt (*25*) referred to the scheme of I-DNA proposed by Bell (*2*), who found that the number and size of cytoplasmic I-DNA-containing particles is greatly altered in the presence of BrdU. However, the evidence that the expression of differentiated function is mediated by I-DNA or that genes for differentiated functions are amplified and transferred into episomes is not convincing enough. Wilt proposed another model, that DNA containing BrdU is sufficiently different in its structure that regulatory controls are abortive and ineffective.

Finally, we want to describe the tumorigenicity of B-6 cell. This interspecific hybrid between mammary carcinoma cell of C3H mouse and nontumorigenic Chinese hamster fibroblast cell exhibited tumorigenicity not only to newborn C3H mouse but also even to the adult, although the survival time of the host was prolonged 2-fold compared with that of the parent mammary carcinoma cell. In addition, even *in vivo*, *i.e.*, in the peritoneal cavity of the host, it retained the capacity to produce hyaluronic acid.

REFERENCES

1. Barski, G., Sorieul, S., and Cornefert, F. Formation of "hybrid" cells in combined cultures of two different mammalian cell strains. J. Natl. Cancer Inst., *26*: 1269–1291, 1961; Barski, G. and Cornefert, F. Characteristics of "hybrid"-type clonal cell lines obtained from mixed cultures *in vitro*. J. Natl. Cancer Inst., *28*: 801–821, 1962.
2. Bell, E. I-DNA: Its packaging into I-somes and its relation to protein synthesis during differentiation. Nature, *224*: 326–328, 1969.
3. Bischoff, R. and Holtzer, H. Inhibition of hyaluronic acid synthesis by BUDR in cultures of chick amnion cells. Anat. Rec., *160*: 317 (Abstr.), 1968.
4. Bischoff, R. and Holtzer, H. Inhibition of myoblast fusion after one round of DNA synthesis in 5-bromodeoxyuridine. J. Cell Biol., *44*: 134–150, 1970.
5. Bregula, U., Klein, G., and Harris, H. The analysis of malignancy by cell fusion, II. Hybrids between Ehrlich cells and normal diploid cells. J. Cell Sci., *8*: 673–680, 1971.

6. Daniel, M. R., Dingle, J. T., and Lucy, J. A. Cobalt tolerance and mucopolysaccharide production in rat dermal fibroblasts in culture. Exp. Cell Res., *24*: 88–105, 1961.
7. Davidson, R. L., Ephrussi, B., and Yamamoto, K. Regulation of pigment synthesis in mammalian cells, as studied by somatic hybridization. Proc. Natl. Acad. Sci. U.S., *56*: 1437–1440, 1966.
8. Ephrussi, B. Hybridization of Somatic Cells, Princeton University Press, New Jersey, 1972.
9. Grossfeld, H. Positive mucin clot test in supernates of cultures of avian embryonic brain. Proc. Soc. Exp. Biol. Med., *96*: 844–846, 1957.
10. Koyama, H. and Ono, T. Effect of 5-bromodeoxyuridine on hyaluronic acid synthesis of a clonal hybrid line of mouse and Chinese hamster in culture. J. Cell Phys., *78*: 265–272, 1971.
11. Koyama, H. and Ono, T. Control of alkaline phosphatase activity in cultured mammalian cells: induction by 5-bromodeoxyuridine, cyclic AMP, and sodium butyrate. *In;* W. Nakahara *et al.* (eds.), Differentiation and Control of Malignancy of Tumor Cells, pp. 325–341, University of Tokyo Press, Tokyo, 1974.
12. Koyama, H., Utakoji, T., and Ono, T. A new cell line derived from newborn Chinese hamster lung tissue. Gann, *61*: 161–167, 1970.
13. Koyama, H., Yatabe, I., and Ono, T. Isolation and characterization of hybrids between mouse and Chinese hamster cell lines. Exp. Cell Res., *62*: 455–463, 1970; Koyama, H. and Ono, T. Initiation of a differentiation function (hyaluronic acid synthesis) by hybrid formation in culture. Biochim. Biophys. Acta, *217*: 477–487, 1970.
14. Lin, S. and Riggs, A. D. *Lac* operator analogues: Bromodeoxyuridine substitution in the *Lac* operator affects the rate of dissociation of the *Lac* repressor. Proc. Natl. Acad. Sci. U.S., *69*: 2574–2576, 1972.
15. Markert, C. L. Neoplasia: A disease of cell differentiation. Cancer Res., *28*: 1908–1914, 1968.
16. Nakano, N. Establishment of cell lines *in vitro* from a mammary ascites tumor of mouse and biological properties of the established lines in a serum-containing medium. Tohoku J. Exp. Med., *88*: 69–84, 1966.
17. Okada, Y. The fusion of Ehrlich's tumor cells caused by HVJ virus *in vitro*. Biken J., *1*: 103–110, 1958.
18. Peterson, J. A. and Weiss, M. C. Expression of differentiated functions in hepatoma cell hybrids: Induction of mouse albumin production in rat hepatoma-mouse fibroblast hybrids. Proc. Natl. Acad. Sci. U.S., *69*: 571–575, 1972.
19. Potter, V. R., Walker, P. R., and Goodman, J. I. Survey of current studies on oncogeny as blocked ontogeny: Isozyme changes in livers of rats fed 3′-methyl-4-dimethyl-aminoazobenzene with collateral studies on DNA stability. GANN Monograph on Cancer Research, *13*: 121–134, 1972.
20. Rowe, W. P., Lowy, D. R., Teich, N., and Hartley, J. W. Some implications of the activation of murine leukemia virus by halogenated pyrimidines. Proc. Natl. Acad. Sci. U.S., *69*: 1033–1035, 1972.
21. Silagi, S. and Bruce, S. A. Suppression of malignancy and differentiation in melanotic melanoma cells. Proc. Natl. Acad. Sci. U.S., *66*: 72–78, 1970.
22. Stellwagen, R. H. and Tomkins, G. M. Differential effect of 5-bromodeoxyuridine on the concentrations of specific enzymes in hepatoma cells in culture. Proc. Natl. Acad. Sci. U.S., *68*: 1147–1150, 1971.

23. Stellwagen, R. H. and Tomkins, G. M. Preferential inhibition by 5-bromodeoxyuridine of the synthesis of tyrosine aminotransferase in hepatoma cell cultures. J. Mol. Biol., *56*: 167–182, 1971.
24. Tomida, M., Koyama, H., and Ono, T. Hyaluronic acid synthetase in cultured mammalian cells producing hyaluronic acid. Oscillatory change during the growth phase and suppression by 5-bromodeoxyuridine. Biochim. Biophys. Acta, *338*: 352–363, 1974.
25. Wilt, F. H. and Anderson, M. The action of 5-bromodeoxyuridine on differentiation. Develop. Biol., *28*: 443–447, 1972.
26. Wu, A. M., Ting, R. C., Paran, M., and Gallo, R. C. Cordycepin inhibits induction of murine leukovirus production by 5-iodo-2′-deoxyuridine. Proc. Natl. Acad. Sci. U.S., *69*: 3820–3824, 1972.

Discussion of Paper by Drs. Ono et al.

DR. JOHNSON: We found that dibutyryl cyclic AMP increases the amount of chondroitin sulfates in the growth medium, but it has no effect on the intracellular levels of these compounds. Have you considered differences between extracellular and intracellular hyaluronate?

DR. ONO: Most of the hyaluronate synthesized by this cell is secreted in the medium. We have measured only the extracellular hyaluronate throughout these experiments.

DR. POTTER: Is it correct that you have concluded, in general terms, that
1) BrdU is accepted as thymidine in cell replication.
2) BrdU is not accepted as thymidine in mRNA production.
3) The BrdU effect is easily seen in the case of enzymes with a short half-life.
4) In mRNA production, BrdU affects the rate of production, not the properties of the enzyme produced.
5) Something is different about the way thymidine or BrdU occurs in DNA sequences that govern the rate of transcription, as compared with the DNA sequences that determine protein structure.

DR. ONO: That is right. We are assuming that the promotor or repressor genes of differentiated functions might be rich in T and more sensitive to the replacement by BrdU. However, we have no definite evidence yet.

DR. WEBER: Since the discussion and the conclusions revolve in part on the behavior of the enzyme hyaluronic acid synthetase it would be relevant for us to know the method employed for the assay of this enzyme. Would you tell us the essentials of your enzyme assay?

DR. ONO: First the harvested cells were homogenized and incubated with ^{14}C-UDP-glucuronic acid and UDP-N-acetylglucosamine. After incubation, the reaction mixture was boiled to stop the reaction and the solution was spotted on a paper strip and subjected to chromatography. Hyaluronate synthesized was calculated from the counts incorporated in the spot of hyaluronic acid.

DR. OSTERTAG: You did mention that some of your cell lines do secrete B-type virus. Is this virus secreted by your cells still capable of inducing tumors in mice? If you add BrdU to your cells, do you get a reduction in the number of virus particles secreted by your cells? If so, are the BrdU-resistant cells still secreting any B-type particles?

DR. ONO: We have not yet tested the oncogenicity of B-type viruses secreted by these cells. We tried to test the induction of B-type particles by BrdU in this cell, but so far the results have been negative.

TUMOR REVERSAL

Expression of Malignancy in Interspecies Cell Hybrids

Georges Barski and Jean Belehradek, Jr.
Tissue Culture and Virus Laboratory, Gustave-Roussy Institute, Villejuif, France

Abstract: The discovery of somatic cell hybridization opened a new approach for the analysis of mechanisms regulating heredity in somatic cells. The relationship between the genetic content of these cells and its phenotypic expression, the mechanisms determining cell differentiation or dedifferentiation and, in particular, the nature of cancerous transformation of cells could be approached in a new fashion.

In the first experiments, which led to the discovery of somatic hybridization, highly tumorigenic mouse cells were fused with syngeneic cells lacking this capacity. The resulting hybrids, cumulating nearly completely the genomes of the 2 parental cells, were malignant. Since then a considerable body of data concerning hybridization between iso- or homologous malignant and nonmalignant cells (whether normal, diploid or transformed, and heteroploid) was reported from our and other laboratories. The results concerning the heritance of malignancy in the hybrids were variable. However some general rules could be established. (1) If at least one of the parental cells originated from a "permanent" line, the hybrid obtained herited this property. (2) If one of the parental lines was not only transformed but also tumorigenic, the hybrid could inherit this faculty or not. (3) In any case the hybrids showed a considerable chromosomal deletion from the "ideal" addition of the 2 parental chromosomal sets.

More recently, the problem of expression of malignancy in cell hybrids was approached in experiments involving crossings between malignant and nonmalignant cells originating from different species, offering an obvious advantage of a fairly easy identification of chromosomal participation from the 2 parent cell genomes. This could also be related to the recognition of corresponding phenotypic characteristics expressed in the hybrids. The intrinsic tumor-producing capacity of the parent and of the hybrid cells was tested by inoculation into immunologically depleted hosts: cortisoned Chinese hamsters (cheek pouches) or embryonated chick eggs (chorioallantoic membrane or brain). These testing methods when previously checked in parallel with inoculations into histocompatible host appeared as quite reliable. Several hybrid lines obtained by crossing highly malignant mouse cells

(R4) with nontumorigenic Chinese hamster (D/AD/Aza heteroploid) or Akodon Urichi (diploid) cells, were developed and studied as were hybrids resulting from crossing nonmalignant (3T3) mouse with malignant (DC3F) Chinese hamster cells. Some general conclusions could be drawn from these experiments:

1) All the R4 × D/AD/Aza hybrids obtained were malignant as were some R4 × Akodon Urichi hybrids. Accordingly, it may be assumed that the genetic or epigenetic factors determining the invasive behavior can come into expression in interspecies hybrids as they do in intraspecies ones.

2) The relatively easy identification of chromosomes in the interspecies hybrids revealed a quite considerable chromosomal variability and the obvious fact that cells presenting a wide range of karyotypic variants were capable to participate in the formation of tumors.

3) On the background of this variability, in the case of R4 × Akodon Urichi hybrids differentiated into tumorigenic and nontumorigenic hybrid lines, no systematic and significant relationship could be established between the tumor-producing capacity and a particular karyological pattern.

4) In all the mouse × Chinese hamster hybrids studied the mouse chromosomes predominated and it appeared significant that the invasive behavior of the hybrids was dependent on the tumorigenic capacity of the mouse parent cell.

5) In all the studied interspecies hybrids both species-specific antigens were expressed on the cell surface. Additionally, TST-type antigens, brought into the hybrid by the tumorigenic parent cell, could be evidenced in different ways. Interestingly enough, this also occurred in the case when the tumor-producing capacity in the hybrids was repressed or not expressed, proving possible dissociation between the presence of this antigen and the capacity to produce invasive growth.

Discovery of Somatic Cell Hybridization

The first experiments which brought us to the discovery, some years ago, of somatic cell hybridization (*2, 3*) were undertaken, as a matter of fact, with the aim of checking the possibility of transfer of genetic or epigenetic information between high- and low-malignant cells in mixed cultures *in vitro*. The cells used originated from subcutaneous tissue of C3H mice developed in long-term cultures by Sanford *et al.* (*32*) and designated in our laboratory as N1 and N2. They were strikingly different in their morphology: N1 presented a criss-cross, noncontact inhibited growth of spindle-shaped or polygonal fibroblasts with numerous cytoplasmic processes whereas the N2 cells were rather round and showed a contact-inhibited behavior in dense cultures. The N1 cells were highly malignant and produced rapidly growing tumors in syngeneic mice and also when inoculated into embryonated chick eggs (chorioallantoic membrane or brain). N2 gave, with very high inocula, only very rare tumors growing very slowly and being scarcely of an invasive nature. The karyotypes of the cells used were also distinctly different: the N1 cells contained, with minor deviations, 50–55 acrocentric chromosomes with one "extra long" marker: the N2 cells had a modal number of 62 chromosomes, 13 of them on average biarmed. The hybrid M cells which appeared in the mixed cultures could

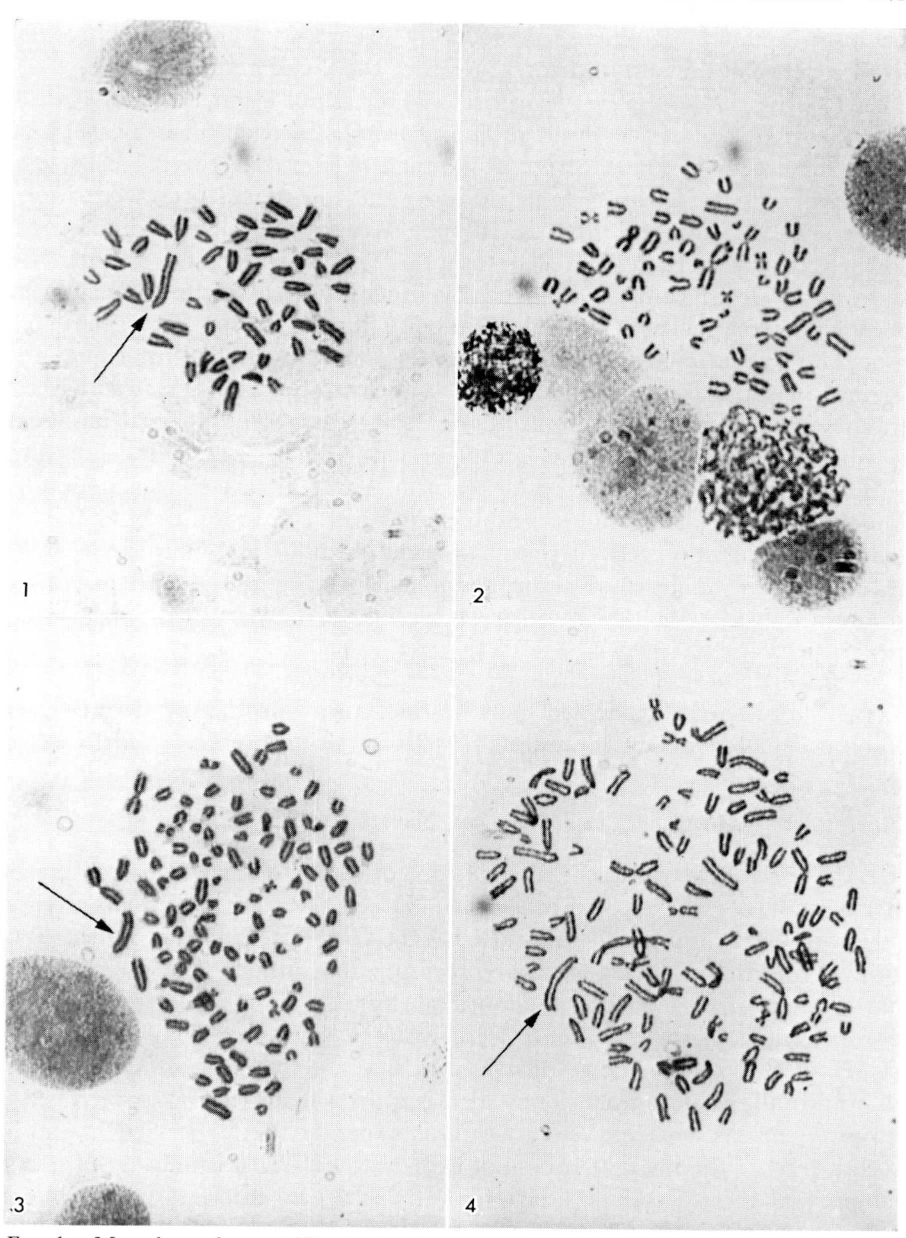

FIG. 1. Metaphase of parent N1 cell: 54 chromosomes, one extra long (arrow).

FIG. 2. Metaphase of parent N2 cell: 61 chromosomes, 12 biarmed.

FIG. 3. Metaphase of an M hybrid cell: 113 chromosomes, 11 biarmed, the extra-long marker present (arrow).

FIG. 4. Metaphase of another M hybrid cell: 122 chromosomes, 14 biarmed, the extra-long marker present (arrow). Figures 1–4 are from G. Barski *et al.*, "hybrid"-type cells in combined cultures of two different mammalian cell strains. J. Natl. Cancer Inst., *26*: 1269–1291, 1961.

be unmistakably recognized as such by their karyotypes which resulted, visibly, from a complete fusion and integration of the 2 cell genomes (Figs. 1–4).

The hybrid cells grew vigorously (which is not a general rule, as we know now) and could be isolated without difficulty from mixed cultures by cloning, especially since these cells had a selective advantage *in vitro* over the N1 cells, which adhere less firmly to the glass. When, in turn, the mixed cultures were inoculated into syngeneic C3H mice the hybrid M cells overgrew the low-malignant N2 cells and formed nearly pure hybrid cell tumors. The hybrid clones isolated from *in vitro* cultures or from tumors, checked by inoculation into syngeneic mice, showed a tumorigenic capacity in the same range as the N1 malignant parent cell.

The hybrid character of the M cells, including their "intermediary" morphological aspect, was maintained for years of continuous culture *in vitro* in spite of some fluctuations in chromosome numbers and shapes and a general tendency for chromosomal loss from the "ideal" additive value of the 2 genomes integrated.

These first findings showed clearly that, in contrast to the generally accepted belief, somatic cells of higher animals are able to fuse not only at the level of their cytoplasms, as observed currently in some pathological conditions, but also at the level of their nuclei and their genomes, forming fairly stable and proliferating somatic hybrid cells, new man-made biological entities created by an experimental procedure.

Since these first experiments, a considerable body of data concerning hybridization between transformed cells having tumor-producing capacity and nonmalignant cells has been accumulated by different authors.

Expression of Malignancy in Homologous Somatic Hybrids

In 1965, Scaletta and Ephrussi (*33*) obtained a successful crossing between the permanent, *in vitro*-adapted malignant N1 cell line, originating, as already mentioned, from C3H mouse and normal CBA-T6T6 mouse cells having the T-6 translocation chromosomal marker, easily recognizable among the 40 normal telocentric chromosomes. Hybrid clones identified by the simultaneous presence of marker chromosomes of the 2 parents were isolated and checked by inoculation into F1 (C3H × CBA) mice. The results showed that the hybrid cells were capable of producing malignant tumors. They also acquired from the N1 parent the capacity to grow *in vitro* as a permanent line. This capacity was expressed in the hybrid cells well ahead of the normal course of rather slow "transformation" of *in vitro*-cultivated normal mouse cell populations and it can be inferred that this capacity was acquired and expressed as an immediate consequence of the fusion with a transformed cell.

Later Silagi (*34*) obtained hybrids by mating *in vitro* cells from a highly malignant C57BL mouse melanoma with an A9 8-azaguanine-resistant variant of L cells which do not normally produce tumors when inoculated into syngeneic C3H mice. The parent A9 cells were eliminated in Littlefield's (*21*) selective HAT medium containing aminopterin, hypoxanthine, and thymidine. Hybrid cell clones, identified by their karyotype, containing chromosomes of both parents, and also by the simultaneous presence of H-2^b (C57BL) and H-2^k (C3H) histocompatibility antigens, were isolated and assayed for malignancy. Five from the total of 6 hybrid clones produced malignant (nonmelanotic) tumors when inoculated into F1 (C3H $\times$C57BL) but not in the parent C3H or C57BL mice. There were no very significant changes in the hybrid karyotypes during an observation period covering nearly 500 generations *in vitro*. The author concluded that "the capacity for progressive growth *in vivo* appears to be "dominant" in the hybrid."

Several attempts were made to hybridize cells converted by an oncogenic virus with nonmalignant cells. Gershon and Sachs (*16*) in 1963 obtained hybrids between polyoma-transformed malignant SWE mouse cells and L cells (of C3H origin, nonmalignant). The hybrids have been shown to be capable of producing tumors when inoculated into F1 (C3H$\times$SWR) mice. They also contained the polyoma virus-induced transplantation antigen.

Similarly, Defendi and his associates (*10*) in 1964 showed that in hybrids produced by mating the low-cancer NCTC 2555 cell line with polyoma-transformed Swiss mouse cells, the polyoma virus-induced complement fixing (CF) and transplantation antigens could be found. Unfortunately, no appropriate histocompatible host was available to check the malignancy of these hybrids *in vivo*.

A more complete and detailed study was made later by Defendi *et al.* (*11*) on hybrids obtained by crossing polyoma-transformed and malignant cells of A/Sn mouse origin with normal, short-term, diploid CBA-T6T6 cells. The hybrids were easily recognized by the presence of marker chromosomes from both parents. They showed positive tests for the transplantation and CF polyoma antigens. This time, assays for malignancy could be performed on appropriate hosts. Most of the hybrid clones proved to be neoplastic when inoculated into F1 (CBA$\times$A/Sn) mice. They gave, as expected, no tumors in parental mice. The authors concluded that the properties of polyoma-transformed parent cells (tumorigenicity and production of polyoma-induced antigens) appear dominant in the hybrid cells. It should not be overlooked that the situation in the reported findings was quite special, since one of the parental cells was also carrying, in addition to the information related to invasive cell behavior, the viral genome, potentially capable, as one can speculate, in new intracellular surroundings created by cell fusion, of initiating *de novo* at least a stage of virus-induced malignant conversion of the hybrid cell. Incidentally, in more recent studies, Klein and Harris (*20*) as well as Meyer *et al.* (*25*) observed polyoma-induced specific antigens not only in the high-tumorigenic, but also in the low-tumorigenic hybrids resulting from matings between polyoma-induced highly malignant cell and normal or low-tumorigenic cells.

On the other hand, Harris *et al.* using the Sendai virus-induced cell-fusion technique devised by Okada (*29, 30*), obtained a number of cell hybrids from crossings between highly malignant, chiefly ascites tumor, cells with nonmalignant (aneuploid or diploid) cells (*9, 19, 37*). These hybrids appeared at first as non- or low-tumorigenic. Their malignancy could still come into expression as a result of *in vitro* or *in vivo* passaging of the hybrids followed by a considerable loss of chromosomes from the "ideal" sum of the integrated 2 parent genomes.

According to the view of these authors, the "suppression" of malignancy in the hybrids resulted from the introduction by the nonmalignant cell partner of a specific chromosomal ingredient missing in the malignant parent cell. The re-expression of the tumorigenic capacity of the hybrid is explained by the secondary loss of the "normalizing" genetic factor, supposedly a specific chromosome. The search for such a "normalizing" chromosome has not so far resulted in its identification. Nevertheless in many experiments, described by Klein *et al.* (*19*), when derivatives of the L cell line, containing a considerable number of biarmed marker chromosomes, were used as a nonmalignant partner for hybridization, massive loss of these recognizable chromosomes accompanied the re-expression of malignancy and was related to it. Other experiments reported by the same group and involving Ehrlich tumors crossed with normal, diploid fibroblasts, were less conclusive in the sense of retention of chromosomes from normal cells being responsible for the suppression of malignancy in the hybrids (*9*).

A similar study was performed by us (*5*) by mating highly malignant N1 cells, of C3H origin, with nonmalignant BALB/c fibroblasts. An attempt was made to establish a possible relationship between the karyological features and the expression of malignancy in the hybrid cell line (HyEN) as well as in 6 clones derived from it. As a rule, the karyotypes of the hybrid line and of its clones showed a marked dispersion of chromosome numbers and modal values, ranging from 80 to 129, the presumed "ideal" sum of the parent cell modes being 122 (Fig. 5).

Among the 6 randomly isolated hybrid clones the only nontumor-producing

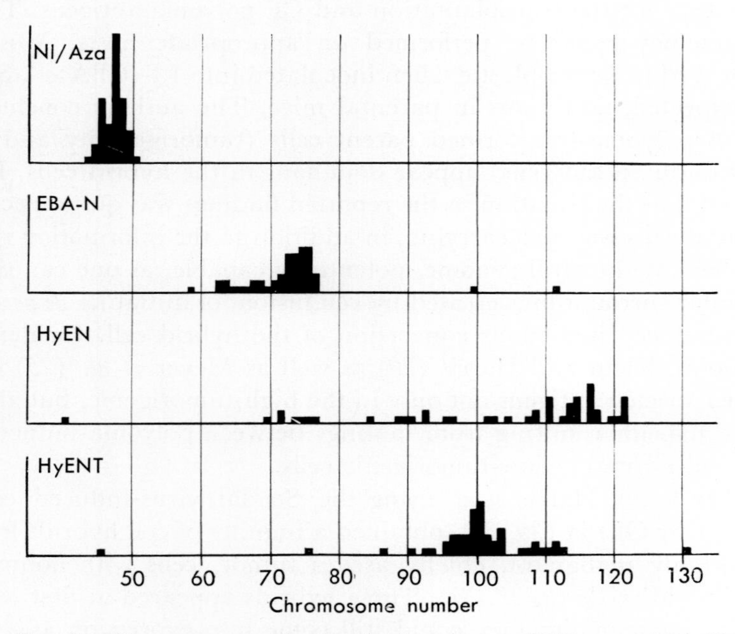

FIG. 5. Chromosome number distribution of parent N1/Aza and EBA, N cells, and of HyEN and HyEN. T hybrid cells. (The HyEN. T line is a tumor derivative of the HyEN line). From J. Belehradek, Jr. *et al.*, karyological patterns and expression of malignancy in some homologous mouse somatic hybrid cells. Int. J. Cancer, *8*: 1–9, 1971.

one had a mode of 112–118 chromosomes, quite similar to those of 2 malignant clones (Fig. 6). Three more clones were malignant and had modes of 80, 98, and 129 chromosomes. Early cultures of cells recovered from hybrid tumors, produced in F1 (C3H × BALB/c) mice, showed on the whole a decrease in chromosomal mode values (Fig. 7). However, in spite of a general tendency toward chromosomal loss, the hybrid cells participating in the formation of tumors represented a wide range of chromosomal variants.

Thus, the considerable background karyological variability of mouse *in vitro*-adapted cell lines in general and of the hybrid lines in particular, makes it quite difficult, if it is even possible, to establish a precise relationship between the presence or absence of specific, individual chromosomes (originating from the one or the other mouse-cell partner) and the expression of malignancy.

So far, the only firmly established conclusion from these studies is that the adaptation of a homologous hybrid cell line to efficient permanent growth *in vitro* and, even more so, to efficient invasive growth *in vivo* is reached through a selection of cells most fit to respond to these growth requirements, and that chromosomal deletion is a currently recognizable sign of this selection.

Whether this phenomenon corresponds to a trivial "alleviation" of the overcharged hybrid cell genome or to a more specific genetic deletion (which may be

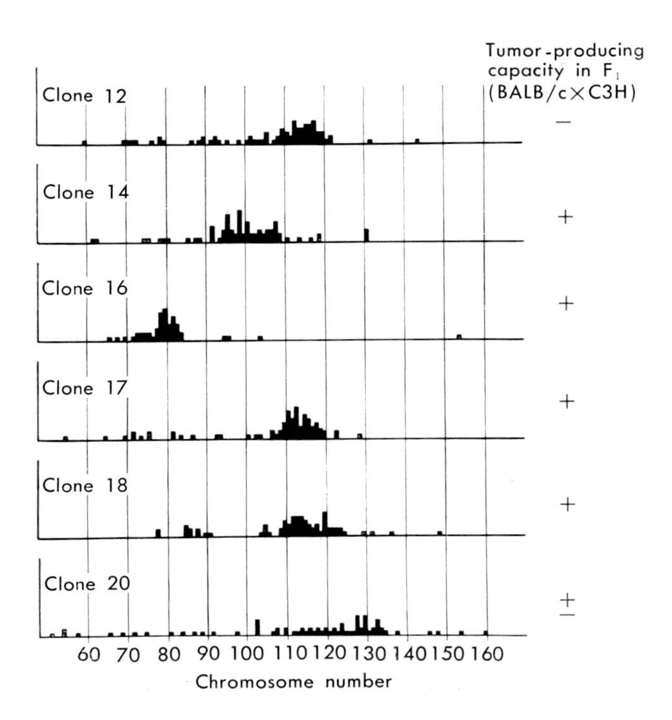

FIG. 6. Chromosome number distribution of several clones (4th passage) derived from the HyEN hybrid line. From J. Belehradek, Jr. *et al.*, karyological patterns and expression of malignancy in some homologous mouse somatic hybrid cells. Int. J. Cancer, *8*: 1–9, 1971.

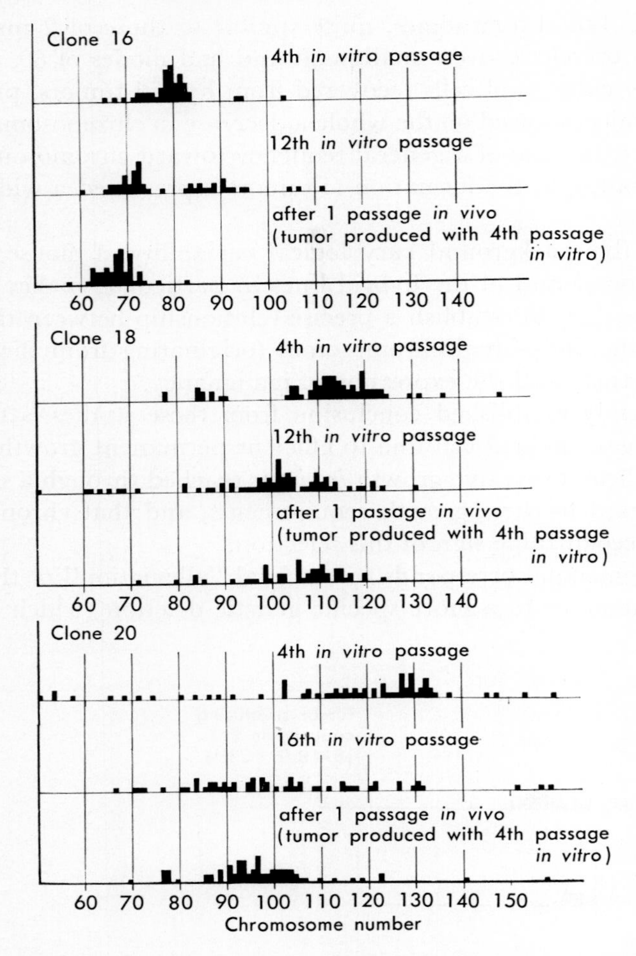

FIG. 7. Evolution of chromosome number distribution of HyEN clones Nos. 16, 18, and 20, following *in vitro* passages or following one animal passage as a tumor. A more-or-less pronounced shift toward lower chromosome numbers is observed. From J. Belehradek, Jr. *et al.*, karyological patterns and expression of malignancy in some homologous mouse somatic hybrid cells. Int. J. Cancer, *8*: 1–9, 1971.

compensated by hybridization and again decompensated) remains still an open question.

Malignant × Nonmalignant Interspecies Cell Hybrids

The problem of expression of tumorigenic properties in somatic hybrids was studied more recently in our laboratory with hybrids obtained by crossing malignant and nonmalignant cells from different species.

Interspecies hybrids were obtained by many authors, since Ephrussi and Weiss (*12*) produced the first viable and proliferating mouse x rat somatic hybrids in 1965.

Numerous successful crosses were obtained since the mating of transformed, aneuploid mouse cells with normal human fibroblasts by Matsuya *et al.* (*24*) or with human blood leucocytes by Nabholz *et al.* (*28*), followed by other authors. Similarly Goldstein and Lin (*17*) hybridized Syrian hamster-transformed heteroploid cells with human "senescent" fibroblasts.

In all these experiments, it was well demonstrated that the long life span characteristic of one of the parent cell could prevail in the hybrids obtained over the properties of the cell partner having a limited *in vitro* life span or even lacking any capacity to proliferate in cultures.

Moreover, the results obtained and especially the karyological data suggest that at least the permanent growth characteristics acquired by the hybrids were related to an adequate dosage of genetic (or epigenetic) factors from the transformed and permanently growing cell parent.

These studies, however, shed no light on the effective expression of invasive properties in interspecies somatic hybrids. We attempted to explore this aspect of interspecies hybrids using several crosses of tumorigenic and nontumorigenic heterologous cells. The following cell material was studied: (1) hybrids between R4/B, bromodeoxyuridine (BrdU)-resistant, BALB/c mouse malignant cells, carrying an overt murine C-type virus infection, crossed with Chinese hamster D/AD/Aza, 8-azaguanine-, and actinomycin D-resistant, nontumorigenic cells* (hybrids designated as HyCS (*4*)), (2) hybrids between 3T3.4E, Swiss mouse, nonmalignant cells, crossed with DC-3F/Aza7, 8-azaguanine-resistant highly malignant Chinese hamster cells (hybrids designated as Hy307 (*6*)), (3) hybrids between malignant R4/B cells and normal Akodon Urichi and (Venezuelian vole mouse) fibroblasts (hybrids designated as HyRA (*7*)).

Evaluation of Tumorigenic Potential in Immunologically Depleted Host Systems

In order to ascertain and to evaluate the tumorigenic capacity of the interspecies hybrids obtained, 2 test systems were used: (1) inoculation into the cheek pouch of cortisone-treated Syrian hamsters and (2) inoculation into embryonated chick eggs at the 11th day of incubation (on dropped chorioallantoic membrane (CAM) or, alternatively by intracerebral route).

In preliminary experiments, several mouse-cell lines having high- or low-tumorigenic capacity were checked comparatively by parallel inoculations into syngeneic mice and into the hamster cheek pouch. Quite concordant results were observed. Thus, for example, highly tumorigenic mouse-cell lines such as N1 (C3H), used in our first hybridization experiments, P4bisT (C57BL), developed in our laboratory, or R4/B (BALB/c), which produced 100% rapidly growing tumors in syngeneic mice inoculated with 10^6 cells, gave nearly 100% tumors in cheek pouches of Syrian hamsters inoculated with $2\text{–}3 \times 10^6$ cells. In contrast, nontumorigenic mouse cell lines, like the 3T3.4E line were entirely negative when

* This line was a derivative of the DC-3F/ADX cells kindly supplied by Dr. J. L. Biedler (Sloan-Kettering Institute, New York).

inoculated in hamsters. According to the same criteria, actinomycin D-resistant Chinese hamster D/AD/Aza cells were nontumorigenic whereas the DC-3F/Aza7 cells were regularly tumorigenic.

Quite similar and generally concordant results were obtained with inoculations into embryonated chick eggs, especially on the chorioallantoic membrane.

A legitimate conclusion from these preliminary studies was that, in confirmation of previous data (*8*, *18*, *23*), the cheek pouch of Syrian hamster as well as the CAM inoculation system, (the last used recently in similar circumstances by Murayama-Okabayashi *et al.* (*27*)) were quite reliable, in the absence of a natural host, for direct checking and evaluation of the tumorigenic capacity of interspecies cell hybrids.

Karyotypes and Expression of Species-specific Antigens in Interspecies Hybrids

In all the three cell crossings studied, hybrid cells isolated and accumulated in selective media (Littlefield's (*21*) HAT medium or HAT supplemented with 1 μg/ml of actinomycin D for the HyCS hybrid), were subjected to karyological analysis. The distinction between the mouse, nearly exclusively telocentric chromosomes, and predominantly (Chinese hamster) or exclusively (Akodon Urichi) biarmed chromosomes was relatively easy.

The analysis showed for the HyCS hybrid a relative stabilisation in the range of 90–100 chromosomes with an average of 10 to 12 biarmed (Fig. 8). Obviously, HyCS cells containing a number of telocentrics, mostly mouse chromosomes, close to the 2S value and a number of biarmed, chiefly Chinese hamster chromosomes, well below the 1S value, had in the long term a selective advantage. This selection operated obviously against the background of an important, constantly observed chromosomal variability. One aspect of this variability was the frequent appearance of chromosomes of a new type nonexistent in the parent cells and originating obviously from intra- or interspecies translocations.

A similar situation was seen with the other hybrid lines studied: HyRA and Hy307, where again the number of telocentric or acrocentric chromosomes indicated in nearly all isolated clones the presence of more than one set of the murine parent line. Similar observations were made by Matsuya *et al.* (*24*) and Sonnenschein (*36*) suggesting that a special mechanism might exist involving preferentially tetraploid or dividing mouse cells in the fusion process.

Concerning Chinese hamster chromosomes participation in the Hy307 hybrids, it varied from less than 1S to nearly 2S. The number of Akodon Urichi chromosomes in the HyRA lines studied was reduced to only a few (Figs. 9 and 10).

In spite of the predominant chromosomal participation of the mouse genome in the hybrids studied and often a quite limited presence of Chinese hamster or Akodon Urichi chromosomes, both parent species-specific antigens could be demonstrated on the surface of the hybrid cells. Thus, for example, the index of specific immunofluorescence (performed according to the technique of Möller (*26*)) was 0.90 for the HyCS cells with anti-mouse sera and 0.98 with anti-Chinese hamster

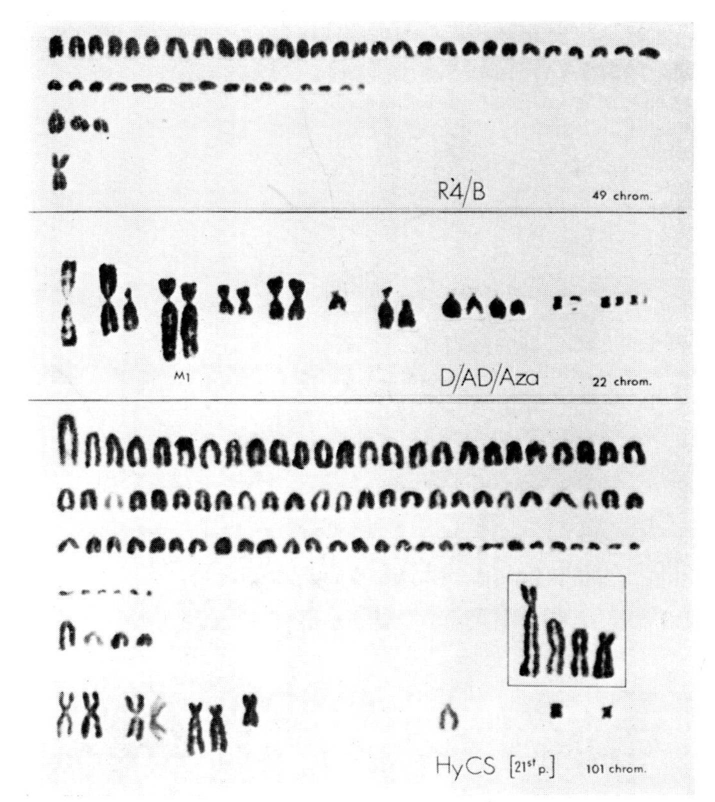

FIG. 8. Ideograms of chromosomes of parent R4/B (mouse), D/AD/Aza (Chinese hamster) cells, and of HyCS interspecies hybrid.

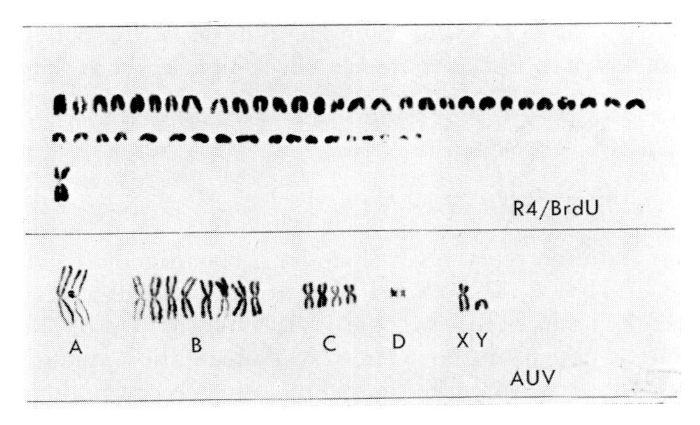

FIG. 9. Ideograms of chromosomes of parent R4/B (mouse) and AUV (Akodon) cells.

sera. Similarly immunofluorescence indices of HyRA hybrids were close or equal to 1.0 with both anti-mouse and anti-Akodon Urichi antisera.

The consequence of this double antigenicity, expressed on the surface of the

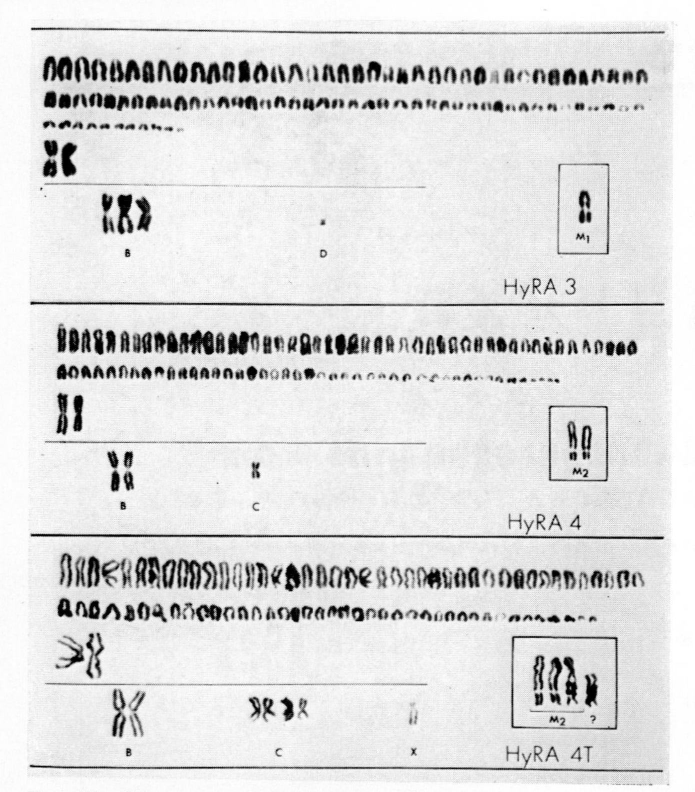

FIG. 10. Ideograms of chromosomes of HyRA3, HyRA4, and HyRA4T interspecies hybrid cells (the HyRA4T line was recovered from a tumor produced in Syrian hamster cheek pouch following inoculation of HyRA4 line).

hybrid cells, was that both the HyCS and the malignant derivatives of HyRA cells, which were tumorigenic in the Syrian hamster cheek pouch, were rejected when inoculated into BALB/c mice, syngeneic with the highly malignant R4/B parental cells.

Intrinsic Malignancy of Interspecies Cell Hybrids

The intrinsic host immune reaction-independent, tumorigenic capacity of the hybrid cell lines HyCS, Hy307, and clonal derivatives, as well as HyRA3 and HyRA4 (the last 2 being clonal sublines of the HyRA hybrid) were checked using the Syrian hamster cheek pouch and also the CAM inoculation system.

The HyCS cell population was tumorigenic to a degree comparable in quantitative terms with that of the malignant mouse parent R4/B cell: 100% takes was obtained in a total of 35 checking inoculations with the 8th, 16th, 18th, and 29th passages of the HyCS cells. The HyCS cells were equally tumorigenic in the CAM system. Concurrently, the invasive character of HyCS cells growth *in vivo* or *in ovo* was ascertained with sections of the tumors produced according to routine histopathological criteria.

Karyological studies revealed that HyCS cells participating in the formation of the tumors were selected from the higher ploidy segment of the hybrid cell population. However, even after this selection, the cell lines developed from the HyCS tumors presented a wide display of chromosomal variants and 2 identical karyotypes could hardly be found among 50 to 60 mitoses routinely examined in detail for each of these lines (see Table 1).

The results were quite different from the Hy307 hybrid series, resulting from fusion of a malignant DC-3F/Aza7, Chinese hamster cell line, with a nonmalignant 3T3. 4E Swiss mouse cell line. As in the case of the HyCS, the Hy307 hybrid clones contained predominantly mouse chromosomes corresponding in mode numbers chiefly to values between 1S and 2S. The participation of Chinese hamster chromosomes varied in different Hy307 clones, reaching 10 to 33 metacentrics, *i.e.*, mode numbers ranging from less than 1S to nearly 2S values.

However, in contrast to what occurred with HyCS, the Hy307 hybrids were

TABLE 1. Karyological Variability in the Parent R4/B (Mouse) and in the Hybrid HyCS (Mouse × Chinese Hamster) Cell Populations

Chromosome No.[a]	Total	Telocentrics			Acrocentrics with satellites	Subtelocentrics	Metacentrics	Estimated participation in HyCS of chromosomes from the parent cells	
		Big and medium	Small	Minutes				R4/B mouse	D/AD/Aza Ch. hamster
R4/B 30th pass.	47	35	4	3	3		2		
	48	36	4	4	2		2		
	49	36–37	4	4	2		1		
	50	34	7	4	3		2		
	50	29	9	7	4		1		
	52	34	6	5	5		2		
	53	40	4	6	2		1		
	56	38	7	8	2		1		
HyCS 5th pass.	34	14	6	5	2	3	4	26	8
	42	23	3	1	4	5	6	31	11
	45	24	6	1	6	3	5	36–35	8–9
	53	31	7	3	5	3	4	46	7
	52	35	4	1	5	1	6	44	8
	56	33	7		4	5	6	44	12
	56	28	5	10	4	3	9	42	14
	59	34	10	4	5	4	2	51	8
	60	42		4	6	4	4	51	9
	60	29	6	4	10	5	4	47–46	13–14
	63	41	8	4	5	3	2	53	8
	64	30	8	6	8	5	6	48	16
	91	60	6	6	4	5	10	79–77	12–14
	107	61	6	3	10	9	15	83	24
	114	68	18	2	8	5	14	93–92	21–22
	121	68	16	8	10	5	14	101	20

[a] Following chromosomal counts on ideograms of individual mitoses analyzed.

devoid of any tumorigenic capacity, which was not unexpected given the nonmalignant character of the 3T3.4E mouse parent cell. In other terms, the "majority" participation of the mouse genome in interspecies hybrids imposed on them either the capacity, in the case of HyCS, or the incapacity, in the case of Hy307, to grow as tumors, depending on whether the mouse parent cell was malignant or not.

"Superhybrids" and Shift of Malignancy

The question then arose as to whether by a "superhybridization" of Hy307 cells fused secondarily with the malignant DC-3F/Aza7 cells a shift toward a tumorigenic capacity could be obtained.

For this fusion we used:

1) A subclone of the Hy307 hybrid cell line containing nearly one set of parent mouse cell chromosomes (a modal number of 57 telocentrics) and 7 to 15 biarmed, chiefly CH chromosomes. This subclone was developed as a BrdU-resistant, thymidine kinase-negative variant of one of the Hy307 hybrid clones, and was designated as Hy307II/B.

2) The Chinese hamster malignant DC-3F/Aza7 subline selected for deficiency in hypoxanthine-guanine phosphoribosyl transferase by prolonged cultivation in presence of 8-azaguanine.

After fusion of cells from these 2 lines with inactivated Sendai virus, "superhybrids" were isolated in Littlefield's (*21*) HAT selective medium. They were designated SHy37, developed as lines and studied.

In terms of karyological constitution, a range of cell variants was found. However, as a rule, in all the SHy37 mitoses checked the number of biarmed Chinese hamster chromosomes was significantly augmented as seen by the increase of modal values from 10 to 47. When inoculated into Syrian hamster cheek pouches, the SHy-37 cells produced tumors. In the cells recovered from these tumors, designated as SHy37T, the shift toward a supremacy of Chinese hamster over mouse chromosomes was indicated by some increase of the first and decrease of the second. Thus, when interpreting these results, both the increase of genetic (or epigenetic) dosage from the malignant parent and its decrease from the nontumorigenic one, have to be taken into consideration as responsible for the expression of the malignancy in the SHy37 "superhybrids."

This situation is quite similar to that described by Murayama-Okabayashi *et al.* (*27*) who obtained in a homologous mouse system by simple or double hybridization between Ehrlich ascites (malignant) and L (nonmalignant) cells a series of cell strains in which gradation of tumor-forming capacity apparently paralleled a corresponding dosage of Ehrlich ascites on L cell genomes.

Comparison of Malignant and Nonmalignant Cell Variants of an Interspecies Hybrid

A quite interesting situation concerning the expression of malignancy was found in HyRA (mouse malignant × Akodon Urichi normal cell) hybrids.

Two clonal HyRA lines, HyRA3, and HyRA4, were selected for more detailed study. In these 2 lines, most of the chromosomes were telocentrics or acrocentrics and their number oscillated around values close to 2 complete chromosome sets (2S) of the mouse parent cell. The biarmed chromosomes, originating chiefly from the Akodon Urichi cell partner, were restricted in number and, with a background of considerable variability, were always well below the value proper to the euploid Akodon Urichi karyotype.

The HyRA3 line when inoculated into cheek pouches of the Syrian hamster never produced tumors in a total of test inoculations exceeding 20, performed so far between the 6th and 30th *in vitro* passages, whereas HyRA4 cells regularly produced tumors in nearly 100% of inoculated animals. These tumors reached a size of about 15 mm in diameter within 2 weeks. Routine histological examinations showed in all cases invasive, undifferentiated sarcomas.

In spite of their relative karyological similarity, some tendency for a lower number of telocentric chromosomes was seen in HyRA3 when compared with HyRA4. Whether this shift in HyRA4 cells can account for the expression of malignancy in HyRA4 but not in HyRA3 hybrid cells remains dubious, since in HyRA4T cells obtained from a HyRA4 tumor, a significant loss of mouse chromosomes was observed.

In the spirit of the rule postulated by Harris and Klein (*9*, *19*, *37*) one would expect that the nonmalignancy of the HyRA3 line should be due to the presence of a particular "malignancy-suppressing" factor from the Akodon Urichi normal karyotype, a factor missing in the HyRA4 line. As a matter of fact, a slightly higher mean number of large metacentric chromosomes, supposedly of Akodon Urichi group B, was observed in the HyRA3 line as compared with the tumor-derivative HyRA4T line. However, this difference was not statistically significant and, in any case, was not absolute and hence, inconclusive.

In other words, in spite of rather favorable conditions of comparison between the karyotypes of the malignant (HyRA4) and nonmalignant (HyRA3) hybrid sister clonal lines (the relatively easy identification of chromosomes as originating from one or other species and the quite limited Akodon Urichi chromosomal participation in the hybrid karyotype) it has, so far, been impossible to establish, against the background of the quite important chromosomal variability in the 2 HyRA lines studied, a precise and constant interrelationship between the presence or absence of determined chromosomes and the expression of tumorigenic capacity.

Fate of the Tumor-specific Antigens and of C-type Murine Virus in Interspecies Hybrids

The interspecies malignant × nonmalignant cell hybridization offered possibilities to study in favorable conditions the interrelationship between the expression of tumorigenic capacity in the hybrids and some other characteristics of the hybrid phenotype; in the first place, the tumor-specific or tumor-related antigens and the fate of C-type murine leukemia viruses if introduced into the hybrids by the mouse parent cell.

In the HyCS hybrid both the mouse and Chinese hamster species-specific cell-surface antigens were expressed, as was the tumor-producing capacity brought in by the mouse R4/B parent cell. However, the Gross-type murine leukemia virus, carried originally by the parent R4/B mouse cell was repressed. This was shown by the absence of C-type virus particles in the hybrid cells and by rather negative results of immunofluorescence tests (index 0.3) with anti-Gross specific antisera. Nevertheless some R4/B cell-specific antigenicity persisted certainly in or on the HyCS hybrid cells since inoculation of these cells into BALB/c mice induced in them a fair degree of immunity against challenge with R4/B cells. An attempt was made to activate the C virus by treating the HyCS cells with BrdU, following the usual procedure (*1*, *22*). The results were inconclusive and no clear activation of the virus could be, so far, obtained under these conditions.

We had to conclude that the presence in the HyCS hybrids of some elements of Chinese hamster genome did not interfere with the expression of either the mouse species-specific antigens or the tumor-producing capacity, but was efficient in interfering with the replication of murine C-type virus and, to some extent, with the expression of virus-induced antigenicity. This conclusion is reminiscent of the observations of Fenyö *et al.* (*14*, *15*) on the repression of murine C-type virus infection in some homologous mouse × mouse hybrids.

Another pattern of interrelationship between malignancy and the expression of tumor-specific or tumor-related antigens was found in the HyRA interspecies hybrid system.

As mentioned already, in spite of only very few Akodon Urichi chromosomes (no more than 5 on average) both mouse and Akodon Urichi species-specific antigens were regularly present at the surface of HyRA cells. This expression was, as far as one could judge according to the specific immunofluorescence index, quite similar for the malignant HyRA4 and nonmalignant HyRA3 hybrids.

Moreover, HyRA4 cells which proved to be tumorigenic in cortisone-treated Syrian hamsters were rejected when inoculated into BALB/c mice, obviously because of the integration in these cells of Akodon Urichi antigenicity.

Additionally, BALB/c mice injected previously with HyRA4 cells resisted subsequent challenge with parent highly malignant R4/B cells, which indicated that the tumor-specific transplantation antigen known to be present in the R4/B cells was also expressed in the HyRA4 hybrids.

This antigen was, supposedly, related to the Gross-type murine leukemia antigen, as demonstrated by positive immunofluorescence tests with specific anti-Gross leukemia serum (index 1.00). This was not unexpected for the HyRA4 line, since C-type particles were abundantly present in the HyRA4 cells and since these cells, when overlayered with XC cells (following the technique of Rowe *et al.* (*31*), produced numerous foci of degenerating polykaryocytes indicating the production of infectious C-type murine leukemia virus. The characteristics of the nonmalignant HyRA3 hybrid line were different in many respects. The HyRA3 cells did not show (on electron microscopic examinations) C-type particles, a conclusion reinforced by negative results obtained in the XC test. Nevertheless, immunofluorescence tests for murine Gross-type leukemia antigen were clearly positive. Furthermore, BALB/c mice preinoculated with HyRA3 cells resisted challenge with R4/B cells.

These concordant results proved that some common antigen(s), alien to the syngeneic BALB/c mice and possibly related to the tumorous nature of the R4/B cells, was shared by the nonmalignant HyRA3 variant of the hybrid cells and the parental R4/B cells. The nature of these antigen(s) remains unknown. However, our results did show that in some circumstances a dissociation between the tumorigenic potential and the tumor-specific antigen can occur in interspecies hybrid cells. This is reminiscent of the observations reported by Klein and Harris (*20*) as well as by Meyer *et al.* (*25*) of a similar dissociation (in homologous mouse-cell hybrids) between the expression of polyoma tumor-specific antigens and tumor-producing capacity. On the other hand, our observations concerning the repression in the HyRA3 hybrids of C-type virus release with maintenance of virus-related cell-surface antigen(s), are close to the data reported by Fenyö *et al.* (*14*) in homologous mouse-cell hybrids.

CONCLUSIONS

1) In confirmation of previously obtained data, crossing of *in vitro*-adapted cells from established lines with diploid, nontransformed cells may result in hybrids and, particularly, interspecies hybrids in which permanent growth characteristics are prevailing. The acquisition of these characteristics appears, as in the case of our HyRA interspecies, mouse × Akodon Urichi hybrids, as an immediate consequence of this kind of cell fusion.
2) If one of the heterologous cell partners involved in hybridization is transformed and malignant, this property can also be expressed in the interspecies hybrid.

This expression can be demonstrated in a reliable way—as shown in comparative control experiments with syngeneic cell-host systems—by using test inoculation in immunologically depleted recipient systems: cheek pouches of cortisone-treated Syrian hamsters and chorioallantoic membrane or brain of chick embryos at the 11th day of incubation. Once expressed in the hybrids, the tumorigenic property remains apparently stable through *in vitro* or *in vivo* passages in spite of karyological changes, which may be important.
3) It is becoming clear that, under these conditions, whatever the genetic or epigenetic determinants of cell malignancy or its repression are, they are expressed in a universally intelligible code which is fully operational in interspecies hybrids.

This is not unexpected in the light of the more general properties of interspecies hybrids in which, according to numerous data, a remarkable coordination and synchronization of diverse synthetic processes, involving nucleic acids, enzymes, and other proteins, is observed.
4) On the basis of this fundamental observation, experiments could be designed in which cell partners for hybridization were chosen having as different and as easily recognizable karyotypes as possible. Such were the mouse × Chinese hamster and the mouse × Akodon Urichi couples.

Under these circumstances, it was found that an important chromosomal segregation occurred in the very first phase of hybrid cell development, resulting in quite asymmetric hybrids in which mouse chromosomes prevailed, frequently reaching numbers close to 2S values, whereas the Chinese hamster and Akodon Urichi chromosomes were usually well below the 1S values. However, both species-specific antigens were strongly expressed in all the hybrids obtained. On the other hand, the tumorigenic capacity of these hybrids appeared to depend on whether the mouse parent was malignant (as in the case of HyCS) or nonmalignant (as in the case of Hy307).

5) These data were completed by the observed shift of the nonmalignant Hy307 hybrid toward malignancy when, by the way of a secondary hybridization, some supplementary genetic (and possibly epigenetic) material from the malignant Chinese hamster cell was introduced in the "superhybrid." The interpretation of this finding was complicated by the fact that in these new hybrids a significant decrease of genetic material from the nonmalignant mouse cell was also noted.

The notion of "genetic dosage" of 2 factors one "promoting" and the other "repressing," the expression of malignancy, as formulated by Yamamoto *et al.* (*38*) and demonstrated by Murayama-Okabayashi *et al.* (*27*) has to be considered as an interpretation of our results. This point of view is supported by generally accepted evidence based on cancer pathology data, clinical as well as experimental, showing that the capacity to produce an invasive growth is not an all-or-none phenomenon but a cell property (or, most probably a set of properties) expressed to different degrees in different cell cites, (including cell membranes), and liable to evolution.

6) Our observations on the expression of malignancy in interspecies cell hybrids are compatible with the 2 proposed concepts concerning the mechanism determining malignant cell transformation: (a) Malignant behavior is determined by positive, new or modified, genetic or epigenetic information which can eventually be matched, in the somatic hybrids in particular, by a so-far hypothetical "repressor" of malignancy, present in nonmalignant cells (whether normal or transformed). (b) The other concept is formulated by Wiener *et al.* (*37*) as follows: "one theory of the origin of malignancy would fit our findings very well; the idea that malignancy results from a specific genetic loss. If malignancy were the result of a genetic deletion or of a defect that produced a nonfunctional gene product, then recessiveness of malignancy in hybrid cells would be expected, for the nonmalignant or normal partner in the hybrid would then be able to complement the defect." It can, by the way, be remarked that in the same article Wiener and his associates (*37*) also admit, implicitly, the validity of the first concept by saying: "whatever heritable character is responsible for the malignancy of the tumor cells, it is not irretrievably lost as a result of cell fusion."

One can add that, considering the second concept in the light of the reported results of interspecies, malignant × nonmalignant cell hybridization, we should accept that the genetic deletion responsible for malignancy might be specifically repaired by a chromosomal compensation from a foreign species.

7) The experiments with HyRA interspecies hybrids offered the possibility of comparing 2 clonal hybrid cell lines developed in parallel, one being malignant and the other nonmalignant. In spite of a fairly satisfactory recognition between the chromosomes introduced into the hybrids by the 2 parent species, it has to be stated that against the background of constant though numerically moderate karyo-

logical variability, no significant and permanent chromosomal pattern could be found establishing a link between the presence or loss of specific chromosomes and expressed and nonexpressed (or repressed) tumorigenic capacity.

As a matter of fact, these observations are also compatible with a hypothesis of epigenetic changes as being responsible for malignant cell transformation, triggered and perpetuated at the same level of cell mechanisms as cell differentiation, for which no specific chromosomal pattern is either found or even looked for. A similar opinion is expressed by Ephrussi (*13*) on the basis of parallelism obtained in some experiments, among others those of Silagi and Bruce (*35*), between extinction of differentiated cell functions and repression of malignancy. Ephrussi concludes:

"The phenomenological similarity between suppression and extinction makes one wonder whether malignancy is not, after all, due to epigenetic changes rather than to genetic ones, *sensu stricto*."

8) In the HyCS as well as in the HyRA interspecies hybrid systems, in addition to the 2 parent, species-specific antigens, tumor-specific antigens capable of immunizing against parent malignant R4/B mouse cells were present and expressed, in particular on the cell surface. Interestingly enough, the same antigens were present and expressed in a similar way in the malignant (HyRA4) and nonmalignant (HyRA3) clonal lines of the HyRA hybrid, showing in this way possible dissociation between tumorigenic capacity and tumor-specific antigenicity. Moreover, Gross virus-related cell-surface antigen was expressed in both the virus-releasing HyRA4 and in the nonreleasing HyRA3 clones, demonstrating the possibility of dissociation between virus production and expression of virus-induced or virus-related antigens.

The murine C-type virus identified as having Gross-type antigenic characteristics, introduced into the hybrids by the R4/B mouse parental cell, was repressed at least in its replicating form in the malignant HyCS and in the nontumorigenic HyRA3 cells. It was fully present in the tumorigenic HyRA4 line. The reason for these differences and the factors (genetic patterns, more especially) determining different degrees of permissivity for the C-type murine virus in the hybrids, are unknown. It is also not clear to what extent the tumor-specific antigenicity of the parent and hybrid cells can be identified with the virus proper or virus-induced antigens.

Thus, as we see, the somatic cell hybridization procedure which has been applied since its discovery in 1960 to the study of the nature of malignant cell conversion and its heritage, while bringing some new approaches in this field propounded many more questions than it did answers. This may be a sign that we are following the right path.

ACKNOWLEDGMENT

The experimental part of this work was supported by grants from the Centre National de la Recherche Scientifique, the Institut National de la Santé et de la Recherche Médicale, and the New York Cancer Research Institute.

REFERENCES

1. Aaronson, S. A., Todaro, G. J., and Scolnick, E. M. Induction of murine C-type viruses from clonal lines of virus-free BALB/3T3 cells. Science, *174*: 157–159, 1971.
2. Barski, G., Sorieul, S., and Cornefert, F. Production dans des cultures *in vitro* de deux souches cellulaires en association de cellules de caractère "hybride." C.R. Acad. Sci. (Paris), *251*: 1825–1827, 1960.
3. Barski, G., Sorieul, S., and Cornefert, F. Formation of "hybrid" cells in combined cultures of two different mammalian cell strains. J. Natl. Cancer Inst., *26*: 1269–1291, 1961.
4. Barski, G., Blanchard, M. G., Youn, J. K., and Leon, B. Expression of malignancy in interspecies, Chinese hamster x mouse cell hybrids. J. Natl. Cancer Inst., in press.
5. Belehradek, J., Jr. and Barski, G. Karyological patterns and expression of malignancy in some homologous mouse somatic hybrid cells. Int. J. Cancer, *8*: 1–9, 1971.
6. Belehradek, J., Jr., Barski, G., and Thonier, M. Unpublished data.
7. Belehradek, J., Jr., Barski, G., and Thonier, M. Interspecies, mouse ×Akodon urichi, somatic hybrids: expression of malignancy. Unpublished data.
8. Biedler, J. L. and Riehm, H. Cellular resistance to actinomycin D in Chinese hamster cells *in vitro*: cross resistance, radioautographic and cytogenetic studies. Cancer Res., *30*: 1174–1184, 1970.
9. Bregula, U., Klein, G., and Harris, H. The analysis of malignancy by cell fusion. II. Hybrids between Ehrlich cells and normal diploid cells. J. Cell Sci., *8*: 673–680, 1971.
10. Defendi, V., Ephrussi, B., and Koprowski, H. Expression of polyoma-induced cellular antigen in hybrid cells. Nature, *203*, 495–496, 1964.
11. Defendi, V., Ephrussi, B., Koprowski, H., and Yoshida, M. C. Properties of hybrids between polyoma-transformed and normal mouse cells. Proc. Natl. Acad. Sci. U.S., *57*: 299–305, 1967.
12. Ephrussi, B. and Weiss, M. C. Interspecific hybridization of somatic cells. Proc. Natl. Acad. Sci. U.S., *53*: 1040–1042, 1965.
13. Ephrussi, B. Hybridization of Somatic Cells, pp. 107–109, Princeton University Press, New Jersey, 1972.
14. Fenyö, E. M., Grundner, G., Klein, G., Klein, E., and Harris, H. Surface antigens and release of virus in hybrid cells produced by the fusion of A9 fibroblasts with Moloney lymphoma cells. Exp. Cell Res., *68*: 323–331, 1971.
15. Fenyö, E. M., Grundner, G., Wiener, F., Klein, E., Klein, G., and Harris, H. The influence of the partner cell on the production of L virus and the expression of viral surface antigen in hybrid cells. J. Exp. Med., *137*: 1240–1255, 1973.
16. Gershon, D. and Sachs, L. Properties of a somatic hybrid between mouse cells with different genotypes. Nature, *198*: 912–913, 1963.
17. Goldstein, S. and Lin, C. C. Rescue of senescent human fibroblasts by hybridization with hamster cells *in vitro*. Exp. Cell Res., *70*: 436–439, 1972.
18. Handler, A. H. and Foley, G. E. Growth of human epidermoid carcinomas (strains KB and Hela) in hamsters from tissue culture inocula. Proc. Soc. Exp. Biol. Med., *91*: 237–240, 1966.
19. Klein, G., Bregula, U., Wiener, F., and Harris, H. The analysis of malignancy by cell fusion. I. Hybrids between tumour cells and L-cell derivatives. J. Cell Sci., *8*: 659–672, 1971.

20. Klein, G. and Harris, H. Expression of polyoma-induced transplantation antigen in hybrid cell lines. Nature, *237*: 163–164, 1972.
21. Littlefield, J. Selection of hybrids from matings of fibroblasts *in vitro* and their presumed recombinants. Science, *145*: 709–710, 1964.
22. Lowy, D. R., Rowe, W. P., Teich, N., and Hartley, J. W. Murine leukemia virus: high frequency activation *in vitro* by 5-iododeoxyuridine and 5-bromodeoxyuridine. Science, *174*: 155–156, 1971.
23. Lutz, B. R., Fulton, G. P., Patt, D. I., Handler, A. H., and Stevens, D. F. The cheek pouch of the hamster as a site for the transplantation of a methylcholanthrene-induced sarcoma. Cancer Res., *11*: 64–66, 1951.
24. Matsuya, Y., Green, H., and Basilico, C. Properties and uses of human-mouse hybrid cell lines. Nature, *220*: 1199–1202, 1968.
25. Meyer, G., Berebbi, M., and Klein, G. Expression of polyoma viral genome without expression of malignancy in a hybrid cell line. Nature, in press.
26. Möller, G. Demonstration of mouse isoantigens at the cellular level by the fluorescent antibody technique. J. Exp. Med., *114*: 415–432, 1961.
27. Murayama-Okabayashi, F., Okada, Y., and Tachibana, T. A series of hybrid cells containing different ratios of parental chromosomes formed by two steps of artificial fusion. Proc. Natl. Acad. Sci. U.S., *68*: 38–42, 1971.
28. Nabholz, M., Miggiano, V., and Bodmer, W. Genetic analysis with human-mouse somatic cell hybrids, Nature, *223*: 358–363, 1969.
29. Okada, Y. The fusion of Ehrlich tumor cells caused by HVJ virus *in vitro*. Biken J., *1*: 103–110, 1958.
30. Okada, Y. Analysis of giant polynuclear cell formation caused by HVJ virus from Ehrlich ascites tumor cells. I. Microscopic observation of giant polynuclear cell formation. Exp. Cell Res., *26*: 98–107, 1962.
31. Rowe, W. P., Pugh, W. E., and Hartley, J. W. Plaque assay techniques for murine leukemia viruses. Virology, *42*: 1136–1139, 1970.
32. Sanford, K. K., Likely, G. D., and Earle, W. R. The development of variations in transplantability and morphology within a clone of mouse fibroblasts transformed to sarcoma-producing cells *in vitro*. J. Natl. Cancer Inst., *15*: 215–237, 1954.
33. Scaletta, L. J. and Ephrussi, B. Hybridization of normal and neoplastic cells *in vitro*. Nature, *205*: 1169–1171, 1965.
34. Silagi, S. Hybridization of a malignant melanoma cell line with L cells *in vitro*. Cancer Res., *27*: 1953–1960, 1967.
35. Silagi, S. and Bruce, S. A. Suppression of malignancy and differentiation in melanotic melanoma cells. Proc. Natl. Acad. Sci. U.S., *66*: 72–78, 1970.
36. Sonnenschein, U., Richardson, I., and Tashjian, A. H., Jr. Loss of growth hormone production following hybridization of a functional rat pituitary cell strain with a mouse fibroblast line. Exp. Cell Res., *69*: 336–344, 1971.
37. Wiener, F., Klein, G., and Harris, H. The analysis of malignancy by cell fusion. III. Hybrids between diploid fibroblasts and other tumor cells. J. Cell Sci., *8*: 681–692, 1971.
38. Yamamoto, T., Rabinowitz, Z., and Sachs, L. Identification of the chromosomes that control malignancy. Nature, *243*: 247–250, 1973.

Discussion of Paper by Drs. Barski and Belehradek, Jr.

DR. WEBER: Dr. Barski, I enjoyed your closely argued, lucid presentation of this interesting and difficult subject. There is an argument that seems to support your thesis that normal karyotype is well-compatible with neoplasia in the observations reported by Drs. P. Nowell and H. P. Morris. Their paper in Cancer Res. showed that in some of the very slow-growing transplantable hepatomas, the karyotype and distribution were similar to those of the normal rat liver. Yet, as you know, this liver tumor is malignant and eventually kills the rat.

DR. BARSKI: Yes, Dr. Weber, I agree with you. Many other examples of normal-looking karyotypes in tumor cells could be quoted. I would, however, add that when we are speaking about "normal" karyotypes of tumor cells, we should rather say "*apparently* normal" since, after all, our karyological technology is still quite a crude instrument of investigation.

DR. HOZUMI: You mentioned some karyotype and antigenic changes in non-malignant and malignant hybrids, HyRA3 and HyRA4. Did you find any other phenotypic differences connecting the malignancy of the cells between nonmalignant and malignant hybrids?

DR. BARSKI: Indeed, the nonmalignancy of the HyRA3 line was accompanied by other phenotypic features such as the absence of C-type particles, as demonstrated by electron microscopy and by XC test. However, we are not ready to draw a conclusion of any cause-effect relationship concerning these phenotypic manifestations.

DR. THOMPSON: You stated that the gene-dose suggestion of Dr. T. Yamamoto and others and Dr. Okada and others was confirmed by your "superhybrid" experiments. The Israeli group particularly has suggested that no single chromosome or even a very few chromosomes are involved in malignancy/benignancy. Do you agree, or do you feel that the issue is still in doubt?

DR. BARSKI: What we did observe is that hybrids obtained by crossing Chinese hamster (malignant) with mouse (nonmalignant) cells were initially nonmalignant. However, following a secondary crossing of these hybrids, again with the Chinese hamster parent malignant cell, at least, several clones of these "superhybrids" appeared as clearly tumorigenic when inoculated into Syrian hamster cheek pouches.

From the karyological point of view, the malignant "superhybrids" showed a shift in the equilibrium of the karyological composition, manifested as an increase in the proportion of biarmed (chiefly Chinese hamster) and a decrease of telocentric (chiefly mouse) chromosomes.

In this sense our observations agree as well with the data of Dr. F. Murayama and Dr. Okada as with those of the Israeli group, except that our findings concern interspecies somatic hybrids in which, of course, the respective chromosomal participations, of the 2 cell parents, were relatively easy to recognize.

Concerning the second part of your question regarding the specific chromosomes, one or several, responsible for malignancy or nonmalignancy, I would prefer to refrain from making a judgment, being not at all sure that tumor-producing capacity or its suppression can, in the present state of our knowledge, be considered as a genetic characteristic. Karyological shift *may be* a direct cause of the expression or nonexpression of malignancy, but is *certainly* a sign of some hereditary events in the cell.

Modification of Ascites Tumor Cells by Hybridization with Fibroblasts

Yoshio OKADA,*[1] Fumiko MURAYAMA,*[2] and Kiyota GOSHIMA*[3]

Research Institute for Microbial Diseases, Osaka University, Osaka, Japan,[1] Tissue Culture Division, Laboratories of Dainippon Pharmaceutical Co., LTD., Osaka, Japan,*[2] and Biological Research Laboratories, Central Research Division, Takeda Chemical Industries, LTD., Osaka, Japan*[3]*

Abstract: 1) Ascites tumor cells, such as Ehrlich ascites tumor cells (E cells) and Friend leukemia (F cells), cells have lost the ability to adhere to a substratum for growth *in vivo* and *in vitro*.
2) On hybridization with fibroblast-like L cells, they showed the ability to adhere to a substratum *in vitro* and formed solid-type tumors in the peritoneum when injected into the abdomen of mice. Their tumor-forming activity was much less than that of ascites tumor cells.
3) The capabilities of adhering to a substratum and forming junctions expressed in the hybrids seem to be derived not only from the fibroblasts but also from the ascites tumor cells, in which these characteristics may be latent.

Modification of parent cell characteristics by hybridization with another kind of cell has been reported at this symposium by Dr. Silagi, Dr. Thompson, and Dr. Barski. Here, I would like to talk about a modification of the surface character of ascites tumor cells by hybridization with fibroblasts.

Hybrid Cell Lines

As shown in Table 1, Ehrlich ascites tumor cells (E cells) and Friend leukemia cells (F cells), obtained from Drs. Ikawa and Sugano, were used as ascites tumor cells. As fibroblasts, L cells and embryo skin fibroblasts from C3H (C cells) or ddO (d cells) mice were used. The hybrids were prepared by fusion of cells with UV-irradiated HVJ (*6, 7*) or with reassembled envelopes of HVJ. First, as a series of hybrids of E and L cells, LE hybrids and LL hybrids were prepared by one-step fusion. LLE and LEE hybrids were then prepared by double hybridization of LL and E or LE and E, respectively. E cells had 2 metacentric chromosomes and L cells had 13 (L_{BrdU}) to 14 (L_{AG})metacentric chromosomes. As shown in Table 2, the

TABLE 1. Hybrid Cell Lines

A. L......L_{AG} & L_{BrdU}	
L_{AG}+E	LE
LE......LE_{AG}	
LE_{AG}+E	LEE
L_{AG}+L_{BrdU}	LL
LL+E	LLE
B. C*+E	CE
d*+E	dE
C. L_{BrdU}+F_{AG}**	LF

C* and d*, embryo skin fibroblasts from normal C3H and ddO mouse, respectively; secondary culture. F**, an established line derived from the spleen of a DDD mouse infected with Friend virus.

TABLE 2. Karyological Characteristics of the Hybrids

Cells	No. of chromosomes	Long metacentric chromosomes (%)
L_{AG}	57	24.8
L_{BrdU}	50	26.7
E	45	4.45
LL	106	25.8
LE	88	16.6
LE_{AG}	83	16.4
LEE	123	11.0
	103	11.2
LLE	108	20.8

karyological features of series of hybrids, such as LLE, LE, and LEE, indicate that the double hybridization was successful.

Biological Characteristics of Hybrid Cells

As shown in Table 3, when the hybrids were injected into the abdomen of mice, all the hybrids grew and formed tumors, but the numbers of cells required for tumor formation varied: L cells formed no tumors, E cells formed tumors efficiently, while many hybrid cells were required for tumor formation. Among the hybrids, the number of cells required increased with increase in the L components

TABLE 3. Biological Characteristics of the Hybrids

Cells	50% tumor-forming dose		Tumor size on CAM	Generation time (hr)
	C3H mice	ddO mice		
L_{AG}	—	—	Small	18
LLE	6.6×10^7	6.0×10^7	Large	10
LE	8.5×10^6	1.7×10^6	Large	12
LEE	4.3×10^5	2.9×10^5	Large	16
E	5.0×10^2	1.0×10^2	Small	11
CE	1.4×10^4	6.8×10^3		
dE	5.0×10^5	4.0×10^4		

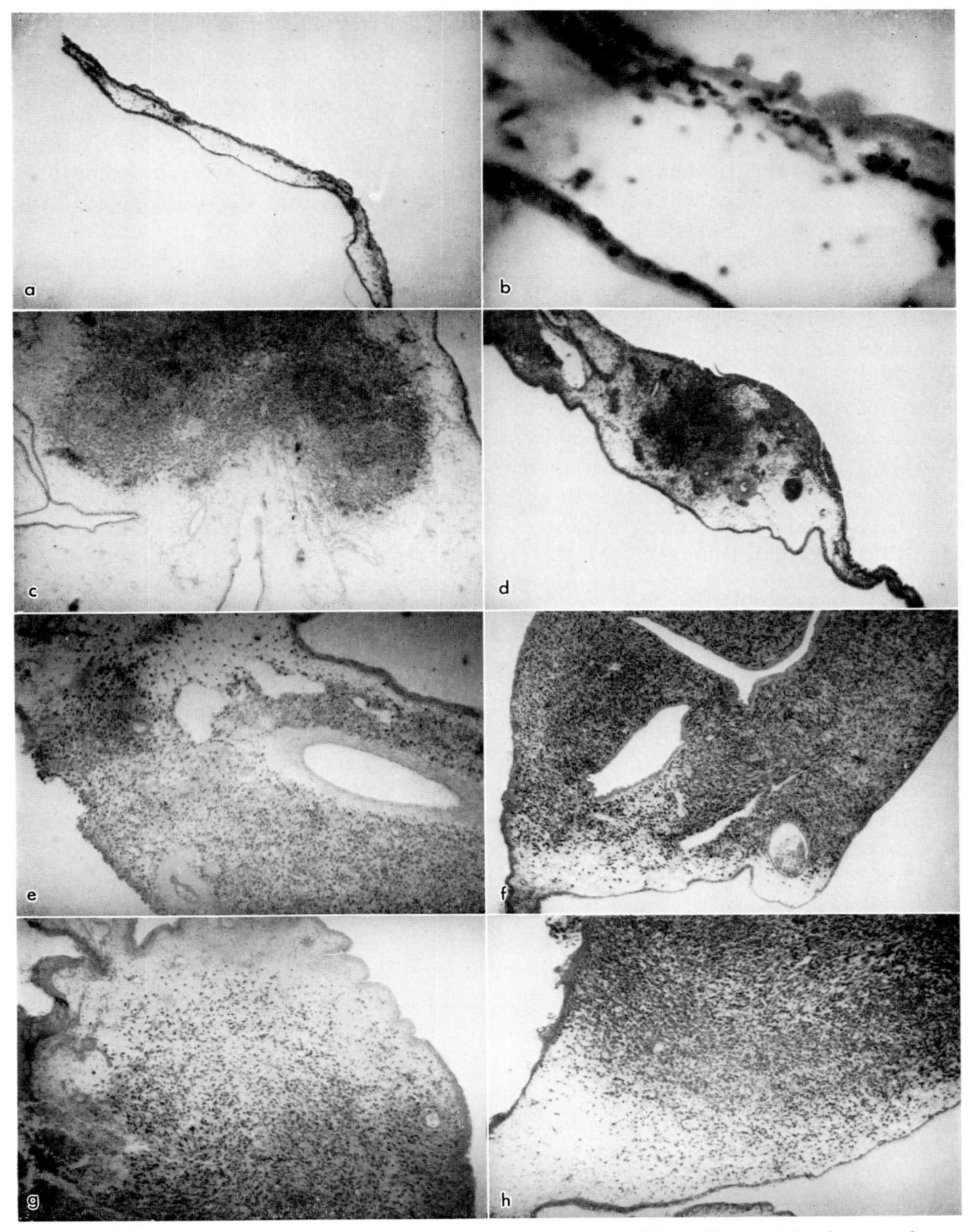

FIG. 1. Tumor mass in CAM. Cell pellets were put on CAM of 9-day-old embryonated eggs and incubated for 7 days at 35.5°C. CAM's with E cells usually appeared rather normal (a and b, high magnification), but small tumor masses were seen in a few (c). L cells formed small tumors in CAM's (d). Hybrids of L and E cells, such as LEE (e), LE (f), and LLE (g), or of E and embryo skin fibroblasts from C3H mouse (CE, h) formed large tumor masses efficiently in CAM's.

in the hybrids. If the number of cells required corresponds simply to the so-called malignancy of cells, it seems that the malignancy of E cells is suppressed by hybridization with L cells. However, the tumors formed by E cells clearly differed histologically from those formed by hybrids. E cells formed an ascites-type tumor and exudation of red blood cells into the ascites was minimal. However, the hybrids formed solid-type tumors embedded in the peritoneum and severe bleeding into the abdomen of mice was often observed. This suggests that the affinity of E cells for the peritoneum of mice is weak and that the cells mainly grow in the abdomen as an ascites form, while the hybrids have strong affinity for the peritoneum and grow embedded in it, and that blood vessels develop in the tumor masses as they grow. The results suggest that the affinity of the hybrids for the peritoneum is a characteristic derived from L cells, while malignancy is derived from E cells.

Similar differences in affinity were observed on growth in chorioallantoic membrane (CAM). When injected onto the CAM of 9-day-old embryonated eggs, E and L cells grew poorly but the all hybrids grew rapidly and formed tumors in 7 to 9 days. As shown in Fig. 1, histological examination showed that CAM's with E cells were usually rather normal in appearance and the added E cells remained on the CAM and degenerated, only occasionally becoming embedded in the CAM and forming small tumors. A few L cells were seen in the mesodermal layer of the CAM and very few of these were mitotic. However, the hybrids formed large tumor mass efficiently in the mesodermal layer of CAM, many mitotic cells were seen, and many blood vessels developed associated with the tumors. These findings suggest that when E cells become attached to the epithelial layer of CAM their ability to penetrate the layer is low while the abilities of L cells and hybrids are high. Moreover, when embedded in the mesodermal layer of the CAM, the growth rate of L cells is very slow while those of the hybrids are very high. Similar tendencies were observed with CE hybrid cells (a hybrid of E cells with embryo skin fibroblasts from a C3H mouse), as shown in Fig. 1, and with dE hybrids (E and embryonated skin fibroblasts from a ddO mouse).

Malignancy of Hybrid Cells

The results under these 2 experimental conditions suggest that the hybrids are less malignant in mouse abdomen, but more malignant on CAM than E cells. However, histological examination showed that this difference was mainly due to a difference in the ability of the cell surface to become attached to a substratum rather than to a difference in so-called malignancy. When cultured *in vitro*, E cells grew in a floating form, L cells grew as monolayers and all 3 hybrids grew on plating as intermediates between the 2 parental types. E cells seem to be a cell line which has lost the ability to adhere to a substratum and they can grow without attachment to a substratum. L cells and the hybrids seem to require attachment to a substratum for growth. The following experiment supports this idea. E, LEE, and LLE cells were injected into the abdomen of mice, and the numbers of free cells in the abdomen were counted every day for 5 days; after this period humoral immunoglobulin against the injected cells may appear. As shown in Fig. 2,

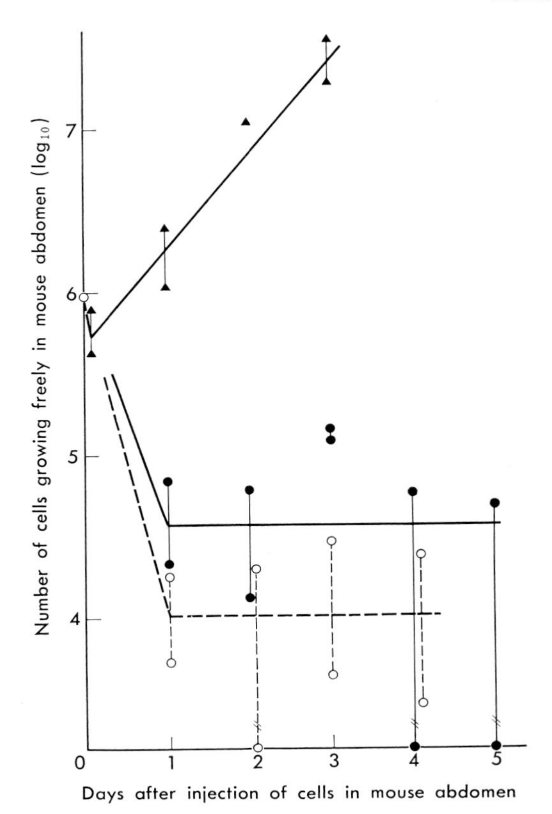

FIG. 2. Number of free cells in mouse abdomen after injection. ▲ E cells; ○ LLE cells; ● LEE cells.

E cells grew logarithmically from immediately after injection, while the numbers of free LEE and LLE cells in the abdomen remained constant during the period. As shown in Table 3, the number of cells of the LE series of hybrids required for tumor formation was 10^3 or more times greater than the number of E cells required. This great difference represents the sum of the effects of immunological rejection, interaction with normal tissue cells and other unknown factors appearing *in vivo*, and cannot be explained simply as due to suppression of so-called malignancy.

What is malignancy? How can we measure the degree of malignancy? These are important problems. However, they cannot be answered exactly, because so many factors may affect the growth of cells *in vivo* and we cannot test each of them separately. Moreover, there is no exact definition of malignancy and the definition seems to differ in different experimental fields. So at present, I will only discuss the correlation between the ability of cells to grow *in vivo* and cell-surface characteristics on the basis of cell-to-cell attachment, excluding immunological reactions.

Cells transformed by viruses or carcinogenic reagents grow piled up on monolayers of nontransformed cells *in vitro*. "Piling up" means loss of ability to adhere to the substratum. We do not know whether the degree of loss corresponds directly to the malignancy of a cell, which would mean that E cells are very malignant, or whether this loss of ability to adhere is independent of any increase in malignancy. However, it may be true that cells which have lost the ability to adhere are not influenced by surrounding cells, while cells which have this ability will be affected by adhering cells. When the latter type of cell is injected into the abdomen, it may be necessary to choose a site to anchor, and this will be on normal tissue cells. We believe that in normal cells there is a kind of system accepting regulatory signals from the animal body, and that cell-to-cell attachment forms part of this system. At least, E cells, which do not have to become attached to normal tissue cells, can grow without receiving regulatory signals from normal cells in the animal. Iijima *et al.* (*5*) showed by electrophysiological examination that the resistance of the junctional membrane between 2 MH 134 cells (an ascites tumor cell line of C3H/He mice) was very high, even in a solid tumor mass formed by subcutaneous injection of the cells. This clearly indicates that ascites tumor cells have no ability of cell-to-cell communication. Recently, we prepared 21 clones of LE hybrids after one-step fusion. Three of them showed the ability to grow as a floating form *in vitro* and readily formed ascites tumors in the abdomen of C3H mice. The other 18 clones grew as plates *in vitro* and when injected into mice as an inoculum of 3 to 4×10^6 cells, 3 of them did not form tumors. The other 15 clones formed solid-type tumors *in vivo*.

Electrical Coupling with Myocardial Cells

The ability of LE hybrids to adhere to the substratum seems to be derived from L cells. However, in collaboration with one of the authors we demonstrated that the cell-surface activity was also derived from E cells. Goshima prepared single myocardial cells from the ventricles of fetal or neonatal mice by trypsinization. He (*1–4*) found that when cultured, these cells become attached to the surface of a Petri dish and beat spontaneously. When cells beating at different frequencies came in contact, their beats became synchronous even when these myocardial cells were connected through intermediate FL cells. When a microelectrode was inserted into an FL cell mediating the synchronized beating, a rhythmical depolarization potential was obtained and the frequency of oscillation of the potential was the same as the frequency of beating of the myocardial cells. As shown in Fig. 3, depolarizing pulses

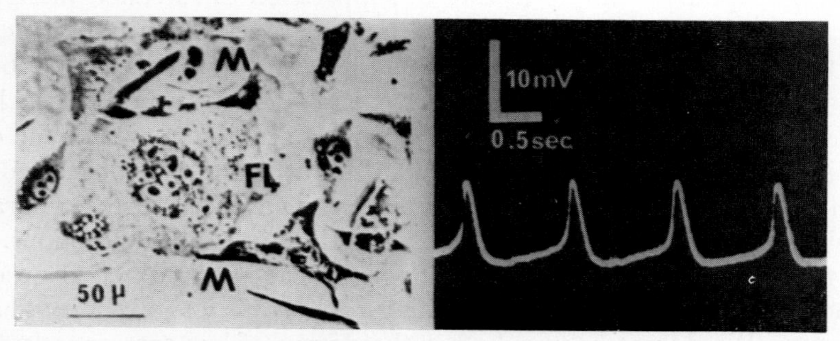

Fig. 3. Mediation of myocardial-cell (M) beating by a fused FL cell and the rhythmical depolarization potential recorded from it.

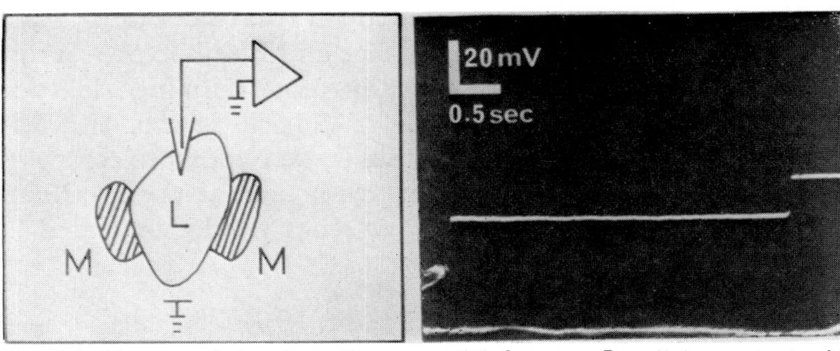

FIG. 4. Absence of depolarization potential from an L cell in contact with myocardial cells (M).

could also be recorded from FL cells fused by HVJ mediating the synchronization. Electron microscopy showed a nexus between the myocardial and FL cells, like that between 2 adjacent myocardial cells. Electrical coupling between cells has been observed in various tissues and a nexus is believed to be a low-resistance junction and to be the site of electrical coupling between cells. Many kinds of strain cells, such as HeLa and 3T3, could also mediate the synchronized beating of mouse myocardial cells, but L cells could not mediate synchronized beating. As shown in Fig. 4, when a microelectrode was inserted into L cells in contact with beating cells, no depolarization potential was recorded from them.

Based on these findings, the 3 lines of hybrids were tested in this system. E cells showed no ability to mediate the beat, because they did not attach or adhere to myocardial cells. However, it was found that L and LLE hybrid cells could not mediate the synchronized beating and LE and LEE cells could, as shown in Table 4. The percentage of cells which could mediate the beating increased on increasing the ratio of E cells to L cells. It seems that E cells retain a latent ability to form electrical coupling with myocardial cells. This indicates that the ability of the hybrids to form cell-to-cell attachments is derived not only from L cells but also from E cells.

TABLE 4. Synchronized Beating of Myocardial Cells Mediated by Hybrids of L and E Cells

Cells	No. of cells	
	Synchronized	Asynchronized
L	0 (0%)	30 (100%)
LLE	0 (0)	43 (100)
LE	19 (17)	93 (83)
LEE	28 (47)	31 (31)

Surface Antigens of Hybrid Cells

The abilities of hybrids, such as LLE, LE and LEE, to form tumors *in vivo* differ. This difference is not due to a difference in the generation times of the cells *in vitro*, as shown in Table 3. The surface antigens expressed on the hybrids were estimated in detail (*8*) and the results showed that the expression corresponds to the contents of chromosomes from both parents in the hybrids. It is known that L cells express a specific antigen(s) on their surface and E cells express a tumor-specific antigen(s). It is possible that the difference in tumor-forming ability mainly corresponds to the degree of immune rejection of the cells *in vivo*. However, it is also possible that LLE cells keep the functional ability to respond to control signals from the animal body through junctions formed with normal tissue cells to a greater extent than LEE cells. Unfortunately, we have not yet demonstrated the presence of this characteristic in the hybrids.

Hybrid of Friend Leukemia Cells

Recently, a hybrid of Friend leukemia cells with L cells was prepared. The Friend leukemia cells grew in a floating form *in vitro*, and as an ascites form in the abdomen of mice, and the mice developed splenomegaly. Just like LE hybrids, the hybrid cells grew as plates *in vitro* and formed solid-type tumors in the abdomen of ddO mice without causing splenomegaly.

REFERENCES

1. Goshima, K. Synchronized beating of and electronic transmission between myocardial cells mediated by heterotypic strain cells in monolayer culture. Exp. Cell Res., *58*: 420–426, 1969.
2. Goshima, K. Formation of nexuses and electronic transmission between myocardial and FL cells in monolayer culture. Exp. Cell Res., *63*: 124–130, 1970.
3. Goshima, K. Synchronized beating of myocardial cells mediated by FL cells in monolayer culture and its inhibition by trypsin-treated FL cells. Exp. Cell Res., *65*: 161–169, 1971.
4. Goshima, K. and Tonomura, Y. Synchronized beating of embryonic mouse myocardial cells mediated by FL cells in monolayer culture. Exp. Cell Res., *56*: 387–392, 1969.
5. Iijima, N., Yamamoto, T., Inoue, K., Soo, S., Matsuzawa, A., Kanno, Y., and Matsui, Y. Experimental studies on dynamic behavior of cells in subcutaneous solid tumor (MH 134-C3H/He). Jap. J. Exp. Med., *39*: 205–221, 1969.
6. Murayama, F. and Okada, Y. Efficiency of hybrid-cell formation from heterokaryons fused by HVJ. Biken J., *13*: 1–9, 1970.
7. Murayama, F. and Okada, Y. Appearance, characteristics and malignancy of somatic hybrid cells between L and Ehrlich ascites tumor cells formed by artificial fusion with UV-HVJ. Biken J., *13*: 11–24, 1970.
8. Murayama-Okabayashi, F., Okada, Y., and Tachibana, T. A series of hybrid cells containing different ratios of parental chromosomes formed by two steps of artificial fusion with UV-HVJ. Proc. Natl. Acad. Sci. U.S., *68*: 38–42, 1971.

Discussion of Paper of Drs. Okada et al.

DR. OSTERTAG: You isolated hybrid cells of Friend and L cells. Do these hybrids still differentiate on addition of dimethylsulfoxide?

DR. OKADA: This is not yet known.

DR. ONO: Do you think that it is possible to get a malignant hybrid by hybridization between normal cells?

DR. OKADA: I don't know. It may be possible to take a segregant from hybrids which keep a malignant characteristic such as "piling up," or being able to grow *in vitro* for a long time.

Natural History of Malignant Stem Cells

G. B. PIERCE, P. K. NAKANE, and J. E. MAZURKIEWICZ

Department of Pathology, University of Colorado Medical Center, Denver, Colorado, U.S.A.

Abstract: The presence of stem cells in neoplasmas has been known since the pioneering work of Makino, but their developmental aspects have been overlooked. Malignant stem cells (embryonal carcinoma) of teratocarcinomas were identified by cloning and shown to be multipotential and capable of differentiating into tissue representing each of the primary embryonic germinal layers. The differentiated progeny of embryonal carcinoma were benign. In addition, differentiation of stem cells of malignant tumors of breast, skin, and cartilage was observed and led us to the conclusion that carcinogenesis was equivalent to differentiation; it had, as its target, normal stem cells which were converted by alteration of genomic controls to malignant stem cells. Then, depending upon the capacity for differentiation the tumor appeared undifferentiated or differentiated. Stem cells are the essence of the malignant process.

A stem cell is capable of mitosis and of differentiation. The normal stem cell of origin of the teratocarcinoma has been identified as the primordial germ cell, the ultimate in an undifferentiated stem cell. The normal counterparts of the other malignant stem cells have not been identified. All stem cells have in common an abundance of organelles associated with the synthesis of cytoplasmic structure at the expense of organelles associated with the production of luxury molecules. The ultrastructural attributes of stem cells vary depending upon the tissue of origin and the propensity for growth at the expense of differentiation.

The phylogenetic position of neoplasms in the scale of biology has not been adequately defined. Neoplasms have been considered a "disease" and for the most part, studies of this disease have employed the philosophy of investigation proved useful in the study of infectious diseases. Unlike infectious diseases, the neoplasm is composed of cells of the body, which are derived by the process of carcinogenesis from normal cells. Although a certain neoplasm may secrete excessive amounts of a hormone or other metabolite, such as gamma globulin, it usually lacks the ability to perform many of the specialized functions of the tissue of origin and is morphologically "undifferentiated."

The lack of special functions or overt manifestations of the differentiated tissue of origin has led to the idea that these special features have been irretrievably lost. In an effort to explain these supposed losses and the stability of the neoplastic process, the mechanism of carcinogenesis has been attributed to a mutation in DNA, leading to dedifferentiation and ultimately, the malignant phenotype. This notion has so dominated thinking in cancer research that little attention has been paid to alternative explanations.

An alternative explanation for the stability and heritability of the malignant phenotype might lie in differentiation, the process by which all tissues in the body develop. Differentiation is the process whereby new phenotypes evolve with stable, heritable properties that differ from those of the progenitor tissue. The mechanism is presumed to be by repression and derepression of parts of genome, modulated by environmental factors. Thus, if carcinogenesis is equivalent to the process of differentiation, then the target cell in carcinogenesis must contain the genomic information required for expression of the malignant phenotype. Then during carcinogenesis, whether mediated by viral, chemical, or physical agents, activation of these genomic loci would occur presumably in a manner similar to the repression and derepression of genomic loci in any differentiation. Since normal cells, at some stage in development, express all of the characteristics which have been attributed to malignancy, it can be assumed that the genome of the normal cell contains the requisite information (*13*).

Target Cell in Carcinogenesis

Two pieces of information aid in identifying the target cell in carcinogenesis. If a chemical carcinogen is applied to the skin and a tumor develops at the site of application, that tumor will be a skin tumor and not one of some other tissue type. Thus, the target cells in carcinogenesis are already determined for a particular differentiation, and they are capable of synthesizing DNA and undergoing mitosis. This is strong presumptive evidence that the cell of origin of neoplasms is either the stem cell of a normal tissue or its partially differentiated progeny still capable of synthesizing DNA.

Accordingly, it is proposed that the target in carcinogenesis is in the stem cells, or the reserve cells responsible for renewal of normal tissues. The net effect of the carcinogenic episode would be repression or derepression of parts of genome with activation of gene loci repressed during embryonic development (fetal antigens, ability to migrate, ability to metastasize, *etc.*) and repression of others (the ability to make certain luxury molecules). Normal stem cells would be converted to malignant stem cells which would operate at a different level of control from the normal. The net result would be the malignant phenotype.

The idea that neoplasia fits into the scheme of biology as an aberration of tissue renewal should not be surprising because tissue reactions such as hyperplasia, metaplasia, and atrophy are mediated by stem cells or reserve cells of normal tissues.

If too many cells are produced in response to an environmental stimulus, hyperplasia results, if too few, atrophy. In the presence of vitamin A, squamous epithelium undergoes mucous metaplasia, not by responses of the adult squamous cells but by the reaction of the stem cells that normally would have undergone squamous differentiation (*5*). If the environment is returned to normal, then these processes are reversed and homeostasis is regained.

Benign and Malignant Neoplasia

When neoplasia is considered as a postembryonic differentiation, superimposed upon the process of cell renewal, for the first time benign neoplasms begin to make sense in relationship to malignant ones. Previously, the relationship of benign and malignant tumors has been unsatisfactorily explained. When a chemically pure carcinogen was painted on the ears of rabbits, many more benign tumors were produced than malignant ones (*23*). This has led to the belief that benign tumors were a stage in the development of malignant ones. Others believed that the processes were unrelated. When polyoma virus was injected into a mouse, benign tumors and malignant tumors developed. It is suggested that the explanation for this variation in outcome probably resides in either the reacting cells or their environment.

It is now well-known that an appropriate environment is essential for differentiations to occur, and when threshold numbers of cells with like propensity are aggregated together, they tend to create the optimum environment for their own tissue reactions (*8*). Probably the reason that benign tumors appear before malignant ones in the carcinogenesis experiments resides in the fact that the benign cells are little altered from normal, and the optimal circumstances for their proliferation are met in the environment of the normal tissue. Environment is also important for optimal proliferation of malignant cells as is well-known in the phenomenon of dormancy, in which small numbers of malignant cells trapped in a hostile environment are incapable of expressing their malignant phenotype. Newly formed and widely separated malignant stem cells may not find optimal growth conditions in normal tissue and it may take a long time to develop the critical number of cells necessary for the expression of the malignant phenotype.

The state of differentiation of the reacting cells is probably of even greater importance in determining whether or not the resulting neoplasm will be benign or malignant (*14*). Since the cells reacting to oncogenic stimuli are involved in tissue renewal, the events of cell and tissue renewal as they might occur in various tissues are schematically reviewed in Fig. 1. The normal stem cell for renewal of tissues such as skin, gastrointestinal tract epithelium, seminiferous epithelium, or bone marrow cells is at the left. It is determined for its particular differentiation and has one function: it can divide into 2 cells. One of the cells undergoes maturational changes and divides again and the cycles of maturation and division are repeated until renewal of functional postmitotic cells has been accomplished. This situation may not hold for other tissues such as liver or some of the endocrines. During liver regeneration, it would appear that functional hepatocytes are able to discontinue

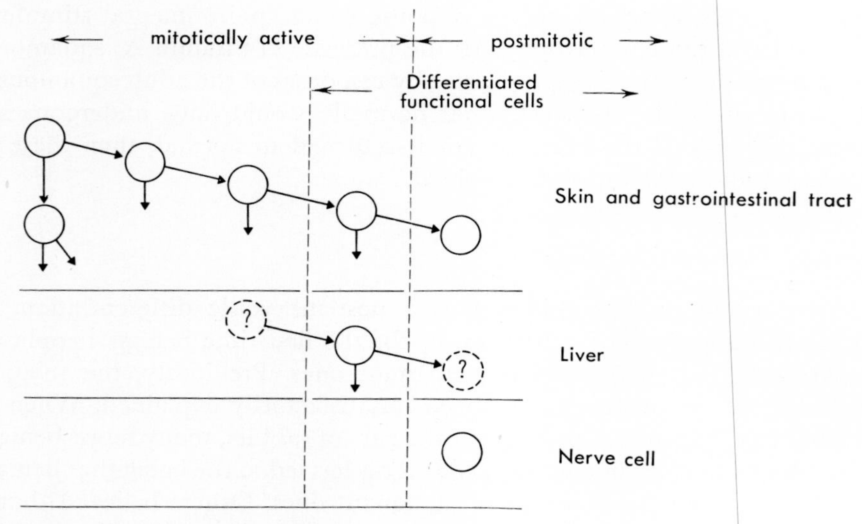

FIG. 1. Models of tissue renewal.

production of luxury molecules in favor of synthesis of DNA and cell replication with restoration of the liver mass. It is not known if liver cells become senescent in the same sense that squamous epithelial cells become senescent. It is also not known whether or not there are immature stem cells in liver lobules involved in the normal renewal processes. Nevertheless, the situation would still hold that carcinogenesis would involve a cell capable of synthesizing DNA. This situation also differs from that in nerve tissue. After embryonic development, there is no renewal of nerve cells, and there are no tumors of these cells.

Although, by definition, stem cells would be only the cells on the left of Fig. 1, for the purposes of this discussion, we will use the term "normal stem cell" to describe any cell capable of synthesizing DNA whose progeny undergo further differentiation as evidenced by synthesis of specific new gene products or assumption of new phenotypic traits. Our concept of carcinogenesis is diagrammed in Fig. 2. It is postulated that the state of differentiation of the cells that respond to the carcinogenic stimulus will in large part determine the state of differentiation of the tumor that develops. Benign tumors would be derived from almost terminally differentiated cells still capable of synthesizing DNA. Aside from progressive and excessive growth, the phenotype of the benign tumor differs little from the normal. In studies of the metabolic requirements of plant tumor cells in tissue cultures, Braun (*2*) has discovered that the principle difference between neoplastic and normal cells lies in the ability of the former to synthesize 2 essential growth ingredients. These are present in embryonic and rapidly growing tissues, but as the result of differentiation, their production is repressed. In the tumor, the genes responsible for their production are apparently derepressed and lead to a level of proliferation that leads to development of a mass. There have been no studies to determine whether or not the essential difference in benign tumors and normal tissues is in the ability of the former to synthesize essential metabolites not present in the normal

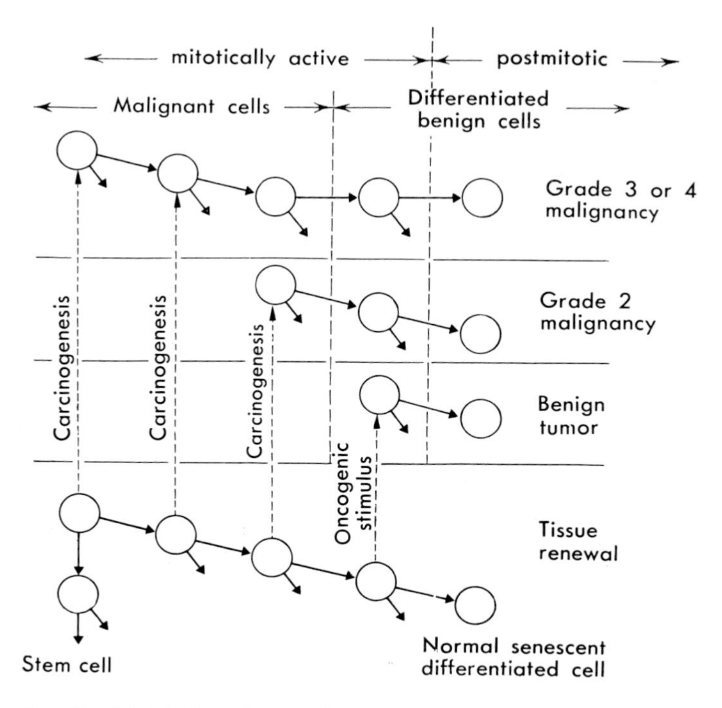

FIG. 2. Model of carcinogenesis.

adult tissue. In this respect, it is of interest that it is easier to grow tumor cells *in vitro* than their normal adult counterpart.

In the scheme illustrated in Fig. 2, malignant tumors would develop when stem cells or their rapidly proliferating and relatively undifferentiated progeny would respond to an oncogenic stimulus. Changes in control of gene loci with repression of certain genes active during embryogenesis and derepression of others which had been repressed during development would occur. It is suggested from the nuclear transplantation experiments that stable chromosomal configurations may occur during differentiation that preclude activation of certain genetic loci (*3, 4*). Thus the highly differentiated target cell may be protected from some of the effects of carcinogenic reactions because of the state of differentiation.

Teratocarcinoma

Data from a series of experiments will now be reviewed that support the notion that the target in carcinogenesis is either a stem cell or its incompletely differentiated progeny. These are converted to malignant stem cells that operate with new levels of control evolving a caricature of the tissue of origin.

Stevens (*18–20*) has identified the primordial germ cell as the cell of origin of teratocarcinomas. Teratocarcinomas are composed of chaotically arranged somatic and extraembryonic tissues intermingled with a highly malignant cancer named

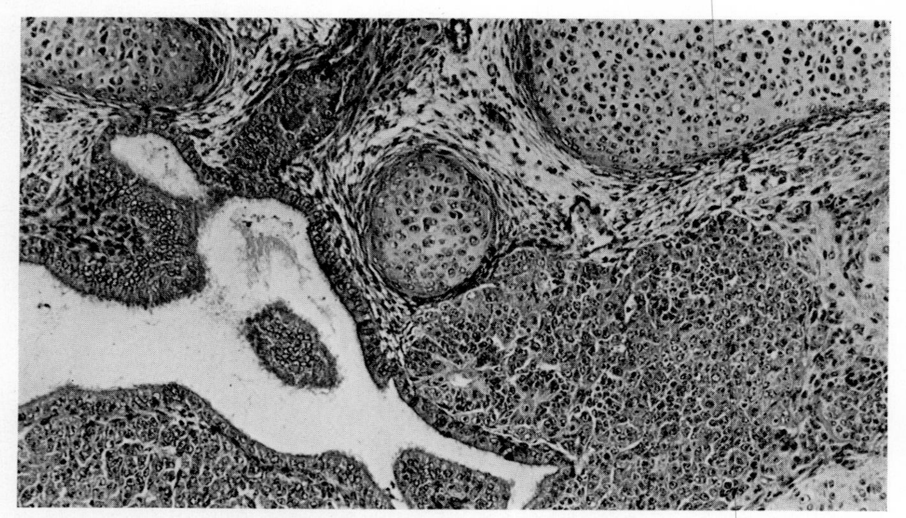

FIG. 3. A typical subcutaneous transplant of a strain 129 teratocarcinoma. Present are glands, cartilage, brain, and a large mass of embryonal carcinoma. (×300)

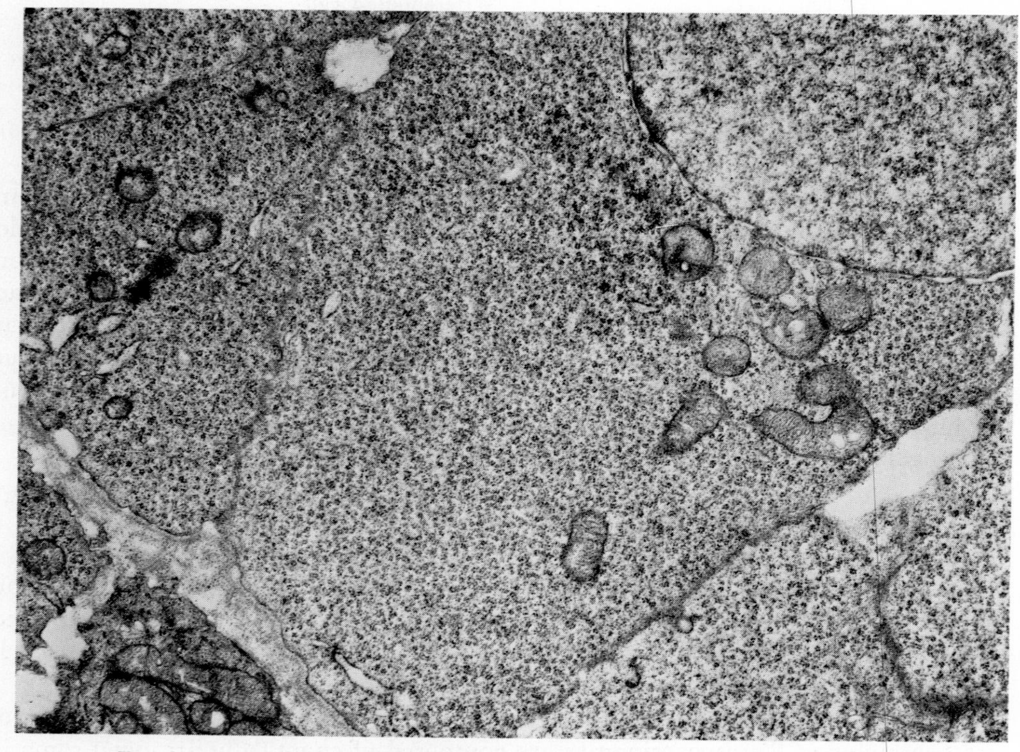

FIG. 4. A typical electron micrograph of an embryonal carcinoma cell of strain 129 mice. Note the mitochondria and paucity of other membranous cytoplasmic organelles. (×13,000)

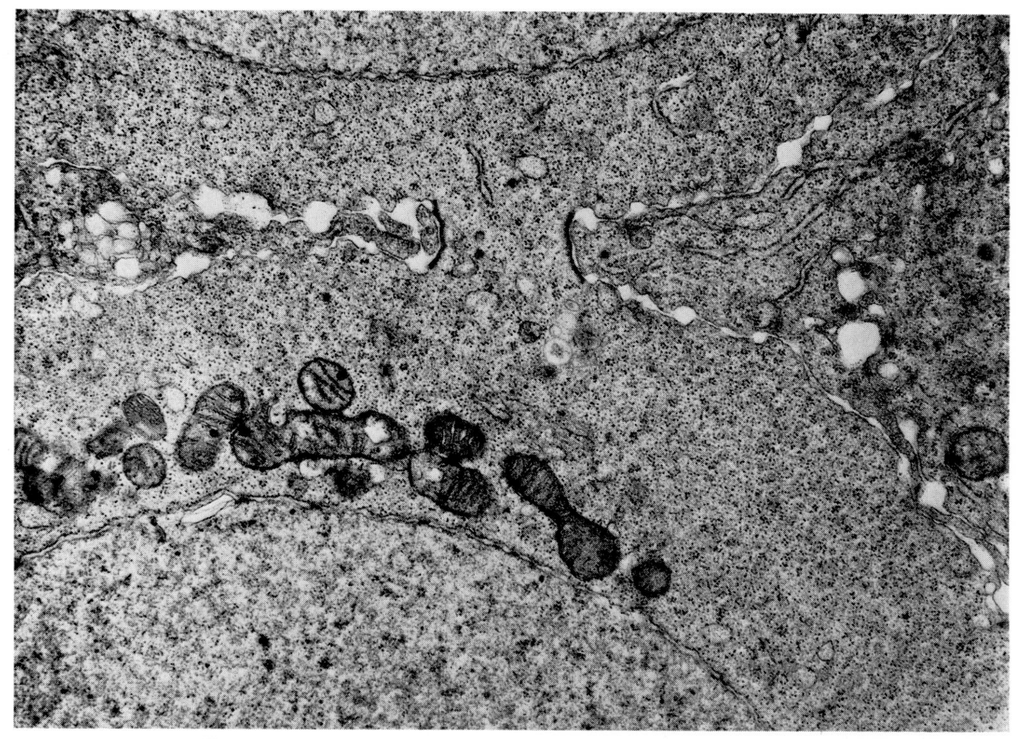

FIG. 5. Two primordial germ cells of strain 129 mice. Nuclei are present at top and bottom and the cytoplasm of the cells are connected by a bridge. Note the mitochondria and lack of other membranous organelles. (× 15,000)

embryonal carcinoma because of its resemblance to primitive embryonic epithelium (Fig. 3). The tumors are most commonly found in the gonads. Stevens (*18*) studied the early development of these tumors in strain 129 mice and found that the most rudimentary could be identified at about 15 days of gestational age. Extrapolating from the data, it was concluded that the carcinogenic event must take place around 12 days of gestational age, and he transplanted genital ridges from 12-day mouse embryos into the testes of adult strain 129 mice (*19*). Half of these grafts differentiated into ovaries, the other half into testes. Eighty percent of the embryonic testes developed teratocarcinomas recognizable 7 days after transplantation. By incorporating genes for the congenital absence of germ cells, the incidence of testicular tumors was reduced to zero (*20*). The incidence could be increased by incorporating other genes. In an electron microscopic study of the process of teratocarcinogenesis from transplanted genital ridges, it became apparent that the primordial germ cell (Fig. 4) and the embryonal carcinoma cell were almost indistinguishable from each other (*17*) (Fig. 5). These experiments established that the target cell for this particular tumor was an undifferentiated cell.

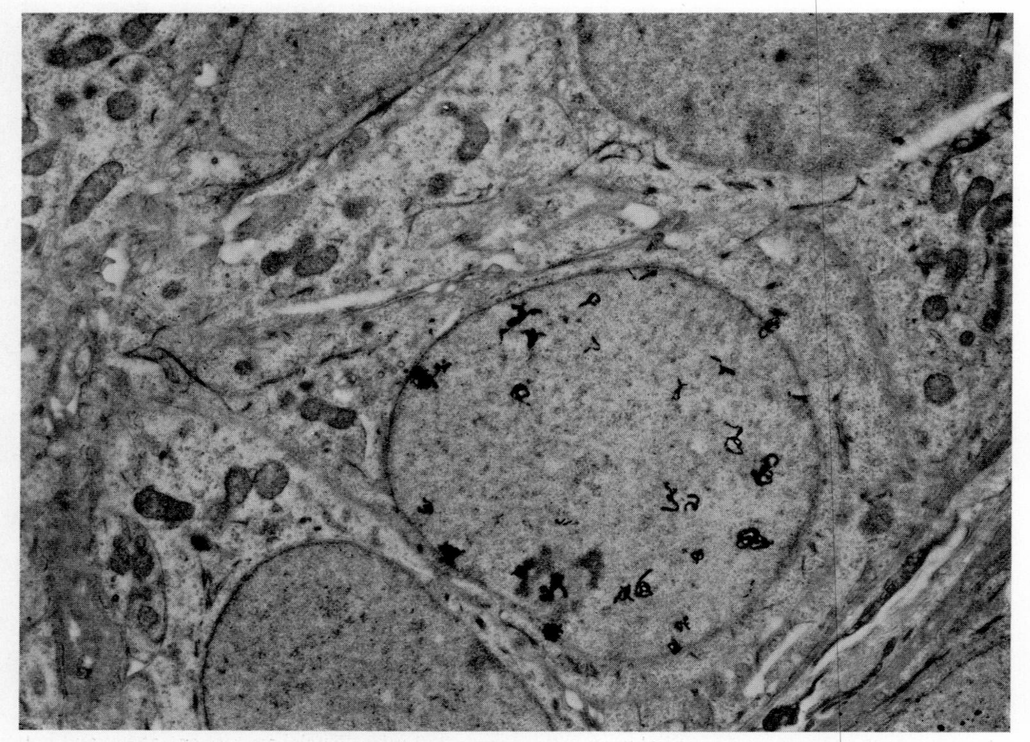

FIG. 6. An autoradiogram with the electron microscope of a stem cell of the squamous cell carcinoma taken 2 hr after a pulse of tritium-labeled thymidine. Note the undifferentiated nature of the cytoplasm in comparison to Figs. 7 and 8. (×10,000)

Embryonal Carcinoma

Kleinsmith and Pierce (*9*) cloned embryonal carcinoma cells and demonstrated that they had a capacity for differentiation into somatic and extraembryonic tissues, many of which were benign (*16*). This established for one kind of tumor a developmental flow starting in undifferentiated normal cells, leading to undifferentiated neoplastic cells and terminating in postmitotic, benign differentiated cells (*12*). The resulting tumor was a caricature of embryogenesis suggesting that the essential alteration in carcinogenesis was at the level of control. In similar studies it was shown that squamous cell carcinomas contained malignant stem cells (Fig. 6) that could differentiate into benign postmitotic squamous elements (Fig. 7) and dark, eccrine sweat gland cells (Fig. 8) which were incapable of forming a tumor upon transplantation (*15*). Thus, this tumor was a caricature of tissue-genesis in a developing system. It has been impossible to establish the cell of origin of this tumor, although from an ultrastructural standpoint, its stem cells have many features in common with basal cells.

The essential ingredient of the neoplastic process is the malignant stem cell, so it behooves us to understand these cells better.

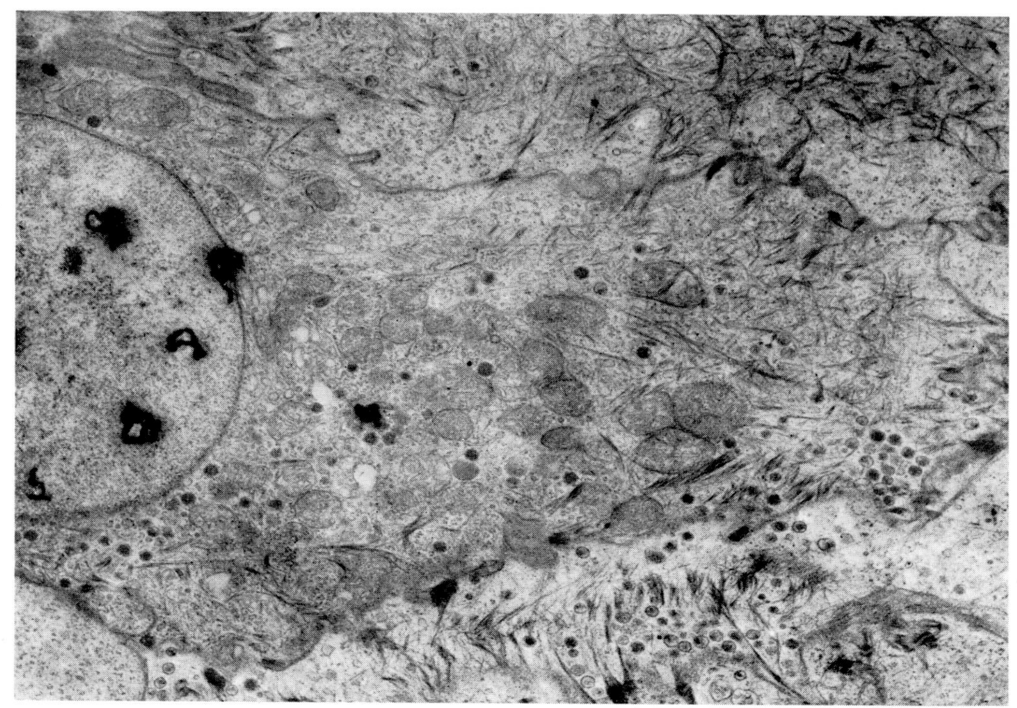

FIG. 7. An autoradiogram with the electron microscope made 96 hr after a pulse of tritium-labeled thymidine. Note the desmosomes with tonofibrils, membrane lining granules, and other aspects of squamous differentiation. (× 14,000)

Organelles in Stem Cells

From an ultrastructural standpoint, stem cells usually have an abundance of organelles associated with the synthesis of cytoplasmic structure at the expense of organelles associated with the production of molecules for export. This is illustrated in the ultrastructure of embryonal carcinoma cells (*11*) (Fig. 4). Notice that the cytoplasm of the embryonal carcinoma cell is dominated by multiple polysomes with few stunted elements of endoplasmic reticulum and an occasional small Golgi complex. Biochemical study of the enzyme systems in this type of cell would be a reflection of those enzymes responsible for cell proliferation. This supports the concepts of Greenstein (*7*) who observed that the biochemistry of malignant cells approaches that of embryonic cells. What it really approached was the biochemistry of a cell organized for replication rather than synthesis of luxury molecules.

It is important to realize that the ultrastructural appearance of a given neoplastic stem cell will be a reflection of the state of differentiation of the normal stem cell. Since normal chondroblasts are well-equipped for production of luxury molecules in relationship to primordial germ cells, it comes as no surprise that the stem cell of chondrosarcoma should contain well-developed organelles for the synthesis of collagen, chondromucoprotein, and glycosaminoglycans (*14*). Figure 9 illustrates the ultrastructural appearance of a stem cell from a chondrosarcoma. Note that this

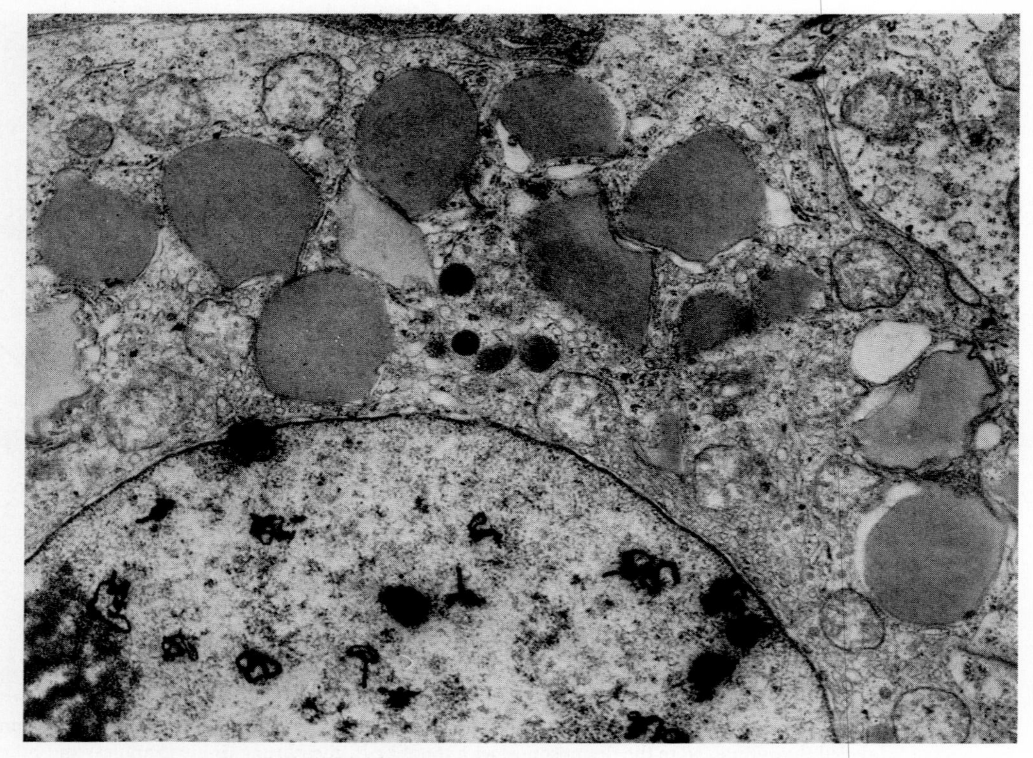

FIG. 8. A "dark eccrine" sweat gland cell that differentiated from the stem cells of the squamous cell carcinoma. (×17,000)

cell synthesized DNA in the 2 hr prior to fixation, as evidenced by the silver grains overlying the nucleus in this autoradiogram. Notice also the abundance of organelles associated with the production of luxury molecules. This malignant cell closely resembles a chondroblast. Aside from the presence of more free polysomes and ribosomes, this cytoplasm is little different from that of chondrosarcoma cells which are not synthesizing DNA. The latter are illustrated in Fig. 10. There may be a few more and better organized elements of endoplasmic reticulum and Golgi complexes, but each cell is well differentiated.

It is of interest that electron microscopy of tumors was first undertaken to resolve the essential differences in appearance between normal and neoplastic cells. None were found, yet the negative information was never used to identify the cell of origin of tumors.

Changes in Spontaneous Tumors

To ensure that we were not dealing with an artefact of transplantable tumors or of manipulation of *in vitro* situations, studies were undertaken in spontaneously arising adenocarcinomas of the breasts of mice to determine whether or not the malignant stem cells of these tumors had a capacity for differentiation (*22*). In auto-

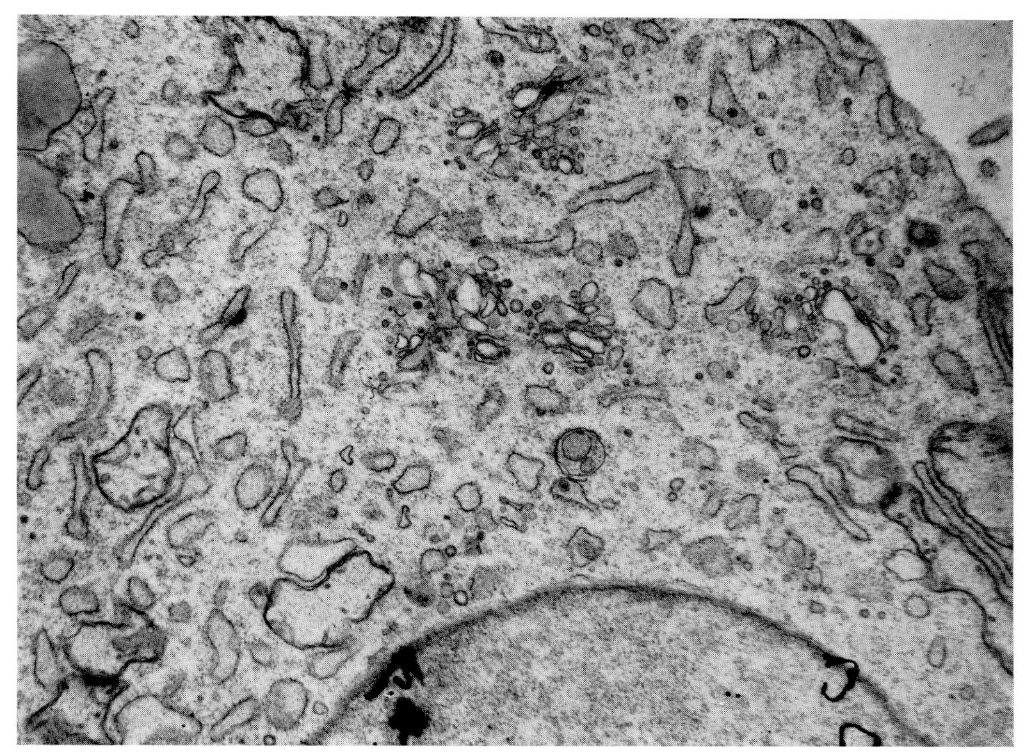

FIG. 9. An autoradiogram of a chondrosarcoma cell capable of synthesizing DNA taken 2 hr after a pulse of tritium-labeled thymidine. Note the degree of differentiation in comparison to that illustrated in Fig. 10. (× 11,000)

radiograms with the electron microscope 2 hr after administration of a pulse of tritium-labeled thymidine, the stem cells were identified as lacking many of the organelles and subcellular components characteristic of differentiated cells of the breast. Animals bearing spontaneous tumors were perfused for 5 days with tritium-labeled thymidine; many of the cells that had not synthesized DNA during this period of time had differentiated to a state incompatible with synthesis of DNA. Although it was not possible to determine whether or not these cells were tumorigenic in direct transplantation studies, the data strongly supported the notion that even in spontaneously occurring tumors, processes of differentiation from undifferentiated stem cells occurred and that these differentiated cells were benign. Of interest in these experiments was the confirmation of Mendelsohn's observation (*10*) that a significant number of the stem cells of these tumors did not synthesize DNA during the 5 day test interval. This would suggest that these stem cells were responding to environmental signals which precluded synthesis of DNA. Concepts of autonomy should be reconsidered in light of this observation.

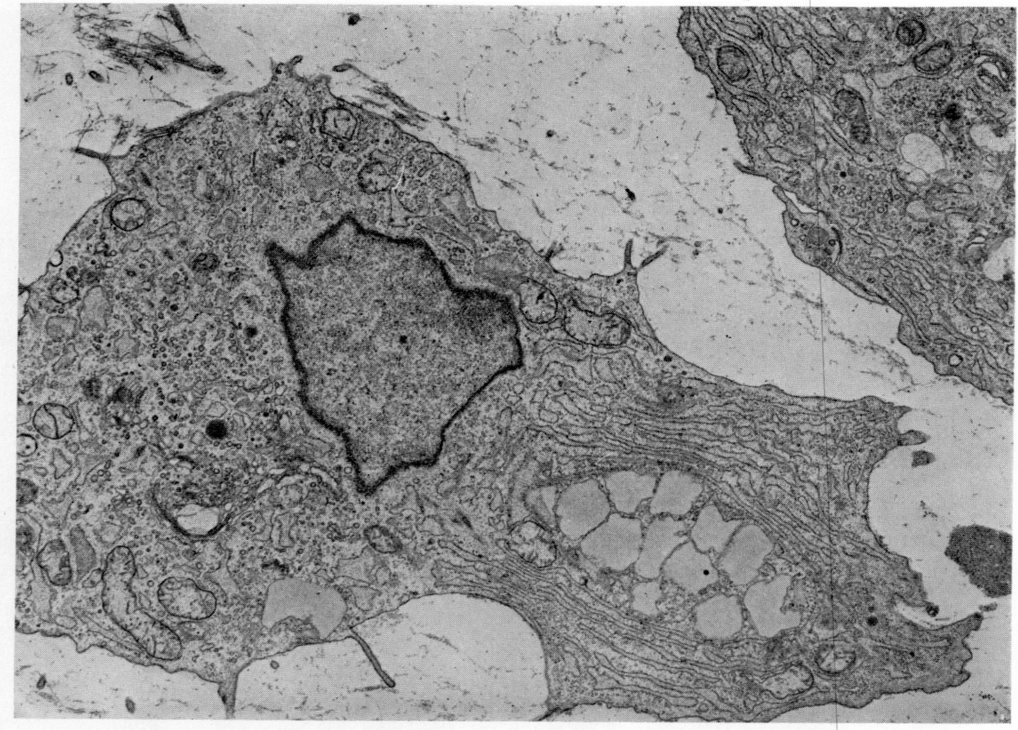

FIG. 10. A cell from the chondrosarcoma to illustrate the greatest differentiation in the tumor. (×10,000)

Pituitary Tumors

The implication from these studies is that enhancement or direction of differentiation in tumors might be a logical alternative to cytotoxic chemotherapy. To this end, we searched for a model system in which a known agent induces a specific response in neoplastic cells. We were attracted to the pituitary tumors isolated by Furth (*6*) and later studied by Bancroft (*1*) and Tashjian *et al.* (*21*). Furth (*6*) showed that if pituitary tumors of rats were cloned, the resulting tumor often synthesized more than one hormone. Either a single tumor cell could synthesize more than one hormone or this was a differentiating system that produced more than one cell type, each capable of synthesizing a particular hormone. Pituitary tumor cells were acquired from Tashjian and a clonal line GH3CL12 was isolated from it. This subclone produced only prolaction as demonstrated immunocytochemically using peroxidase-labeled antibodies to the pituitary hormones and by immunoassay. A significant number of the cells contained no hormone and were called uncommitted. Prolactin could be demonstrated in the others. When vasopressin was added to these cultures production of adrenocorticotropic hormone (ACTH) occured in the cultures detectable 10 hr after administration of the stimulus. Thus, these cultures synthesized prolactin and ACTH, but the hormones were never found in the same cell.

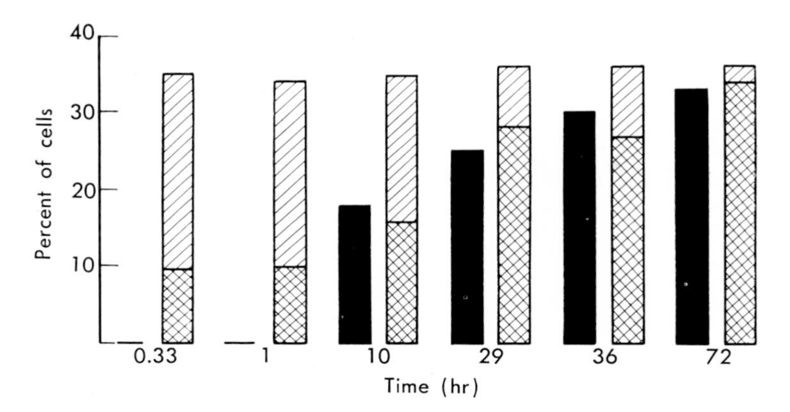

FIG. 11. A chart illustrating the relationship of hormone contest to mitotic activity in monolayer cultures of pituitary cells. ▨ prolactin positive; ▩ prolactin positive + ^{3}H-labeled nuclei; ■ ACTH positive + ^{3}H-labeled nuclei.

The population doubling time for the subclone was determined with and without vasopressin to be approximately 29 hr. Subcultures were continuously exposed to vasopressin and tritium-labeled thymidine for 72 hr and samples were processed at time intervals for immunocytochemistry and autoradiography. Uptake of tritium-labeled thymidine was not dependent upon the presence of vasopressin. When the cells were classified according to their hormone content, and presence of absence of labeled nuclei, prolactin positive cells remained constant, but the percent of ACTH positive cells increased. Throughout the experiment, about one-third of the cells contained prolactin. After 72 hr of exposure to labeled thymidine, all cells containing prolactin had labeled nuclei. After 10-hr exposure to vasopressin, about 20% of the cells were ACTH positive, ultimately about 30% were ACTH positive. All of the ACTH positive cells had labelled nuclei. Concomitant with the increase in ACTH positive cells was a decrease in the uncommitted cells and it was concluded that prolactin producing cells were not affected by exposure to vasopressin, and that vasopressin induced ACTH synthesis in the uncommitted cell population (Fig. 11). Once ACTH synthesis was initiated, it continued in the absence of vasopressin.

These studies confirm the notion that malignant stem cells do respond to environmental stimuli and are not the autonomous creastures classically believed. Those of the pituitary tumor respond to environmental stimuli by a specific differentiation.

CONCLUSION

In summary, it is proposed that the neoplastic process is an aberration of tissue renewal most likely paralleling in mechanism the process of differentiation. Normal stem cells are converted to neoplastic stem cells and these in turn are often productive of significant numbers of benign cells. From a clinical standpoint, there is a preponderance of proliferating malignant cells which invariably leads to the demise of the host unless the tumor is eradicated. Rather than kill malignant stem cells, it is proposed that studies should be undertaken to direct their differentiation to the benign state.

ACKNOWLEDGMENTS

The authors wish to acknowledge the excellent technical support of Miss B. Wilson and Mr. Alan Jones. These studies were supported in part by grant #DT-14-0 from the American Cancer Society and #AM-15663 and #AI-09109 from the National Institutes of Health.

REFERENCES

1. Bancroft, F. C., Levine, L., and Tashjian, A. H. Control of growth hormone production by a clonal strain of rat pituitary cells. J. Cell Biol., *43*: 432–441, 1969.
2. Braun, A. C. The usefulness of plant tumor systems for studying the basic cellular mechanisms that underlie neoplastic growth generally. *In;* R. Harris, P. Allin, and D. Viza (eds.), Cell Differentiation, pp. 115–119, Munksgaard, Copenhagen, 1972.
3. DiBerardino, M. A. and Hoffer, N. Origin of chromosomal abnormalities in nuclear transplants—a re-evaluation of nuclear differentiation and nuclear equivalence in amphibians. Develop. Biol., *23*: 185–209, 1970.
4. DiBerardino, M. A. and King, T. J. Development and cellular differentiation of neural nuclear transplants of known karyotype. Develop. Biol., *15*: 102–108, 1967.
5. Fell, A. B. and Mellanby, E. Metaplasia produced in cultures of chick ectoderm of high vitamin A. J. Physiol., *119*: 470–488, 1953.
6. Furth, J. Pituitary cybernetics and neoplasia. The Harvey Lectures, *63*: 47–71, 1967.
7. Greenstein, J. P. Biochemistry of Cancer, 2nd ed., Academic Press, New York, 1954.
8. Grobstein, C. and Zwilling, E. Modification of growth and differentiation of chorioallantoic grafts of chick blastoderm pieces after cultivation at a glass-clot interface. J. Exp. Zool., *122*: 259, 1953.
9. Kleinsmith, L. J. and Pierce, G. B. Multipotentiality of single embryonal carcinoma cells. Cancer Res., *24*: 1544, 1964.
10. Mendelsohn, M. L. Autoradiographic analysis of cell proliferation in spontaneous breast cancer of C3H mouse. III. The growth fraction. J. Natl. Cancer Inst., *28*: 1015–1028, 1962.
11. Pierce, G. B. Ultrastructure of human testicular tumors. Cancer, *19*: 1963, 1966.
12. Pierce, G. B. Teratocarcinoma: Model for a developmental concept of cancer. *In;* A. A. Moscona and A. Monroy (eds.), Current Topics in Developmental Biology, vol. 2, pp. 223–246, Academic Press, New York, 1967.
13. Pierce, G. B. and Johnson, L. D. Differentiation and Cancer. In Vitro, *7*: 140–145, 1971.
14. Pierce, G. B. and Nakane, P. K. Unpublished.
15. Pierce, G. B. and Wallace, C. Differentiation of malignant to benign cells. Cancer Res., *31*: 127–134, 1971.
16. Pierce, G. B., Dixon, F. J., Jr., and Verney, E. L. Teratocarcinogenic and tissue forming potentials of the cell types comprising neoplastic embryoid bodies. Lab. Invest., *9*: 583, 1960.

17. Pierce, G. B., Stevens, L. C., and Nakane, P. K. Ultrastructural analysis of the early stages of development of teratocarcinomas. J. Natl. Cancer Inst., *39*: 755–773, 1967.
18. Stevens, L. C. Testicular teratomas in fetal mice. J. Natl. Cancer Inst., *28*: 247, 1962.
19. Stevens, L. C. Experimental production of testicular teratomas in mice. Proc. Natl. Acad. Sci. U.S., *52*: 654, 1964.
20. Stevens, L. C. Origin of testicular teratomas from primordial germ cells in mice. J. Natl. Cancer Inst., *38*: 549–552, 1967.
21. Tashjian, A. H., Bancroft, F. C., and Levine, L. Production of both prolactin and growth hormone by clonal strains of rat pituitary tumor cells. J. Cell Biol., *47*: 61–70, 1970.
22. Wylie, C. V., Nakane, P. K., and Pierce, G. B. Degrees of differentiation in non-proliferating cells of mammary carcinoma. Differentiation, *1*: 11–20, 1973.
23. Yamagiwa, K. and Ichikawa, K. Experimental study of the pathogenesis of carcinoma. J. Cancer Res., *3*: 1–21, 1918.

Discussion of Paper by Drs. Pierce et al.

Dr. Nakahara: I am afraid I cannot quite see eye to eye with Dr. Pierce as to the malignant stem cells. I can see very well how he arrived at the conclusion of the existence of malignant stem cells, working as he did with embryonic teratocarcinoma, but generally speaking in chemical carcinogenesis any somatic cell capable of mitosis can be made to respond by becoming malignant. There seems to be no necessity for assuming malignant stem cell as a general rule for carcinogenic processes. It is true that Dr. S. Makino used the term "stem cell" but that stem cell referred to the main cell line responsible for the maintaining of the cell line specific characters in transplanted tumor strain.

Dr. Pierce: I agree, Dr. Nakahara, that Dr. S. Makino used the term "stem cells" to denote multiple lines of cells in leukemias, each recognizable by a chromosomal marker. Each line bred true in the sense that it always gave rise to progeny carrying its chromosomal marker. Our work and that of the leukemia people expand Dr. S. Makino's idea by showing that some of the progeny of the stem cells can differentiate. Unfortunately, for good health and long life, not enough differentiate and the malignant ones kill the host.

As for the first part of your question, I believe our differences may be semantic. Rather than coin new words, I have employed the terms "normal stem cells and their partially differentiated progeny." All of these cells are capable of synthesizing DNA and, as you point out, carcinogenesis affects cells capable of synthesizing DNA. Those that respond become the malignant stem lines of the tumor.

Finally, it is possible that some malignant tumors may not contain stem cells in the sense that I have used the term. We know that teratocarcinomas, squamous cell carcinomas, breast carcinomas, and choriocarcinomas contain them, but whether or not chondrosarcomas or hepatomas contain stem cells is not known. In the case of chondrosarcoma, we have shown that the progeny cells capable of synthesizing DNA undergo further maturation. So it, too, is a developing system.

Dr. Paul: Is it not true that benign tumors may progress to malignant ones and that malignancy often progresses? Can you reconcile this with your view that the degree of malignancy is correlated with the degree of differentiation of the target cell?

Dr. Pierce: Although some pathologists believe that certain benign tumors may be a stage in the development of malignant ones, there is no hard evidence to support the view. Others contend that they are unrelated. Progression as a phenomenon of the natural history of neoplasms was proposed by Dr. L. Foulds as the mechanism to explain changes in phenotype of tumors with time. It is now apparent from the work of Dr. G. Klein on ascites tumors and our work with melanoma that tumors contain heterogeneous populations of stem cells with differing attributes, which may be selected for under the conditions of the environment. If a stem line with particular characteristics has selective advantage over the others, it will eventually overgrow the others and the malignancy will have progressed in the sense that its particular characteristics will now dominate.

DR. SUGIMURA: I completely agree with your statement that the reaction of chemical carcinogen with protein and RNA should be carefully considered as well as DNA. N-methyl-N′-nitro-N-nitrosoguanidine, a potent mutagen and carcinogen, can readily modify proteins by producing nitroamidination on ε-amino group of lysine residue of protein. The magnitude of this reaction was bigger than that methylation of DNA by this mutagen.

Factor(s) Stimulating Differentiation of Mouse Myeloid Leukemia Cells Found in Ascitic Fluid

Motoo Hozumi, Kenji Sugiyama, Michiko Mura, Haruo Takizawa, Takashi Sugimura, Taijiro Matsushima, and Yasuo Ichikawa*

*National Cancer Center Research Institute, Tokyo, Japan and Institute for Virus Research, Kyoto University, Kyoto, Japan**

Abstract: Mouse myeloid leukemia line cells, Ml, could be induced to differentiate *in vitro* into macrophage-like cells by a factor(s) in ascitic fluids of animals bearing the ascites tumors Yoshida ascites hepatoma AH-130, AH-108A, and AH-7974, and mouse mammary carcinoma FM3A/B. Although the Ml cells were cytologically identified as myeloblastic leukemic cells, their cytological appearance was in general similar to that of macrophages of normal hematopoietic tissues. Differentiation of Ml line cells was induced by ascitic fluid in liquid media and also in soft agar layered on firmer agar containing ascitic fluid. The rate of induction of differentiation was roughly proportional to the concentration of ascitic fluid in the medium and the length of exposure of the cells. The factor(s) in ascitic fluid stimulating differentiation was nondialyzable and heat-labile, and retained activity after lyophilization. An active fraction of the ascitic fluid was precipitated by up to 50% saturation of ammonium sulfate and the molecular weight of the material with activity was estimated to be about 100,000.

Cultures of the leukemic cell line, Ml, from a spontaneous myeloid leukemia of an SL strain mouse, can be induced to develop into macrophages and granulocytes, losing their tumorigenicity for isologous mice, by a factor(s) present in conditioned medium from cultured embyo cells of various animals (*1–3, 5*). The chemical nature of this factor(s) is unknown, but it was demonstrated to be nondialyzable and thermolabile and to be a protein (*1–4*). This factor(s) should have potential value in the control of leukemia. We recently found that a factor(s) stimulating differentiation of Ml cells was also present in ascitic fluid of animals bearing various ascites tumors. This paper reports some properties of this factor and its mode of action.

Differentiation of Ml Cells into Phagocytic Cells with Ascitic Fluid

The cells used in this experiment were a clone, Ml clone 34, from a spontaneous myeloid leukemia of an SL mouse (*2*). The cells were cultured in Eagle's minimum essential medium with twice the normal concentrations of amino acids and vitamins and supplemented with 10% calf serum. Most cells are known to differentiate into macrophage-like cells, but seldom into granulocytes in the presence of conditioned media of embryo cells of various mammals.

Ascitic fluid was obtained using 3 strains of transplantable Yoshida ascites rat hepatoma, AH-108A, AH-130, and AH-7974 (*9*). Ascitic fluid was also obtained using the transplantable mouse mammary carcinoma FM3A/B (*7*). Eight days after intraperitoneal inoculation of 0.1 ml of ascites of the hepatomas or mammary carcinoma into female Donryu rats and C3H/He mice, respectively, ascites were harvested aseptically. The ascites were centrifuged at $800\times g$ for 30 min to remove the tumor cells. The cell-free supernatants were centrifuged again at $10{,}000\times g$ for 30 min and clear supernatants were obtained. Averages of 3 and 10 ml of clear ascitic fluid were obtained per mouse and rat, respectively. The protein contents of the ascitic fluids of rat hepatomas AH-130, AH-108A, AH-7974, and mammary carcinoma FM3A/B were 79.0, 73.6, 67.2, and 70.5 mg/ml, respectively, but values may vary in different preparations. The ascitic fluid was stored at -10°C until use. Conditioned medium was prepared from secondary embryo-cell cultures from ICR mice by the method of Ichikawa (*2*).

Cells of Ml clone 34 were incubated at a concentration of $1–5\times10^5$ cells/ml with the ascitic fluid or conditioned medium. The cell nutrients, amino acids, vitamins, and glucose were adjusted to the same concentrations in medium containing ascitic fluid as in Eagle's minimum essential medium with or without calf serum. Undiluted ascitic fluid was used without any additions. After 2 to 3 days, the cells were washed and incubated for 4 hr with a suspension of polystyrene latex particles (one drop/20 ml of Eagle's minimum essential medium; average diameter 1 μm, Dow Chemical, U.S.A.). The cells were vigorously washed, and the percentage of

TABLE 1. Differentiation of M1 Cells to Phagocytic Cells with Ascitic Fluids of Animals Bearing Various Tumors

Source of factor stimulating differentiation	Concentration (%, v/v)	Phagocytic cells/ total cells[a] (%)
Ascitic fluid of hepatoma-bearing rat		
AH-130	50	67
	100	75
AH-108A	50	65
	100	73
AH-7974	50	40
	100	70
Ascitic fluid of mammary carcinoma-bearing mouse		
FM3A/B	50	40
	100	70
Conditioned medium from a mouse-embryo cell culture	50	32
	100	60
Without ascitic fluid	—	0.5

[a] Three days after addition of ascitic fluid in culture medium without serum.

phagocytic cells among at least 200 viable cells was calculated by the method described previously (*2, 5*).

Three days after addition of 50 to 100% ascitic fluid from 3 strains of rat hepatomas and mouse mammary carcinoma to cultured Ml cells, 40 to 75% of the total viable cells had become phagocytic cells (Table 1). Under the same experimental conditions, 50 to 100% conditioned medium from secondary cultures of mouse embryo cells induced 30 to 60% phagocytic cells.

Cytological Observations on Cells Differentiated from Ml Cells with Ascitic Fluid

The clonal line, Ml clone 34, was established from a single colony of Ml cells in soft agar. Ml Clone 34 cells were morphologically homogeneous and were round when grown in suspension (Fig. 1a). However, on addition of ascitic fluid, some of the Ml cells adhered to the surface of the culture dishes and the number of adhering cells increased with the concentration of ascitic fluid and incubation time. The cells adhering on to a cover slip were fugiform, elongated, triangular, or rectangular and were distinctly different from the round Ml cells in suspension (Fig. 1b).

After incubation in medium containing polystyrene latex particles, many of the cells treated with ascitic fluid and almost all the cells adhering to the surface of the culture dishes showed active phagocytosis of polystyrene latex particles. The untreated cells did not take up any particles (Figs. 2a, b).

The morphology of cells treated with ascitic fluid was examined after staining the cells with May-Grünwald-Giemsa solution. Untreated Ml cells had a large round nucleus and a little, strongly basophilic cytoplasm (Fig. 3a). However, many of the cells treated with ascitic fluid, especially these adhering to the cover slip, had

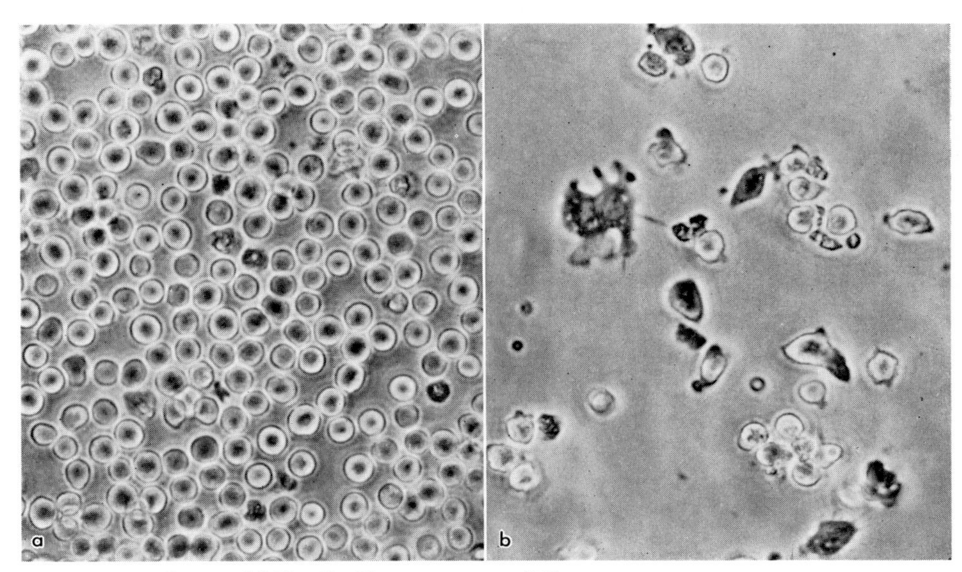

FIG. 1a. Untreated M1 cells. Phase contrast. ×250

FIG. 1b. M1 cells treated for 2 days with 50% ascitic fluid of a rat bearing hepatoma AH-130. Phase contrast. ×250 a smaller nucleus, eccentrically located in the cytoplasm and their cytoplasm was enlarged and weakly basophilic. The cytoplasm contained many phagocytized particles of India ink after incubation of the cells with the latter for 4 hr (Fig. 3b).

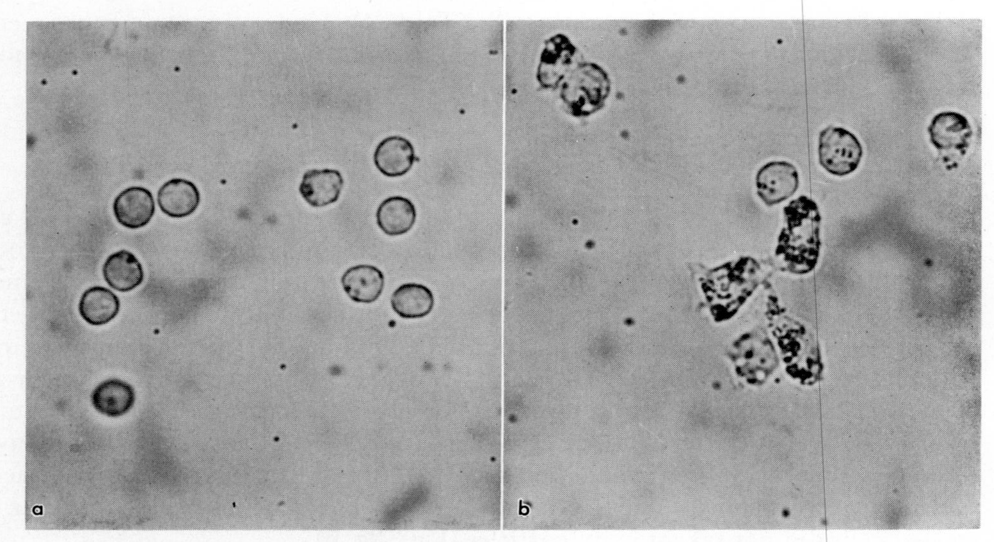

FIG. 2a. Untreated M1 cells showing inability to phagocytize polystyrene latex particles. Living cells. ×500

FIG. 2b. M1 cells treated for 2 days with 50% ascitic fluid of a rat bearing hepatoma AH-130. The cells phagocytized polystyrene latex particles. Living cells. ×500

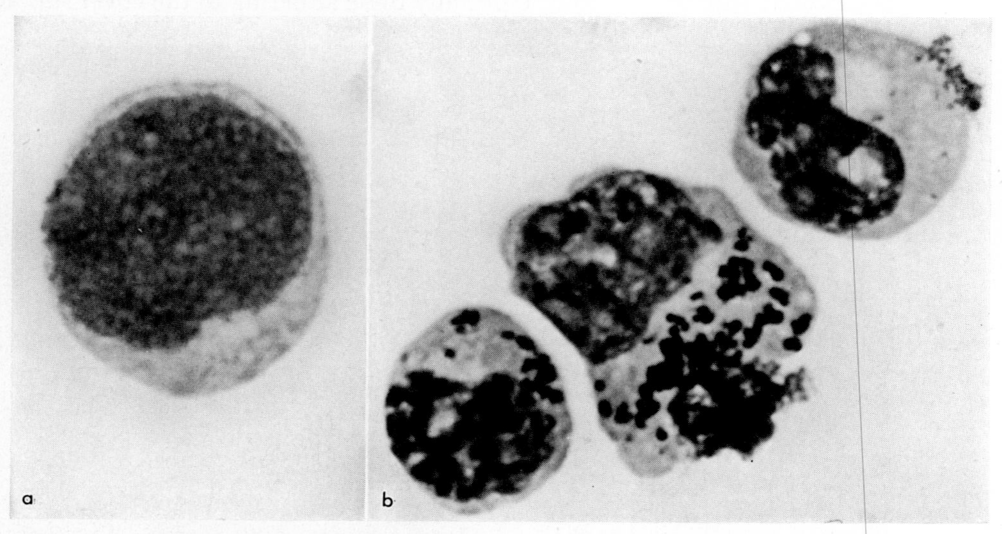

FIG. 3a. M1 cells stained with May-Grünwald-Giemsa. ×1,000

FIG. 3b. M1 cells treated for 2 days with 50% ascitic fluid of a mouse bearing mammary carcinoma FM3A/B. The cells contained phagocytized carbon (Indian ink) particles after incubation with serum-free medium containing 0.5% Indian ink for 4 hr. May-Grünwald-Giemsa. ×1,000

After treatment with ascitic fluid most cells appeared similar to macrophages from normal hematopoietic organs, although a few had ring-shaped nuclei, like those frequently seen in normal mouse granulocytes.

Morphology of Colonies of Ml Cells in Soft Agar with Ascitic Fluid

Ml clone 34 cells were seeded, at an inoculum of 400 cells/6 cm glass Petri dish, in soft agar (0.33%) overlayed on firmer agar (0.5%) containing various concentrations of ascitic fluid of AH-130. The cultures were incubated at 37°C in a humidified atmosphere of 5% CO_2 in air. The numbers and types of colonies were checked under an inverted microscope after 12 days. The colonies were classified into 3 types (I, II, or III) depending on to the grade of dispersion of the cells in the colony (*2*). Without ascitic fluid, only compact, type I colonies were formed, at a plating efficiency of 78% (Fig. 4a, Table 2). However, with ascitic fluid, dispersed colonies were formed (Fig. 4b, Table 2). The plating efficiency of total colonies was slightly stimulated with lower concentrations of ascitic fluid, but markedly reduced with

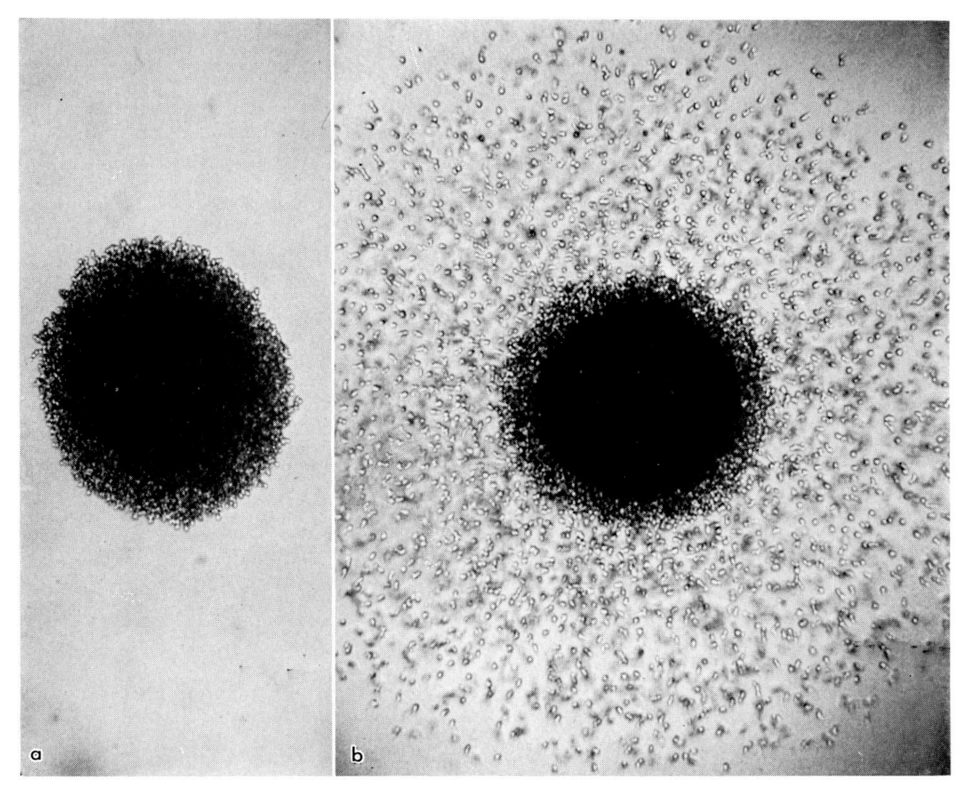

FIG. 4a. Colony of M1 cells in soft agar culture medium without ascitic fluid. Twelve days after seeding the cells. ×55
FIG. 4b. Colony of M1 cells in soft agar culture medium with ascitic fluid of rat bearing hepatoma AH-130. Twelve days after seeding the cells. ×55

TABLE 2. Effect of Concentration of Ascitic Fluid of Rats Bearing Hepatoma AH-130 on Morphology and Plating Efficiency of Colonies

Concentration of ascitic fluid[a] (%, v/v)	No. of colonies per plate[b]			Plating efficiency (%)	Dispersed colonies (II+III)/total (%)	No. of cells /colony (×10²)
	I[c]	II[d]	III[e]			
0	390	0	0	78.0	0	4.2
3.12	465	4	0	93.8	0.9	17.8
6.25	266	140	1	80.1	34.6	27.6
12.5	71	203	3	55.4	74.4	22.6
25	12	51	1	12.6	81.3	6.4

[a] Dialyzed and lyophilized ascitic fluid was used. The concentration is expressed as the final concentration of untreated original ascitic fluid in the total volume of culture medium. [b] Twelve days after seeding 500 cells per dish. [c] Cells all adhered together and formed a very compact colony. [d] Cells at the periphery of the colony spread out into the agar but those in the center remained compact. [e] All cells spread out in the agar and not even the center of the colony was compact.

higher concentrations. The number of cells per colony with 3.12 to 12.5% ascitic fluid was 5 to 7 times that without ascitic fluid. With 25% ascitic fluid, colony formation was much reduced, but the percentage of dispersed colonies increased to 81.3% of the total colonies. Most of the dispersed colonies were type II colonies, in which cells located at the periphery of the colony spread out over the agar while those in the center remained compact (Fig. 4b). A few colonies were completely dispersed, type III colonies. The dispersed cells at the periphery of colonies often had a twisted form, suggesting motility, and frequently had much phagocytized agar in their cytoplasm. The nuclei of these dispersed cells were smaller and eccentrically located and the cytoplasm was enlarged and rather acidophilic. In general, the cytological appearance of dispersed cells in the colonies was similar to that of the macrophage-like cells described above, produced by treatment with ascitic fluid (Figs. 1b, 2b, and 3b).

Some Factors Affecting Differentiation of Ml Cells to Phagocytic Cells with Ascitic Fluid

Ascitic fluid from rats bearing hepatoma AH-130 had high activity to induce differentiation of mouse leukemic cells, and it was relatively easy to obtain a large amount of ascitic fluid. The properties of the factor(s) in the ascitic fluid inducing differentiation were investigated.

Effect of the concentration of ascitic fluid: The percentage of phagocytic cells seen after treatment of Ml cells with ascitic fluid increased progressively with the concentration of ascitic fluid added (Fig. 5). Almost 50% of the cells were phagocytic cells on treatment with 50% ascitic fluid, and 70% were differentiated on treatment with 100% ascitic fluid. The growth rate of the cells decreased with increase in the concentration of ascitic fluid and with 50% ascitic fluid the number of cells was half that without ascitic fluid (Fig. 5). This relationship between differentiation and growth of the cells was closely related to the observation that increase in the number of dispersed-type colonies in soft agar with increase in concentration of ascitic fluid was always accompanied by a decrease in the plating efficiency (Table 2). This

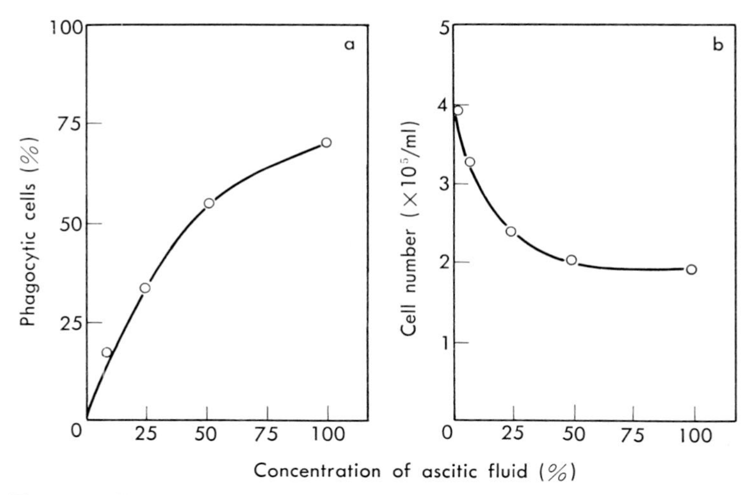

FIG. 5. Effect of the concentration of ascitic fluid of rats bearing hepatoma AH-130 on the differentiation of M1 cells to phagocytic cells. The cells were cultured for 2 days in serum-free culture medium. a: differentiation to phagocytic cells; b: cell growth.

decrease in cell growth might be partly due to formation of nonmitotic differentiated cells or to some factors in ascitic fluid retarding cell growth.

Time course of differentiation of Ml cells to phagocytic cells with ascitic fluid: After addition of ascitic fluid, the percentage of phagocytic cells in the medium without calf serum increased linearly with time to almost 75% after 3 days (Fig. 6). In medium with calf serum the induction of phagocytic cells also increased linearly, but the rate of induction was approximately 30% less than without calf serum (Fig. 6). Cell growth was significantly more with than without calf serum.

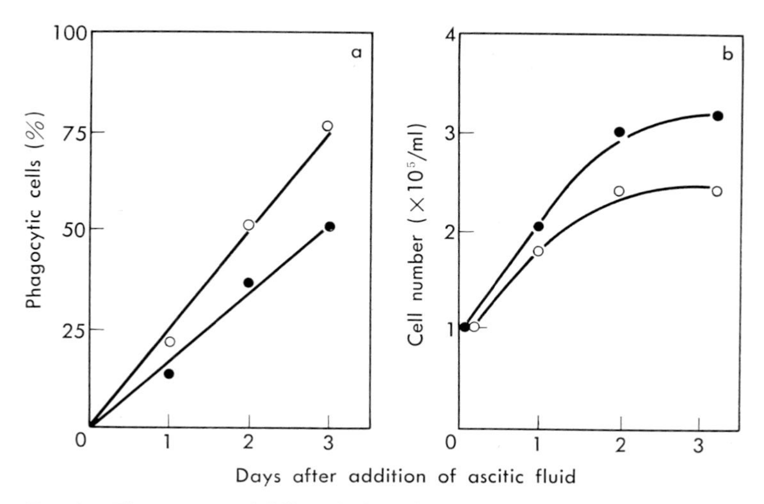

FIG. 6. Time course of differentiation of M1 cells to phagocytic cells with 50% ascitic fluid of rats bearing hepatoma AH-130. The cells were cultured with (●) or without (○) serum. a: differentiation to phagocytic cells; b: cell growth.

TABLE 3. Effect of Heating of Ascitic Fluid of Rats Bearing Hepatoma AH-130 on the Activity Causing Differentiation of M1 Cells to Phagocytic Cells

Treatment of ascitic fluid	Cell No. ($\times 10^5$/ml)	Phagocytic cells/ total cells[a] (%)
Untreated	4.5	30
56°C, 30 min	6.0	13
70°C, 10 min	6.0	0
Without ascitic fluid	9.0	0.2

[a] Two days after addition of 50% (v/v) ascitic fluid to culture medium with serum.

TABLE 4. Differentiation of M1 Cells to Phagocytic Cells with Dialyzed and Lyophilized Ascitic Fluid of Rats Bearing Hepatoma AH-130

Ascitic fluid	Concentration (%, v/v)	Cell No. ($\times 10^5$/ml)	Phagocytic cells / total cells[a] (%)
Untreated	10	4.2	29
	50	2.3	38
Dialyzed	10	3.0	28
	25	2.2	37
	50	1.6	53
Dialyzed and lyophilized	10	3.3	23
	25	2.2	35
	50	1.9	52
Without ascitic fluid	—	4.7	0.2

[a] Two days after addition of ascitic fluid to culture medium with serum.

Some Properties of Factor(s) in Ascitic Fluid Stimulating Differentiation

Heat stability: The activity of the factor(s) in the ascitic fluid stimulating differentiation was found to be heat-labile (Table 3). The activity decreased to less than 50% on heating at 56°C for 30 min and was completely lost on heating at 70°C for 10 min.

Dialysis and lyophilization: The ascitic fluid from a rat bearing hepatoma AH-130 was dialyzed in a cellophane tube against distilled water for 24 hr and then lyophilized. Its activity to induce differentiation did not decrease on either dialysis or lyophilization (Table 4).

Precipitation with ammonium sulfate: Forty milliliters of ascitic fluid from rats bearing hepatoma AH-130 was treated with successively increasing concentrations of ammonium sulfate and the activities of the precipitates to induce differentiation were assayed after dialysis and lyophilization. Most of the activity was recovered in the fraction precipitated at 0–50% saturation of ammonium sulfate and the specific activity of this fraction was almost 3 times higher than that of the original ascitic fluid (Table 5).

Sephadex G-200 gel filtration: The above fraction precipitated at up to 50% saturation of ammonium sulfate (equivalent to 7 ml of the original ascitic fluid) was

TABLE 5. Ammonium Sulfate Fractionation of Ascitic Fluid of Rats Bearing Hepatoma AH-130

% saturation of ammonium sulfate	Yield of protein (mg)	Phagocytic cells/total cells/ml equivalent ascitic fluid[a] (%)
Total ascitic fluid	1,013.9	31.0
0–30	111.0	29.4
30–40	74.5	21.0
40–50	157.0	18.5
50–60	23.4	1.4
60–70	497.0	4.9
>70	182.0	4.0
Without ascitic fluid	—	0.1

[a] Two days after addition of ascitic fluid to culture medium with serum.

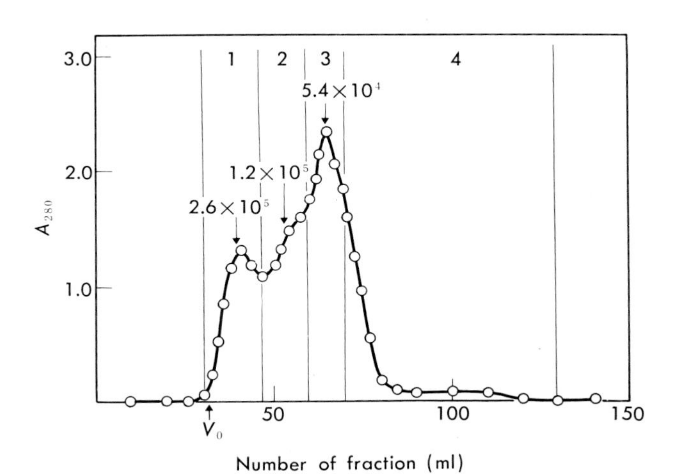

FIG. 7. Sephadex G-200 gel filtration of ascitic fluid of rats bearing hepatoma AH-130. Fractions of ascitic fluid of rats bearing hepatoma AH-130 precipitated with up to 50% saturation of ammonium sulfate (equivalent to 7 ml of the original ascitic fluid) were combined, dialyzed against distilled water, and lyophilized. The lyophilized material was dissolved in phosphate-buffered saline and insoluble material was removed by centrifugation. Then the preparation was loaded on a column of Sephadex G-200 (2×43 cm). The column was eluted with phosphate-buffered saline. Fractions of 1 ml were collected. The numerals indicated by arrows in the figure are molecular weights estimated from the positions of marker substances (gamma globulin, bovine albumin, and chymotrypsinogen). V_0 indicates the void volume of the column.

loaded on a Sephadex G-200 column and eluted with phosphate-buffered saline. The elution of protein was measured as absorbance at 280 nm (Fig. 7). Two peaks of protein were eluted with a shoulder before the second peak. The elution profiles of the marker substances (gamma globulin, bovine albumin, chymotrypsinogen) were obtained under the same conditions and the molecular sizes of the materials in the 2 peaks and the shoulder, designated as fractions 1, 2, and 3, were calculated as 2.6×10^5, 1.2×10^5, and 5.4×10^4, respectively (Fig. 7).

These 3 fractions and a low-molecular-weight fraction were condensed in a

TABLE 6. Differentiation of Ml Cells to Phagocytic Cells with Fractions of Ascitic Fluid of Rats Bearing Hepatoma AH-130 Obtained by Sephadex G-200 Gel Filtration

Fraction[a]	Yield of protein (mg)	Phagocytic cells/total cells/ml equivalent ascitic fluid[b] (%)
Ammonium sulfate	62.0	31.6
Fraction loaded		
1	3.2	18.7
2	17.4	26.9
3	19.3	10.8
4	1.5	12.4
Without ascitic fluid	—	0.2

[a] See Fig. 7 for explanation of fraction numbers. [b] Two days after addition of ascitic fluid to culture medium with serum.

Diaflo membrane and their activities to induce differentiation were determined as shown in Table 6. Fraction 2, with an estimated molecular weight of 10^5, had the highest activity but the other fractions also had fairly high activities.

DISCUSSION

A factor(s) stimulating differentiation of myeloid leukemia Ml line cells was found in conditioned media from cultured cells of various species (*2*, *5*), but it was unknown whether such a factor(s) occurs *in vivo*. The present experiments clearly demonstrate that ascitic fluid of animals bearing various tumors contains the factor(s) and suggest that such a factor(s) might play a role in the regulation of differentiation of leukemia cells *in vivo*. Our preliminary experiments showed that ascitic fluid from SL strain mice inoculated intraperitoneally with Ml cells had no ability to stimulate differentiation of the same Ml cells in culture. Therefore, the properties of the ascitic fluid of Ml cells may differ qualitatively from those of other tumor cells. This difference may explain why Ml cells grow in the ascites without differentiation.

The origin of the factor(s) in the ascitic fluid stimulating differentiation is uncertain. Ascitic fluid is thought to be formed by accumulation of transudates or exudates of body fluids including blood plasma. The sera from normal rats and those bearing ascites tumors had no ability to stimulate differentiation of Ml cells, so the active factor(s) must originate from sources other than the serum.

A factor(s) stimulating growth and differentiation of normal hematopoietic cells to macrophages and granulocytes has been found not only in conditioned medium from cultures of various cells, but also in various tissues and human urine (*6*, *8*). Fibach *et al.* (*1*) reported that one of these factors also stimulated the differentiation of Ml cells to macrophages and granulocytes. The general properties of the factor(s) in the ascitic fluid stimulating differentiation were closely related to those of the factor(s) in conditioned medium, although in recent studies on the purification of this factor(s) indicated some differences (*3*). The factor(s) stimulating differentiation of both normal hematopoietic cells and Ml cells was found to be a protein(s) with a molecular weight of 65,000–70,000 (*4*). Further purification and characterization of this factor(s) and the factor(s) in ascitic fluid will cast light on the interrelatiohs between them and their roles in the regulation of differentiation of hematopoietic cells and leukemia cells.

REFERENCES

1. Fibach, E., Landau, T., and Sachs, L. Normal differentiation of myeloid leukemic cells induced by a differentiation-induction protein. Nature New Biol., *237*: 276–278, 1972.
2. Ichikawa, Y. Differentiation of a cell line of myeloid leukemia. J. Cell Physiol., *74*: 223–234, 1969.
3. Ichikawa, Y. Differentiation of leukemic myeloblasts. GANN Monograph on Cancer Research, *12*: 215–229, 1972.
4. Landau, T. and Sachs, L. Characterization of the inducer required for the development of macrophage and granulocyte colonies. Proc. Natl. Acad. Sci. U.S., *68*: 2540–2544, 1971.
5. Maeda, M. and Ichikawa, Y. Production of growth- and differentiation-stimulating factors for mouse leukemia cells by different cell species. Gann, *64*: 257–263, 1973.
6. Metcalf, D. and Moore, M. A. S. Regulation of growth and differentiation in hemopoietic colonies growing in agar. *In;* G. E. W. Wolstenholme (ed.), Hemopoietic Stem Cells (Ciba Foundation Symposium), vol. 13, pp. 157–182, Associated Scientific Publishers, Amsterdam, 1973.
7. Nakano, N. Establishment of cell lines *in vitro* from a mammary ascites tumor of mouse and biological properties of the established lines in a serum-containing medium. Tohoku J. Exp. Med., *88*: 69–84, 1966.
8. Sachs, L. *In vitro* control of growth and development of hematopoietic cell clones. *In;* A. S. Gordon (ed.), Regulation of Hematopoiesis, vol. 1, pp. 217–233, Appleton-Century-Crofts Co., New York, 1970.
9. Yoshida, T. Comparative studies of ascites hepatomas. *In;* H. Busch (ed.), Methods in Cancer Research, vol. 6, pp. 97–157, Academic Press, New York, 1971.

Discussion of Paper by Drs. Hozumi et al.

DR. PRASAD: I think these are very important findings. Is it specific to myeloid leukemia or can you induce similar differentiation in other type of leukemia?

DR. HOZUMI: So far, I have not tested the effect of the factors on differentiation in other type of leukemia. However, Dr. Y. Ichikawa observed that differentiation-stimulating factors present in conditioned medium from cultures of embryo cells were specifically active to myeloid leukemia cells. I want to study this problem in the near future.

DR. WEBER: Is it possible that your culture of myeloid leukemia cells might be contaminated with normal leukocytes?

DR. NAKAHARA: Dr. Hozumi is working on a clone isolated as a single cell, and there is no chance of the culture being contaminated with any normal leukocytes.

DR. TOMKINS: Have you examined the effects of cyclic AMP or glucocorticoids on differentiation?

DR. HOZUMI: No, I have not tested the effects of these chemicals. I want to try this in the near future.

DR. DUBE: Do you find any C-type virus in these cells?

DR. HOZUMI: We have not checked the presence of the C-type virus in the myeloid leukemia cells, yet. However, Dr. R. C. Gallo's group in the National Cancer Institute in the United States reported recently that they found high activity of reverse transcriptase in leukemia cells, suggesting the possible presence of some viruses in the leukemia cells.

DR. RUTTER: Does the ascites preparation contain growth factor for your leukemia cells?

DR. HOZUMI: I'm not sure about the growth factor in the ascites. However, the formation of the colonies and the number of cells per colony in soft agar were stimulated at low concentrations of ascites, suggesting that there are some growth-stimulating factors for the leukemic cells.

DR. WEINHOUSE: I wonder whether the sera from rats or mice have such activity for differentiation of your leukemia cells?

DR. HOZUMI: I tested rat serum, but I could not find any activity.

Viral Involvement in the Differentiation of Erythroleukemic Mouse and Human Cells

W. Ostertag,*[1] T. Cole,*[1] T. Crozier,*[1] G. Gaedicke,*[2] J. Kind,*[1] N. Kluge,*[1] J. C. Krieg,*[1] G. Roesler,*[1] G. Steinheider,*[1] B. J. Weimann,*[1] and S. K. Dube*[1]

Max-Planck-Institut für Experimentelle Medizin, Göttingen, West Germany[1] and Molekularbiologisch-Hämatologische Arbeitsgruppe, Universitäts-Kinderklinik, Hamburg-Eppendorf, West Germany*[2]*

Abstract: The factors involved in transformation and differentiation of spleen focus-forming virus (SFFV)-transformed erythroid precursor cells in culture are examined. Two types of cell clones were isolated. The first cell clone, F4—after 5 years of tissue culture—is still releasing transforming SFFV and LLV helper virus. This cell line is also normal in respect to thymidine kinase (TK) activity. A second cell clone which was made resistant to bromodeoxyuridine (BrdU), is TK$^-$ and does not release C-type particles. Both cell lines can be induced to differentiate to hemoglobin synthesizing late erythroblasts. In step with the appearance of globin mRNA, we find a 10–150 fold increase in virus release (SFFU) in F4 cells. In B8 cells globin mRNA is synthesized but no virus release is found. BrdU-resistant B8 cells however show an increase in intracisternal A particles on induction of globin synthesis with dimethylsulfoxide (DMSO), BrdU (!), and azidothymidine. The A particles possibly correspond to the transforming SFFV. The coordinated increase of A particles, or C-type viruses in F4 with globin mRNA suggests a common regulatory mechanism.

Human cell lines from patients with polycythemia vera contain ribonucleoprotein (RNP) particle bound reverse transcriptase activity and a few A-type particles.

Spleen focus-forming virus (SFFV)-transformed cells offer a unique system to study cellular transformation by a tumor virus, differentiation which is blocked by the transformation event, and release of the block by aprotic solvents (*18*, *41*, *49*) or by steroid hormones (*19*).

In another chapter (*9*) we have tried to describe cellular changes which might reflect normal mechanisms of differentiation of erythroid cells. In this chapter we will try to approach the relationship of the transformational event and of induced differentiation to viral parameters. In the second part of this chapter we would like to draw attention to possibly related phenomena in human cells. The latter have been maintained in tissue culture. The cells are derived from patients with proliferative disorders of the erythropoietic compartment.

Transformation by the Friend Virus (FV) Complex and Possible Involvement in Differentiation

Three different RNA tumor viruses are known to transform erythroid cells in mice, the Rauscher (*45*), Stansly (*61*), and FV complex (*17*). All of these viruses are of obscure origin but their specificity and complexity is similar. Whether any of these viruses are derived from a naturally occurring mouse erythroleukemia is unclear. All of these viruses, however, induce erythroleukemia in mice. The course of this disease is acute. Only a few weeks are needed to manifest the disease after injection of the virus in young or adult mice.

If the Friend or Rauscher virus is passaged through C57 B1 mice or rats, the disease picture changes and lymphatic leukemia can be obtained (*7*, *8*). The virus recovered by this kind of passage induces lymphatic leukemia in mice which previously could be used to obtain erythroleukemia (*39*). This suggested that there are 2 components in the FV complex, a virus that induces erythroleukemia and which can be eliminated by passage through rats or C57 B1 mice, and another virus which is able to transform lymphatic cells. Proof of this dual nature of the FV complex was obtained by separation of the 2 transforming activities by end-point dilution (*63*) and even more strikingly by obtaining transformed erythroid cells which do contain erythroid-transforming, but defective viral information, and lack the lymphatic leukemia virus (*16*). The usual association of the 2 virus components, as far as erythroid viral transformation is concerned, is due to the defectiveness of the erythroid-transforming (SFFV) virus. All the known viral-specific proteins of the FV complex are coded by LLV-F. The LLV-F virus can be easily replaced as a helper virus by other lymphatic leukemia viruses (*16*, *33a*). The pseudo-types of SFFV cannot be distinguished serologically from the parent helper virus serotype. We were able to show (see below) that even endogenous leukosis virus, which can be induced by bromodeoxyuridine (BrdU), is also capable of acting as a helper virus for SFFV (*43*).

The resistance of rats and C57 B1 mice to the transforming activity of SFFV was shown to be a simple Mendelian trait. The gene FV-2 which governs the transforming response of SFFV (*38*) is located on chromosome 9, linkage group II (*32*). The resistance is recessive and absolute. Steeves (*65*) has obtained some viral variants which permit erythroid transformation at a low rate in C57 B1 mice.

Another pair of mouse genes which are involved in the susceptibility of mice to SFFV transformation are W and S1. Mouse mutants homozygous for these genes are anemic (*54*). Both genes are also known to modify erythroid differentiation. Homozygotes are resistant to SFFV transformation (*4*, *62*). A study of these mouse mutants and of viral mutants able to transform previously resistant mice might be of great value in studying the process leading to erythroid transformation.

Another gene, FV-1, is responsible for susceptibility of resistance to the LLV helper virus (*31*, *33*). FV-1 is located on chromosome 4, linkage group VIII (*53*) next to the Gpd-1 locus (*51*). The resistance is a dominant Mendelian trait. A group of inbred mouse strains, DBA2, NIH Swiss, and DDD are N-type and can be infected by N-tropic FV (LLV-F) and another group of mice, such as Balb/C and

C57 Bl are resistant to N-tropic LLV-F. The resistance is not absolute. By passage of the original N-tropic FV complex in Balb/C mice NB-tropic FV was obtained (*31*). FV complex with NB host range infects and transforms N or B mice with equal efficiency.

The nature of the resistance or sensitivity to LLV or SFFV is at present unknown. The host range, however, is not a property of the mouse cell membrane (*26*, *73*).

On investigating the karyotype of cells transformed by SFFV and infected by LLV we have obtained several cell lines which show a translocation of the 2 homologues of chromosome 4, which is known to carry the FV1 gene (*41* and Schnedl, personal communication).

The ability of SFFV to transform erythroid target cells in the spleen of sensitive mice can be used to titrate the number of SFFV. It is assumed that a single SFFV transforms a single erythroid spleen cell, which is then able to form a clone of transformed cells. This clone can be recognized as a spleen focus (*2*, *44*). The spleen focus-forming assay for SFFV is the only clonal and *in vivo* transformation assay in mice. The unavailability of biologically active FV (SFFV) in tissue cultures was a major obstacle to a study of the nature of the FV complex. We have obtained several cell lines which are transformed by and which still liberate significant amounts of biologically active SFFV (FV) (*42*).

1. In vivo transformation by FV obtained from permanent cell culture

In our chapter on the differentiation of erythroleukemia cell cultures (*9*) we have listed and characterized a series of cell strains. Cell clones FSD1/F4 and the BrdU-resistant subclones B8, B8/3, and B16 are those which have been characterized best.

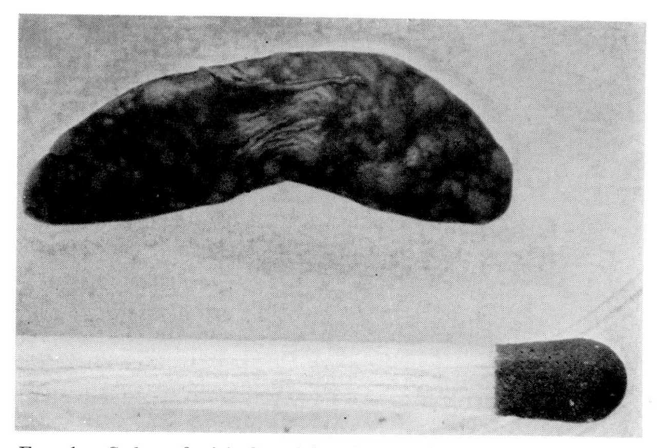

FIG. 1. Spleen foci induced by tissue-culture supernatants of cell clone F4. F4A cells were grown in modified Eagle's medium (*41*) to a density of 1×10^6 cells/ml. The cellular supernatant was passed through a 0.45 μ filter. The cell-free supernatant was then diluted 1: 10 and 0.1 ml of this suspension injected into the lateral tail vein of DBA-2 mice. The mice were killed 9 days later and the spleen was fixed in 70% ethanol.

TABLE 1. Friend Virus (LLV+SFFV) Release in SFFV Transformed Cell Cultures

Cell line	Mouse origin	Clone of:	Differentiation with DMSO	Virus release			Previous treatment
				SFFU assay	XC assay	Electron microscopy	
FSD1	DBA-2	—	Low	0	3×10^5	⧻	—
F4 sublines							
N	„	FSD1	Yes	500–1,000	3×10^5	⧻	—
A	„	„	Yes	10,000–20,000	2.5×10^5	⧻	—
O	„	„	No	10,000–20,000	3×10^5	⧻	—
CD	„	„	Low	0	n. t.	⧻	Grown in calf serum, treated with amphotericin B
1	„	F4N	Yes	0	~100	n. t.	—
B8	„	F4A	Yes	1–5	5×10^2	(±)	BrdU 200 μg/ml
B8/3	„	B8	Yes	0–2	1.4×10^3	n. t.	Agar clone of B8
B8/7	„	„	Yes	500–700	n. t.	n. t.	Agar clone of B8
B8/1	„	„	Yes	0	n. t.	n. t.	Agar clone of B8
B8/22	„	„	Low	0	10	n. t.	Agar clone of B8 Kiang and Krooth
B16	„	F4A	Moderate	500–1,000	n. t.	+	BrdU 200 μg/ml
F4-AG	„	F4A	Yes	10,000	n. t.	n. t.	Azaguanine-resistant
A1	„	F4AG	Yes	1,000	n. t.	n. t.	Spleen clone of F4AG
FSD 2	Balb/C	—	Yes	n. t.	n. t.	n. t.	Isolated from spleen
FLD 2	Balb/C	—	Yes	n. t.	n. t.	n. t.	Isolated from liver

n. t., not tested.

If the supernatant of F4 cells is injected into the tail vein of Balb/C or DBA2 mice, we obtain spleen foci, as shown in Fig. 1. We can distinguish several sublines of F4 (Table 1). One of these sublines, F4CD, does not release any C-type particles capable of forming spleen foci, F4N releases 1,000–5,000 spleen focus-forming units (SFFU)/ml and F4A 10,000–20,000. At present we do not know how cell lines of the same clone diverge. Cell clones B8 and B16, which are derived from F4A (*40*) are low in overall C-type particle release (*43*).

2. *In vitro XC assay*

If the XC tissue-culture assay for the LLV helper virus is used we obtain approximately the same virus titers ($3\times10^5/10^6$ cells) in all sublines of clone F4 except in F4/1. The XC assay and the number of SFFU cannot be compared (Table 1). A correlation of the transforming titer and the XC titer is, however, obtained for the BrdU-resistant subclones of B8, for B8 ,and for B16 as compared with the parent clone F4A. B8 shows very low or zero virus titers in both assays. On recloning of B8 we were able to obtain subclones which are heterogeneous in their SFFU and XC titers. Some subclones no longer show detectable SFFU, such as B8/1 or B8/22 (unpublished results).

3. *Electron microscopy*

If C-type particles are counted by electron microscopy (Tables 2 and 3; Figs. 2 and 3) we essentially are able to confirm the XC assay data. The number of virus particles is low or zero in B8 and B16 and is high in F4N or F4A. No difference between F4N and F4A has been detected.

TABLE 2. Virus Release as Assayed by Electron Microscopy

	C particles/cell section	
F4 (All sublines)	11.3	BrdU, azaguanine-sensitive, TK^+
B8	0.03–0.45	TK^-, BrdU-resistant
B13	0.7	TK^-, BrdU-resistant
B16	6.2	Partial TK^-, partially resistant to BrdU

TABLE 3. C-type Particle Release and A Particles as Assayed by Electron Microscopy Following Treatment with Thymidine Analogues

Cell type	Type of treatment	Days	C particles	Intracist. A particles
F4:	Untreated	—	11.3	1.4
	BrdU 6×10^{-4}M	1	10.2	3.4
	AT 1×10^{-4}M	1.4	2.3	8.0
	AT ,,	2	3.2	20.9
	AT ,,	8	1.5	9.2
	AT 2.5×10^{-4}M	1	1.9	9.0
B8 :	Untreated	—	0.03–0.50	4.9
	BrdU 6×10^{-4}M	1.4	0.65	6.6
	BrdU 6×10^{-4}M	2	4.2	10.7
	DMSO 1.5%	2	0.1	17.3
	DMSO 1.5% } BrdU 6×10^{-4}M }	2	9.0	16.7
	AT 2.5×10^{-4}M	1.4	0.5	13.7
	AT 2.5×10^{-4}M	8	0.45	14.4
	NH_2T 1×10^{-4}M	1.4	0.4	7.6
	NH_2T 2.5×10^{-4}M	1.4	0.4	11.2
	TdR 2.5×10^{-4}M	1.4	0.4	6.7
B16:	Untreated	—	6.2	27.6
	BrdU 6×10^{-4}M	1.4	1.9	49

4. *Effect of DMSO*

If 1 or 1.5% DMSO is added to cell clone F4N, it induces differentiation and a dramatic 10- to 100-fold increase in the biological activity of FV using the spleen focus-forming assay (Fig. 4). Only a moderate, at most 10-fold, increase can be obtained by particle counting (*42*). If direct virus assays were used a 5- to 10-fold stimulation of virus release was found. For that, we measured virus release by electron microscopy, by OD, and by ^{32}P-labeling of viral particles (Tables 4 and 5). This is similar to the observations of Sato *et al.* (*57*). The increase in SFFU is in line with data reported by Ikawa (*23*) in this volume using the XC assay. A subline

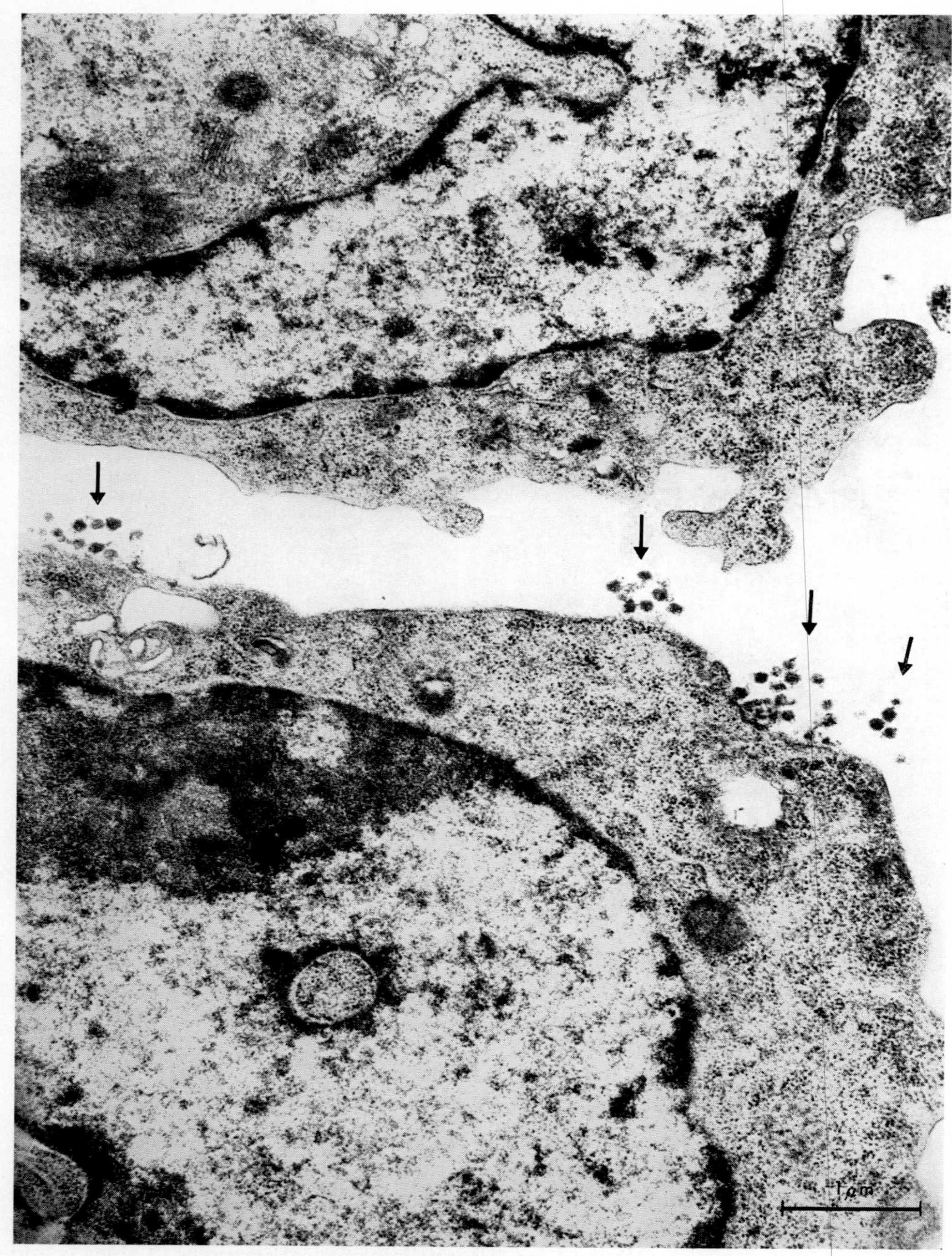

FIG. 2. Thin section of a pellet of F4A cells. Arrows: C-type viruses in close contact with the surface of the cell. ×25,000.

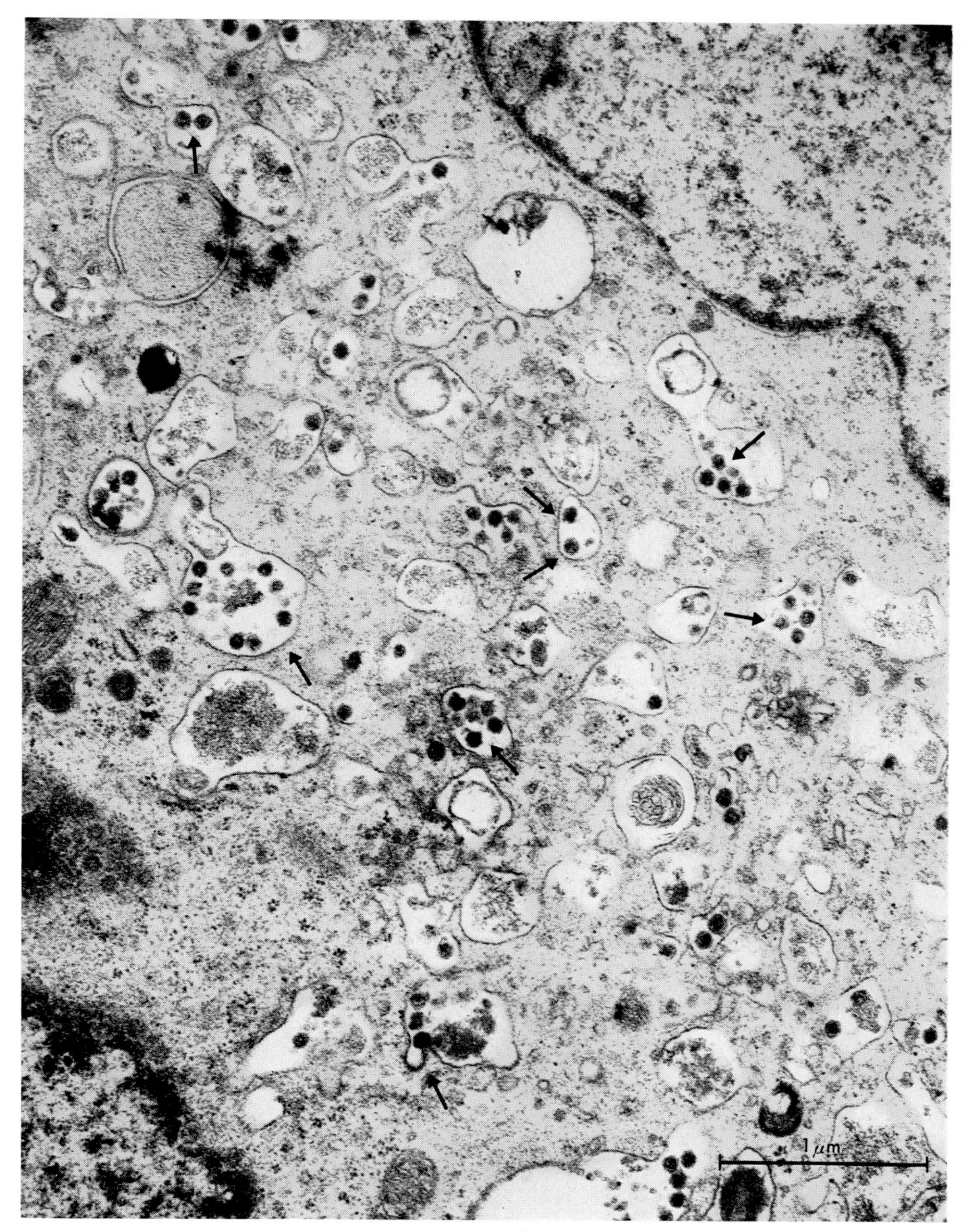

FIG. 3. Thin section of a pellet of B8 cells (uninduced). Arrows: numerous intracisternal A-type particles, some of them budding. ×40,000.

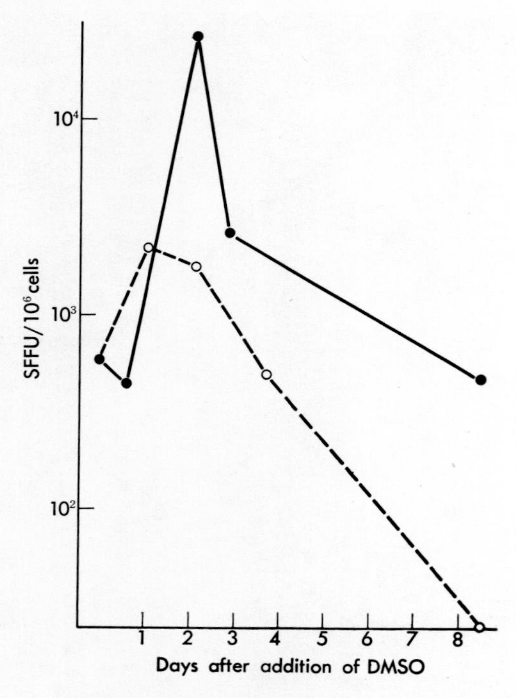

FIG. 4. Increase in released SFFU in F4N cells at day 1–2 of DMSO-induced differentiation. Note: decrease of virus release in surviving cells. ● 1% DMSO; ○ 1.5% DMSO.

TABLE 4. Increase of FV Release by DMSO-stimulated F4N Cells (^{32}P Label in Virions)

	^{32}P (cpm)/hr	Relative values
Control (4 hr)	70,000	1
1.5% DMSO (20–24 hr)	75,000	1.07
1.5% DMSO (24–36 hr)	600,000	8.6

F4N cells were labeled with 0.5 mCi/ml in PO_4-free medium for 6 hr. Virus released in the indicated time periods was collected. Viruses were purified by 2 sucrose density gradients and then counted.

TABLE 5. Increase of FV Release by DMSO-stimulated F4N Cells (OD_{260} of Viral RNA)

	OD viral RNA/hr	Relative values
Control (7 hr)	2	1
1% DMSO (0–16 hr)	3.5	1.7
1% DMSO (16–25 hr)	9	4.5
1% DMSO (25–38 hr)	13	6.5

F4N cells (10L) were grown to a density of 2×10^6 cells/ml in Roller cultures then were spun down. New medium (5L) was added and incubation was continued for 7 hr. After 7 hr the cells were pelleted. The supernatant was removed to collect the virus. Incubation of the cells was continued with fresh medium containing 1% DMSO for time periods with medium exchange as indicated in column 1. Virus was separated by the usual methods and viral RNA prepared.

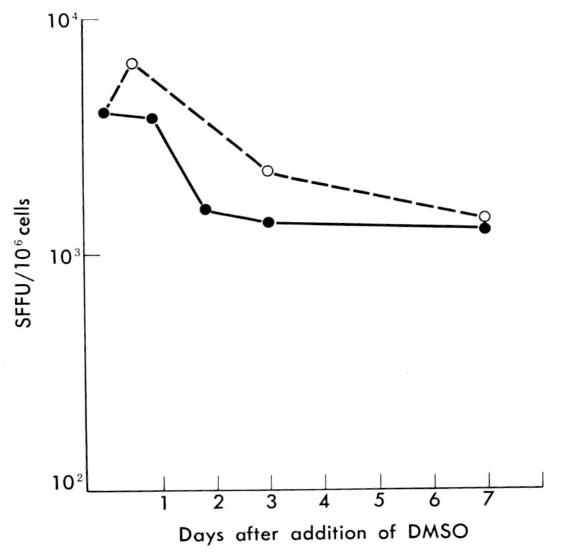

FIG. 5. No increase in release of SFFU in F4AO (=F4A) cells. These cells do not differentiate in the presence of 1 or 1.5% DMSO. ● 1% DMSO; ○ 1.5% DMSO.

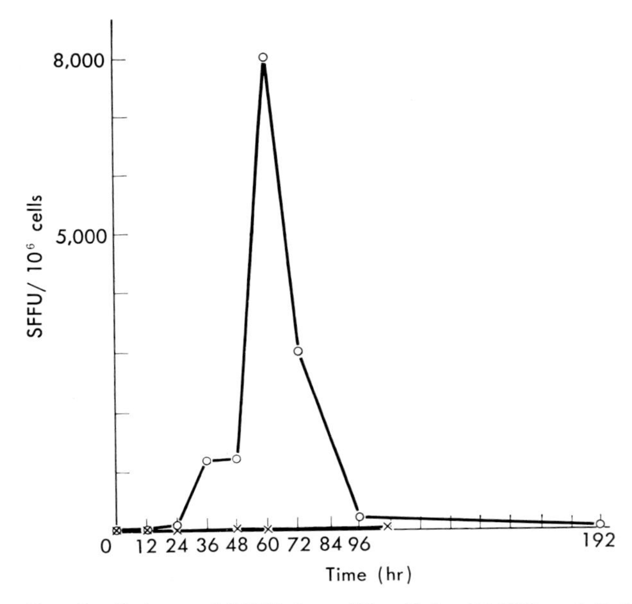

FIG. 6. Release of SFFV from B8 cells by DMSO and BrdU. B8 cells show no release of SFFV even in the presence of DMSO (1.5%). Inducibility of SFFU is seen on adding 200 μg/ml BrdU. × DMSO; ○ BrdU (200 μg/ml).

of F4A (F4AO) which does not differentiate on adding DMSO shows no increase of virus release (Fig. 5), but only an overall decline. We would therefore like to conclude that the 10- to 100-fold increase in virus release is due to an effect of DMSO correlated with erythroid differentiation. However, if DMSO is added to B8 or to sublines of B8, no increase in virus release is obtained, although differentiation can be induced (Fig. 6). Virus release is not obligatory for differentiation in the virus-negative cell line B8.

The increase in virus release in F4 1–2 days after DMSO induction is correlated with the appearance of globin mRNA in the cytoplasm of the differentiating cell (*9*). RNA synthesis and viral release decline below uninduced levels 3 days after addition of DMSO (Fig. 4). Some of the cells survive and grow continuously even in the presence of 1% or 1.5% DMSO. These cells show no erythroid differentiation. The release of FV is 1,000-fold lower than it was before DMSO exposure of the cells.

To conclude, we observe several effects of DMSO on virus release in F4 cells: (1) Correlated with the time of increase of globin mRNA in the cytoplasm, we observed a 60- to 100-fold increase of virus release as judged by the transforming capacity in the spleen focus-forming test, but only a 10-fold increase in virus particles. (2) This increase in virus release is dependent on cellular differentiation. (3) The increase in virus release is not obligatory for cellular differentiation. (4) Surviving and nondifferentiating cells after induction of differentiation show much reduced capacity to release virus particles as compared to cells which have not been treated with DMSO.

5. Viral RNA

C-type viruses contain 60–70S RNA, rRNA, some ribosomal RNA, or ribosomes, and bits of RNA of other size classes (*11*, *13*, *37*, *47*, *48*, *69*). The 60–70S RNA is the viral genome, and the other RNA species, whose function is unknown at the present time, are thought to be derived from cellular RNA. Extensive degradation of the 60–70S RNA is found if virions are not harvested within a few hours of release. If, however, the virions are harvested immediately after release, RNA predominantly in the 30–35S class is recovered after treating 60–70S RNA with either 99% DMSO or at 100°C for 45 sec (*11*). The 30–35S RNA species are considered to be viral RNA subunits or the viral chromosomes. Each virion would thus contain about 2–4 pieces of 30–35S RNA in the 60–70S RNA. In a series of experiments, Duesberg and Vogt (*12*, *29*) were able to correlate the size of the 30–35S with some properties of the virions: the largest 30–35S RNA pieces are obtained if RNA of non-defective chick sarcoma viruses is separated on acrylamide gels. Replication-defective sarcoma viruses have a smaller sized 30–35S, as do chick leukosis viruses. We were able to show that a similar correlation is found for the FV genome (*35*). After ^{32}P-labeling of transforming FV from tissue-culture supernatants and separation of the RNA on acrylamide gels, several RNA species are obtained. The 2 largest subunits are larger than 28S, *ca.* 30–35S, while the smaller RNA species correspond in size to 28S, 18S, 7S, and 4S RNA (Fig. 7). The autoradiography shown in Fig. 7 was obtained by using virions of DMSO-stimulated cells. No RNA could be obtained in the region of globin mRNA. By electrophoresis of LLV RNA on acrylamide gels only the larger of the two 30–35S RNA species can be observed. We therefore conclude that at least the larger of the 2 RNA species in the 30–35S region is the genome for the helper virus (LLV-F) of the FV complex. By exclusion, the smaller species of 30–35S RNA subunit should therefore contain the SFFV RNA.

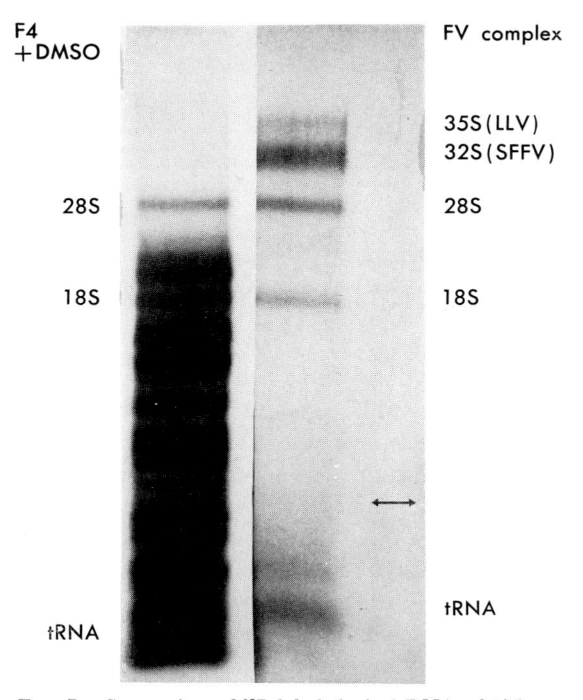

FIG. 7. Separation of ^{32}P-labeled viral RNA of FV complex released by DMSO-stimulated F4A cells. At the left, cellular RNA for comparison. F4A cells were grown and stimulated to differentiate by adding increasing concentrations (1–1.5%) of DMSO. Cells were labeled with ^{32}P (0.5 mCi/ml) in phosphate-free medium with 1.5% DMSO 24–36 hr after adding 1% DMSO. Viruses were separated by 2 sucrose density centrifugation. Viral RNA was extracted, heated to 100°C for 45 sec, cooled to 4°C, and applied to 2.2% acrylamide slab gels. Cellular RNA was prepared similarly, treated, and applied to a different slot of the same acrylamide slab. Note: the viral RNA consists of 2 large subunits (35S: possibly LLV RNA, 32S: possibly SFFV RNA) and intact 28 and 18S, as well as low-molecular-weight RNA in the tRNA region. No labeled RNA could be detected in the region where globin mRNA should move! The cellular ribosomal RNA in differentiating cells is degraded to 22S and 16–17S RNA.

A handle would than be available for the first time to study the function of the transforming SFFV. It is surprising that most of the label of the two 30–35S RNA species is not in the larger (LLV) subunit. This could indicate that viruses released by transformed erythroid cells in tissue culture predominantly carry SFFV genetic information.

6. Viral RNA during DMSO-induced differentiation

We isolated viral RNA of virions released by F4 cells before and during stimulation with DNSO. We observed a big discrepancy in the relative amounts of intact 30–35S RNA (Table 6). Only 3% of the total RNA in the virions is found in the 30–35S region in virions released in 12 hr by control cells, whereas 31% intact 30–35S RNA can be found in the virions liberated by differentiating cells. We have also checked SFFV-transformed cell lines which do not differentiate in the

TABLE 6. Relative Distribution of Total ^{32}P-labeled Viral RNA before and during DMSO Stimulation

		30–35S	4–29S
Virus of control cells	^{32}P	2,560	78,000
	%	3.2	97
Virus of DMSO-stimulated cells	^{32}P	3,200	7,200
	%	31	69

F4N cells were labeled with 0.5 mCi/ml ^{32}P in PO_4-free medium for 6 hr. Virus released over a period of 12 hr was collected and total viral RNA was extracted, then heated to 105°C for 45 min, cooled to 4°C, and then separated by acrylamide gel electrophoresis. DMSO stimulation was carried out for 2 days with 1% DMSO.

presence of DMSO. The viruses released by such cells treated with 1.5% DMSO show very little intact 30–35S RNA. This difference in the intactness of viral RNA is probably the explanation for the discrepancy of biological activity and particle release during DMSO-induced differentiation. A decrease in the activity of cellular ribonuclease which can be observed during erythroid differentiation could lead to a decreased incorporation of ribonuclease into the virions released by differentiating cells. As a consequence, less viral RNA breakdown would occur. This could lead to a longer half-life of the biological activity of the virions.

The change in the pattern of 30–35S RNA breakdown explains only part of the 10- to 100-fold increase of biological activity. It does not account for the 5- to

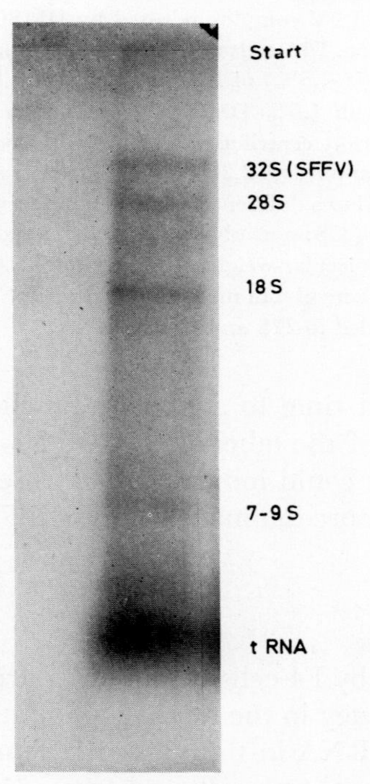

Fig. 8. Autoradiograph of ^{32}P-labeled viral RNA released by F4N cells after 2.2% acrylamide gel electrophoresis. Details as in the legend to Fig. 7. Note: very little RNA which moves in the region of the larger viral RNA subunit can be detected.

10-fold increase in the actual number of viral particles which can be observed during differentiation.

In cell line F4A we have observed that 20–30% of the viral RNA is in the larger subunit of the 30–35S pieces, and 70–80% is in the smaller one. In F4N, a subline of F4A, we obtain very little of the large 35S subunit and more than 95% of the smaller subunit (Fig. 8). F4A has a relatively high titer of SFFU (10–30,000/ml tissue-culture supernatant), whereas F4N has a much lower titer (1,000–3,000/ml). The XC titer in both cell lines is identical. We would therefore speculate that attenuation of viruses in tissue culture is due to loss of genomic material of LLV, which leads to inability of the virions to replicate *in vivo*. The attenuated LLV, however, can infect and multiply in the fibroblasts which are used for the XC assay (*42*). For this reason, some of the RNA of the smaller subunit of the 30–35S class might be deleted LLV RNA.

7. *Structure of viral RNA*

In a previous section we have shown that the genomic RNA of FV has a sedimentation coefficient of about 70S and upon heating resolves into 2 bands of 30–35S on polyacrylamide gels. By analogy with other RNA tumor viruses we have assumed that the 2 bands represent the so-called *a* and *b* subunits (*11, 12, 14, 15*). In

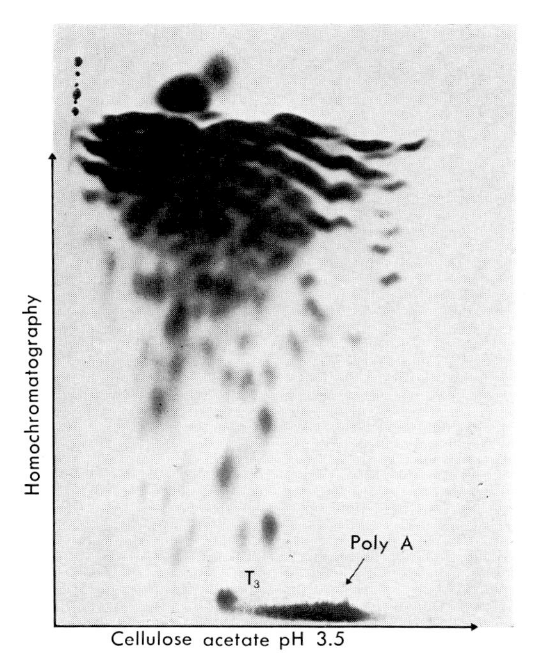

FIG. 9. A two-dimensional fractionation of a ribonuclease T_1 digest of FV RNA. ^{32}P-labeled viral RNA was digested with RNase T_1 for 30 min at 37°C (0.01M Tris buffer, pH 7.4, enzyme-to-substrate ratio of 1 to 20), and fractionated on a two-dimensional system using high-voltage electrophoresis in the first dimension and homochromatography in the second (*6, 10, 55*). The chromatogram was autoradiographed. The direction of mobility is shown by arrows. Radioactivity at the origin is poly A.

the case of Rous sarcoma virus (RSV), where the subunit structure of viral RNA was first demonstrated, the question of whether the subunits are identical or not has been studied in detail. There are arguments favoring both possibilities. The biological properties of the virus, namely, that transforming RSV's contain only *a* subunits whereas transformation-defective variants contain only *b* subunits, suggest that the subunits are identical and that deletion of one *a* subunit gives rise to a population of *b* subunits (*15*). On the other hand, hybridization data (*68*) indicate that the subunits are not identical. In the case of FV, we have attempted to gain some insight into the structure of viral RNA by subjecting it to fingerprint analysis. The results of these studies, which are at a preliminary stage, are described below. Figure 9 shows a T_1 ribonuclease fingerprint of ^{32}P-labeled 70S viral RNA. The RNA, purified on sucrose gradients (see last section) was digested with T_1 RNase at 37°C using an enzyme-to-substrate ratio of 1 to 20. The digest was fractionated on a two-dimensional system using high-voltage electrophoresis on a cellulose strip at pH 3.5 for the first dimension and homochromatography on a DEAE-cellulose thin layer plate for the second (*6, 55*). The chromatogram was autoradiographed using Kodak H film. Under the conditions used for fractionation, oligonucleotides of chain length up to 25 nucleotides are well-separated from one another.

A, T_1 RNase fingerprint of uniformly ^{32}P-labeled RNA provides a direct assay

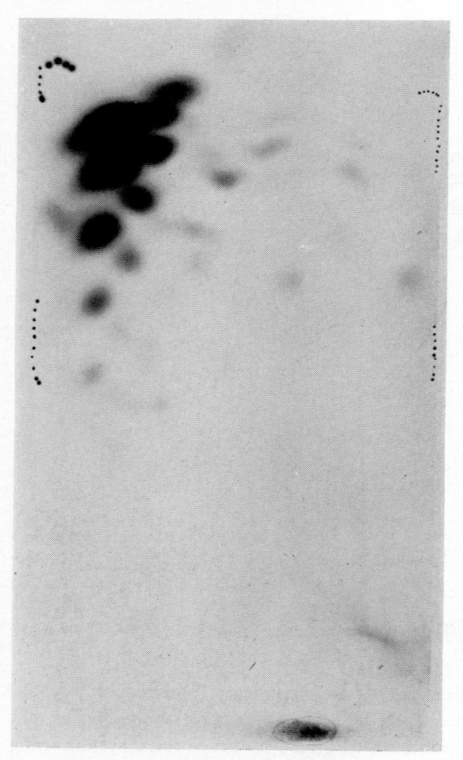

FIG. 10. A two-dimensional fractionation of a combined RNase T_1 and pancreatic ribonuclease (RNase A) digest of ^{32}P-labeled FV RNA. Details as in Fig. 9, except that RNA was digested in buffer containing 0.3M NaCl.

for the chain length of RNA because the relative yields of various oligonucleotides (also of mononucleotides, *e.g.*, Gp) can not only be estimated by inspection, but determined quantitatively by counting the radioactivity in the oligonucleotide spots. For converting cpm in the spot to cpm per phosphate we have used spot T_3 (Fig. 10) as the standard. Further analysis using RNases A and U_2, and T_2 and also alkaline hydrolysis, has shown that T_3 contains 22 nucleotides. Dividing cpm in T_3 by 22 gave cpm per phosphate.

Using this value we have estimated the molar yield of various oligonucleotides per mole of RNA. Adding these values for all T_1 products, the molecular weight of the RNA comes out to about 2.5×10^6. It should be emphasized that the estimated molecular weight represents the minimum molecular weight of the RNA, because we have assumed that T_3 is present in a single copy, which may not be the case. Also, we have not taken into account many spots which were present in low yields because the radioactivity was too low for accurate determination. Since one of the 30–35S bands on polyacrylamide gels accounts for about 20% of the 70S viral RNA, it is quite conceivable that most of the low-yield spots originate from this RNA subunit, which is presumably LLV (see previous section). One of the possibilities our results appear to suggest is that the viral RNA is composed of 4 subunits, of which 3 are identical or have extensive homologies.

8. Poly A in viral RNA

We have analysed the radioactive material at the origin on the fingerprint (Fig. 10) by further hydrolysis with RNases A and U_2, and alkali. Most of the material was resistant to RNase A but sensitive to U_2 and alkali. The U_2 and alkaline hydrolysis product was predominantly A, indicating that the FV RNA contains poly A, like other viral RNA's (*28*). We have also hydrolyzed total viral RNA with a mixture of RNases A and T_1. This procedure is useful for obtaining poly A and tracks of oligo A. Since RNases A is known to split after A residues in low salt concentrations (*5*) the hydrolysis was carried out in a high-salt buffer (0.01 M Tris-HCl, pH 7.4, 0.002 M EDTA. 0.3 M KCl). The digest was fractionated on a two-dimensional system, as described in the previous section. The result of this experiment, in addition to showing that the viral RNA contains poly A, also shows that the RNA contains a number of internal oligo A tracks of the type (A) 8–10 C, (A) 8–10 U, and (A) 8–10 G.

9. Effect of BrdU and azidothymidine (AT) on BrdU-sensitive cells

It is known that BrdU interferes with C-type virus replication (*66*). BrdU induces mutations in C-type tumor viruses (*3*). BrdU or iododeoxyuridine (IdU) can also be employed to induce endogenous C-type virus which presumably exists as a provirus in the cellular genome before BrdU treatment (*34, 52*). This latter effect of BrdU is especially marked in some cell lines which are already transformed (*1, 24, 30, 60*).

AT is another thymidine analogue which has an N^3 group replacing the 3′-OH group in the deoxyribose. AT can be phosphorylated by thymidine kinase. The efficiency of phosphorylation of AT in *in vitro* assays is 2–3 times

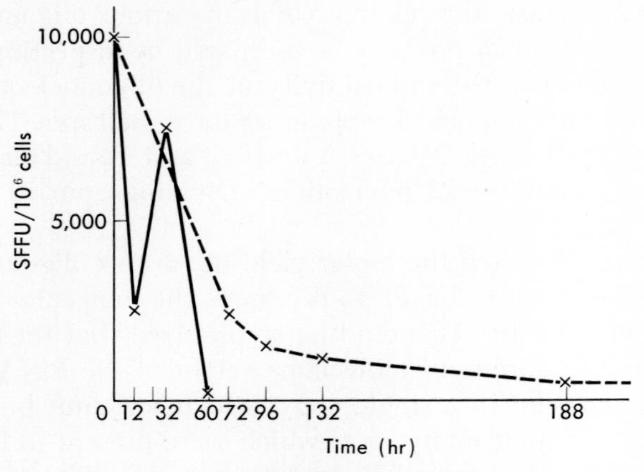

FIG. 11. Decrease of SFFU after treatment of F4A cells with BrdU. — BrdU (200 μg/ml); --- BrdU (2.5 μg/ml).

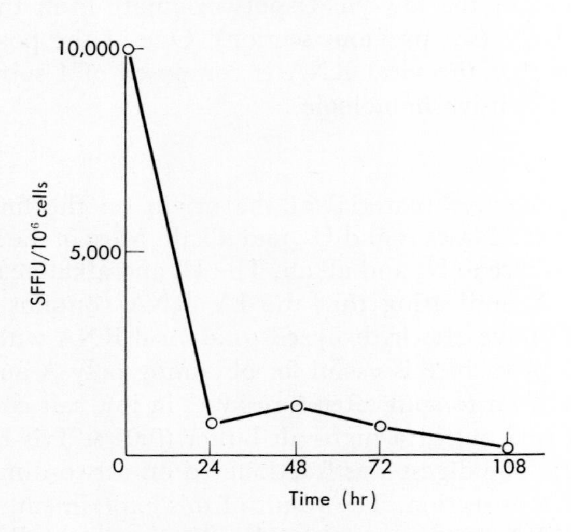

FIG. 12. Decrease of SFFU after treatment of F4A cells with AT (2.5×10^{-4}M).

lower than that of thymidine. AT probably interferes with the chain elongation of DNA. The N^3 group in the 3′-OH position does not allow linkage to the next phosphate (*43*).

After addition of either 10^{-5} M BrdU or 2.5×10^{-4} M AT to TK+ cells (F4A) we observe a pronounced decrease of SFFU (Figs. 11 and 12). If higher doses (2×10^{-3} M) BrdU are added we also find an overall decrease of SFFU. However, after the first or second day of addition of BrdU we also obtain a short transient increase in SFFU. This is an indication of induction of C-type virus in BrdU-sensitive erythroleukemic cells in culture. BrdU-resistant cells can also be induced (see below). AT differs from BrdU in its lack of toxicity. Cells grow perfectly well in the presence of up to 5×10^{-4} M AT, whereas doses above 10^{-5} M BrdU are toxic. AT could thus possibly be used as a substance to inhibit DNA virus replication without affecting cellular viability.

To summarize: BrdU and AT interfere with the release or possibly with the replication of FV complex. The interference with virus replication is probably not related to internal incorporation of thymidine analogs into cellular DNA. Possibly cellular repair enzymes can remove terminally incorporated AT in the growing DNA strand. This would explain the low toxicity of AT.

10. Virus-negative transformed erythroleukemic cells

The effect of BrdU in reducing virus release was used to obtain cells deficient in C-type virus release. F4 (TK^+) cells were made resistant to 10^{-3} M BrdU. The resistant cell population was cloned by spleen or agar cloning (*40*, *43*). Virus titers were determined by either the XC or SFFU assay.

The resistant cells show reduced thymidine kinase activity. B16 has only moderately decreased TdR kinase levels and an intermediate level of C-type virus release in the SFFU assay whereas in B8 TdR kinase is hardly detectable, but the virus level is more than 2,000-fold lower than in F4 (TK^+). The difference is even more pronounced in the XC assay. The virus level in clone B8 is reasonably stable with time if the cells are grown in medium without BrdU. The level of virus release in B16 recovers after omission of BrdU. The recovery takes about 6–12 months. B8 cells were recloned and several subclones were obtained. Some of these (B8/22 obtained by Drs. Krooth and Kiang, Columbia University; B8/1 and others) show no detectable SFFU. B8/22 releases very few virions, as shown by the XC assay.

To summarize: we can obtain various BrdU-resistant erythroleukemic cell clones. In these cell clones thymidine kinase activity is reduced. The level of TdR kinase is usually correlated with the number of C-type particles which are released by the cells. This and other results suggest (see below) that C-type viruses interact with TdR kinase as has been reported for DNA tumor viruses (*20*, *22*, *25*). Resistant B8 (TK^-) cells revert at a very low frequency to TK^+ phenotype. Increased viral release is correlated with increased TdR kinase activity in cell clone B16. B16 reverts at a high frequency to TK^+ phenotype and regains the ability to liberate high titers of FV.

11. Increase of virus release is not a prerequisite for differentiation

SFFV transformation arrests differentiation. Transformed cells in tissue culture are in a proerythroblast stage of differentiation. Spontaneous differentiation is observed in only very few cells. We also showed that there is an actual increase of released viral particles, not only an increase in biological activity of C-type viruses in transformed cells during DMSO-induced differentiation (*42*, *57*). This raises the interesting possibility that the control of differentiation exerted by the virus is analogous to that observed in *Escherichia coli* lysogenically infected with bacteriophage. If DNA is incorporated as a prophage genome in the *E. coli* chromosome, a repressor protein is formed which prevents the expression of lytic replication of the phage. If the SFFV genome also is integrated during transformation as a provirus it might be possible that a repressor protein is formed by the SFFV genome similar to that of λ and *E. coli*. This might prevent expression of the independent replicative form of SFFV. Two cellular programs exist in eukaryotes, one of which is BrdU-sensitive, the other being BrdU-resistant (*70*). The BrdU-sensitive program leads to the synthesis of "luxury" molecules for further differentiation. We suggest that tumor viruses use part of the BrdU-sensitive program for their autonomous replication. One way to avoid uncontrolled virus multiplication would be to interfere with the BrdU-sensitive program, that is, with further differentiation. Possibly transforming C-type viruses form a repressor in the provirus stage which arrests differentiation in order to avoid uncontrolled replication of their own viral genome.

The correlation of low virus yield in unstimulated Friend cells and the 7-fold higher yield (by OD) in differentiating Friend cells would support such a model. To test this hypothesis we have developed the virus-negative cell line B8, which can be initiated by DMSO to differentiate (*9*) without virus release (Fig. 6). Similarly, by addition of AT and DMSO to BrdU-sensitive cells we observe even better differentiation (*9*), but no virus release (Tables 3 and 7). Interferon also reduces virus yield from F4 cells more than 100-fold, but does not interfere with DMSO-induced differentiation (*67*). We conclude that the normally observed increase in C-type virus release with differentiation is not a prerequisite for differentiation.

TABLE 7. Correlation of Differentiation with Increase in Intracisternal A Particles in B8 and Lack of Correlation in F4

		% globin synthesis	A particles/cell
F4:	Untreated	1.1	1.4
	+Azido TdR 2.5×10^{-4}M	4.0	15
	+DMSO 1.5%	22.5	3.5
	+Azido TdR +DMSO	26.0	n. t.
B8:	Untreated	2.6	4.8
	+DMSO 1.5%	23.5	17.3
	+BrdU 6×10^{-4}M	5	10.7
	+DMSO +BrdU	30.4	16.7

Since the C-type virus proteins are all contributed by the nontransforming LLV which is present as a helper, we can only conclude that the presence or release of LLV helper virus is not conditional for differentiation. To check for a correlation between differentiation of the cells and SFFV functions, we have to measure SFFV functions in the cell. This has not yet been done. The virus-negative cell line B8 still contains SFFV information (see below) and shows very little LLV release. It can be induced to differentiate. It is therefore a good cell line to test for SFFV involvement in differentiation.

12. BrdU-resistant cells can be induced to release endogenous and SFFV by treatment with BrdU

Transformed cell lines sometimes liberate active C-type virus if exposed to high concentration of BrdU (*1, 24, 30, 60*). After spleen cloning of B8 which is BrdU-resistant and virus-negative the cell line was kept for 2 years without BrdU. The base level of release of SFFU is about 5 units/10^6 cells. By treatment of those cells with BrdU at a concentration of 200 μg/ml a more than 1,000-fold transient

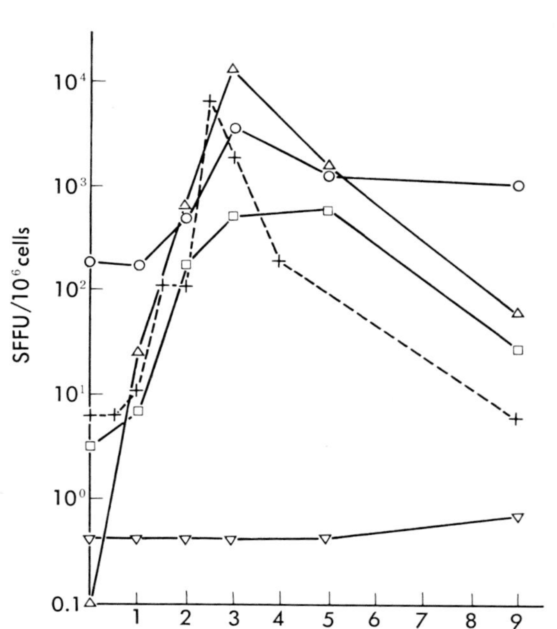

FIG. 13. Inducibility of endogenous virus and SFFV by adding 200 μg/ml to B8 cells or subclones of B8. N-type DBA-2 mice were used as recipients. The host range of induced virus is N. ○ B8/7; △ B8/3; □ B8/6; +B8; ▽ B8/1. B8/1, B8/4, B8/22, with SFFU noinduction detectable.

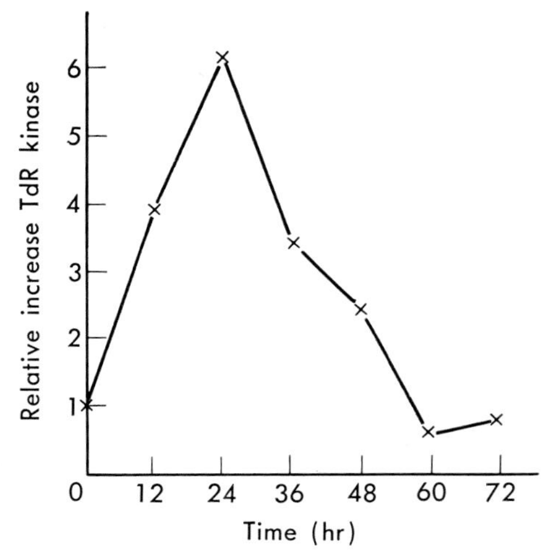

FIG. 14. Increase of thymidine kinase activity in B8 cells (resistant cells) on treatment of the cells with 200 μg/ml BrdU.

increase of SFFU (Figs. 6 and 13) and at the same time a correlated increase of TdR kinase is observed (Fig. 14). The BrdU-inducible virus does not grow or is very deficient in B-type cells as shown by the XC assay using A31 cells of Balb/C origin or by injection of the virus in B-type mice (*43*). The LLV-F/SFFV complex which was originally used to obtain the transformed cell lines F4 and B8 is of the NB type. The N host range specificity of the induced virus is a proof that the BrdU-inducible virus is an endogenous virus and not a mutated BrdU-resistant LLV (Fig. 15). We can conclude that the B8 cells do contain SFFV and that an endogenous C-type virus is induced by BrdU which supplies the helper function for the SFFV in cell line B8. The endogenous virus, unlike the LLV or our cell lines, has only N, not NB, host range properties.

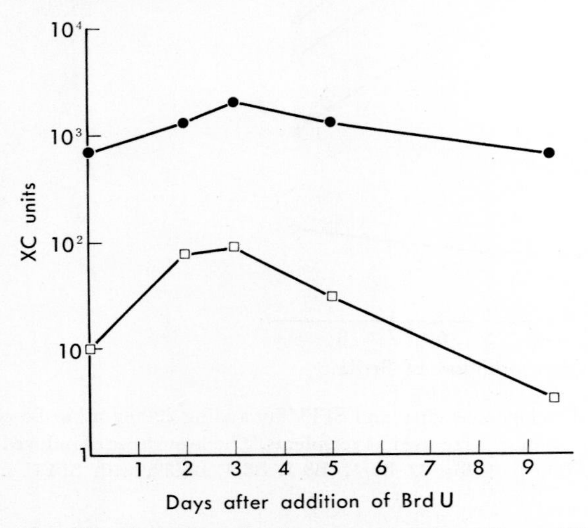

FIG. 15. Induction of endogenous virus from B-type (3T3, Balb/C) indicator cells by BrdU. B-type indicator cells in the XC assay permit only a low rate of plaque formation with BrdU-induced endogenous virus. The host range of endogenous virus is not B or NB! ● B8; □ B8/22.

13. Are intracisternal A particles involved in transformation and differentiation of erythroleukemic mouse cells?

The only morphological equivalent to virus particles inside the SFFV-transformed cells are intracisternal A particles which are found abundantly in BrdU-resistant and much less so in BrdU-sensitive cells (Fig. 3). This suggests that some or all A particles are morphological equivalents of SFFV and that some genomes of SFFV normally get incorporated into LLV protein coats. They might appear as external C-type virus if they acquire the LLV protein envelope. We would then expect a compensatory increase of A particles if C-type virus formation is inhibited.

Intracisternal A particles are known to be serologically different from C particles (*27*). They do contain 60–70S "genomic" RNA (*72*) and a genuine reverse transcripase activity (*71*) similar to extracellular C particles. They are sometimes observed during normal differentiation and are found more often in transformed cells. Their biological function remains obscure.

Intracisternal A particles in SFFV-transformed cells show the following properties: (1) A particle replication is not inhibited by BrdU, AT or interferon. With all of these agents, C particle release is depressed by more than 2 orders of magnitude

(Figs. 11 and 12). This suggests a different level of integration of intracisternal A-particle function than of C-type viruses; (2) if BrdU-sensitive cells are treated with AT or interferon we do get a compensatory increase of A particles; (3) if A particles

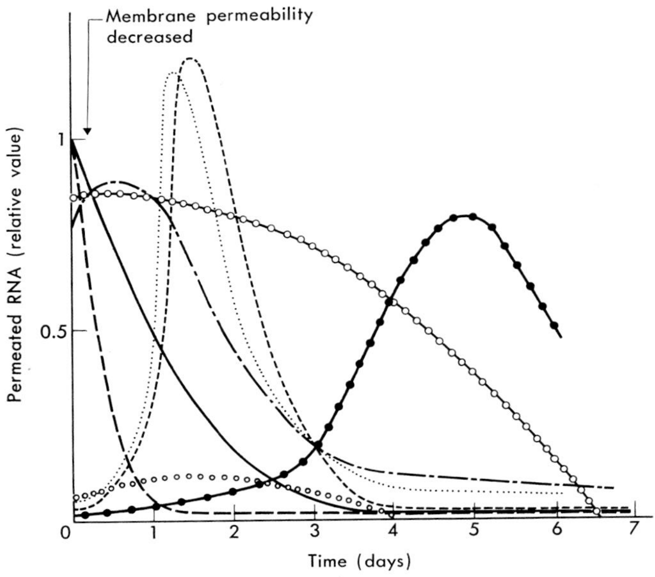

FIG. 16. Schematic drawing of the series of events during DMSO-induced differentiation in cell clone FSD 1/F4. — synthesis of cytoplasmic RNA; — — synthesis of nuclear RNA; - - - cytoplasmic globin mRNA synthesis; -○- DNA synthesis; ······ C-type virus (LLV+SFFV); — - — TdR kinase; ○ A particles (intracisternal) (synthesis); ● globin synthesis.

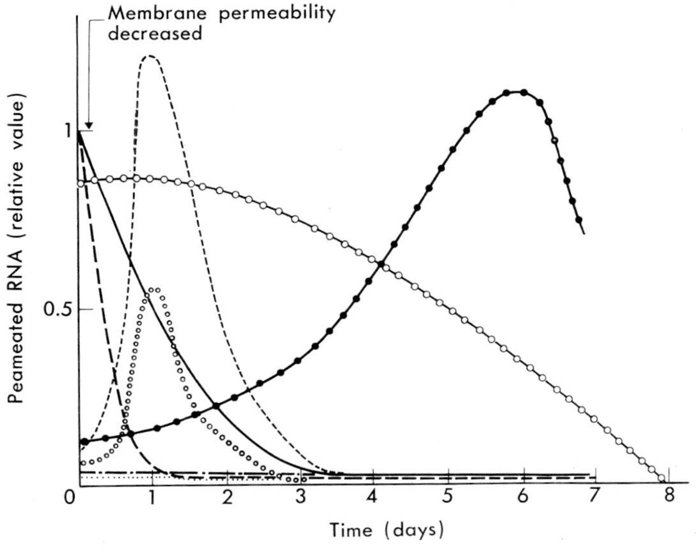

FIG. 17. Schematic drawing of the series of events during DMSO-induced differentiation in cell clone B8 (BrdU-resistant). Symbols, see legend for Fig. 16.

TABLE 8. Effect of Various Substances on Differentiation (Globin Synthesis), C-type Virus Release (SFFU), and A-type Particles

	Control	DMSO	BrdU	Azidothymidine	Thymidine	Amphotericin B	Interferon
F4A:							
Globin synthesis	0.5–1%	20–25%	+DMSO 0.5%	3–4%	0.5%	+DMSO 0.5–1%	+DMSO 20%
FV (SFFV)	10,000 medium and large 1,000 large	Not done 120,000	200	150	10,000–29,000	1,100	50
A-type particles	1.4	3.5	3.4	9–20	n. t.	n. t.	6
Membrane permeability	Normal	Decreased	n. t.	n. t.	n. t.	Increased	n. t.
B8:							
Globin synthesis	2.5%	25–30%	4–5%	4–5%	2%	+DMSO 1–2%	26% +DMSO
FV (SFFV)	5	5	I: 8,000 B: 6	0.2	n. t.	n. t.	0.2
A-type particles	4–6	17.3	6–11	13.7 NH_2-TdR: 11.2	6.7	n. t.	7

n. t., not tested; I, BrdU-induced; B, permanently in BrdU.

are genetically SFFV and if they are involved in the process of differentiation we would expect in the virus-negative line B8 an increase of A particles during DMSO stimulation similar to that found for C particles in the sensitive cell line F4. This is actually observed ! (Tables 3 and 7).

We were also able to show that A particles in SFFV-transformed cells have the density described for intracisternal A particles and possess 60–70S RNA (Weimann, unpublished data).

To summarize the evidence reported here on viral involvement during differentiation (Figs. 16 and 17; Table 8): C-type virus release is not conditional for differentiation but intracisternal A-type particles or their possible equivalent in the C-type population, SFFV, might be required for arrest and induction of differentiation in these erythroleukemic cells. Naturally it is tempting to suggest that the transforming SFFV itself is involved in differentiation.

Particle-bound Reverse Transcription: Transformed Cells of Human Origin with Possibly Erythroid Differentiation

One approach to determine if any information derived from C-type RNA tumor viruses or of intracisternal A-type particles is present in human leukemic cells is to isolate subcellular cytoplasmic fractions containing RNA tumor virus-like particles from these cells, to identify proteins, and compare the property of these particles with those from known C-type leukosis viruses.

Cellular DNA polymerase isolated from virus-infected cells can be distinguished from the viral DNA polymerase by biochemical and immunological criteria (*21, 46, 50, 59*).

An enzyme similar to the viral reverse transcriptase has been found and partially purified from human leukemic cells (*56*). We have established a cell line derived from the bone marrow of a patient which polycythemia vera. We have isolated a DNA polymerase from a subcellular cytoplasmic particle fraction of these polycythemic human cells in cell culture. The particles have a density of 1.16–1.19 g/cm³, characteristic of that of RNA tumor viruses of animals.

The DNA polymerase is distinct and different in its chromatographic properties on phosphocellulose from DNA polymerases N and C, adopting the nomenclature of McCaffrey *et al.* (*36*).

In addition, the enzyme activity, unlike polymerase T, has no terminal transferase activity when tested with oligo dT and oligo dG. It can use poly rC-oligo dG as a template/primer and transcribes heterologuous regions of natural RNA's.

Since RNA tumor viruses contain 2 outstanding diagnostic features, the RNA-dependent DNA polymerase (reverse transcriptase) and 70S RNA, it is possible to demonstrate their presence by simultaneous detection assay according to the method of Schlom and Spiegelman (*58*).

This assay is based on the observation that the initial DNA product is complexed *via* hydrogen bonding to the 70S RNA template and travels with the 70S RNA on sedimentation analysis.

To synthesize DNA on the endogenous template a high-speed pellet of the cytoplasm from PV cells was centrifuged through a linear sucrose gradient and fractions with a density of 1.16–1.19 g/cm³ were taken. DNA was synthesized with a detergent-disrupted preparation of the subcellular fraction in the presence of actinomycin D to inhibit DNA-directed DNA synthesis.

After deproteinisation, the probe was subjected to a sedimentation analysis in glycerol gradients. The newly synthesized product is found in the 70S region. On preincubation with ribonuclease (RNase) no radioactivity can be found in this region.

The product banding in the 70S region was isolated and examined by equilibrium centrifugation in Cs_2SO_4. Most of the product bands at a density of 1.62–1.66 g/cm³ are characteristic of RNA, although some radioactivity is found in the DNA region.

After alkaline hydrolysis of the material from the RNA region and recentrifugation in Cs_2SO_4, the labeled product is shifted into the region with a density of 1.42 g/cm³, characteristic of DNA. From these results we conclude that human polycythemia vera cells in cell culture contain particles in the cytoplasm with a density similar to that of RNA tumor viruses in animals. These particles are associated with a DNA polymerase which catalyses an endogenous ribonuclease-sensitive DNA synthesis.

ACKNOWLEDGMENTS

We thank A. Rohmann and B. Neu for assistance. This work was supported by a grant from the Deutsche Forschungsgemeinschaft (W. Ostertag, G. Gaedicke, J. Kind, J. C. Krieg, and G. Steinheider).

REFERENCES

1. Aaronson, S. A. Chemical induction of focus-forming virus from non-producer cells transformed by murine sarcoma virus. Proc. Natl. Acad. Sci. U.S., *68*: 3069–3072, 1971.
2. Axelrad, A. A. and Steeves, R. A. Assay for Friend leukemia virus: Rapid quantitative method based on enumeration of macroscopic spleen foci in mice. Virology, *24*: 513–518, 1964.
3. Bader, J. P. and Brown, N. R. Induction of mutations in an RNA tumour virus by an analogue of a DNA precursor. Nature, *234*: 11–12, 1971.
4. Bennett, M., Steeves, R. A., Cudkowicz, G., and Mirand, E. A. Mutant Sl alleles of mice affect susceptibility to Friend spleen focus-forming virus. Science, *162*: 564–565, 1968.
5. Beers, R. F. Hydrolysis of polyadenylic acid by pancreatic ribonuclease. J. Biol. Chem., *235*: 2393–2398, 1960.
6. Brownlee, G. G. and Sanger, F. Chromatography of ^{32}P-labelled oligonucleotides on thin layers of DEAE-cellulose. Eur. J. Biochem., *11*: 395–399, 1969.
7. Dawson, P. J., Rose, W. M., and Fieldsteel, A. H. Lymphatic leukaemia in rats and mice inoculated with Friend virus. Brit. J. Cancer, *20*: 114–121, 1966.
8. Dawson, P. J., Tacke, R. B., and Fieldsteel, A. H. Relationship between Friend virus and an associated lymphatic leukaemia virus. Brit. J. Cancer, *22*: 569–576, 1968.
9. Dube, S. K., Gaedicke, G., Kluge, N., Weimann, B. J., Melderis, H., Steinheider, G., Crozier, T., Beckman, H., and Ostertag, W. Hemoglobin-synthesizing mouse and human erythroleukemic cell line as model systems for the study of differentiation and control of gene expression. *In;* W. Nakahara *et al.* (eds.), Differentiation and Control of Malignancy of Tumor Cells, pp. 103–136, Univ. of Tokyo Press, Tokyo, 1974.
10. Dube, S. K., Marcker, C. A., Clark, B. F. C., and Cory, S. The nucleotide sequence of N-formyl-methyl-transfer RNA. Products of complete digestion with ribonuclease Tl and pancreatic ribonuclease and derivation of their sequences. Eur. J. Biochem., *8*: 244–255, 1969.
11. Duesberg, P. H. Physical properties of Rous sarcoma virus RNA. Proc. Natl. Acad. Sci. U.S., *60*: 1511–1518, 1968.
12. Duesberg, P. H., Lai, M. M. C., and Maisel, J. Tumor virus RNAs and tumor virus genes. *In;* R. Neth, R. Gallo, S. Spiegelman, and F. Stohlman (eds.), Modern Trends in Human Leukemia, pp. 134–144, Grune and Stratton, New York, 1974.
13. Duesberg, P. H. and Robinson, W. S. Nucleic acid and proteins isolated from the Rauscher mouse leukemia virus (MLV). Proc. Natl. Acad. Sci. U.S., *55*: 219–227, 1966.
14. Duesberg, P. H. and Vogt, P. K. Differences between the ribonucleic acids of transforming and non-transforming avian tumor viruses. Proc. Natl. Acad. Sci. U.S., *67*: 1673–1680, 1970.
15. Duesberg, P. H. and Vogt, P. K. Gel electrophoresis of avian leukosis and sarcoma viral RNA in formanilide: Comparison with other viral and cellular RNA species. J. Virol., *12*: 594–599, 1973.
16. Fieldsteel, A. H., Kurahara, C., and Dawson, P. J. Moloney leukaemia virus as a helper in retrieving Friend virus from a non-infectious reticulum cell sarcoma. Nature, *223*: 1274, 1969.

17. Friend, C. Cell-free transmission in adult Swiss mice of a disease having the character of a leukemia. J. Exp. Med., *105*: 307–318, 1957.
18. Friend, C., Scher, W., Holland, J. G., and Sato, T. Hemoglobin synthesis in murine virus-induced leukemic cells *in vitro*: Stimulation of erythroid differentiation by dimethyl sulfoxide. Proc. Natl. Acad. Sci. U.S., *68*: 378–382, 1971.
19. Gaedicke, G., Abedin, Z., Dube, S. K., Kluge, N., Neth, R., Steinheider, G., Weimann, B. J., and Ostertag, W. Control of globin synthesis during DMSO induced differentiation of mouse erythroleukemic cells in culture. *In;* R. Neth, R. Gallo, S. Spiegelman, and F. Stohlman (eds.), Modern Trends in Human Leukemia, pp. 278–287, Grune and Stratton, New York, 1974.
20. Glaser, R., Ogino, T., Zimmerman, J., and Rapp, F. Thymidine kinase activity in Burkitt lymphoblastoid somatic cell hybrids after induction of EB virus. Proc. Soc. Exp. Biol. Med., *142*: 1059–1062, 1973.
21. Goodman, N. C. and Spiegelman, S. Distinguishing reverse transcriptase of an RNA tumor virus from other known DNA polymerases. Proc. Natl. Acad. Sci. U.S., *68*: 2203–2206, 1971.
22. Hampar, B., Derge. J. G., Martos, L. M., and Walker, J. L. Persistence of a repressed Epstein-Barr virus genome in Burkitt lymphoma cells made resistant to 5-bromodeoxyuridine. Proc. Natl. Acad. Sci. U.S., *68*: 3185–3189, 1971.
23. Ikawa, Y., Ross, J., Gielen, J., Packman, S., Leder, P., Ebert, P., Hayashi, K., and Sugano, H. Erythrodifferentiation of cultured Friend leukemia cells. *In;* W. Nakahara *et al.* (eds.), Differentiation and Control of Malignancy of Tumor Cells, pp. 515–547, Univ. of Tokyo Press, Tokyo, 1974.
24. Klement, V., Nicolson, M. O., and Huebner, R. J. Rescue of the genome of focus-forming virus from rat non-productive lines by 5′-bromodeoxyuridine. Nature New Biol., *234*: 12–14, 1971.
25. Klemperer, H. C., Haynes, G. R., Shedden, W. T. H., and Watson, D. H. A virus-specific thymidine kinase in BHK21 cells infected with herpes simplex virus. Virology, *31*: 120–128, 1967.
26. Krontiris, T. G., Soeiro, R., and Fields, B. N. Host restriction of Friend leukemia virus. Role of the viral outer coat. Proc. Natl. Acad. Sci. U.S., *70*: 2549–2553, 1973.
27. Kuff, E., Lueders, K. K., Older, H. L., and Wivel, N. Some structural and antigenic properties of intracisternal A particles occurring in mouse tumors. Proc. Natl. Acad. Sci. U.S., *69*: 218–222, 1972.
28. Lai, M. M. C. and Duesberg, P. H. Adenylic acid rich sequences in RNAs of Rous sarcoma virus and Rauscher mouse leukaemia virus. Nature, *235*: 383–386, 1972.
29. Lai, M. M. C., Duesberg, P. H., Horst, J., and Vogt, P. K. Avian tumor virus RNA: A comparison of three sarcoma viruses and their transformation-defective derivatives by oligonucleotide finger printing and DNA-RNA hybridization. Proc. Natl. Acad. Sci. U.S., *70*: 2266–2270, 1973.
30. Lieber, M. M., Livingston, D. M., and Todaro, G. J. Superinduction of endogenous type C virus by 5-bromodeoxyuridine from transformed mouse clones. Science, *181*: 443–444, 1973.
31. Lilly, F. Susceptibility to two strains of Friend leukemia virus in mice. Science, *155*: 461–462, 1967.
32. Lilly, F. Fv-2: Identification and location of a second gene governing the spleen focus response to Friend leukemia virus in mice. J. Natl. Cancer Inst., *45*: 163–169, 1970.

33. Lilly, F. Mouse leukemia: A model of a multiple-gene disease. J. Natl. Cancer Inst., *49*: 927–934, 1972.
33a. Lilly, F. and Steeves, R. B tropic Friend virus: A host range pseudotype of SFFV. Virology, *55*: 363–370, 1973.
34. Lowy, D. R., Rowe, W. P., Teich, N., and Hartley, J. W. Murine leukemia virus: High-frequency activation *in vitro* by 5-iododeoxyuridine and 5-bromodeoxyuridine. Science, *174*: 155–156, 1971.
35. Maisel, J., Klement, V., Lai, M. M. C., Ostertag, W., and Duesberg, P. H. Ribonucleic acid components of murine sarcoma and leukemia viruses. Proc. Natl. Acad. Sci. U.S., *70*: 3536–3540, 1973.
36. McCaffrey, R., Smoler, D. F., and Baltimore, D. Terminal deoxynucleotidyl transferase in a case of childhood acute lymphoblastic leukemia. Proc. Natl. Acad. Sci. U.S., *70*: 521–525, 1973.
37. Obara, R., Bolognesi, D., and Bauev, H. Ribosomal RNA in avian leukosis virus particles. Int. J. Cancer, *7*: 535–546, 1971.
38. Odaka, T. Inheritance of susceptibility to Friend mouse leukemia virus. V. Introduction of a gene responsible for susceptibility in the genetic complement of resistant mice. J. Virol., *3*: 543–548, 1969.
39. Odaka, T. and Yamamoto, T. Inheritance of susceptibility to Friend mouse leukemia virus. Japan. J. Exp. Med., *32*: 405–413, 1962.
40. Ostertag, W., Crozier, T., Kluge, N., Melderis, H., and Dube, S. K. Action of 5-bromodeoxyuridine on the induction of haemoglobin synthesis in mouse leukaemia cell resistant to BUdR. Nature New Biol., *243*: 203–205, 1973.
41. Ostertag, W., Melderis, H., Steinheider, G., Kluge, N., and Dube, S. K. Synthesis of mouse haemoglobin and globin mRNA in leukemic cell culture. Nature New Biol., *239*: 231–234, 1972.
42. Ostertag, W., Kluge, N., Gaedicke, G., Steinheider, G., Dube, S. K., and Pragnell, I. Induction of endogenous and spleen focus-forming virus (SFFV) dependent on differentiation of SFFV transformed mouse embryo leukemic cells. Unpublished.
43. Ostertag, W., Roesler, G., Krieg, J. C., Kind, J., Cole, T., Crozier, T., Gaedicke, G., Steinheider, G., Kluge, N., and Dube, S. K. Interference of BrdU with Friend virus replication; induction of endogenous virus and of TdR kinase by BrdU in FV transformed cells. Proc. Natl. Acad. Sci. U.S., 1974, in press.
44. Pluznick, D. H. and Sachs, L. Quantitation of a murine leukemia virus with a spleen colony assay. J. Natl. Cancer Inst., *33*: 535–546, 1964.
45. Rauscher, F. J. A virus-induced disease of mice characterized by erythrocytopoiesis and lymphoid leukemia. J. Natl. Cancer Inst., *29*: 515–543, 1962.
46. Robert, M. S., Smith, R. G., Gallo, R. C., Sarin, P. S., and Abrell, J. W. Viral and cellular DNA polymerases: Comparison of activities with synthetic and natural RNA templates. Science, *176*: 798–800, 1972.
47. Robinson, W. S. and Baluda, M. A. The nucleic acid from avian myeloblastosis virus compared with the RNA from the bryan strain of Rous sarcoma virus. Proc. Natl. Acad. Sci. U.S., *54*: 1686–1692, 1965.
48. Robinson, W. S., Pitkanen, A. P., and Rubin, H. The nucleic acid of the Bryan strain of Rous sarcoma virus: Purification of the virus and isolation of the nucleic acid. Proc. Natl. Acad. Sci. U.S., *54*: 137–144, 1965.
49. Ross, J., Ikawa, Y., and Leder, P. Globin messenger-RNA induction during erythroid differentiation of cultured leukemia. Proc. Natl. Acad. Sci. U.S., *69*: 3620–3623, 1972.

50. Ross, J., Scolnick, E. M., Todaro, G. J., and Aaronson, S. A. Separation of murine cellular and murine leukemia virus DNA polymerase. Nature New Biol., *231*: 163–167, 1973.
51. Rowe, W. P. and Sato, H. Genetic mapping of the Fv-1 locus of the mouse. Science, *180*: 640–641, 1973.
52. Rowe, W. P., Lowy, D. R., Teich, N., and Hartley, J. W. Some implications of the activation of murine leukemia virus by halogenated pyrimidines. Proc. Natl. Acad. Sci. U.S., *69*: 1033–1035, 1972.
53. Rowe, W. P., Humphrey, J. B., and Lilly, F. A major genetic locus affecting resistance to infection with murine leukemic virus. III. Assignment of the FV-1 locus to linkage group VIII of the mouse. J. Exp. Med., *137*: 850–853, 1973.
54. Russell, E. S. Abnormalities of erythropoiesis associated with mutant genes in mice. *In;* A. S. Gordon (ed.), Regulation of Hematopoiesis, pp. 649–675, Appleton-Century-Crofts, New York, 1970.
55. Sanger, F., Brownlee, G. G., and Barrell, B. A two-dimensional fractionation procedure for radioactive nucleotides. J. Mol. Biol., *13*: 373–398, 1965.
56. Sargadharan, M. G., Sarin, P. S., Reitz, M. S., and Gallo, R. C. Purification from human leukaemic cell, response to AMV 70S RNA and characterization of DNA product. Nature New Biol., *240*: 67–72, 1972.
57. Sato, T., DeHarven, E., and Friend, D. Increased virus budding from Friend erythroleukemic cells treated with DMSO, DMF and/or BrdU *in vitro*. VIth Int. Symp. on Comparative Leukemia Research, 1974, in press.
58. Schlom, J. and Spiegelman, S. Simultaneous detection of reverse transcriptase and high molecular weight RNA unique to oncogenic RNA viruses. Science, *174*: 840–843, 1972.
59. Scolnick, E. M., Parks, W. P., Todaro, G. J., and Aaronson, S. A. Immunological characterization of reverse transcriptase. Nature New Biol., *235*: 35–40, 1972.
60. Silagi, S., Beju, D., Wrathall, J., and Deharven, E. Tumorigenicity, immunogenicity and virus production in mouse melanoma cells treated with 5-bromodeoxyuridine. Proc. Natl. Acad. Sci. U.S., *69*: 3443–3447, 1972.
61. Stansly, P. G. and Soule, H. D. Transplantation and cell free transmission of a reticulum-cell sarcoma in BALB/c mice. J. Natl. Cancer Inst., *29*: 1083–1106, 1962.
62. Steeves, R. A., Bennett, M., Mirand, E. A., and Cudkowicz, G. Genetic control by the W locus of susceptibility to (Friend) spleen focus-forming virus. Nature, *218*: 372–374, 1968.
63. Steeves, R. A., Eckner, R. J., Bennett, M., Mirand, E. A., and Trudel, P. J. Isolation and characterization of a lymphatic leukemia virus in the Friend virus complex. J. Natl. Cancer Inst., *46*: 1209–1218, 1971.
64. Steeves, R. A., Eckner, R. J., Mirand, E. A., and Priore, R. L. Rapid assay of murine leukemia virus helper activity for Friend spleen focus-forming virus. J. Natl. Cancer Inst., *46*: 1219–1228, 1971.
65. Steeves, R. A., Mirand, E. A., Bulba, A., and Trudel, P. J. Spleen foci and polycythemia in C57Bl mice infected with host-adapted Friend leukemia virus. Int. J. Cancer, *5*: 346–356, 1970.
66. Stephenson, J. R., Reynolds, R. K., and Aaronson, S. A. Isolation of temperature-sensitive mutants of murine leukemia virus. Virology, *48*: 749–756, 1972.
67. Swetly, P. and Ostertag, W. Effect of interferon on release of Friend virus and on DMSO induced stimulation of hemoglobin synthesis in erythroleukemic cells in culture. Nature, 1974, in press.

68. Taylor, J. M., Varmus, H. E., Faras, A. J., Levinson, W. E., and Bishop, J. M. Evidence for non-repetitive subunits in the genome of Rous sarcoma virus. J. Mol. Biol., *84*: 217–221, 1974.
69. Watson, J. D. The structure and assembly of murine leukemia virus: Intracellular viral RNA. Virology, *45*: 586–597, 1971.
70. Weintraub, H., Campbell, G. L. M., and Holtzer, H. Identification of a developmental program using bromodeoxyuridine. J. Mol. Biol., *70*: 337–350, 1972.
71. Wilson, S. H. and Kuff, E. A novel DNA polymerase activity found in association with intracisternal A-type particles. Proc. Natl. Acad. Sci. U.S., *69*: 1531–1536, 1972.
72. Yang, S. S. and Wivel, N. Analysis of high-molecular-weight ribonucleic acid associated with intracisternal A particles. J. Virol., *11*: 287–298, 1973.
73. Yoshikura, H. Host range conversion of the murine sarcoma-leukemia complex. J. Gen. Virol., *19*: 321–327, 1973.

Discussion of Paper by Drs. Ostertag et al.

DR. NAKAHARA: Did I understand you correctly that BrdU-resistant cells lose their ability to produce C-type particles? Can such cells, without C-type particles, transmit Friend disease in susceptible mice?

DR. OSTERTAG: Some BrdU-resistant cells have largely lost their ability to produce C-type particles. Some other resistant clones release BrdU-resistant C-type viruses. The cells of cell line B8, which release 10^{-3}–10^{-4} times as much C-type particles as F4, if injected into susceptible mice grow as tumor cells similar to the virus negative cell lines of Fieldsteel *et al.* (*16*). The few C-type particles in the cell free supernatant of B8 cells induce erythroleukemia in DBA-2 mice at a very low rate.

DR. TOMKINS: Why are TK^- cells unable to induce C-type particles?

DR. OSTERTAG: In TK^- cells BrdU induces TK activity and release of endogenous virus. The activation of TK leads to incorporation of BrdU into cellular DNA. This seems to be a prerequisite for the induction of endogenous viruses. The lack of DMSO to induce an increase in virus release in TK^- cells is possibly due to the lack of TK activity in these cells or due to complete loss of the LLV helper virus.

DR. HOLTZER: Could you please state what happens to C-type viruses after: 1) BrdU alone, 2) DMSO alone, and 3) BrdU+DMSO. What happens to the FV under the above conditions?

DR. OSTERTAG: BrdU decreases virus release in TK^+, BrdU-sensitive cells. In TK^- BrdU-resistant cells C-type viruses can be induced by BrdU. DMSO alone does increase virus release in TK^+ and in partially TK^- (B16) cells. It does not induce C-type particles in the TK^- lines B8, B4, or B13. In all BrdU-resistant lines an increase in intracisternal A particles can be observed after adding BrdU. BrdU and DMSO if added together, induce endogenous virus in TK^- cells at the same rate as if BrdU is added alone. The experiment has not been done for TK^+ cells. The virus release was measured by the spleen focus-forming assay. Spleen focus formation is the characteristic property of the FV complex. These results were confirmed by the XC assay and by electron microscopy.

Erythrodifferentiation of Cultured Friend Leukemia Cells

Yoji IKAWA,*[1] J. ROSS,*[2] P. LEDER,*[2] J. GIELEN,*[2] S. PACKMAN,*[2] P. EBERT,*[3] K. HAYASHI,*[4] and Haruo SUGANO*[1]

Cancer Institute, Tokyo, Japan,[1] and National Institute of Child Health and Human Development,*[2] National Cancer Institute,*[3] and National Institute of Dental Research,*[4] NIH, Bethesda, Maryland, U.S.A.*

Abstract: In an attempt to investigate leukemogenesis by Friend leukemia virus (FLV), we found that the virus arrested erythrocytic maturation of erythropoietin-induced proerythroblasts to proliferate neoplastically at an immature stage. One of the clonal proerythroblastoid leukemia lines (T-3-Cl-2), isolated from the focal splenic lesions of DDD mice, does not undergo spontaneous differentiation, but does undergo erythrodifferentiation when treated with dimethylsulfoxide (DMSO). Kinetics of erythrodifferentiation of T-3-Cl-2 cells was studied in globin mRNA, globin, erythrocyte membrane-specific antigens, δ-aminolevulinic acid synthetase, and hemoglobin. Uninduced T-3-Cl-2 cells contain little or no detectable globin mRNA. Within 32 hr after addition of 1% DMSO to the culture medium, however, the RNA, hybridizable with ^{3}H-DNA transcribed from globin mRNA of reticulocytes, appears. This RNA has a sedimentation coefficient of 9S in sucrose gradients, and reaches a constant level within 76 hr. Considering the non-reiterating character of the globin genes, their expression during differentiation is best explained in terms of transcriptional activation. Actinomycin D and cycloheximide caused abrupt cessation of globin mRNA accumulation, which suggested *de novo* synthesis of globin mRNA at every moment. During differentiation, infectious FLV increased up to 200 times. Tumorigenicity of differentiated T-3-Cl-2 cells decreased, while inducibility of classical Friend disease increased in the transplantation study. Since erythroblastosis-inducing capacity of FLV depends upon its host erythroid cells, something carried out by the virions from those erythroid cells appears responsible in leukemogenicity, which prompted us to examine the FLV virions from induced cells. In those virions, globin mRNA was found associated with 60S viral RNA. Whatever the function, our observations raise at least the formal possibility that host mRNA's can be conveyed by the RNA tumor viruses.

There exist methodological difficulties in the study of carcinogenesis. One of them lies in the characteristics of cancer occurrence; low efficiency of cancerization at the cell level. Cancer causes tremendous problems in our life, although detectable malignant conversion of cells in our body is rarely occurring—very infrequent, especially when we account for the number of the cells we have. Malignant transformation of a single cell out of billions of cells can result in killing the host. It is considered methodologically impossible to know every factor to cause malignant transformation in such a particular single cell among many other cells, most of which would never become malignant within the rest of the life span of the host.

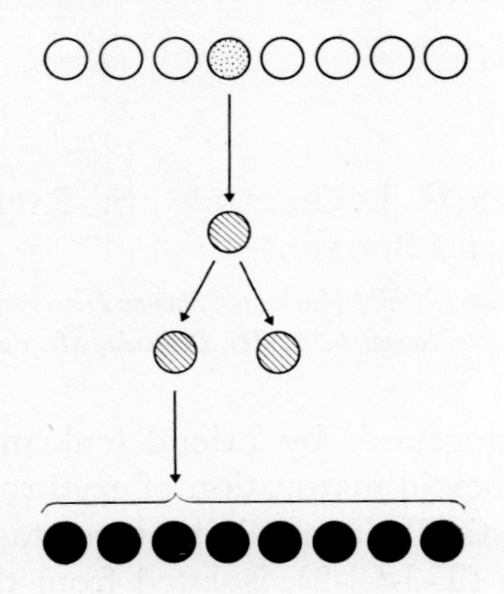

Suppose the white circles in the illustration are normal cell, and the black circles are malignant. It appears unreasonable to compare the white ones with the black ones to seek for certain etiological factors to cause malignant conversion, because almost all the white ones would never become malignant, and only the particular dotted circle, supposedly premalignant, among many white ones may be expressed as a malignant cell. The white ones are too far from the black ones to compare with. Accordingly, such investigations as electron microscopic or biochemical study of the process of cancerization are methodologically unwarranted. Suppose we have an experimental system, in which we can obtain cells of a single clone out of one million cells one month after initiation of a certain carcinogenic treatment. To analyze the mechanism of transformation in those cells deriving from the transformed single cell clone by comparing them with the other cells, most of which would not transform, does require premises that the particular cell to be transformed is exactly equivalent to other cells before transformation, and its principal nature is preserved during multiplication.

In this respect, it was necessary in the study of carcinogenesis to establish a system, in which transformation takes place in high efficiency within a short period. The system of Friend virus could bring a reconciliation to this need to some extent. The virus could initiate neoplastic transformation of erythrocyte precursors within a day or two to result in a fatal erythroblastosis.

Historical Review of Friend Virus

In 1957 Friend (*10*) described the recovery of a virus from the enlarged spleen of a Swiss mouse which had been inoculated with a cell-free extract of Ehrich ascites tumor cells. The rationale of this experiment was based on Selby's electron microscopic particles in Ehrlich ascites tumor cells which were later identified as reovirus particles. The electron micrographs of Friend virus particles shown by de Harven in 1958 (*7*) were consistent with classical type C particles, 100 nm in diameter, with dense nucleoids.

Friend virus, routinely prepared as 10% homogenate of the virus-infected, enlarged spleens, induces splenomegaly in susceptible mice about a week after intraperitoneal inoculation. This splenic lesion is followed by the appearance of blastic cells in the peripheral blood, and diffuse infiltration of the blastic cells in the liver, terminating the hosts due to hemorrhagic tendency.

The original Friend virus was kindly sent as a lyophilyzed sample in 1960 by Dr. C. Friend to Dr. Waro Nakahara. The Swiss mice in our hands were not highly susceptible to the virus, but Kasuga found that ddOM mice were highly responsive to the virus. The virus was passed through ddOM mice until 1965, and then through DDD mice at the Cancer Institute, Tokyo, by routine intraperitoneal injection of supernatant of the infected spleen homogenate.

Friend Disease: Reticulum Cell Sarcoma, or Erythroid Neoplasm?

Friend virus causes splenomegaly in DDD mice within a week or two after inoculation. The DDD mouse is an inbred albino mouse strain established at the Institute of Medical Science, University of Tokyo. Following splenomegaly the blastic cell count in the peripheral blood increases up to 500,000/mm^3 in association with an increased erythroblast count.

The description of the above disease is detailed in previous reports by Friend (*10*), Metcalf *et al.* (*32*), and Kasuga and Oota (*29*). The enlarged spleen consisted of large polygonal cells. Hepatomegaly occurred in the splenectomized mice after infection with the virus. Until 1965, the disease was interpreted as an undifferentiated reticulum cell sarcoma associated with erythroblastosis.

The nature of the Friend virus-induced disease as an undifferentiated reticulum cell sarcoma was also supported by the morphologic character of the transplantable tumors derived from the splenic and hepatic lesions of Friend disease. Since the establishment of the transplantable tumors, histological studies have been concerned primarily to those transplantable tumors (*6*). Many classical pathologists have raised questions regarding the neoplastic character of the primary splenic lesions, and have forced the researchers to study on the transplantable tumors as qualifying neoplasms.

Only Zajdela (*54*) reported in 1962 that Friend virus induced an erythroid neoplasia, because the lesions lacked reticulin fibers, and incorporated radio-iron. Prior to this, Mirand (*33*) reported increase of radio-iron uptake in the Friend virus-infected mice, but this finding was considered due to associated erythroblastosis. Amano also suggested in 1962 an erythroid orgin (personal communication).

Soule *et al.* in 1966 (*46*) observed *in vitro* erythrocytic maturation of the cells of a similar leukemia induced by Stansly leukemia virus. Rauscher in 1962 (*39*) described a murine leukemia virus which has also similar properties of Friend virus.

In Rauscher disease the primary splenic lesions were the principal materials for the study of its histogenesis. No transplantable nor cultured lines have ever been isolated from Rauscher virus-induced erythroid lesions. In 1966 Yokoro and Thorell (*52*) determined hemoglobin concentration on impression smears of blastic cells from the hepatic lesions of Rauscher disease by a microspectrophotometric technique. Dunn and Green (*8*) described the disease as reactive proliferation of erythroid and myeloid cells, emphasizing its involvement of only hemopoietic organs. Pluznik and Sachs in 1966 (*36*) showed a high incidence of Rauscher virus-induced disease in anemic mice which had a high population of immature erythroblasts.

Friend *et al.* in 1965 (*11*) isolated a Friend leukemia cell line which differentiated along the erythrocytic line. Rossi and Friend in 1966 (*42*) reported the increase of radio-iron uptake in the lethaly irradiated mice inoculated with cultured Friend leukemia cells. Patuleia and Friend (*35*) demonstrated in 1967 erythrocytic maturation of a Friend leukemia cell line in the soft agar medium. Friend *et al.* (*12*) further reported in 1971 an enhancement of erythrocytic maturation of the same cell line by adding dimethylsulfoxide (DMSO) to the cultured medium. These serial findings by Friend and her colleagues indicate that a cell line derived from Friend virus-induced leukemia in DBA/2 mice has a potentiality to mature towards erythrocytes.

Friend Disease as Malignant Erythroblastosis

Ikawa *et al.* (*17*) reported in 1966 on the early focal splenic lesions of Friend disease induced by inoculation with the diluted virus material. Prior to this, Axelrad and Steeves (*3*), and Pulznik and Sachs (*37*) applied these splenic foci for the assay of Friend virus and Rauscher virus, respectively. The number of splenic foci was proportionate to the quantity of the virus inoculated. Ikawa *et al.* (*18*) found that these spleen foci were good materials for the study of histogenesis, and described that they appeared in the red splenic pulp a few days after infection.

Electron microscopic studies of early spleen foci of Friend disease revealed two types of blastic cells, one simulating proerythroblasts, and the other, basophilic erythroblasts, both shedding viruses (*20*). Infrequent budding of virus particles was also observed in the cells morphologically corresponding to normoblasts. Impression smears of spleen foci of Friend disease stained with Giemsa solution revealed classical proerythroblasts with occasional small normoblasts.

Pulse labeling of the spleen focus cells with tritiated thymidine showed a labeling index higher than 75%, showing their short generation time comparable to that of erythroblasts, allowing the early lesions of Friend disease to grow rapidly to replace the entire spleens (*25*).

In 1967 Ikawa and Sugano (*19*) observed early lesions of Friend disease in the bone marrow of splenectomized mice and concluded that the hepatomegaly in splenectomized mice was secondary to the bone marrow lesions.

Ikawa and Sugano established in 1966 (*16*) six transplantable Friend leukemia strains after hundreds of attempts of transplanting 7- to 12-day spleen foci to syngeneic mice. Those mice successfuly taking the transplants showed no splenomegaly, suggesting minimal infectious virus in the transplants. Those mice rejecting the

TABLE 1. High Heme Synthesis in Transplantable Friend Ascites Tumor Cells

Cells	No.	^{59}Fe incorporation into heme per mouse cpm (mean)		^{59}Fe incorporation into heme per milliliter cell cpm (mean)	
SFAT-1-15	2	2,830	(1,847)	5,440	(3,360)
SFAT-1-15	4	865		1,280	
SFAT-3-13	6	23,500	(32,300)	62,700	(86,900)
SFAT-3-13	7	49,100		117,000	
SFAT-3-13	8	24,300		81,000	
SFAT-4-12	9	15,350	(12,095)	45,200	(31,100)
SFAT-4-12	10	8,840		17,000	
SFAT-5-10	14	20,800	(22,300)	80,000	(68,350)
SFAT-5-10	15	23,800		56,700	
SFAT-6-6	17	26,120	(21,510)	15,300	(28,750)
SFAT-6-6	20	16,900		42,200	

DDD mice, 1 week after inoculation of 10^7 SFAT cells, were injected intraperitoneally with $^{59}FeCl_3$, 1 μCi in 0.5 ml saline. Four hours later, SFAT cells were collected, washed repeatedly, and assayed for *de novo* synthesized heme, crystalized by addition of carrier hemoglobin. (By courtesy of Drs. S. Sasa and F. Takaku)

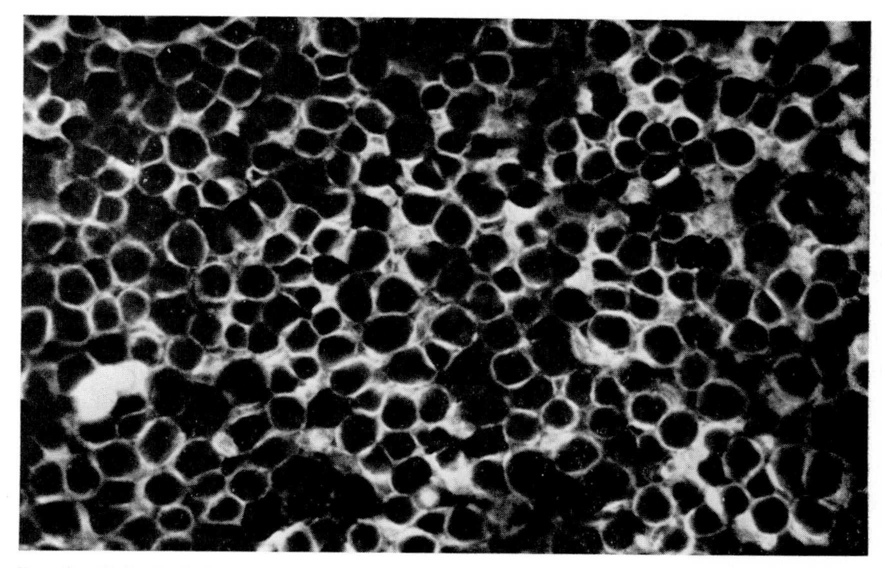

FIG. 1. Splenic lesions of FLV-induced leukemia stained with fluorescein isothiocyanate (FITC)-labeled antisera against erythrocyte membrane-specific antigens. Most Friend cells fluoresced.

transplants died of classical Friend disease. The subcutaneously transplanted tumors were called SFST-1 to -6 (spleen focus-derived solid tumor-1 to -6). All of them were converted to ascites forms, SFAT-1 to -6. SFAT-1, -2, -4, and -5 showed uniform monocytoid cells, whereas SFAT-3 and -6 consisted of immature erythroid cells at different stages. Sassa *et al.* (*43*) demonstrated a high heme synthesis in SFAT-3 and SFAT-6 cells (Table 1). Kasuga (*30*) showed the presence of globin in SFAT-3 cells by an immunofluorescent technique.

All of these serial experiments on Friend virus-induced leukemia cells which had been carried out by Ikawa and Sugano (*21*) proposed a hypothesis that Friend disease was a malignant erythroblastosis.

Recently, Ikawa *et al.* (*26*) obtained conclusive evidence that the cells neoplastically proliferating in Friend disease were of an erythroid nature. Adachi and Furusawa (*1*) purified normal erythrocyte membranes and obtained an erythrocyte membrane-specific antibody by immunizing guinea pigs. This specific antibody reacts only to erythrocytic cells from immature proerythroblasts to mature erythrocytes. Almost all the cells of both early, focal and advanced, diffuse splenic lesions of Friend virus-induced leukemia of DDD mice reacted to this specific antibody against normal erythrocyte membrane (Fig. 1). This result indicates that the cell membrane of those proliferating cells in Friend disease have the same proteins as normal erythroid cells.

Maturation Arrest of Target Cells by Friend Virus

The proliferating cells in the spleen after infection with Friend virus are thus found erythroid. A question raised here; "which cells in the red splenic pulp were forced to proliferate and cause fatal erythroblastosis; stem cells or proerythroblasts?" The similar question was also raised by Pulznik *et al.* (*36*).

Repeated syngeneic red blood cell transfusion resulted in polycythemia or a decrease in the population of erythroblasts in the spleen. An intraperitoneal injection of six cobalt units of erythropoietin (supplied by Dr. T. Asano, Teikokuzoki Pharmaceutical Co., Ltd., Tokyo) induced synchronous proliferation of proerythroblasts along the splenic trabeculae in the red splenic pulp 24 hr after inoculation. This was followed by the appearance of smaller erythroblasts. Seventy-two hours after injection of erythropoietin into the polycythemic mice, the erythropoietic situation becomes almost normal in the spleens. If erythropoietin is injected into normal mice, the population of erythroblasts in the spleen increases, resulting in a moderately enlarged spleen of about 0.5 g (normal spleen weight; 0.12 g).

When Friend virus was injected into mice with different degrees of erythropoiesis, splenomegaly paralleled with the quantity of proerythroblasts at the time of infection, which suggested that the erythropoietin-induced proerythroblasts are one of the target cells of Friend virus (*22*, *24*, *25*).

The proerythroblasts synchronously induced by erythropoietin in the spleens of DDD mice were labeled with tritiated thymidine and were then followed autoradiographically with and without Friend virus. Approximately 90% of the spleen focus cells, 24 and 48 hr after Friend virus infection, showed silver grains and were found to be formed by the proliferation of the labeled erythropoietin-induced proerythroblasts (Table 2). Without Friend virus, the labeled proerythroblasts differentiated into smaller erythroblasts which were positive for silver grains autoradiographically.

Approximately 70% of erythropoietin-induced proerythroblasts differentiate into smaller erythroblasts, while 30% of them remain as reserve cells, some of which remain undifferentiated even one week after stimulation by erythropoietin. When infected with freshly harvested Friend virus in a high titre, approxi-

TABLE 2. Autoradiographic Study of ^{3}H-thymidine-labeled, Erythropoietin (EP)-induced Proerythroblasts with or without Friend Virus Infection

Spleens (post infection)	Materials	No. of cells checked	No. of cells labeled (over 5 grains)	Labeling index
EP only	Proerythroblasts	228	210	92.1
	Other cells in red splenic pulp	586	24	4.1
FV after labeling (24 hr)	FV-induced SF cells	351	302	86.3
	Other cells in red splenic pulp	1,012	68	6.7
FV after labeling (48 hr)	FV-induced SF cells	311	283	91.0
	Other cells in red splenic pulp	982	83	8.5

FV: Friend virus. SF: spleen foci.

TABLE 3. Forty-eight-hour Friend Virus-induced Spleen Foci in Polycythemic Mice with Pulse-labeled, EP-induced Proerythroblasts

Mouse strain	No. of foci checked	Large proerythroblastoid cells[a]		Small erythroblasts[a]	
		No. of cells checked	No. of cells labeled	No. of cells checked	No. of cells labeled
DDD	46	2,622	2,361	111	104
BALB/c	17	796	653	393	325

a) In a section of the spleen material.

90% of erythropoietin-induced proerythroblasts remained undifferentiated (Table 3).

The maturation arrest of erythropoietin-induced proerythroblasts by infection with Friend virus was less conspicuous in BALB/c mice (Table 3), which might be one of the reasons for unsuccessful transplantation of focal splenic lesions of Friend disease in BALB/c mice. This finding may correspond to the remarkable tendency of Rauscher virus-induced erythroblastoid cells to differentiate into erythrocytes.

Recently Tambourin and Wendling (*50*) reported that Friend virus transforms erythropoietin responsive stem cells.

Establishment of Cultured Friend Leukemia Lines

Since January, 1968, Ikawa and Sugano (*16*) have maintained *in vitro* a Friend leukemia line, TSFAT-3 (or T-3), which originated from SFAT-3, an ascites Friend leukemia line. The cells grew in suspension in HAM F-12 medium (GIBCO) supplemented with 10% heat-inactivated calf serum. This T-3 cell line was cloned in soft agar medium according to the technique described by Patuleia and Friend (*35*), establishing T-3-Cl-1 and -2, both having an approximate doubling time of 18 hr. Although the cells are all round-shaped and are growing in suspension as shown in Fig. 2, stationary cultures give the best growth rate. Intermittent shaking disturbs multiplication of the cells (Y. Inoue and Y. Ikawa, unpublished data). A few fine intercellular bridges are occasionally observed between the cells, which may be the remnants of the cell division, and may communicate the cells.

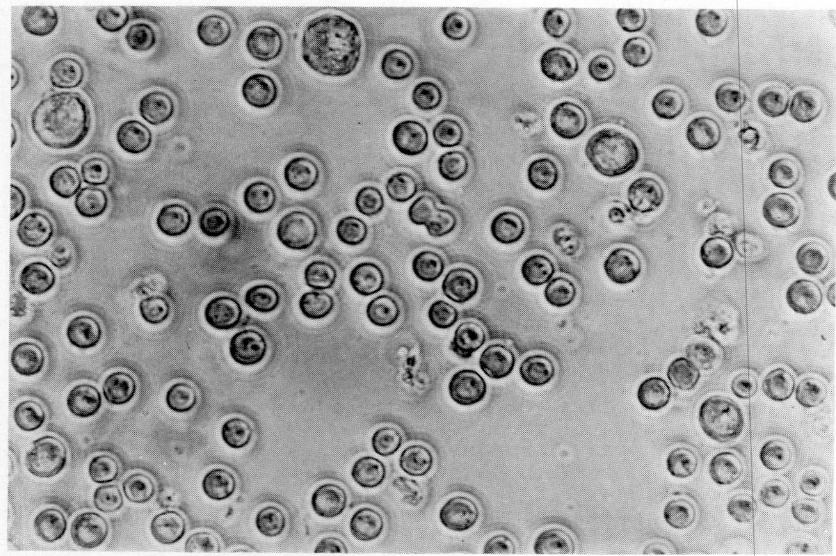

FIG. 2. T-3-Cl-2 cells growing in suspension.

When 10^5 cells of T-3-Cl-1 or -2 lines were inoculated intraperitoneally into syngeneic DDD mice, the cells grow in suspension as hemorrhagic ascites fluid. The Giemsa-stained smears showed erythroblastoid cells at various stages of maturation, which suggested a possibility of T-3-Cl-2 cells differentiating towards erythrocytes in the peritoneal cavity. Since T-3-Cl-1 or -2 cells still propagate infectious Friend leukemia virus (FLV), erythroid cells induced by the virus in the spleen can infiltrate into the peritoneal cavity. Accordingly the need to cultivate T-3-Cl-1 or -2 cells *in vivo* apart from the systemic body fluid occurred, which was realized by an intraperitoneal diffusion chamber culture.

SFAT-3 cells were occasionally positive for erythrocyte membrane-specific antigens, but T-3-Cl-1 or -2 cells were negative for them in the routine passage even after digestive enzyme treatment, including neuraminidase.

Erythrocytic Maturation of T-3-Cl-1 or -2 Cells in Peritoneal Diffusion Chambers (14)

Samples of T-3-Cl-1 or -2, 5×10^4 or 2×10^5 cells, were placed in a diffusion chamber, 13 mm in diameter, securely sealed with Millipore membrane filter of 0.45 μm pore size, and cultured in the peritoneal cavity of syngeneic or allogeneic mice for 4 to 9 days. The chambers were then taken out, cleaned by physiological saline, and one side of the filter membranes was broken with a needle. In each chamber a gelatinous mass was found, and the fluid content inside was yellowish or pinkish. The masses were imprinted on the glass slides for cytological examination, and then fixed for electron microscopy.

The imprinted cells stained with May-Grünwald Giemsa solution showed several stages of erythrocytic maturation. Most of the various-sized erythrocyte-

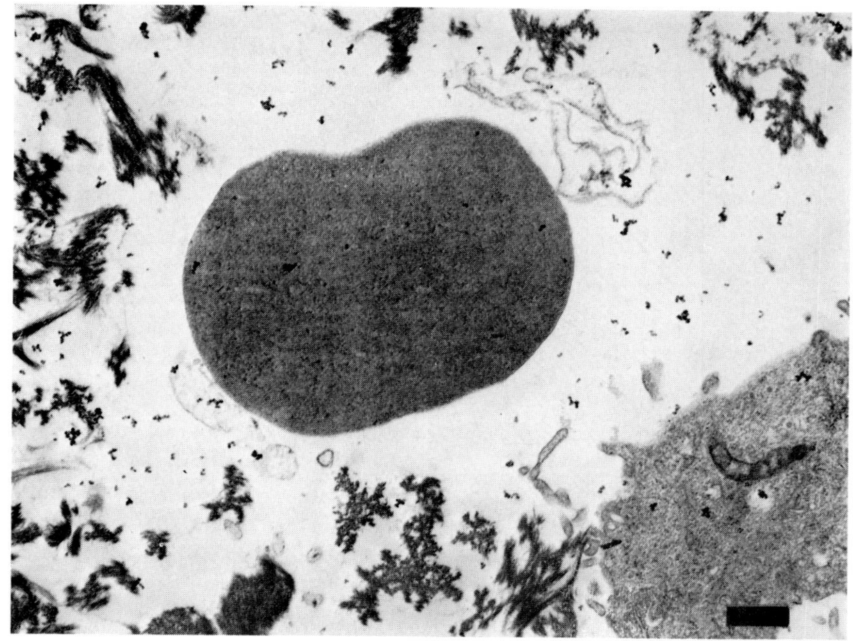

FIG. 3. Erythrocyte-like corpuscle induced from a T-3-Cl-1 cell in an intraperitoneal diffusion chamber. Electron-dense cytoplasm should be compared with that of the other cell fragment shown right lower. Fibrin fibers are also shown.

like corpuscles fluoresced when fixed and stained with the fluorescence-labeled antibody against normal erythrocyte membrane (*1*, *13*). Electron microscopy also revealed electron-dense erythrocyte-like corpuscles in the matrix containing fibrin fibers. These fibers appeared to infiltrate from the outside of the chamber (Fig. 3). The Friend leukemia cells remaining undifferentiated in the chamber showed virus particles budding and accumulating in the intracytoplasmic inclusions (*26*, *48*).

As to the shape of the erythrocyte-like corpuscles thus obtained in the diffusion chamber cultures, they are various in size, mostly round-shaped or spheroid with occasional tails. Biconcave corpuscles were infrequent.

Phagocytic activity of the Friend leukemia cells remaining undifferentiated was also observed with fibrin fibers and erythrocyte-like corpuscles as inclusions (*15*).

Erythrocytic Maturation in Vitro of T-3-Cl-2 Cells

After the success in inducing erythrodifferentiation of T-3-Cl-1 or -2 cells in the intraperitoneal diffusion chamber culture, several attemps were made to induce erythrodifferentiation *in vitro*. However, none of the attemps were successful including addition of erythropoietin, activated Vitamin B_{12}, *etc.*, to the culture medium.

Friend *et al.* (*12*) then succeeded in inducing erythrocytic maturation of their Friend leukemia cell lines by addition of DMSO into the culture medium. Ap-

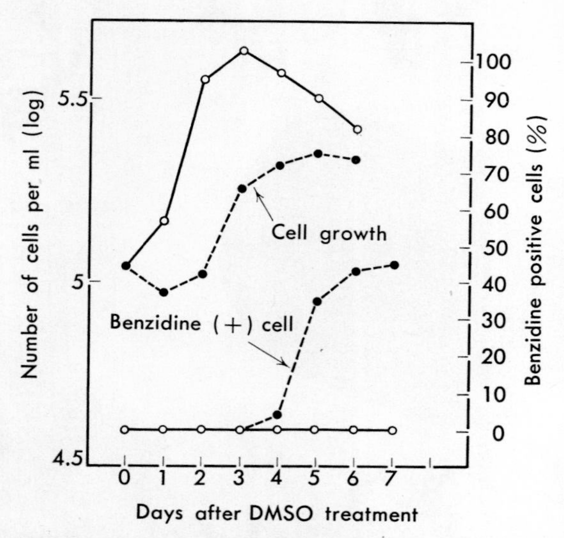

FIG. 4. Cell growth and hemoglobin production in DMSO-treated T-3-Cl-2 cells. ○ control; ● 2% DMSO.

plying their technique, T-3-Cl-1 or -2 cells were cultured in the medium containing 2% DMSO. Erythrocytic maturation was observed as Friend *et al.* (*12*) described. The cell pellets on day 6 had apparent hemoglobin color, and about 40% of the cells became positive for Ralph's hemoglobin staining (Figs. 4–7). Binucleated cells with intense benzidine-positivity (Ralph's staining) were frequently observed (Fig. 8). There were infrequent erythrocyte-like corpuscles (Fig. 9). Most of the differentiation-induced T-3-Cl-1 or -2 cells became positive for erythrocyte membrane-specific antigens as observed by an immunofluorescent technique (Fig. 10). Electron microscopically, on day 2 to 3 after addition of DMSO, frequent budding of type C virions were observed at the surface of some leukemia cells (*26*). E. de Harven also observed increased budding of type C particles from the surface of DMSO or dimethylformamide (DMF)-treated Friend leukemia cells by a scanning electron microscope (personal communication). On day 6 to 7, mature normoblast-like cells were frequently observed with a small number of type C particles budding, and accumulation of mature type C virions were observed intracysternally in occasional cells (*26*) (Fig. 11). Addition of DMSO at the concentration of 2% into the culture medium decreased the growth of Friend leukemia cells, but the titre of attenuated but infectious FLV in the culture fluid after sonication of the cells increased up to 200 times by a focus assay on sarcom virus-positive, leukemia virus-negative cells (*26*). This was also confirmed by inoculating the harvested virus to syngeneic DDD mice, which shortened the latent period.

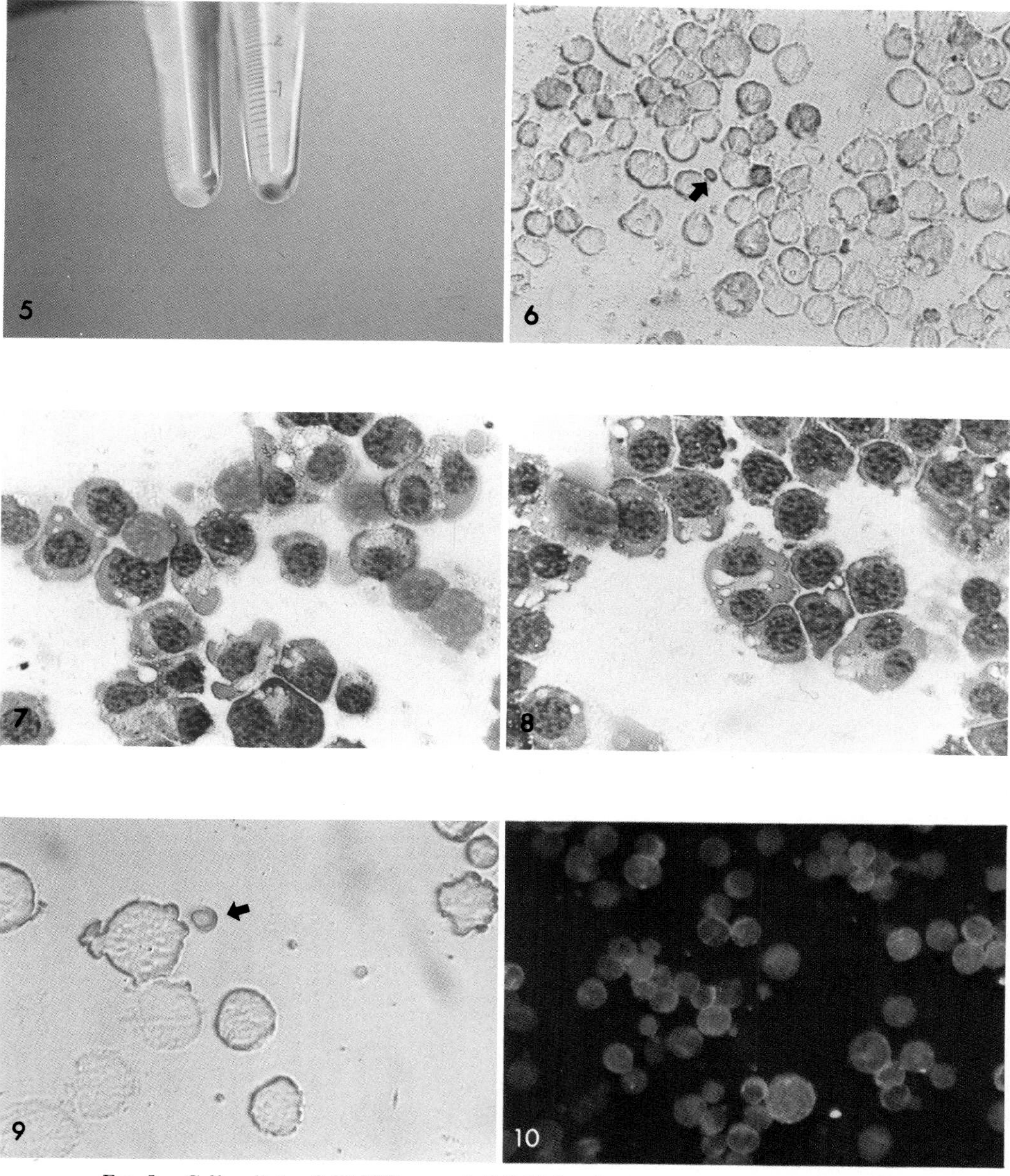

FIG. 5. Cell pellets of DMSO-treated T-3-Cl-2 cells showing hemoglobin color (right). The left whitish cell pellet is that of control cell.

FIG. 6. T-3-Cl-2 cells, 6 days after initiation of 1.5% DMSO treatment. Approximately 40% of the cells positive for hemoglobin (Hb) staining. An erythrocyte-like corpuscle (arrow) is shown.

FIG. 7. Normoblastoid cells, DMSO-induced, in T-3-Cl-2. Yellow-or gray-colored cells positive for Hb.

FIG. 8. A binucleated T-3-Cl-2 cell, intensely stained for Hb.

FIG. 9. A thick, refractile, erythrocyte-like corpuscle (arrow) induced.

FIG. 10. DMSO-treated T-3-Cl-1 cells stained with FITC-labeled antisera against erythrocyte membrane-specific antigens. Smaller cells intensely fluoresced.

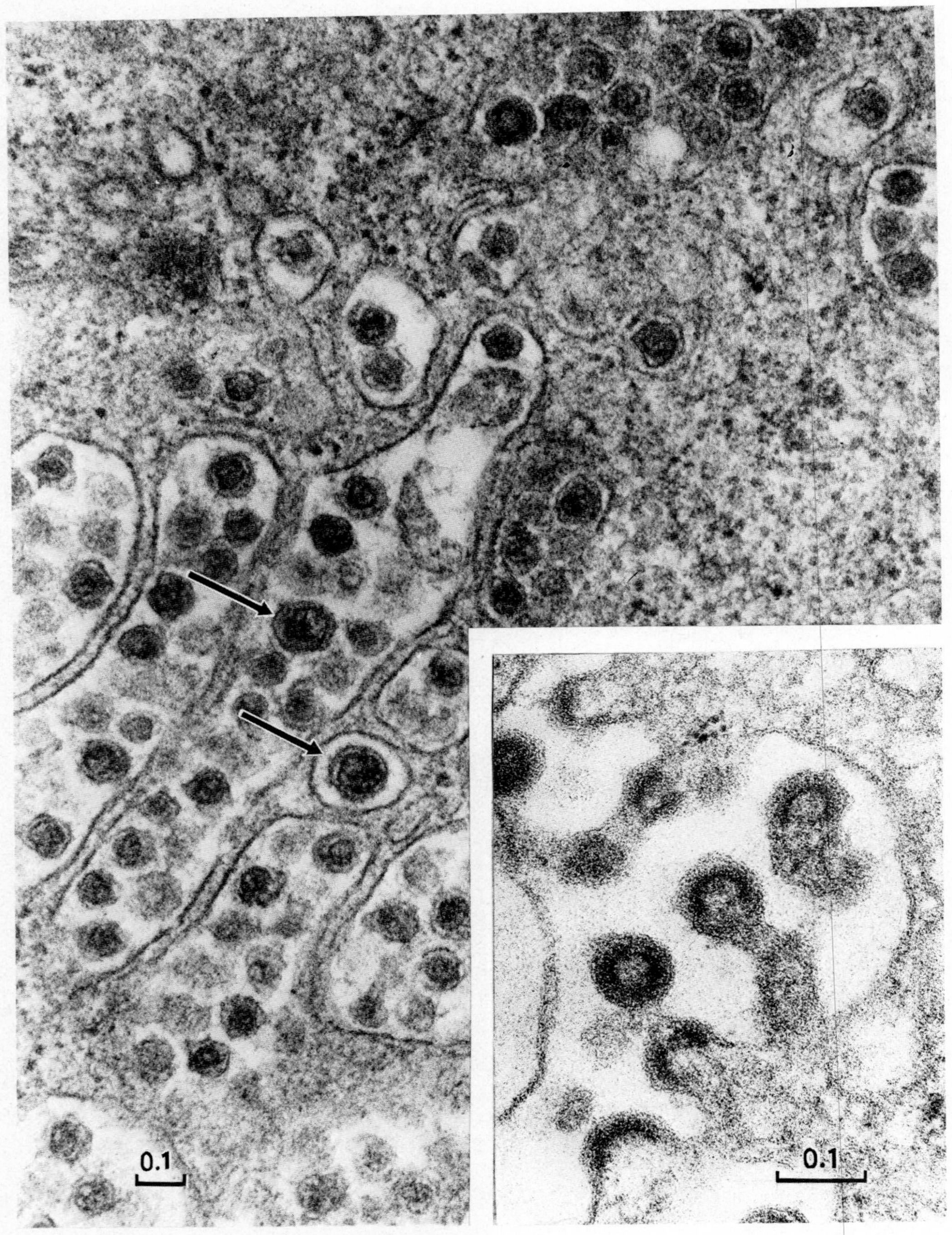

FIG. 11. Virions in differentiation-induced T-3-Cl-2 cells, so-called "maturation" of type C particles in the intracysternal spaces of a differentiation-induced T-3-Cl-2 cell. Large particles (arrows) are shown.

TABLE 4. Substances and Conditions Inducing Erythrocytic Maturation of Cultured Friend Leukemia Cells (T-3-Cl-1 Cells)

Substances or conditions	Effects[a]	Most effective doses	Range
DMSO	+++	2 %	0.5–5.0 %
N-dimethyl-rifampicin	+++	100 μg/ml	50–200 μg/ml
Bleomycin	+++	3 μg/ml	0.5–50 μg/ml
Ara-C	++	1×10^{-6} M	1×10^{-8}–1×10^{-5} M
Mitomycin C	++	1 μg/ml	0.5–5.0 μg/ml
Glycerol	++	5 %	2–10 %
Nitrogen mustard	+	1×10^{-6} M	1×10^{-7}–1×10^{-5} M
Vincristine	+	1×10^{-9} M	1×10^{-10}–1×10^{-9} M
5-Fluorouracil (5-FU)	+	1×10^{-5} M	1×10^{-6}–1×10^{-5} M
Culture at 32°C	+		
Culture without serum	+		
5-Bromodeoxyuridine (BrdU)	–		1×10^{-7}–1×10^{-3} M
Actinomycin D	–		0.5–50 ng/ml
X-ray irradiation	–		200–900 R
Puromycin	–		0.01–0.1 μg/ml
Cycloheximide	–		0.1–1.0 μg/ml
Vitamin B_{12}	–		0.5–5.0 μg/ml
Cyclic AMP and derivatives	–		0.1–100 μg/ml
Concanavalin A	–		10.0–50.0 μg/ml
Nonidet	–		0.001–0.1 %
Addition of excess serum	–		

a) As revealed by intensity of fluorescence when stained with FITC-labeled antisera against erythrocyte membrane-specific antigens.

Substances and Conditions Inducing Erythrocytic Maturation and Malignancy of Induced Cells

As reported previously, erythrocyte membrane-specific antigen, detectable by an immunofluorescence technique, was used as a marker of erythrocytic cells (*26*). Since this antigen is already present at an immature erythroblastic stage, and is virtually undetectable in stationary cultured T-3-Cl-1 or -2 cells, this marker was used for screening substances and conditions enhancing erythrocytic maturation of those cultured leukemia cells (*47*).

Among those substances which effectively induce erythrodifferentiation of T-3-Cl-1 cells are DMSO, N-dimethyl-rifampicin and Bleomycin. The former two chemicals induced favorably natural erythrodifferentiation with most intensive fluorescence of erythrocyte membrane-specific antigen on day 4 and 5 after initiation of addition of these chemicals; whereas Bleomycin induced the same antigen as early as on day 2 (data, not shown). Other substances and conditions to enhance erythrocyte membrane-specific antigen on T-3-Cl-1 cells less intensely are listed in Table 4.

T-3-Cl-1 cells have been maintained *in vitro* for years, but 83 % of syngeneic weanling male DDD mice which are intraperitoneally inoculated with 10^6 cells die either of retention of voluminous ascites or of splenomegaly induced by FLV

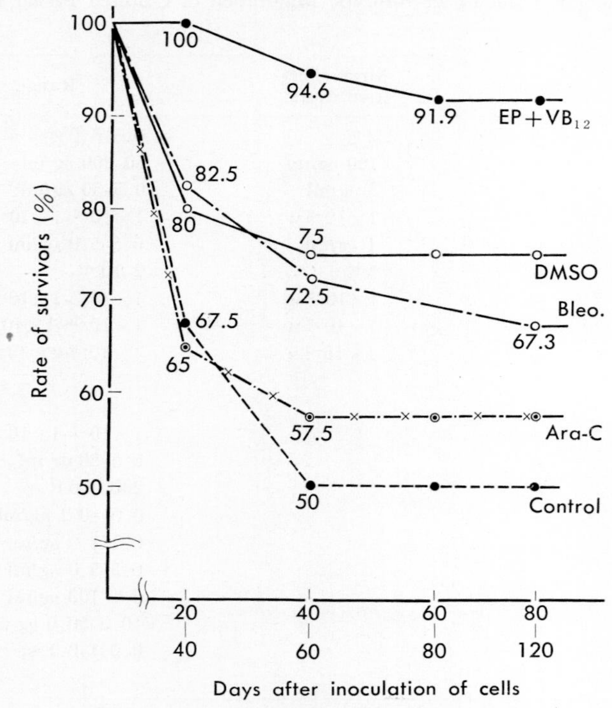

FIG. 12. Survival rate (in terms of ascites death) of DDD mice inoculated with T-3-Cl-1 cells after respective treatment of Ara-C, Bleomycin (Bleo.), DMSO, and EP+Vitamin B_{12} (VB_{12}).

realeased from the inoculated cells. Once adopted *in vivo*, T-3-Cl-1 line is taken in 100% of the syngeneic hosts.

Cultured T-3-Cl-1 cells were treated at most effective doses shown in Table 4 by cytosine arabinoside (Ara-C), Bleomycin, DMSO, or a combination of erythropoietin and Vitamin B_{12}, and the cells were harvested on day 4, washed repeatedly with culture medium, and the same number (10^6) of the viable cells in each treatment was inoculated respectively into the syngeneic hosts.

Those mice inoculated with the differentiation-induced cells showed higher survival rates in terms of ascites death—direct death of the inoculated cells (Fig. 12). Instead, they showed higher incidence of splenomegaly or Friend disease, suggesting increased induction of FLV from those differentiation-induced cells.

Accordingly the overall survival rate of the mice inoculated with the differentiation-induced T-3-Cl-1 cells is not markedly altered compared to that of control T-3-Cl-1 cells (Fig. 13). It may be noteworthy that a combination of erythropoietin, grade III, 10 unit/ml, and Vitamin B_{12}, 0.5 μg/ml resulted in long survivors, 35% of the recipient mice, although it did not damage T-3-Cl-1 cells *in vitro* at all.

Since these Friend leukemia cells are still propagating potent FLV, induction of differentiation in those cells did not alter the survival rate of the recipient mice.

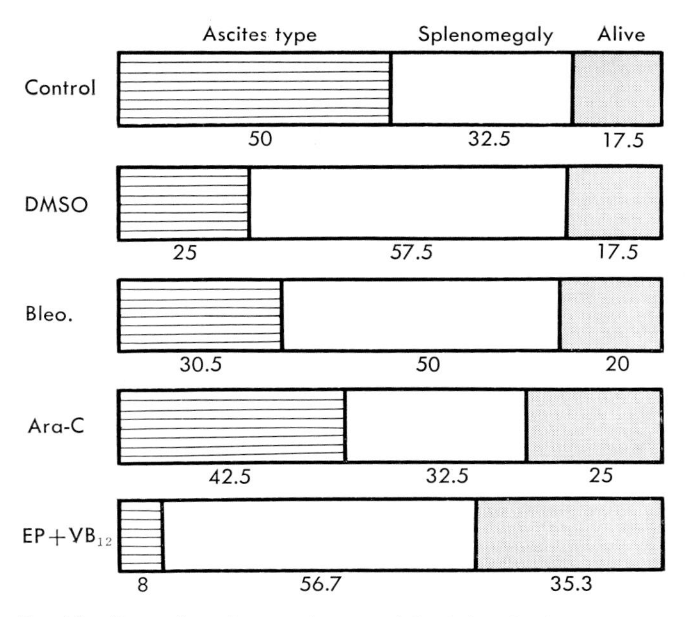

FIG. 13. Rate of survivors and cause of death in mice inoculated with treated T-3-Cl-1 cells.

If the leukemia cells are induced spontaneously or chemically without viral involvement, the similar attempt may result in marked improvement of the survival rate. SLA-5 cells, originating from *n*-butyl-nitrosourea-induced erythroid leukemia in a Sprague Dowley rat, were also treated with 2% DMSO and 10^6 viable cells were inoculated into the syngeneic hosts, which gave us more than 60% of long survivors (Y. Ikawa *et al.*, unpublished observations).

Kinetics of Erythrodifferentiation Detected by Several Erythrocytic Markers

As described previously, erythrodifferentiation of T-3-Cl-1 and -2 cells is detected in their morphology, erythrocyte membrane-specific antigens, and several markers relating to hemoglobin synthesis. Erythrocyte membrane-specific antigens can be analyzed by intensity of fluorescence when the cells are stained with fluorescein isothiocyanate (FITC)-labeled antisera against erythrocyte membrane-specific antigens (by courtesy of Dr. M. Furusawa, Osaka City University, and of Dr. K. Hayashi, NIDR, NIH, U.S.A.). Several pretreatment could not differentiate erythrocyte membrane-specific antigens of induced T-3-Cl-2 cells from erythrocyte membrane-specific antigens of normal erythrocytes (Table 5). ^{125}I-labeled antisera against erythrocyte membrane-specific antigens were also used for quantitation of erythrocyte membrane-specific antigens (Fig. 14). This technique, however, required skillfulness, since ^{125}I-labeled antisera reacted to some extent nonspecifically to DMSO-treated cells. DMF (0.5%)-treatment, another aprotic solvent to induce erythrodifferentiation of cultured Friend leukemia cells, was useful as a control.

As to the markers relating to hemoglobin production, "heme" and "globin"

TABLE 5. Effects of Pretreatment on Erythrocyte Membrane-specific Antigens Induced in T-3-Cl-2 Cells as Detected by FITC-labeled Antisera (Intensity of Fluorescence)[a)]

Pretreatment	T-3-Cl-2 cells	Normal erythrocyte
Acetone (5′ at r.t.)	+++	+++
CCl_4 (5′ at r.t.)	+	±
Ethanol (2′ at r.t.)	+++	+++
Methanol (2′ at r.t.)	−	±
Butanol (5′ at r.t.)	Not done	++++
2% formaldehyde (5′ at r.t.)	+++	+++
2.5% glutaraldehyde (5′ at r.t.)	+++	+++
DMSO (5′ at r.t.)	++	+++
Ether (5′ at r.t.)	++++	++++
Chloroform (5′ at r.t.)	+++	+++
Heating, 56°C, 30′	+++	+++
Air dried (control)	++++	++++

[a)] F/P molar ratio: 0.75. Staining titer=1:4 (by UG-1 exciter system), 1:8 (by interference filter) Inhibition tests by unlabeled antibody: 1 step, inhibited; 2 step, inhibited

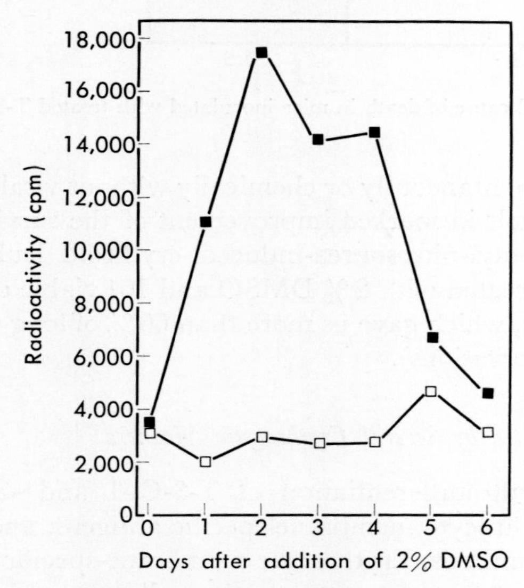

FIG. 14. Erythrocyte membrane-specific antigens enhanced in differentiation-induced T-3-Cl-2 cells. Cell line: T-3-Cl-2. 2% DMSO: ■ (+); □ (−).

should be examined. As to the heme synthesis, Sassa *et al.* (*43*) reported a high radio-iron incorporation into heme crystals synthesized by SFAT-3 cells, an ancestor ascitic line of T-3-Cl-1 and -2 cells. Hemoglobin production (Ralph's hemoglobin staining) in DMSO-treated cells was already introduced (Fig. 4).

δ-Aminolevulinic acid (δ-ALA) synthetase, one of the enzymes involving heme synthesis, was also enhanced in T-3-Cl-2 cells by addition of DMSO to the culture medium. This was carried out in collaboration with Dr. P. Ebert, NCI, Bethesda, Md., U.S.A. (*9*). One of the examples is shown in Fig. 15. Similar studies have been done in T-3-Cl-1 cells (*49*).

TABLE 6. Reaction of Friend Leukemia Cells to FITC-anti-mouse Globin Antibody (8×Dilution)

1% DMSO / Days after initiation of treatment	T-3-Cl-2		DBA/Fr	
	+	−	+	−
1	±	±	+ (10–20)	+ (10–20)
2	+ (1–5)	±	+ (20)	+ (20)
3	+ (10–20)	± (10)	⧺ (40)	+ (30–40)
4	⧺ (30–40)	± (20)	⧻ (80)	⧺ (80)
5	⧺ (80–90)	+ (20–30)	⧻ (80–90)	⧺ (60)
6	⧺ (70–80)	+ (20–30)	⧺ (60–70)	⧺ (60)

E/P=2.0; −∼⧻, intensity of fluorescence. Figures in parentheses indicate % positive cells.

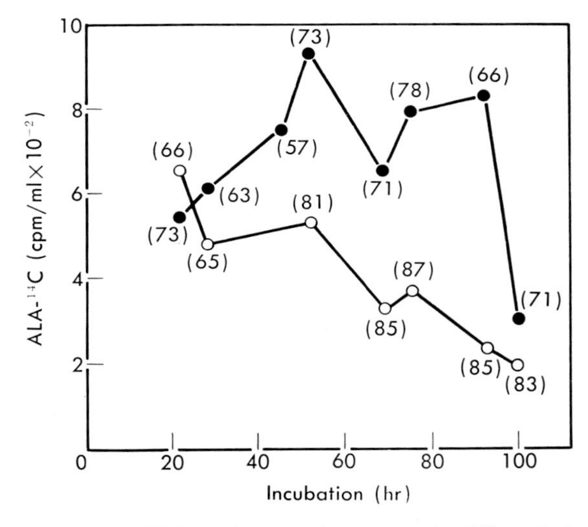

FIG. 15. δ-ALA synthetase enhancement in differentiation-induced T-3-Cl-2 cells (% viable cells). ● 1.5% DMSO; ○ control.

As to globin synthesis, fluorescein-labeled antisera against purified mouse globin was used on both T-3-Cl-2 cells and DBA/Fr cells, another Friend leukemia line originating from a DBA/2 mouse, always showing erythrocytic maturation. Globin synthesis reached to the maximum on day 5 of DMSO treatment (Table 6). As to the phasing of these markers, δ-aminolevulinic acid synthetase increased first, then erythrocyte membrane-specific antigens, and finally globin.

Prior to globin synthesis, globin mRNA should be increased in the treated cells. During Ikawa's fellowship tenure in NCI, NIH, Bethesda, Md., U.S.A., he met Dr. H. Aviv, NICHD, NIH (*2*), who isolated and purified biologically active globin mRNA from mouse reticulocytes by oligo-dT cellulose column, and Dr. J.

Ross, NICHD, NIH, (*40*), who obtained tritiated DNA complimentary to mouse globin mRNA by applying *in vitro* reverse transcriptase from avian myeloblastosis virus. This mouse globin complementary DNA (cDNA) was just prepared to be applied to the differentiation-inducible T-3-Cl-2 cells.

Globin mRNA Induction in T-3-Cl-2 Cells

In order to determine whether the erythroid change in T-3-Cl-2 cells induced by DMSO is accompanied by an accumulation of globin mRNA, total cellular RNA was prepared from DMSO-treated cells at various times after induction. In those cellular RNA probes, the amount of globin mRNA was quantitated in their hybridizability with ^{3}H-DNA complementary to globin mRNA from mouse reticulocytes (Fig. 16). Uninduced cells contain little or no detectable globin mRNA (*41*). Detailed studies indicate that globin mRNA can be detected about 32 hr after the adition of DMSO. The relative amount of globin mRNA continues to accumulate for about 4 days and reaches a constant level. This RNA can be digested with RNase, similarly to polysomal RNA of mouse reticulocytes (*41*). Additional studies (J. Ross *et al.*, unpublished data) indicate that these globin mRNA sequences can be absorbed to oligo-dT cellulose, suggesting that they contain poly (A) sequences. Sucrose gradient analysis indicate that globin sequences have a sedimentation coefficient of 9S (*41*).

In addition, hybridization kinetics analyses can be used to quantitate the amount of globin mRNA in these cells as compared to mature mouse reticulocytes.

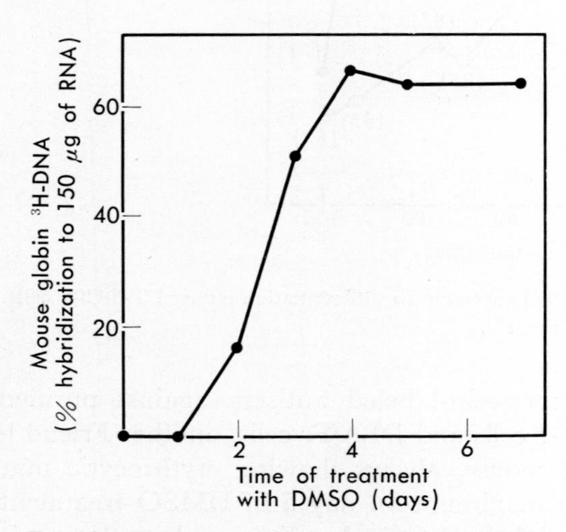

FIG. 16. Time course of induction of globin mRNA after treatment with DMSO. Logarithmically growing T-3-Cl-2 cells were passed to fresh growth medium at a concentration of approximately 0.5×10^5 cells per ml. Twenty-four hours later (day 0), 1.0% (v/v) DMSO was added. At daily intervals thereafter aliquots of cells were harvested, and total RNA was prepared from them. Approximately 150 μg of RNA was incubated to globin ^{3}H-cDNA for 24 hr. The amount of DNA hybridized was measured by determining the percent of the input counts which were resistant to the single-strand specific deoxyribonuclease S_1.

TABLE 7. Quantitation of Globin mRNA in Mouse Reticulocytes and DMSO-induced T-3-Cl-2 Cells

Source of RNA	$C_0t_{1/2}$	Fraction globin mRNA	Total RNA/cell (pg)	Globin RNA/cell (pg)	Globin mRNA/cell (molecules)
9 S globin mRNA (standard)	1.6×10^{-3}	1.0	—	—	—
Reticulocyte (total)	5.7×10^{-2}	0.028	0.25	7×10^{-3}	21,000
T-3-Cl-2 cell (total, induced)	7.7×10^{0}	0.0002	11.1	2.2×10^{-3}	6,600

Quantitation of globin sequences in the various RNA preparations was calculated using the $C_0t_{1/2}$'s derived by analyses described by Leder *et al.* (*31*). The total RNA content of each cell was determined as described by Ross *et al.* (1973). A molecular weight of 200,000 was assumed for the globin mRNA.

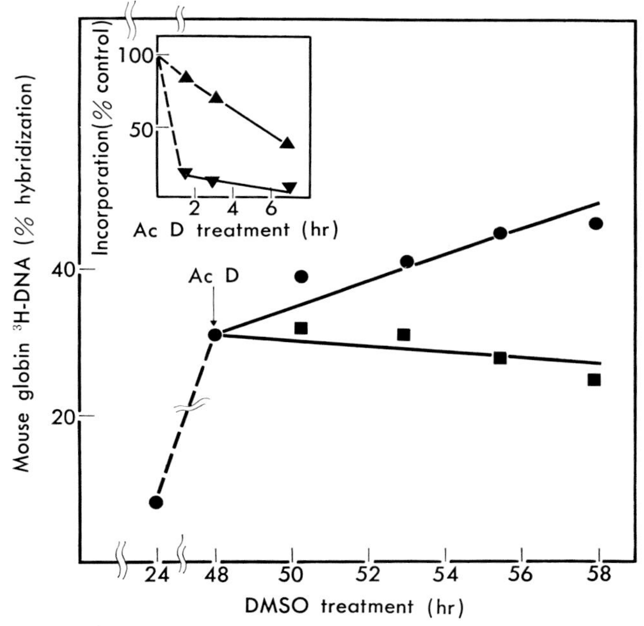

FIG. 17. The effect of actinomycin D (Ac D) on the continued accumulation of globin mRNA in DMSO-treated T-3-Cl-2 cells. T-3-Cl-2 cells were grown logarithmically for 48 hr in the presence of 1% (v/v) DMSO. At that time actinomycin D, dissolved in 50% (v/v) ethanol, was added to cultures at a final concentration of 0.3 μg/ml. An equivalent volume of ethanol (0.03 ml) was added to control cultures. At intervals thereafter cells were harvested, and cytoplasmic RNA was prepared. In control cultures the cell number increased approximately 30% between 48 and 58 hr; in actinomycin-treated cultures the cell number remained stationary for the first 7.5 hr and then decreased by about 25% over the subsequent 2.5 hr (J. Ross and J. Gielen, unpublished observations). Greater than 90% of the cells remained viable in both treated and untreated cultures (J. Ross and J. Gielen, unpublished observations). Forty micrograms of RNA was incubated with globin cDNA for 22 hr, and the percent hybridization was determined. The arrow indicates the time of addition of actinomycin D. ● no actinomycin D; ■ actinomycin D, 0.3 μg/ml. Inserts: incorporation of ^{14}C-amino acids and ^{3}H-uridine into acid-precipitable material. At the times indicated after actinomycin D addition, duplicate 4 ml aliquots of cells were pulse-labeled with ^{3}H-uridine and ^{14}C-amino acids. The amount of precursor incorporated into perchloric acid-precipitable material was determined as previously described (*38*). ▲ amino acid incorporation; ▼ uridine incorporation.

The data are summarized in Table 7. Approximately 6,000 globin mRNA molecules are present per induced T-3-Cl-2 cell, an amount one-third that in circulating mouse reticulocytes (*31*).

This accumulation of globin mRNA sequences could occur as a result of transcriptional activation of globin genes, or it could result from stabilization of *de novo* synthesized globin mRNA, which, in the undifferentiated cells, might be degraded shortly after it had been synthesized. In any case, globin mRNA could be synthesized *de novo* during erythrodifferentiation. Alternatively, undifferentiated cells might contain a pool of globin mRNA undetectable by the hybridization technique —double-stranded, for example.

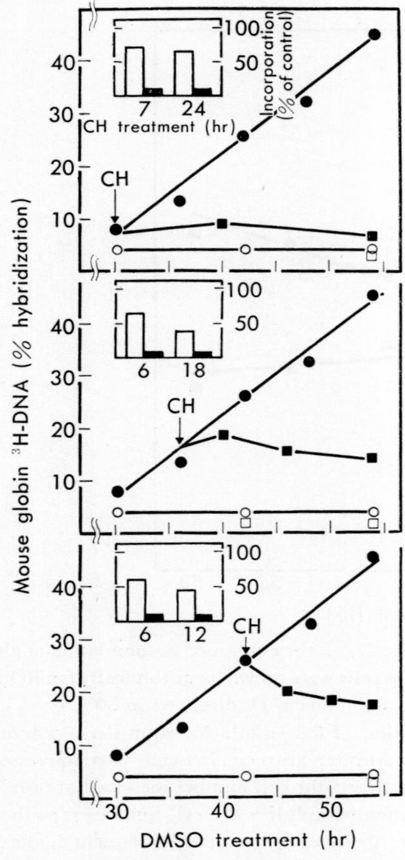

FIG. 18. The effect of cycloheximide (CH) on the accumulation of globin mRNA sequences in DMSO-treated T-3-Cl-2 cells. DMSO (1.5%, v/v) was added to cultures at time zero and cycloheximide was added as noted in the figure. The concentrations of cycloheximide used were as follows: at 30 hr, 1 μg/ml; at 36 hr, 1.25 μg/ml; and at 42 hr, 1.5 μg/ml. The cells were harvested at the times noted, and total cellular RNA was extracted and hybridized to globin cDNA as described (*41*). ● DMSO-treated cells, no cycloheximide added; ■ DMSO-treated cells, cycloheximide added; ○ untreated cells, no cycloheximide added; □ untreated cells, cycloheximide added. Inserts: open bar, the incorporation of ^{3}H-uridine into acid-precipitable material; closed bar, incorporation of ^{14}C-amino acids into acid-precipitable material.

If globin mRNA accumulation requires *de novo* synthesis, then it should be inhibited by actinomycin. Actinomycin D was added to cells, 0.3 μg/ml, which inhibited 90% of uridine incorporation into RNA during the first 2 hr treatment with very little effect on protein synthesis (insert, Fig. 17). Cells were exposed to actinomycin D for only 10 hr, causing no alteration in cell viability. This addition of actinomycin D caused an abrupt cessation of globin mRNA accumulation. Accordingly, globin mRNA accumulation in hemoglobin producing T-3-Cl-2 cells requires *de novo* RNA synthesis.

These results can be most simply explained in terms of transcriptional activation of the globin genes. Similar conclusions have been reached using another line of mouse erythroid leukemia cells (*34*) and erythroid precursor cells in the embryonic liver of the mice (*51*).

The expression of the globin genes must be closely correlated with other biochemical events necessary for maturation of red blood cells. Such cells are required to produce equivalent amount of both alpha- and beta-globin chains as well as heme. It is thought that heme is, in fact, necessary for the continued synthesis of globin (*5*). In order to determine whether globin mRNA synthesis could be uncoupled from protein biosynthesis in these cells, they were treated at various times after induction with cycloheximide (Fig. 18) (*31*). Cycloheximide treatment abruptly caused the cessation of globin mRNA accumulation. This occurs at concentrations of cycloheximide that inhibit 50% of total RNA synthesis. Globin mRNA appears particularly sensitive to cycloheximide. Globin mRNA accumulation may depend upon continued protein synthesis. There may be a certain protein necessary for either translation of the globin message or transcription of the globin gene.

In order to determine further whether the appearance of globin mRNA in T-3-Cl-2 cells is an transcriptional event, we have attempted to determine if the number of globin genes is altered as the cells accumulate hemoglobin. T-3-Cl-2 cells were grown logarithmically for 4 days with and without 1% (v/v) DMSO. Total cellular DNA was extracted and hybridized in vast excess to globin ^{3}H-cDNA added as a hybridization probe (*38*), and the rate of re-association of globin sequences was measured as a function of the parameter C_0t (*4*). The results show that the rate of re-association is virtually identical for DNA prepared from differentiation-induced and -uninduced T-3-Cl-2 cells (Fig. 19). The half C_0t for induced T-3-Cl-2 DNA is 930; that for uninduced T-3-Cl-2 DNA is 880. Since the half C_0t of the unique fraction of C3H mouse embryo DNA assayed under these conditions is 3,000, the estimated reiteration frequency of the globin genes per haploid genome of these cells is approximately 3.4 to 3.2 (Table 8) (*31*). Since almost the same number of globin genes exists in a wide variety of animal tissues, including erythroid cells, mouse liver and spleen cells (S. Packman, unpublished observation) as well as mouse embryo cells, we may conclude that the number of globin genes remained constant during hemoglobin induction in T-3-Cl-2 cells, and that the number of globin genes in T-3-Cl-2 cells, mouse erythroid cells, is similar to the number of globin genes in several tissues of certain avian and mammalian spieces.

The above finding also supports the transcriptional activation of the globin

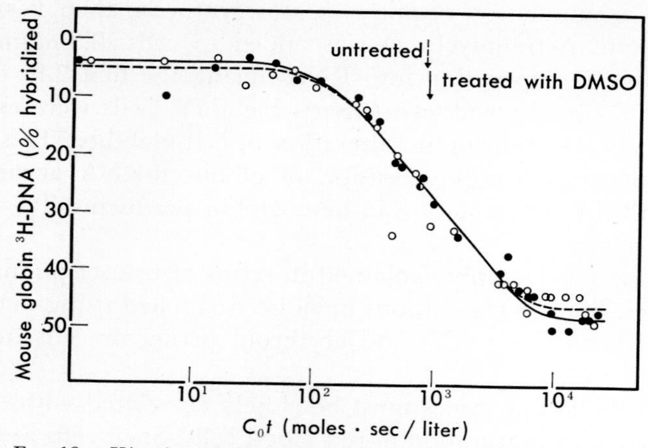

FIG. 19. Kinetics of annealing of radioactive mouse globin cDNA to a vast excess of unlabeled DNA from T-3-Cl-2 cells grown in the absence (untreated) or presence (treated) of 1% (v/v) DMSO. The T-3-Cl-2 cells were grown for 4 days in the presence or absence of DMSO (1%, v/v), during which time they replicated 3 to 4 times. Total cellular DNA was extracted, purified, and sonicated to a molecular weight of approximately 200,000. Each 3.0 ml hybridization reaction mixture contained cellular DNA (1.8 mg/ml untreated, 2.0 mg/ml DMSO-treated) and 4.2×10^{-7} mg/ml mouse globin ^{3}H-cDNA (9.1×10^6 cpm/μg) in buffer containing 1.02 M NaCl and 0.02 M Tris-Cl, pH 7.0. Incubation was at 66°C, and duplicate 0.05 ml aliquots were withdrawn at intervals to determine the percent of the labeled DNA which had annealed. C_0t values were calculated as (A_{260} units of DNA/ml $\times$ hours)/2 (L. Gelb *et al.*, 1971) and were standardized to the value that would obtain at 0.18 M sodium (*14*). ○ untreated $C_0t_{1/2}=880$; ● treated $C_0t_{1/2}=930$.

TABLE 8. Reiteration Frequency of Mouse Globin Sequences in Induced and Uninduced T-3-Cl-2 Cells

DNA	$C_0t_{1/2}$	Reiteration frequency
Mouse globin ^{3}H-DNA+T-3-Cl-2 (untreated) DNA	880	3.4
Mouse globin ^{3}H-DNA+T-3-Cl-2 (treated with DMSO) DNA	930	3.2
Control: mouse embryo ^{14}C-DNA, unique sequence fraction	3,000	1

$C_0t_{1/2}$ determinations were made using the data indicated in Fig. 19. The $C_0t_{1/2}$ for mouse embryo ^{14}C-DNA was determined in control experiments using the technique employed to determine the $C_0t_{1/2}$ shown in Fig. 19.

genes during erythrodifferentiation of T-3-Cl-2 cells. However, the possibility of post-transcriptional stabilization of newly synthesized globin mRNA has not been ruled out yet. In fact, detailed decay studies carried out following actinomycin treatment indicate that the amount of globin mRNA remains constant for at least 5 hr after actinomycin addition, suggesting the stability of globin mRNA (Fig. 17). A number of sudies have shown that the properties of globin mRNA synthesized by induced T-3-Cl-2 cells are indistinguishable from those of globin mRNA extracted from mouse reticulocytes—both having sedimentation coefficient of 9S, containing poly A, and showing a long chemical half life (J. Ross *et al.*, unpublished). Since the amount of hemoglobin in induced cells paralleled with the amount of globin mRNA induced, and the nature of thus induced hemoglobin is indistinguishable from hemoglobin of normal red blood cells (Y. Inoue and Y. Ikawa, unpublished observation), it is more likely that both retained and newly synthesized globin mRNA's are favorably stable.

An Association between Globin mRNA and 60S RNA Derived from FLV

One of the authors (Y.I.) has long suspected that something carried by FLV within the virions is responsible for induction of neoplastic proliferation of the erythroid target cells. When FLV is passed through a fibroblastic cell line, an attenuated FLV variant is easily obtainable after tens of subcultures (*53*). The attenuated FLV has the same FMR antigen, but has no longer capacity to induce Friend disease, or malignant erythroblastosis. When the large quantity of attenuated FLV, yet incapable of inducing erythroblastosis, was infected to the mice with erythropoietin-induced hyper-erythroblastosis, infectious erythroblastosis-inducing FLV was recovered from the plasma as early as 4 days; whereas the diluted potent FLV, limitingly diluted but yet sufficient to in Friend disease in 3 weeks, did not show up in the plasma that early (*23*). The latter finding suggested that something carried by FLV virions from the host target cells is responsible for erythroblastosis induction.

It is also well known that FLV is not efficiently recovered in the lethaly irradiated mice with a low quantity of erythroid hemopoietic cells, which will be cured by inoculation of hemopoietic cells. This fact also suggests the host cell dependency of leukemogenic activity of FLV.

When potent FLV is infected onto the fibroblasts, if the virus has been passed by inoculation of the harvested virions, the fibroblastic cells would easily lose potent FLV; whereas the FLV virions can maintain its erythroblastosis-inducing capacity for a long period, if they are harvested from the potent FLV-producing clone of the originally infected fibroblasts (H. Yoshikura *et al.*, unpublished observation). Once erythroblastosis-inducing factor within the FLV virions is successfully incorporated into the genetic materials of the recipient fibroblasts, the latter cells may propagate the potent FLV virions for a longer period. However, this portion of FLV (some call as 'SFFV') will easily be aborted in the transferring process from one cell to another, or will be readily repressed in the non-erythropoietic recipient cells. A short half-inactivation time of FLV also suggested that the erythroblastosis-inducing factor within FLV virions is from the host erythroid cells.

This hypothesis of presence of some host cell factor in the potent FLV virions has been long harbored by Y.I. The first step to establish this hypothesis is to obtain evidence that FLV virions can convey some cellular factors. Y.I. thought of the application of differentiation-inducible T-3-Cl-2 cells for this purpose, because T-3-Cl-2 cells in the ordinary passages do not contain globin mRNA, but the same cells do synthesize globin mRNA after addition of DMSO. T-3-Cl-2 propagated the type C virions in both conditions. Furthermore, Ikawa *et al.* (*26*) showed an increase of infectious FLV in the DMSO-treated T-3-Cl-2 cells.

These prompted us to investigate the presence of globin mRNA in the virions from induced T-3-Cl-2 cells. This was partly reported previously (*27*).

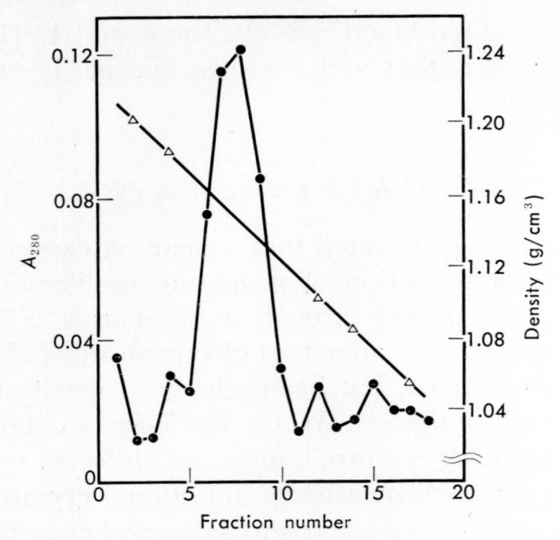

FIG. 20. Equilibrium density gradient centrifugation of partially purified FLV. Viruses which had been once banded by equilibrium centrifugation in a sucrose gradient made in 0.1 M NaCl, 0.01 M Tris-Cl, pH 7.4, 0.001 M EDTA. Centrifugation was for 12 hr at 35,000 rpm in the Spinco SW41 rotor. Fractions were collected dropwise from the bottom of the tube. Density was calculated from the retractive index. △ density (g/cm³); ● A_{280}.

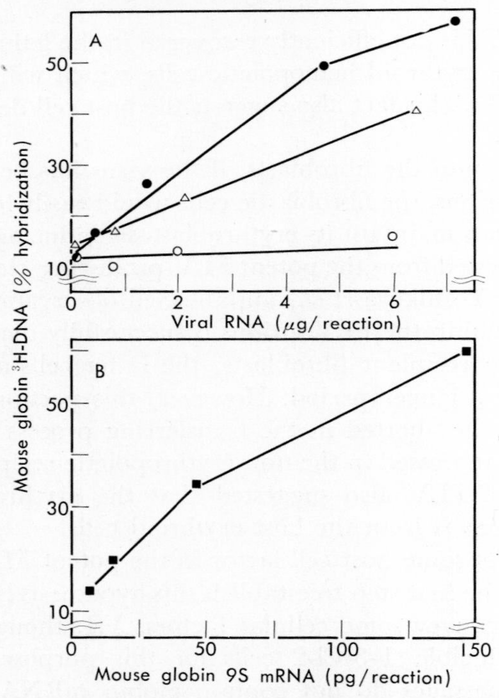

FIG. 21. Detection of globin sequences associated with FLV RNA by hybridization with ³H-globin cDNA. Total viral RNA was extracted from purified virions as described previously (*41*). No carrier RNA was added during the extraction procedure. Incubation was for 90 hr in 0.03 ml at 45°C in buffer containing 0.45 M NaCl and 33% (v/v) formamide. Percent hybridization was determined by the S_1 nuclease assay. A: incubation with FLV RNAs. ○ FLV from T-3-Cl-2 cells, no DMSO (non-induced); ● FLV from T-3-Cl-2 cells, 1% DMSO (induced); △ FLV from Cl 745 cells, no DMSO (constitutive). B: incubation with purified 9S mouse globin mRNA standard.

TABLE 9. Estimation of Globin mRNA Content of Friend Leukemia Virions

Source of virions	Percent globin mRNA in total viral RNA	Number of globin molecules per viral 60S molecule
T-3-Cl-2		
(DMSO-induced)	0.0018	0.0012
Cl-745	0.0010	0.0007

These estimates are calculated from the data in Fig. 16. To determine percent globin mRNA, the amount of purified 9S mouse globin mRNA required to achieve 30% hybridization (H_{30}) was divided by the amount of viral RNA which gave 30% hybridization. H_{30} values are as follows: 9S globin mRNA= 3.9×10^{-5} μg; T-3-Cl-2=2.2 μg; Cl-745=3.9 μg. The number of globin molecules per 60S viral RNA molcule was calculated as follows: assuming a molecular weight of 10^7 daltons for the 60S FLV RNA, the mass of one molecule of 60S RNA=1.7×10^{-5} pg. If 60S RNA=75% of total viral RNA (cf. Fig. 22), then there would be one molecule of 60S RNA per 2.3×10^{-5} pg of total viral RNA. A quantity of 2.3×10^{-5} pg of total viral RNA from T-3-Cl-2 cells should contain 4.1×10^{-10} pg ($2.3\times10^{-5}\times1.8\times10^{-5}$) of globin mRNA. Assuming a molecular weight of 2×10^{-5} daltons for globin mRNA, the mass of one globin mRNA molecule is 3.3×10^{-7} pg. Therefore, the number of globin mRNA molecules in an amount of total viral RNA containing one 60S molecule should be $\frac{4.10\times10^{-10}}{3.3\times10^{-7}}=1.2\times10^{-3}$. Similar calculations were used to determine the number of globin mRNA molecules per 60S RNA of C-l745 virions.

The virus particles used for these studies were purified by two equilibrium density centrifugations in sucrose gradients. An example of the sedimentation properties of virus isolated by a single equilibrium sedimentation is shown in Fig. 20. The virus band is located at a density of about 1.16 g/cm^3—in both samples from induced and uninduced cells.

The RNA prepared from these virions was hybridized with ^{3}H-DNA complementary to mouse globin mRNA. As shown in Fig. 21, RNA from the virions of induced cells contained hybridizable globin mRNA sequences, whereas RNA from those of uninduced cells did not. To assure that the association observed between FLV virions and globin mRNA was not due to addition of DMSO to the culture medium, virus and RNA were prepared as a control from a line of FLV-induced erythroleukemia cells, Cl-745, originally established by Friend and her colleagues (*44*), which constantly produce hemoglobin (consequently globin mRNA) without addition of DMSO. Some of Cl-745 cells are spontaneously showing erythrocyte membrane antigen. Viral RNA derived from these globin producing cells also contains globin mRNA sequences.

The globin mRNA associated with the viral RNA can be roughly quantitated by a comparison of its hybridization properties to that of purified mouse globin mRNA (Table 9). Globin mRNA sequences account for a rather small proportion of the total FLV RNA, approximately $1\text{–}2\times10^{-3}$%, an amount equivalent to about one globin mRNA molecule per thousand 60S viral RNA molecules.

In an attempt to determine the nature of the association between globin mRNA and these viral particles, intact, purified virions containing globin mRNA were incubated together with bovine pancreatic RNase (Table 10). The RNA prepared from these virions and that from similarly treated, control virions from uninduced cells to which mouse reticulocyte polysomes or 9S globin mRNA had

TABLE 10. Insensitivity to Ribonuclease of Globin mRNA in FLV

Source of virus	Additions	RNase	Percent hybridization of ^{3}H-cDNA to RNA derived from virus mixtures	
			FLV ^{3}H-cDNA	Globin ^{3}H-cDNA
Treated cells	None	–	22	6
		+	25	5
Untreated cells	Mouse reticulocyte	–	36	75
	polysomes	+	44	0
Untreated cells	9S mouse globin	–	40	77
	mRNA	+	43	0

Approximately 5 A_{260} units of FLV purified from DMSO-treated T-3-Cl-2 cells were incubated in a 1.0 ml reaction with or without 2 μg of RNase A for 45 min at 37°C. Controls contained FLV from untreated T-3-Cl-2 cells and either 5 μg of mouse reticulocyte polysomes or 16 ng of purified 9S mouse globin mRNA. RNA was extracted and concentrated. Hybridization to mouse globin ^{3}H-cDNA was in 0.15 ml for 96 hr, while hybridization to FLV cDNA was in 0.15 ml for 19 hr. Background (S_1 nuclease resistance cDNA alone) of 7% hybridization has been subtracted from the percent hybridization in the presence of RNA.

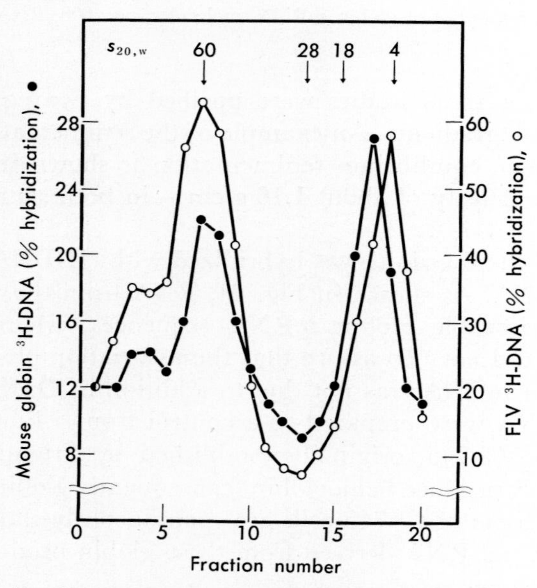

FIG. 22. Sedimentation analysis of globin mRNA associated with FLV. RNA was extracted from FLV purified from T-3-Cl-2 cells grown in the presence of 1% (v/v) DMSO. Centrifugation was carried out for 3.5 hr in a linear 15–30% (w/v) sucrose gradient in buffer containing 0.1 M NaCl. Fractions were collected and concentrated to a final volume of 0.035 ml as described previously (*27*). For hybridization to FLV RNA, a 0.002 ml aliquot was incubated with 200 cpm (approximately 0.04 ng) of FLV ^{3}H-cDNA for 60 min in a 0.06 ml reaction volume. For hybridization to mouse globin mRNA, a 0.03 ml aliquot was incubated with 375 cpm (approximately 0.05 ng) of globin ^{3}H-cDNA for 72 hr in a final volume of 0.06 ml.

been added were assayed for RNase-resistant globin mRNA and FLV RNA. In each case, the amount of hybridizable FLV RNA was not reduced by RNase treatment. Globin mRNA associated with the isolated virions was also resistant

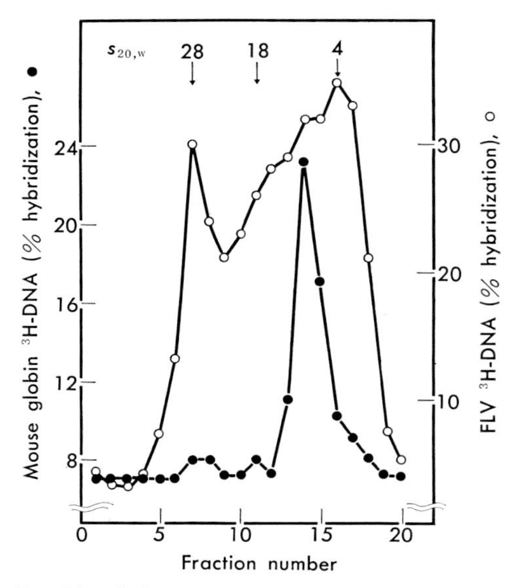

FIG. 23. Sedimentation analysis of globin mRNA associated with FLV following denaturation. RNA from FLV grown in DMSO-treated T-3-Cl-2 cells was heated and rapidly cooled in low-salt buffer (*27*). The heat-denatured RNA was then centrifuged for 11 hr at 4°C in a 15–30% sucrose gradient made in low-salt buffer. Fractions were collected and concentrated to 0.035 ml. A 0.002 ml aliquot was incubated with FLV ^{3}H-cDNA (210 cpm) for 1 hr, while 0.03 ml aliquot was incubated with mouse globin ^{3}H-cDNA (430 cpm) for 78 hr in 0.06 ml reaction volumes.

to RNase treatment, whereas exogenously added globin mRNA was completely digested by the same treatment. These results suggest that the globin mRNA is enclosed within the outer membrane of the virus.

This suggestion is further supported by the sedimentation properties of globin mRNA prepared from purified viral particles (Fig. 22). FLV RNA as detected by hybridization to FLV ^{3}H-cDNA is made up of the characteristic 60S and 4S components. Globin mRNA, as detected by hybridization with globin ^{3}H-cDNA, also sediments as two major components, one coincident with the heavy FLV RNA complex, the other with the 9S RNA fraction. The basis of this association between globin mRNA and 60S FLV complex was further examined by denaturing the viral RNA complex under mild conditions which resulted in the conversion of the 60S viral complex into primarily 30S, intermediate, and 4S components, and cause the degradation of the 60S globin mRNA into the 9S fraction indicating that it was not covalently incorporated into a large sequence (Fig. 23).

The megakaryocytes adjacent to the FLV-induced spleen foci often showed large, atypical type C particles (*20*). DMSO-treated T-3-Cl-2 cells often showed large type C particles (Fig. 11, arrows). The association of globin mRNA in those virions has not been studied yet. A possible L-cell variant propagates type C like particles, non-infectious, with 110S RNA which is readily denatured into smaller fragments (T. Yamaguchi, personal communication).

The possibility that globin mRNA became associated with the 60S RNA com-

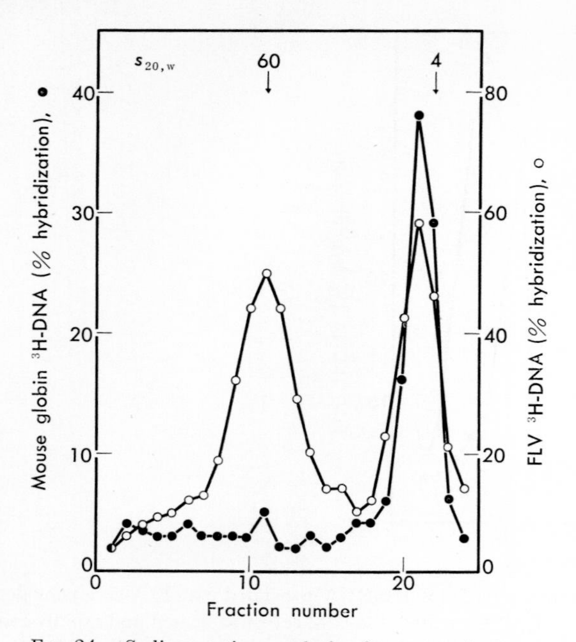

FIG. 24. Sedimentation analysis of mouse reticulocyte polysomal RNA added exogenously to FLV particles. Approximately 5 μg of mouse reticulocyte polysomes in 0.03 ml was added to an equivalent volume of FLV (containing approximately 0.3–0.5 A_{260} units) which had been purified from T-3-Cl-2 cells grown without DMSO. The virus-polysome mixture was mixed gently and allowed to stand at room temperature for 10 min. RNA was then extracted as described previously (*27*) and centrifuged in a 15–30% sucrose gradient under conditions identical to those described in Fig. 20. Aliquots of 0.05 ml taken directly from the gradient fractions were incubated for 18 hr in a reaction volume of 0.15 ml with either FLV or globin ^{3}H-cDNA.

plex in the course of RNA extraction was examined by adding reticulocyte polyzomes to virions purified from the medium of uninduced cells. RNA was extracted from this mixture and subjected to sucrose gradient centrifugation under non-denaturing condition (Fig. 24). Under these conditions, the globin mRNA migrated entirely with 9S RNA fraction.

Although we first described the nature of RNA in the virions being resistant to RNA as an evidence of the RNA existing within the virions, the RNA closely attached to the virions is often subjected to be resistant to RNase. Most of the previous studies showing RNA tumor viruses containing cellular ribosomal and transfer RNA's are along this line. Therefore, the finding of 60S aggregation of 9S globin mRNA is very important in proving that the cellular mRNA is present within the virions. Viral 60–70S RNA is also consisting of several constituents, and can integrate several sequences from the host cells (*45*). There may be a certain mechanism inside the type C virions to conform RNA's into 60–70S. This mechanism as well as the mechanism of infection may act in the virus recombining certain genetic materials from the host cells. Alternatively, mRNA's containing reiterated host nucleoid sequences might provide nucleation sites for the linkage of viral RNA subunits during formation of the 60S viral RNA complex, or similarly serve as primers for the reverse transcriptase enzyme. Perhaps, these host components directly influence the fate of infectious virions (*23*).

As previously described, one globin mRNA sequence is present per 500–1,000 viral genomes (Table 8). Since globin sequences constitute 1.8×10^{-3}% of the total FLV RNA, it is not surprising that a reverse transcript made using FLV from induced T-3-Cl-2 cells fails to hybridize significantly to globin mRNA (J. Ross, unpublished). One from FLV from hypererythroblastosis mice might have a small but detectable amount of hybridizable RNA.

Globin mRNA is only one of many cellular mRNA's similarly incorporated into the virions. Electron microscopy often showed the type C virions containing possible ribosomal particles. Further, the FLV particles used in these studies contain a minimum proportion of globin sequences since a portion of these virions were shed prior to the initiation of globin mRNA accumulation. Globin mRNA starts to increase on day 2 of DMSO treatment, and reaches a constant level on day 3 of treatment. The virions used were harvested on day 4. So three-fourths of the virions may be from uninduced cells. Further, the hemoglobin-positive cells on day 5 were still less than 40% of the treated cells. If we harvest the virions more properly from the induced cells, we might show a higher incidence of globin mRNA in the virions.

Whatever the function of globin mRNA in the virions from differentiation-induced T-3-Cl-2 cells, our observation raise at least the formal possibility that host mRNA's can be conveyed from one cell to another *via* the RNA tumor viruses. The presence of an enzyme in the virions capable of transcribing RNA into DNA makes this possibility an especially interesting one.

ADDENDUM

Repeated cloning has been carried out to isolate favorably stable phenotypic variants of T-3-Cl-2 line, some super-inducible and others induction-resistant clones (*28*). Some clones are minimally virus-producing, other clones differentiate further to shed spherocyte-like corpuscles by DMSO treatment.

ACKNOWLEDGMENTS

Acknowledgments are due to Miss Mariko Aida for the isolation and maintenance of T-3-Cl-1 and -2 variants, Mr. Tokuichi Kawaguchi for immunofluorescence microscopy, Mr. Tsugukatsu Nashiro for electron microscopy, Mr. Yoshio Nagashima for preparation of photographs, and Mr. Koichi Makino for the animal care.

This study was partly supported by the grants from the Ministry of Education of Japan, and from the Ministry of Health and Welfare of Japan. This study was also supported by grants from the Princess Takamatsu Cancer Research Fund, and from the Japanese Society for Promotion of Cancer Research.

One of the authors (Y.I.) was supported by USPHS International Research Fellowship when he was at Viral Leukemia and Lymphoma Branch, National Cancer Institute, Bethesda, Maryland, U.S.A.

REFERENCES

1. Adachi, H. and Furusawa, M. Immunological analysis of the structural molecules of erythrocyte membrane in mice. I. Analysis of the aqueous phase molecules obtained by butanol fraction of erythrocyte membrane. Exp. Cell Res., *50*: 490–496, 1968.
2. Aviv, H. and Leder, P. Purification of biologically active globin mRNA by chromatography on oligothymidylic acid cellulose. Proc. Natl. Acad. Sci. U.S., *69*: 1408–1412, 1972.
3. Axelrad, A. A. and Steeves, R. A. Assay for Friend leukemia virus: Rapid quantitative method based on enumeration of macroscopic spleen foci in mice. Virology, *24*: 513–518, 1964.
4. Britten, R. J. and Kohne, D. E. Repeated sequences in DNA. Science, *161*: 529–540, 1968.
5. Bruns, G. P. and London, I. M. The effect of hemin on the synthesis of globin. Biochem. Biophys. Res. Commun., *18*: 236–239, 1965.
6. Buffet, R. F. and Furth, J. A transplantable reticulum cell sarcoma variant of Friend's viral leukemia. Cancer Res., *19*: 1063–1069, 1959.
7. de Harven, E. and Friend, C. Electron microscopic study of a cell-free induced leukemia of the mouse. A preliminary report. J. Biophys. Biochem. Cytol., *4*: 151–159, 1958.
8. Dunn, T. B. and Green, A. W. Morphology of BALB/c mice inoculated with Rauscher virus. J. Natl. Cancer Inst., *36*: 987–1001, 1966.
9. Ebert, P. and Ikawa, Y. Induction of δ-aminolevulinic acid synthetase during erythroid differentiation of cultured leukemia cells. Proc. Soc. Exp. Biol. Med., *146*: 601–604, 1974.
10. Friend, C. Cell-free transmission in adult Swiss mice of a disease having the character of a leukemia. J. Exp. Med., *105*: 307–319, 1957.
11. Friend, C., Patuleia, M. C., and de Harven, E. Erythrocytic maturation *in vitro* of murine (Friend) virus-induced leukemic cells. Natl. Cancer Inst. Monogr., *22*: 505–522, 1966.
12. Friend, C., Scher, W., Holland, J. G., and Sato, T. Hemoglobin synthesis in murine virus-induced leukemic cells *in vitro*. Simulation of erythroid differentiation by dimethyl sulfoxide. Proc. Natl. Acad. Sci. U.S., *68*: 378–382, 1971.
13. Furusawa, M. and Adachi, H. Immunological analysis of the structural molecules of erythrocyte membrane in mice. II. Staining of erythroid cells with labelled antibody. Exp. Cell Res., *50*: 497–504, 1968.
14. Furusawa, M., Ikawa, Y. and Sugano, H. Development of erythrocyte membrane-specific antigen(s) in clonal cultured cells of Friend virus-induced tumor. Proc. Jap. Acad., *47*: 220–224, 1971.
15. Furusawa, M., Ikawa, Y., and Sugano, H. Phenotypic changes in Friend tumor cells. GANN Monograph on Cancer Research, *12*: 231–239, 1972.
16. Ikawa, Y. and Sugano, H. An ascites tumor derived from early splenic lesion of Friend's disease. Gann, *57*: 641–643, 1966.
17. Ikawa, Y. and Sugano, H. Transplantation of spleen foci of Friend disease. Nippon Byori Gakkai Kaishi (Trans. Soc. Pathol. Japan), *55*: 139, 1966 (in Japanese).
18. Ikawa, Y., Sugano, H., and Oota, K. Spleen focus of Friend's disease. A histological study. Gann, *58*: 61–67, 1967.

19. Ikawa, Y. and Sugano, H. Focal lesions of Friend's disease in bone marrow. Gann, *58*: 305–307, 1967.
20. Ikawa, Y. and Sugano, H. Spleen focus of Friend's disease. An electron microscopic study. Gann, *58*: 155–160, 1967.
21. Ikawa, Y. and Sugano, H. Histogenesis of Friend virus-induced leukemia (malignant erythroblastosis) and its short latency. Annu. Rept. Coop. Res. of Ministry of Education of Japan (Cancer), 317–328, 1967 (in Japanese).
22. Ikawa, Y. and Sugano, H. Target cell analysis of Friend virus by using erythropoietin. Proc. Jap. Cancer Assoc., 26th Annu. Meet, p. 224, 1967 (in Japanese).
23. Ikawa, Y., Yoshikura, H., and Sugano, H. Dependency of leukemogenicity of Friend virus on host cells. Proc. Jap. Cancer Assoc., 27th Annu. Meet., pp. 114–115, 1968 (in Japanese).
24. Ikawa, Y. and Sugano, H. Maturation arrest of erythropoietin-induced proerythroblasts in DDD mice caused by infection with Friend leukemia virus. Abstr. Papers, 10th Int. Cancer Congr., p. 155, 1970.
25. Ikawa, Y., Sugano, H., and Furusawa, M. Pathogenesis of Friend virus-induced leukemia in mice. GANN Monograph on Cancer Research, *12*: 33–45, 1972.
26. Ikawa, Y., Sugano, H., and Furusawa, M. Erythrocyte membrane specific antigens in Friend virus-induced leukemia cells. Bibl. Haemat., *39*: 955–967, 1973.
27. Ikawa, Y., Ross, J., and Leder, P. An association between globin messenger RNA and 60S RNA derived from Friend leukemia virus. Proc. Natl. Acad. Sci. U.S., *71*: 1154–1158, 1974.
28. Ikawa, Y., Aida, M., Inoue, Y., and Sugano, H. Erythrocyte-like corpuscle formation in cultured Friend leukemia cells. Proc. Jap. Cancer Assoc., 33rd Annu. Meet., p. 11, 1974 (in Japanese).
29. Kasuga, T. and Oota, K. Pathological characteristics of Friend disease. Gan-no-Rinsho (Cancer Clinic), *8*: 251–265, 1962 (in Japanese).
30. Kasuga, T. Presence of globin in Friend ascites tumor cells: An immunofluorescence study. Proc. Jap. Cancer Assoc., 26th Annu. Meet., p. 223, 1966 (in Japanese).
31. Leder, P., Ross, J., Gielen, J., Packman, S., Ikawa, Y., Aviv, H., and Swan, D. Regulated expression of mammalian genes. Cold Spring Harbor Symp. Quant. Biol., *38*: 753–761, 1973.
32. Metcalf, D., Furth, J., and Burret, R. F. Pathogenesis of mouse leukemia caused by Friend virus. Cancer Res., *19*: 52–59, 1959.
33. Mirand, E. A., Prentice, T. C., Hoffman, J. G., and Grace, J. T., Jr. Effect of Friend virus in Swiss and DBA/1 mice on ^{59}Fe uptake. Proc. Soc. Exp. Biol. Med., *106*: 423–426, 1961.
34. Ostertag, W., Melderis, H., Steinheider, G., Kluge, N., and Dube, S. Synthesis of mouse haemoglobin and globin mRNA in leukaemic cell cultures. Nature New Biol., *239*: 231–234, 1972.
35. Patuleia, M. and Friend, C. Tissue culture studies on murine virus-induced leukemia cells. Isolation of single cells in agar-liquid medium. Cancer Res., *27*: 726–730, 1967.
36. Pluznik, D. H. and Sachs, L. The mechanism of leukemogenesis by the Rauscher leukemia virus. Natl. Cancer Inst. Monogr., *22*: 3–14, 1966.
37. Pluznik, D. H. and Sachs, L. Quantitation of a murine leukemia virus with a spleen colony assay. J. Natl. Cancer Inst., *33*: 535–546, 1964.
38. Packman, S., Aviv, H., Ross, J., and Leder, P. A comparison of globin genes in duck

reticulocytes and liver cells. Biochem. Biophys. Res. Commun., *49*: 813–819, 1972.

39. Rauscher, F. J. A virus-induced disease of mice characterized by erythrocytopoiesis and lymphoid leukemia. J. Natl. Cancer Inst., *29*: 515–543, 1962.
40. Ross, J., Aviv, H., Scolnick, E., and Leder, P. *In vitro* synthesis of DNA complementary to purified rabbit globin mRNA. Proc. Natl. Acad. Sci. U.S., *69*: 264–267, 1972.
41. Ross, J., Ikawa, Y., and Leder, P. Globin messenger-RNA induction during erythroid differentiation of cultured leukemia cells. Proc. Natl. Acad. Sci. U.S., *69*: 3620–3623, 1972.
42. Rossi, G. B. and Friend, C. Erythrocytic maturation of (Friend) virus-induced leukemic cells in spleen clones. Proc. Natl. Acad. Sci. U.S., *58*: 1373–1380, 1967.
43. Sassa, S., Takaku, F., Nakao, K., Ikawa, Y., and Sugano, H. Heme synthesis in Friend ascites tumor cells. Proc. Soc. Exp. Biol. Med., *127*: 527–529, 1968.
44. Scher, W., Prieslar, N. D., and Friend, C. Hemoglobin synthesis in murine virus-induced leukemic cells *in vitro*. J. Cell Physiol., *81*: 63–70, 1973.
45. Scolnick, E. M., Stephenson, J. R., and Aaronson, A. Isolation of temperature sensitive mutants of murine sarcoma virus. J. Virol., *10*: 653–666, 1972.
46. Soule, H. D., Albert, S., Wolf, P. L., and Stansly, P. G. Erythropoietic differentiation of stable cell lines derived from hematopoietic organs of mice with virus-induced leukemia. Exp. Cell Res., *42*: 380–383, 1966.
47. Sugano, H., Furusawa, M., Kawaguchi, T., and Ikawa, Y. Induction of erythrocyte membrane specific antigens in Friend leukemia cells. Bibl. Haemat., *39*: 939–954, 1973.
48. Sugano, H., Furusawa, M., Kawaguchi, T., and Ikawa, Y. Differentiation of tumor cells. Recent Results Cancer Res., *44*: 30–44, 1974.
49. Takahashi, E., Nagasawa, T., Sato, S., Matsushima, T., Sugimura, T., and Ohashi, A. Induction of δ-aminolevulinic acid synthetase in cultured Friend leukemia cells by dimethyl sulfoxide and human placental extract. Gann, *65*: 261–268, 1974.
50. Tambourin, P. and Wendling, F. Malignant transformation and erythroid differentiation by polycythemia-inducing Friend virus. Nature New Biol., *234*: 230–233, 1971.
51. Terada, M., Cantor, L., Metafora, S., Rifkind, R. A., Bank, A., and Marks, P. A. Globin messenger RNA activity in erythroid precursor cells and the effect of erythropoietin. Proc. Natl. Acad. Sci. U.S., *69*: 3575–3579, 1972.
52. Yokoro, K. and Thorell, B. Cytology and pathogenesis of Rauscher virus disease in spleenectomized mice. Cancer Res., *26*: 536–543, 1966.
53. Yoshikura, H., Hirokawa, Y., Ikawa, Y., and Sugano, H. Infectious but nonleukemogenic Friend leukemia virus obtained after prolonged cultivation *in vitro*. Int. J. Cancer, *4*: 636–640, 1969.
54. Zajdela, F. Contribution à l'étude de la cellule de Friend. Bull. Assoc. Franç. Cancer, *49*: 351–358, 1962.

Discussion by Paper of Drs. Ikawa et al.

DR. BARSKI: Do you have any evidence in your investigations of the "double" nature of the Friend virus?

DR. IKAWA: The virus from T-3-Cl-2 cells can induce malignant erythroblastosis, but if it is infected to the resistant mice, it can induce myelomonocytic leukemia or thymic lymphoma.

DR. BARSKI: Did you find any evidence of genetic modification *in vitro* of the Friend virus concerning its transforming capacity?

DR. IKAWA: Erythroblastosis-inducing capacity has been decreasing gradually for the past 6 years, but is still maintained.

DR. TOMKINS: Could you capture SV-40 specific RNA in C-type particles induced with BrdU in SV-3T3 cells?

DR. IKAWA: I have never tried that kind of experiment, but there is a possibility of BrdU-induced C-type particles carrying SV-40 RNA.

DR. OSTERTAG: If I understand you right, you like the hypothesis that globin mRNA in the virions serves to specify the type of transformation (erythroid) by the Friend virus. If this is correct, then your unstimulated cells should not transform erythroid cells since your unstimulated cells do not synthesize globin mRNA. Other Friend cell lines such as our clone F4 which do have some globin synthesis (0.5%) then might synthesize virus which does transform erythroid cells. Did you check that point?

DR. IKAWA: I like the hypothesis partly because of such a short latency and such a rapid inactivation at room temperature. Something unstable which has been carried out within the virions may be active only in the erythroid target cell. If we dilute murine sarcoma virus (Harvey), we cannot induce erythroblastosis, but we can induce lymphoma. So the quantity of the virus is also involving the type of disease induced by the virus. In the present report, I simply want to correlate the capacity of FLV carrying out some mRNA to the transforming activity. The FLV from unstimulated cells should have the same capacity, although there is no globin mRNA to carry out.

Closing Remarks

Dr. Waro Nakahara

It seems that it has become customary for the Chairman of the Organizing Committee to make closing remarks at the end of the scientific session of the Symposium.

A remarkable feature of this Fourth Symposium was the great emphasis on purely molecular biological subjects: gene expression and its regulation, transcriptional and translational mechanisms, *etc*. This, I think, was very fortunate, since without the understanding of these mechanisms, there will be no real progress in cancer research. By the time when these now fashionable subjects are outmoded we will have learned what cancer cells really are.

So many very interesting and exciting papers were presented at this Symposium that any attempt at the integration of the material involved is certainly beyond the scope the closing remarks. I shall therefore limit myself to just commenting on some of the topics upon which many of the discussions converged. I trust that nobody will object if I take up the subjects of decancerization, altered transplantability and tumor reversal for my theme.

The fundamental concept underlying that of the so-called decancerization is that carcinogenesis may not always involve irreversible alteration in cell genetic material (genes), but may represent abnormal differentiation determined only by changes in gene expression; some genes being "switched off" and others "switched on."

Data supporting this concept originated from the isozyme studies on various tumors and normal tissues, and on alteration of isozyme patterns in the course of experimental carcinogenesis. More specifically speaking, it is based on observations on definite but selected biochemical parameters in some tumors, which may or may not be applicable to the generality of tumors. At any rate, further piling up of data seems highly desirable.

The spontaneous or induced loss of transplantability of transplantable animal tumors during cultivation *in vitro* has been well documented. This phenomenon may well be due to some alteration in membrane antigen and/or to the change in histocompatibility to the host animal. Non-transplantable clones isolated from transplantable tumor cell cultures may be mutant clones. Transplantability in itself, moreover, is not necessarily a criterion of cell malignancy, contrary to the assumption among some people. Many non-malignant cells are transplantable.

We know by experience that there is a very wide range of variation in the time of appearance of palpable tumors from the autografts of spontaneous as well as carcinogen-induced tumors, that is, in strictly autochthonous tumor-host system. Transplantability in more complicated systems, as involved in the case of cell culture, would introduce many factors, which may render difficult the correct evaluation of the transplantation results.

It has been axiomatic in cellular pathology that the cell state of malignancy is autonomous (independent from growth regulating influences) and is irreversible (not reversible to the normal cell state). This may be true insofar as the cells that have completed their full cancerization, but the summation concept of carcinogenic mechanism implies that there must be many intermediate cells that stand between normal and fully malignant cells. The cell cancerization process is not a single jump from normal to malignant. It is completed by the summation of many, probably numerous small jumps.

In chemical carcinogenesis, once the cell is transformed into fully malignant state, presence of the inciting agent is no longer needed. Established cancer cells are independent of the carcinogen, and maintain their malignant properties autonomously.

The case may be different in viral leukemia. The causative virus is generally there with the leukemic cells. This fact makes it difficult to exclude the possibility that the leukemia virus may act as stimulator for cell division, the cell proliferation then being virus dependent in one way or another.

It is conceivable that most leukemic cell populations contain apparent leukemic cells which are morphologically not distinguishable from true leukemia cells, partially committed cells, progenitor cells, and even normal blast cells. Some of these apparent "leukemic" cells may respond to normal differentiating influences, *i.e.*, they have not acquired full autonomy. These cells may become non-dividing and proceed to differentiate (mature) into erythroblasts, granulocytes, or macrophages, depending on the nature of the mother cells from which they were derived.

I am inclined to this interpretation for the well known erythroid differentiation *in vitro* of Friend leukemia cells. The same may apply to the differentiation *in vitro* of a myeloid leukemia cell line into mature granulocytes and macrophages.

These cases of induction of leukemic cell differentiation are associated with the stopping of cell multiplication, and apparently support Potter's "oncogeny is blocked ontogeny" idea. However, Potter's thesis has much wider meaning than can be explained away by the simple examples of "reversal" of leukemic cells.

There are many transplantable sarcomas derived from Friend leukemia spleen, which do not spontaneously undergo erythroid differentiation. This proves that not all the Friend leukemia cells are destined to erythroid differentiation but that some are apparently fully and irreversibly cancerized. However, even from such a sarcoma a clonal cell line can be isolated which will undergo erythroid differentiation when dimethylsulfoxide is added to the cell culture (Ikawa).

English dictionaries define "reversion" in biological sense as "return to previous state, especially to ancestral type." Tumor reversal would mean reversion of tumor cells back to normal cells of origin. If tumor cells in general can really be brought back to normal cells it would mean a complete breakdown of the authodox idea of the irreversibility of cancerous changes. That would be a revolution in the pathological concept of neoplasia.

If such a revolution is inevitable, the revolution will have to be accepted whether one likes it or not. It is important, however, to critically scrutinize the significance of new findings, and not to fall into a pitfall which is often very attractively constructed.

In this connection attention should be called to the fact that most of the data supporting the concept of tumor reversal are derived from experiments *in vitro*. How far could the *in vitro* phenomena be extrapolated to the *in vivo* condition is always a moot question. Even if tumor reversal or decancerization takes place *in vivo* also, the process must occur on a whole sale scale, involving all the cancer cells; otherwise cancerous growth will be maintained by the cancer cells which escape decancerization. To propose a search for means of promoting the reversion process as one way of approach to the cure of cancer at this stage of our knowledge may seem too naive. But it must be remembered that any approach to cancer chemotherapy may have to be purely empirical.

At this point, I feel like recalling what Isaac Newton said in his presidential address to the Royal Society of London. He said "We are like children playing on the seashore, picking up pretty shells and pebbles here and there, while the boundless ocean of truth lies undiscovered before us." Although Newtonian physics has since given way to atoms, electrons, and theory of relativity, the words just cited remain generally applicable, especially to cancer research. But let us not be afraid of the vastness of the ocean. Let us consider it as just a much greater field to explore and to make discoveries.

I now close the scientific session of the Fourth International Symposium of the Princess Takamatsu Cancer Research Fund with the feeling that it did build another milestone in the progress of cancer research, as did each of the past three symposia of the Fund.

Esteemed Colleagues, I cannot thank you enough for the active participation and kind cooperation.

Epilogue

Prof. VAN R. POTTER

Your Imperial Highness, on behalf of the privileged visitors who have participated in this conference I wish to thank you for the wonderful interest and thoughtfulness you have shown in making this event possible. Your interest in science maintains a high tradition in Japan, a tradition that if properly nurtured and guided can only result in tremendous benefit to the whole human race.

I have attended and participated in cancer conferences of one kind or another nearly every year since 1945 and I can honestly say that this one has surpassed any previous conference that I have attended.

First the size has been near optimum, and secondly, the continuity and content has been undiluted and unbroken, while the mix of disciplines has been just right. Finally, the hospitality, facilities, and international fellowship—even the season has been ideal.

With all these expressions of confidence and satisfaction, I am sure you wonder where, after more than 30 years of international effort, where does it all lead? I think you are entitled to a realistic answer to that question. And it is one that I undertake with genuine humility.

First, I will say that a comprehensive and penetrating understanding of the molecular and biological nature of cancer is very near at hand, so near that when we look back 4 or 5 years from now it may be difficult to say exactly which discoveries were most important in gaining this understanding. Later perhaps a clearer perspective will be gained. But I must caution you that understanding is not synonymous with conquering the disease. I do not believe that such a goal is realistic, although we must continue to work for maximum results. What we can look forward to on the basis of better understanding is an increased ability to prevent cancer, to evaluate the genetic and environmental components of the disease, and finally to treat that various special manifestations of the disease. But I repeat, in my opinion, it will never be possible to prevent all cases or to treat all cases, once they occur.

However, I would like to emphasize that cancer research impinges on more than the cancer problem. It is the symbol of man's eternal struggle to find order in the midst of a world in which order and disorder are strangely mixed. It is the symbol of man's struggle against death and suffering. We know that death and suffering can never be eliminated but we can hope to die with some preservation of human dignity.

The cancer problem is not the most important problem in the world. Human survival, peace, population control, an equitable distribution of natural resources, and an attempt to maintain and improve the quality of human life, all these demand our attention.

We whose main professional mission is to understand and conquer cancer are not unmindful of these other problems. As citizens of the world we gain much by international conferences such as the one we have just enjoyed. We gain a better understanding of cancer but we also gain a better understanding of each other. Again, your Imperial Highness, your continued interest is greatly appreciated and is most vital to our continued effort, and to international insight.